IITA
Research to Nourish Africa

ISPA

Mycotoxins

Detection Methods, Management, Public Health and Agricultural Trade

Integration of Mycotoxin and Toxigenic Fungi
MYCO-GLOBE
Research for Food Safety in Global System

Mycotoxins

Detection Methods, Management, Public Health and Agricultural Trade

Edited by

John F. Leslie

Ranajit Bandyopadhyay

Angelo Visconti

www.cabi.org

John F. Leslie is Professor and Head of the Department of Plant Pathology at Kansas State University, Manhattan, Kansas, USA.
Ranajit Bandyopadhyay is a senior research scientist at the International Institute of Tropical Agriculture (IITA) in Ibadan, Nigeria.
Angelo Visconti is the Director of the Institute of Sciences of Food Production (ISPA) of the Italian National Research Council (CNR) in Bari, Italy and was the Coordinator of the MycoGlobe Research Network.

CABI is a trading name of CAB International

CABI Head Office
Nosworthy Way
Wallingford
Oxfordshire OX10 8DE
UK
Tel: +44 (0)1491 832111
Fax: +44 (0)1491 833508
E-mail: cabi@cabi.org
Website: www.cabi.org

CABI North American Office
875 Massachusetts Avenue
7th Floor
Cambridge, MA 02139
USA
Tel: +1 617 395 4056
Fax: +1 617 354 6875
E-mail: cabi-nao@cabi.org

A catalogue record for this book is available from the British Library, London, UK.

Library of Congress Cataloging-in-Publication Data

Mycoglobe Conference "Reducing Impact of Mycotoxins in Tropical Agriculture" (2005 : Accra, Ghana)
Mycotoxins : detection methods, management, public health, and agricultural trade / [edited by] John F. Leslie, Ranajit Bandyopadhyay, Angelo Visconti.
p. ; cm.
"This publication is based on contributions from participants at the Mycoglobe Conference 'Reducing Impact of Mycotoxins in Tropical Agriculture' from 13-16 September 2005 in Accra, Ghana"--T.p. verso.
Includes bibliographical references and index.
ISBN 978-1-84593-082-0 (alk. paper)
1. Mycotoxins--Congresses. 2. Mycotoxins--Africa--Congresses. 3. Mycotoxins--Mediterranean Region--Congresses. 4. Fungal diseases of plants--Congresses. 5. Food contamination--Congresses. I. Leslie, John F., 1953- II. Bandyopadhyay, Ranajit, 1953- III. Visconti, Angelo. IV. C.A.B. International. V. Title.
[DNLM: 1. Mycotoxins--toxicity--Congresses. 2. Food Contamination--prevention & control--Congresses. 3. Tropical Medicine--Congresses. 4. World Health--Congresses.
QW 630.5.M9 M9941m 2008]

RA1242.M94M923 2008
615.9'5295--dc22

2008001269

ISBN-13: 978 1 84593 082 0

Printed and bound in the UK by Cromwell Press, Trowbridge, from copy supplied by the editors.

Table of Contents

Preface

Mycotoxins have scourged mankind for thousands of years causing death, hallucination and misery. Since the 1960s with the identification of aflatoxin the toll extracted by these compounds from human populations in developed countries has continuously decreased with increasingly effective government regulations and routine monitoring of the food supply reducing recognized problems by orders of magnitude. The reality in less-developed countries could not be in starker contrast. Food insecurity, few if any enforced regulations, and harsh environments that favor fungal growth and toxin production combine to make chronic mycotoxin contamination and its associated health problems another cruel fact in an already difficult life. The lack of appreciation of these problems amongst policy makers and the general populace limits the public demand for a scientific solution to little more than lip service except when mycotoxins threaten the pocketbook, usually as a non-tariff trade barrier. The EU recognition of this problem led to the establishment of the MycoGlobe research network with participating scientists in the public and private sector from every continent in the world coordinating their efforts to determine the range of potentially contaminated products.

In sub-Saharan Africa the aflatoxin, fumonisin and ochratoxin mycotoxins are common, economically important, and health hazards for humans and domesticated animals due to both chronic and acute toxicological manifestations that include cancer, mutagenicity, birth defects and estrogenic, gastrointestinal, urogenital, vascular, kidney and nervous system disorders. The compromised immune response that accompanies chronic exposure to some mycotoxins reduces resistance to infectious disease and may be the most important and the most underappreciated of the health problems associated with mycotoxins. The reduction of growth and development of children exposed to aflatoxins may reduce quality of life and limit an individual's ability to reach their potential. In 1996, the UN Commission on Sustainable Development approved a work program on indicators of sustainable development that included mycotoxins in food as one of the components/indicators related to protection and promotion of human health.

In 2004, the European Commission approved the "Integration of Mycotoxin and Toxigenic Fungi Research for Food Safety in Global System (MycoGlobe) Specific Support Action" within the Sixth Framework Programme. The aim of the MycoGlobe Specific Support Action (contract number FOOD-CT-2004-007174) was to implement the outcomes of a wide range of European research projects in the area of mycotoxins and toxigenic fungi by supporting, stimulating and facilitating cooperation between countries in the European Union and other countries that have bilateral scientific and technological cooperation agreements with the European Union. As a part of the MycoGlobe Specific Support Action, the International conference "Learning from EC: Reducing Impact of Mycotoxins in Tropical Agriculture with Emphasis on Health and Trade in Africa" was organized in Accra, Ghana, 13-16 September 2005. The conference brought European and US investigators to Africa to interact with local scientists to identify key mycotoxin constraints in tropical Africa and the Mediterranean, to explore challenges and opportunities to mitigate health and trade problems associated with mycotoxins, and to sensitize policy makers and opinion-leaders to the urgency of reducing mycotoxin-related problems. The goal of the conference was to recommend priority actions in the areas of technical, institutional and policy options for improving public health and trade through management of mycotoxins from "field to fork".

Representatives from various development organizations, international research and development organizations, advanced research institutes in Europe and the United States and national research organizations in Africa, public health ministries, regulatory agencies, and farmers' organizations shared their experience on various issues related to mycotoxins. The chapters in this book touch on issues including health, trade, ecology, epidemiology, occurrence, detection, management, awareness and policy. A chapter based on structured discussions at this conference identifies priorities for mycotoxin management from technical, institutional and policy perspectives that can serve as a guide for development of inter-institutional, cross-disciplinary programs to reduce the severity of the current problems.

We thank the European Commission for funding the MycoGlobe project through which this conference was organized. The conference was organized by the International Institute of Tropical Agriculture (IITA), Ghana's Council of Scientific and Industrial Research (CSIR), and Institute of Sciences of Food Production (ISPA) – CNR, and co-sponsored by the Food and Agriculture Organization of the United Nations (FAO), the African Agriculture Technology Foundation (AATF), and two USAID-funded projects: the International Sorghum and Millet Collaborative Research Program (INTSORMIL) and the Peanut Collaborative Research Support Program (Peanut-CRSP). We thank all of them for their sponsorship and for their support of research presented at the conference.

This book is an outcome of the MycoGlobe conference in Accra. Most of the chapters are based on invited oral presentations made at the conference. The content of a number of poster presentations is generally missing. All chapters were reviewed for content and have been edited for style by one or more of the editors of the volume. We thank the corresponding authors for responding to our inquiries and suggestions in a timely manner. The opinions expressed in these chapters are those of the authors and not necessarily those of the editors, the European Union, or any of the sponsoring institutions.

We hope that this book will serve as a source of information on the occurrence and impact of mycotoxins on everything from trade and health to agricultural production in addition to suggesting opportunities for their management in Africa and elsewhere by researchers, policy makers and development investors.

JOHN F. LESLIE
Department of Plant Pathology
Kansas State University (KSU)
Manhattan, Kansas, USA

RANAJIT BANDYOPADHYAY
International Institute of Tropical Agriculture (IITA)
Ibadan, Nigeria

ANGELO VISCONTI
Institute of Sciences of Food Production (ISPA)
National Research Council (CNR)
Bari, Italy

Addresses of Contributors (* corresponding author)

Evans Afriyie-Gyawu
College of Veterinary Medicine
Faculty of Veterinary Toxicology
Texas A&M University
College Station, Texas 77843-4458
USA

Kwaku O. Agyemang
Department of Agricultural Economics, Agribusiness and Extension
Faculty of Agriculture
Kwame Nkrumah University of Science and Technology (KNUST)
Kumasi
Ghana

David Aldred
Applied Mycology Group
Cranfield Health
Cranfield University
Bedford MK43 0AL
United Kingdom

Nii-Ayi Ankrah
Noguchi Memorial Institute for Medical Research
University of Ghana
Legon, Accra
Ghana

Richard T. Awuah*
Department of Crop Science
Faculty of Agriculture
Kwame Nkrumah University of Science and Technology (KNUST)
Kumasi
Ghana
e-mail: awuahrt@yahoo.com

Tomohiro Ban
Kihara Institute for Biological Research
Yokohama City University
641-12 Maioka
Yokohama 244-0813
Japan

Ranajit Bandyopadhyay*
International Institute of Tropical Agriculture (IITA)
PMB 5320, Oyo Road
Ibadan
Nigeria
e-mail: r.bandyopadhyay@cgiar.org

Samuel A. Bankole
Department of Biological Sciences
Olabisi Onabanjo University
PMB 2002
Ago-Iwoye, Ogun State
Nigeria

Marianne Banziger
Centro Internacional de Mejoramiento de Maíz y Trigo (CIMMYT)
Apdo. Postal 6-641
06600 Mexico, D.F.
Mexico

Paola Battilani*
Institute of Entomology and Plant Pathology
Faculty of Agriculture
Università Cattolica del Sacro Cuore
via Emilia Parmense 84
29100 Piacenza
Italy
e-mail: paola.battilani@unicatt.it

David Bergvinson
Centro Internacional de Mejoramiento de Maíz y Trigo (CIMMYT)
Apdo. Postal 6-641
06600 Mexico, D.F.
Mexico

Deepak Bhatnagar*
USDA – Agricultural Research Service
Southern Regional Research Center
1100 Robert E. Lee Boulevard
New Orleans, Louisiana 70124
USA
e-mail: dbhatnag@srrc.ars.usda.gov

Stanford Blade*
Alberta Agricultural Research Institute
Suite 500, Phipps McKinnon Building
10020-101A Avenue
Edmonton, Alberta T5J 3G2
Canada
e-mail: stan.blade@gov.ab.ca

Carlo Brera
GMO and Mycotoxins Unit
National Center for Food Quality and Risk Assessment
Italian National Institute for Health (ISS)
viale Regina Elena, 299
00161 Rome
Italy

Robert L. Brown
USDA – Agricultural Research Service
Southern Regional Research Center
1100 Robert E. Lee Boulevard
New Orleans, Louisiana 70124
USA

Jeffrey W. Cary
USDA – Agricultural Research Service
Southern Regional Research Center
1100 Robert E. Lee Boulevard
New Orleans, Louisiana 70124
USA

Zhi-Yuan Chen
Department of Plant Pathology and Crop Physiology
Louisiana State University Agricultural Center
Baton Rouge, Louisiana 70803
USA

Dyborn Chibonga
The National Smallholder Farmer's Association of Malawi (NASFAM)
P. O. Box 30716
Lilongwe 3
Malawi

Thomas E. Cleveland
USDA – Agricultural Research Service
Southern Regional Research Center
1100 Robert E. Lee Boulevard
New Orleans, Louisiana 70124
USA

Peter J. Cotty*
USDA-ARS
Laboratory for Aflatoxin Reduction in Crops
Division of Plant Pathology and Microbiology
Department of Plant Sciences
University of Arizona
Tucson, Arizona 85721
USA
e-mail: pjcotty@email.arizona.edu

Ousmane Coulibaly*
International Institute of Tropical Agriculture
Biocontrol Centre for Africa
B.P. 08-0932
Cotonou
Benin
e-mail: o.coulibaly@cgiar.org

Jacqueline d'A Hughes
AVRDC – The World Vegetable Center
P. O. Box 42
Shanhua
Tainan 74199
Taiwan, Republic of China

Barbara de Santis
GMO and Mycotoxins Unit
National Center for Food Quality and Risk Assessment
Italian National Institute for Health (ISS)
viale Regina Elena, 299
00161 Rome
Italy

Ineke M. de Vries
Plant Research International
P. O. Box 16
6700 AA Wageningen
The Netherlands

Francesca Debegnach
GMO and Mycotoxins Unit
National Center for Food Quality and Risk Assessment
Italian National Institute for Health (ISS)
viale Regina Elena, 299
00161 Rome
Italy

Stella Denloye
National Agency for Food & Drugs Administration and Control (NAFDAC)
3/4 Oshodi/Apapa Expressway
Lagos
Nigeria

Bamory Diarra
Institut d'Economie Rurale
B. P. 258
Rue Mohamed V
Bamako
Mali

William O. Ellis
Department of Biochemistry
Kwame Nkrumah University of Science and Technology
Kumasi
Ghana

Pascal Fandohan*
Program on Agricultural and Food Technology
National Institute of Agricultural Research
B. P. 128
Porto-Novo
Benin
e-mail: lta@intnet.bj

Simon C. Fialor
Department of Agricultural Economics, Agribusiness and Extension
Faculty of Agriculture
Kwame Nkrumah University of Science and Technology (KNUST)
Kumasi
Ghana

Johanna Fink-Gremmels*
Division of Veterinary Pharmacology, Pharmacy and Toxicology
Faculty of Veterinary Medicine
University of Utrecht
3508 TD Utrecht
The Netherlands
e-mail: J.Fink@uu.nl

Richard A. Frederiksen
8248 Raintree Drive NE
Albuquerque, New Mexico 87122
USA

Wentzel C. A. Gelderblom
PROMEC Unit
Medical Research Council
P. O. Box 19070
Tygerberg 7505
South Africa

Martin Gilmour
Masterfoods
Slough SL1 4JX
United Kingdom

Yun Yun Gong
Molecular Epidemiology Unit
Centre for Epidemiology and Biostatistics
Leeds Institute of Genetics, Health and Therapeutics
Faculty of Medicine and Health
University of Leeds
Leeds LS2 9JT
United Kingdom

Andrew J. Hall
Infectious Disease Epidemiology Unit
London School of Hygiene & Tropical Medicine
University of London
London WC1E 7HT
United Kingdom

Kerstin Hell*
International Institute of Tropical Agriculture (IITA)
Biocontrol Centre for Africa
B.P. 08-0932
Cotonou
Benin
e-mail: k.hell@cgiar.org

Said Hounkponou
International Institute of Tropical Agriculture
Biocontrol Centre for Africa
B.P. 08-0932
Cotonou
Benin

Henry J. Huebner
College of Veterinary Medicine
Faculty of Veterinary Toxicology
Texas A&M University
College Station, Texas 77843-4458
USA

Ramon Jaime-Garcia
Laboratory for Aflatoxin Reduction in Crops
Division of Plant Pathology and Microbiology
Department of Plant Sciences
University of Arizona
Tucson, Arizona 85721
USA

Braima James
International Institute of Tropical Agriculture (IITA)
PMB 5320, Oyo Road
Ibadan
Nigeria

Dan Jeffers
Dow AgroSciences de Mexico
Torre Corey, Av. Vallarta 6503, Piso 8
45010 Guadalajara, Jalisco
Mexico

Yi Jiang
Department of Epidemiology
School of Public Health
University of Alabama at Birmingham
1665 University Blvd., RPHB 217
Birmingham, Alabama 35294-0022
USA

Curtis M. Jolly
Department of Agricultural Economics and Rural Sociology
Auburn University
Auburn, Alabama 36849
USA

Pauline E. Jolly*
Department of Epidemiology
School of Public Health
University of Alabama at Birmingham
1665 University Blvd., RPHB 217
Birmingham, Alabama 35294-0022
USA
e-mail: jollyp@uab.edu

Richard Jones
International Crops Research Institute for the Semi-Arid Tropics (ICRISAT)
P. O. Box 39063
Nairobi 00623
Kenya

Gert H. J. Kema
Plant Research International
P. O. Box 16
6700 AA Wageningen
The Netherlands

Ondié Kodio
Institut d'Economie Rurale
B. P. 258,
Rue Mohamed V
Bamako
Mali

Kafui A. Kpodo*
Food Research Institute
Council for Scientific & Industrial Research (CSIR)
P. O. Box M.20
Accra
Ghana
e-mail: kpodofri@ghana.com

Jurgen Köhl
Plant Research International
P. O. Box 16
6700 AA Wageningen
The Netherlands

P. Lava Kumar
International Institute of Tropical Agriculture (IITA)
PMB 5320, Oyo Road
Ibadan
Nigeria

John F. Leslie
Department of Plant Pathology
Kansas State University
4024 Throckmorton Hall
Manhattan, Kansas 66506-5502
USA
e-mail: jfl@ksu.edu

Marianne Lindblom*
Kraft Foods
Upplands Väsby SE-194 86
Sweden
e-mail: mlindblom2@krafteurope.com

Antonio F. Logrieco*
Institute of Sciences of Food Production (ISPA)
National Research Council (CNR)
via Amendola 122/O
70126 Bari
Italy
e-mail: antonio.logrieco@ispa.cnr.it

Naresh Magan*
Applied Mycology Group
Cranfield Health
Cranfield University
Bedford MK43 0AL
United Kingdom
e-mail: n.magan@cranfield.ac.uk

Kehinde Makinde
Agro-Dealer Development Program
Alliance for a Green Revolution in Africa
FARA Secretariat
PMB CT 173 Cantonments
Accra
Ghana

Walter F. O. Marasas*
PROMEC Unit
Medical Research Council
P. O. Box 19070
Tygerberg 7505
South Africa
e-mail: marasasrika@telkomsa.net

ABEBE MENKIR*
International Institute of Tropical Agriculture (IITA)
PMB 5320, Oyo Road
Ibadan
Nigeria
e-mail: a.menkir@cgiar.org

MARINA MIRAGLIA*
GMO and Mycotoxins Unit
National Center for Food Quality and Risk Assessment
Italian National Institute for Health (ISS)
viale Regina Elena, 299
00161 Rome
Italy
e-mail: miraglia@iss.it

ANTONIO MORETTI
Institute of Sciences of Food Production (ISPA)
National Research Council (CNR)
via Amendola 122/O
70126 Bari
Italy

JIRO MURAKAMI
Centro Internacional de Mejoramiento de Maíz y Trigo (CIMMYT)
Apdo. Postal 6-641
06600 Mexico, D.F.
Mexico

IRENE N. NAWA
School of Agricultural Sciences
University of Zambia
P. O. Box 32379
Lusaka
Zambia

SHYAM N. NIGAM
International Crops Research Institute for the Semi-Arid Tropics (ICRISAT)
Patancheru, Andhra Pradesh 502 324
India

BONNY R. NTARE
ICRISAT
B. P. 320
Bamako
Mali

DAVID OFORI-ADJEI
Noguchi Memorial Institute for Med. Res.
University of Ghana
Legon, Accra
Ghana

MWANCHA N. OKIOMA*
Member of Parliament
Kenya National Assembly (Parliament)
Box 41842-00100 GPO
Nairobi
Kenya
e-mail: mwancha@yahoo.com

SARAH OLEMBO
Directorate for Rural Economy and Agriculture of the African Union
P. O. Box 3243
Addis Ababa
Ethiopia

MONICA OLSEN
National Food Administration
P. O. Box 622
SE-75126 Uppsala
Sweden

RODOMIRO ORTIZ*
Centro Internacional de Mejoramiento de Maíz y Trigo (CIMMYT)
Apdo. Postal 6-641
06600 Mexico, D.F.
Mexico
e-mail: r.ortiz@cgiar.org

GUNER ÖZAY*
Food Institute
TUBITAK Marmara Research Center
P. O. Box 21
41470 Gebze Kocaeli
Turkey
e-mail: Guner.Ozay@mam.gov.tr

Hayrettin Ozer
Food Institute
TUBITAK Marmara Research Center
P. O. Box 21
41470 Gebze Kocaeli
Turkey

Elena Pannunzi
GMO and Mycotoxins Unit
National Center for Food Quality and Risk Assessment
Italian National Institute for Health (ISS)
viale Regina Elena, 299
00161 Rome
Italy

Michelangelo Pascale*
Institute of Sciences of Food Production (ISPA)
National Research Council (CNR)
via Amendola 122/O
70126 Bari
Italy
e-mail: michelangelo.pascale@ispa.cnr.it

Giancarlo Perrone
Institute of Sciences of Food Production (ISPA)
National Research Council (CNR)
via Amendola 122/O
70126 Bari
Italy

Timothy D. Phillips*
College of Veterinary Medicine
Faculty of Veterinary Toxicology (VAPH)
Texas A&M University
College Station, Texas 77843-4458
USA
e-mail: TPhillips@cvm.tamu.edu

Maya Piñeiro*
Nutrition and Consumer Protection Division (AGNS)
Food Quality and Standards Service
FAO
Viale delle Terme di Caracalla
0100 Rome
Italy
e-mail: Maya.Pineiro@fao.org

Claudia Probst
Laboratory for Aflatoxin Reduction in Crops
Division of Plant Pathology and Microbiology
Department of Plant Sciences
University of Arizona
Tucson, Arizona 85721
USA

Kanniah Rajasekaran
USDA – Agricultural Research Service
Southern Regional Research Center
1100 Robert E. Lee Boulevard
New Orleans, Louisiana 70124
USA

S. Veera Reddy
International Crops Research Institute for the Semi-Arid Tropics (ICRISAT)
Patancheru, Andhra Pradesh 502 324
India

Brian G. Shelton*
Pathogen Control Associates, Ltd.
270 Scientific Drive, Suite No. 3
Norcross, Georgia 30092
e-mail: bshelton@pathcon.com

Gordon S. Shephard
PROMEC Unit
Medical Research Council
P. O. Box 19070
Tygerberg 7505
South Africa

Moses Siambi
International Crops Research Institute for the Semi-Arid Tropics (ICRISAT)
P. O. Box 1096
Lilongwe
Malawi

Bupe A. Siame*
Department of Biology
Trinity Western University
7600 Glover Road
Langley, British Columbia V2Y 1Y1
Canada
e-mail: anthony.siame@twu.ca

Richard Sikora
Soil Ecosystem, Phytopathology & Nematology
Universität Bonn
Nussallee
953115 Bonn
Germany

Aoua Traoré
ICRISAT
B. P. 320
Bamako
Mali

Paul C. Turner
Molecular Epidemiology Unit
Centre for Epidemiology and Biostatistics
Leeds Institute of Genetics, Health and Therapeutics
Faculty of Medicine and Health
University of Leeds
Leeds LS2 9JT
United Kingdom

Hari D. Upadhyaya
International Crops Research Institute for the Semi-Arid Tropics (ICRISAT)
Patancheru, Andhra Pradesh 502 324
India

Theo A. J. van der Lee
Plant Research International
P. O. Box 16
6700 AA Wageningen
The Netherlands

Frans Verstraete*
European Commission
Health and Consumer Protection Directorate-General
200, rue de la Loi – Office F101-4/ 56
B-1049 Brussels
Belgium
e-mail: frans.verstraete@ec.europa.eu

Angelo Visconti*
Institute of Sciences of Food Production (ISPA)
National Research Council (CNR)
via Amendola 122/O
70126 Bari
Italy
e-mail: angelo.visconti@ispa.cnr.it

Hester F. Vismer
PROMEC Unit
Medical Research Council
P. O. Box 19070
Tygerberg 7505
South Africa

Cees Waalwijk*
Plant Research International
P. O. Box 16
6700 AA Wageningen
The Netherlands
e-mail: cees.waalwijk@wur.nl

Farid Waliyar*
International Crops Research Institute for the Semi-Arid Tropics (ICRISAT)
Patancheru, Andhra Pradesh 502 324
India
e-mail: f.waliyar@cgiar.org

JIA SHENG WANG
Department of Environmental Toxicology
Division of Human Health Sciences
Texas Tech University
Box 4163
Lubbock, Texas 79409-1163
USA

CHRISTOPHER P. WILD*
Molecular Epidemiology Unit
Centre for Epidemiology and Biostatistics
Leeds Institute of Genetics, Health and Therapeutics
Faculty of Medicine and Health
University of Leeds
Leeds LS2 9JT
United Kingdom
e-mail: c.p.wild@leeds.ac.uk

JONATHAN H. WILLIAMS*
Peanut CRSP Management Entity
College of Agricultural and Environmental Sciences
University of Georgia
1109 Experiment Street
Griffin, Georgia 30223
USA
e-mail: twillia@griffin.uga.edu

XIUDE XU
Institute of Plant Protection
Liaoning Academy of Agricultural Sciences
Shenyang 110161
People's Republic of China

JOHN M. YOHE*
Management Entity
Sorghum, Millet & Other Grains CRSP
113 Biochemistry Hall
University of Nebraska
Lincoln, Nebraska 68583-0748
e-mail: jyohe@unlnotes.unl.edu

JIUJIANG YU
USDA – Agricultural Research Service
Southern Regional Research Center
1100 Robert E. Lee Boulevard
New Orleans, Louisiana 70124
USA

Overview and Introduction

1

The EU MycoGlobe Project: Global Integration of Mycotoxin and Toxigenic Fungal Research for Enhanced Food Safety

Angelo Visconti* and Giancarlo Perrone

Abstract

The MycoGlobe project is a Specific Support Action (SSA) of the European Commission (FOOD-CT-2004-007174) that builds on the outcome of a wide range of European research projects in the area of mycotoxins and toxigenic fungi by supporting, stimulating and facilitating the participation and cooperation with countries that have signed bilateral scientific and technological cooperation agreements with the EU. In particular, MycoGlobe aims to: i) create a consortium of international experts in the area of mycotoxins and toxigenic fungi; ii) identify priorities of common interest by sharing results of the European research with similar major international networks; iii) evaluate research policy in the SSA area through conferences focused on advanced research tools; and iv) improve collaboration in the SSA area by transferring knowledge to developing countries in Africa, Asia and South America and by sharing information with advanced third countries, *e.g.*, the United States and Australia through bilateral workshops and short-term exchanges.

Introduction

Huge quantities of food are wasted every year because they are invaded by toxic fungi or contaminated by fungal metabolic products. Such waste occurs most prominently in hotter countries where food shortages may already be a problem. One estimate (CAST, 2003) is that mycotoxins affect a quarter of the world's food crops, including many basic foodstuffs and animal feed, as well as crops, *e.g.*, coffee, with high economic value. Mycotoxins are toxic secondary metabolites of fungi that usually belong to one of three genera – *Aspergillus*, *Penicillium* or *Fusarium*. Mycotoxins may be detrimental to the health of humans and/or animals and may be produced on a wide range of agricultural commodities under a diverse range of conditions. Some of the mycotoxins, such as aflatoxin are among the most potent mutagenic and carcinogenic substances known (CAST, 2003). Mycotoxins are associated with many chronic health risks, including the induction of cancer, immune suppression, and digestive, blood and nerve defects (CAST, 2003, Shephard, 2006). Mycotoxins occur more frequently under tropical conditions. Diets in many developing countries are based on crops susceptible to mycotoxins, leading to high levels of chronic health problems in tropical, developing countries.

Since the discovery of aflatoxins in the early 1960s (Blount, 1961) researchers worldwide have made significant progress identifying and understanding the major classes of mycotoxins and the fungi that produce them. The economic consequences of mycotoxin contamination are profound, as crops contaminated with high levels of a mycotoxin often must be destroyed. Contaminated crops are sometimes diverted to animal feed which can reduce growth rates, lead to illness of animals consuming contaminated feeds and result in meat and milk containing toxic residues or biotransformation products. In the United States the mean economic annual costs of crop losses from the mycotoxins are estimated to be in the millions of US dollars (Robens and Cardwell, 2003). Regulations exist worldwide against the sale of contaminated commodities, and the economic effect of these regulations has been estimated (Wu, 2004). In spite of four decades of research worldwide establishing the extent of mycotoxin contamination, recurrent episodes of chronic and acute toxicity following mycotoxin contamination, and the extensive economic costs associated with this exposure, mycotoxin contamination is not considered an epidemic. Reports of outbreaks and deleterious effects worldwide often are contradictory. Diagnosis of mycotoxicoses in animals is difficult as the symptoms induced may be similar to other diseases with other causes. Moreover, the effects of mycotoxins often are synergistic (CAST, 2003).

Mycotoxins also affect European crops and recent EC Framework Programs (V and VI) have included projects to study toxic fungi in food crops. In particular 10 projects were merged into a cluster, the Mycotoxin Prevention Cluster, and involved more than 60 research groups from all EU Member States and several candidate countries (www.mycotoxin-prevention.com). This European network was strengthened by the Cost Action 835 on "Agriculturally important toxigenic fungi" (Logrieco and Visconti, 2004). To tackle the mycotoxin menace, a united effort worldwide is required, with better cooperation amongst researchers, particularly those in developed countries, so that developing countries can benefit from a coordinated effort to minimize the mycotoxin menace. The MycoGlobe Specific Support Action was set up to share the results of EU research with other countries that have bilateral science and technology agreements with the EU, within the wider framework of a global information system on mycotoxins and toxigenic fungi.

MycoGlobe Objectives

The MycoGlobe project (FOOD-CT-2004-007174) was funded as a Specific Support Actions (SSA) within Priority 5, "Food Quality and Safety," of the VIth Framework Program. MycoGlobe utilizes the outcome of a wide range of European research projects in the area of mycotoxins and toxigenic fungi by supporting, stimulating and facilitating the participation of and cooperation with countries that have signed bilateral scientific and technological cooperation agreements with the EU. The specific objectives of the project are: i) to create a consortium of major international experts in the area of mycotoxins and toxigenic fungi; ii) to identify priorities of common interest by sharing results of the European research with major international networks in the field; iii) to evaluate research policy in the SSA area by conferences focused on advanced research tools; and iv) to improve cooperation in the SSA area by transfer of knowledge to developing countries and sharing information with advanced third countries through bilateral workshops and short mobility missions.

Public anxiety about the quality and safety of food production in Europe has increased in recent years. MycoGlobe supports actions that transfer knowledge on agriculture systems in difficult environmental conditions and developed a working group of experts whose expertise encompasses the main knowledge of the food production and distribution chain in Europe and elsewhere, especially EC importer/exporter countries. European research from the Mycotoxin Prevention Cluster includes new detection techniques, hazard and risk assessment strategies, fungal infection prevention processes, biological control agents, and novel test kits.

Previous interactions with US mycotoxin researchers, both in Italy and in The Netherlands (Mycotoxin World Forum), confirmed that European researchers could benefit greatly by better understanding the approach and progress made in the United States. MycoGlobe also enabled collaboration between the EU and other countries to obtain information on mycotoxin crop contamination, to map mycotoxin occurrence in the most widely cultivated and exported crops, and to develop networks that include training, conferences and seminars.

MycoGlobe Structure

The MycoGlobe project consists of six Work packages (WPs), whose activities and objectives are detailed below and involve eight participants (Table 1, Fig. 1):

WP1. Steering Committee (SC).

WP2. Conference on "Integration of Mycotoxin and Toxigenic Fungi European Research for Food Safety in Global System" (October 22, 2004, Brussels).

WP3. Conference "Learning from the EC: Reducing Impact of Mycotoxins in Tropical Agriculture with Emphasis on Health and Trade in Africa" (September 13-16, 2005, Accra, Ghana).

WP4. Conference "Advances in Genomics, Biodiversity and Rapid System for Detection of Toxigenic Fungi and Mycotoxins" (September 26-29, 2006, Monopoli, Bari, Italy).

WP5. Workshop/Training course on "Mycotoxins and Toxigenic Fungi Detection Methods" (October 2-6, 2006, Bari, Italy).

WP6. Bilateral workshops with Australia and USA, respectively, Sydney (February 15-17, 2006) and New Orleans (July 5-7, 2005).

The Steering Committee has a key-role in: i) the selection of speakers and session chairpersons for the conferences; ii) preparation and production of the final report; and iii) ensuring the consistency of the scientific programs of the conferences and documents delivered as a result of the project.

All the conferences follow a similar outline: a series of presentations including keynote lectures, oral presentations grouped in different sessions and poster sessions. In each case, the goal is to coordinate research and exchange information at an international level on research in mycotoxins and toxigenic fungi. Conference sessions and discussions are targeted towards European scientists and key international scientists in the public and private sector, decision-makers and stakeholders. Special emphasis is placed on the new EU Member States, candidate countries and countries that have signed exchange agreements with the EU, as well as international organizations and research centers.

The kick-off conference in Brussels focused on the dissemination of the outcome of the Mycotoxin Prevention Cluster and other related projects of FP5 and FP6. Selected speakers from all continents provided a worldwide summary of research on mycotoxins and toxigenic fungi. The meeting was held in conjunction with the Mycotoxin Cluster dissemination meeting that was held the day before at the same location. A CD with pdf files con-

Table 1. Partners' expertise and role in the MycoGlobe project

Participant	Expertise	Role in the project
ISPA-CNR Italy	Mycotoxins, toxigenic fungi, analytical methods, mycology, genetic and genomics, new detection methods	Coordinator, organize WP4 Conference and WP5 Training course; coordinate SC activities
Cranfield United Kingdom	Ecophysiology of spoilage and toxigenic fungi, HACCP for improving food chain safety, mycology	Organize WP2 Conference and collaborate with ISPA in WP5 training course; participate in SC
IITA Nigeria	Food security, agricultural production, management of natural resources, aflatoxin problems	Organize WP3 Conference, improve the dissemination of results in Africa; participate in SC
USDA United States	Food and feed safety, aflatoxin genomics and functional genomics, biochemistry	Organize WP6 bilateral workshop in USA; participate in SC; contribute to WP5 training course
KSU United States	Plant pathology, fungal genetics and genomics	Organize WP6 bilateral workshop in USA and *Fusarium* Laboratory Workshop in Italy; participate in SC
ICRISAT India	Crop productivity, food security, management of environment, aflatoxin research (resistance, storage, agronomics, biological control)	Disseminate results in Asian countries; participate in conferences and SC
RBG Australia	Taxonomy, ecology and phylogenetics of fungi (in particular *Fusarium*)	Organize WP6 bilateral workshop in Australia; participate in SC contribute to *Fusarium* Laboratory Workshop in Italy
UNRC Argentina	Microbiology, chemistry, agronomy, toxigenic fungi and mycotoxins in plant products	Organize conference in Argentina; disseminate results in South American countries; participate in SC

ISPA-CNR: Institute of Sciences of Food Production, CNR, Bari, Italy; **Cranfield:** Applied Mycology Group, Cranfield University, Silsoe, UK; **IITA:** International Institute of Tropical Agriculture, Ibadan, Nigeria; **USDA:** United States Department of Agriculture, New Orleans, LA, USA; **KSU:** Department of Plant Pathology, Kansas State University, Manhattan, KS, USA; **RBG:** Royal Botanic Garden and Domain Trust, Sydney, Australia; **UNRC:** Department of Microbiology and Immunology, The National University of Rio Cuarto, Rio Cuarto, Argentina; **ICRISAT:** International Crops Research Institute for the Semi-Arid Tropics, Patancheru, India.

taining the main results of the research in the area of mycotoxins and toxigenic fungi in relation to food safety was made and distributed to partners and other interested persons.

The African conference in Accra (Ghana) focused on the transfer of information for reducing the impact on mycotoxins in tropical agriculture, whose proceedings are summarized in the present volume. The conference was a medium to exchange information on current situation of mycotoxins in food and export crops, detection methods, institutional agreements and policies associated with public health and trade. The goal of the conference was to recommend priority actions in the areas of technical, institutional and policy options for improving public health and trade through management of mycotoxins in pre-harvest

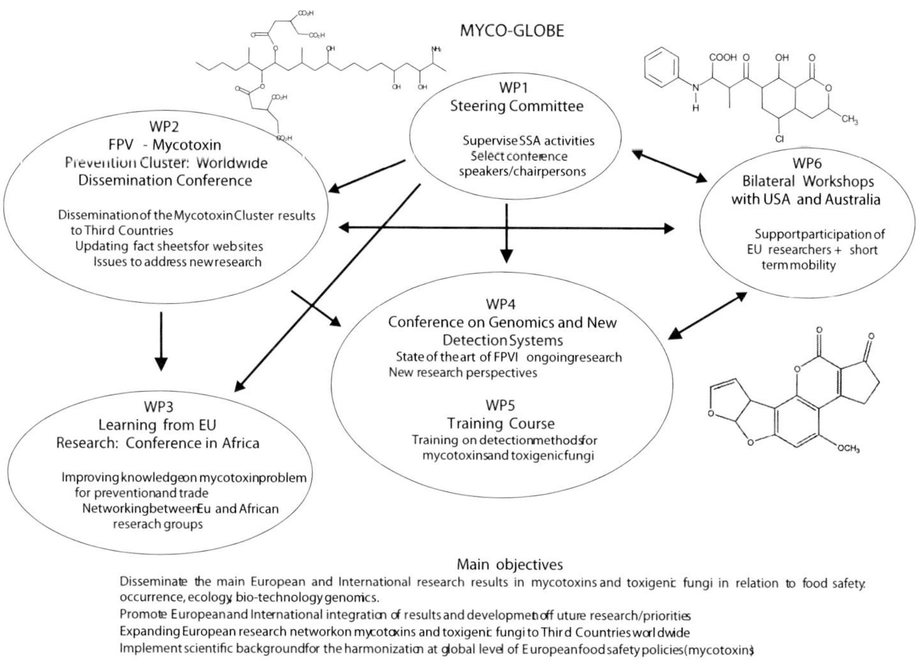

Figure 1. Graphical presentation of work packages

and post-harvest conditions (see Chapter 3). One hundred and nine participants from 28 countries in Africa, Europe, Asia, North America and South America attended the conference. Participants included scientists, ministers and parliamentarians, heads of institutions, policy makers, trade specialists and health experts from government and non-governmental organizations and the private sector.

The conference in Monopoli (Bari) focused on advances in genomics, biodiversity and rapid systems for detection of toxigenic fungi and mycotoxins, with the goal of reducing mycotoxin contamination and increasing food security, food safety, and the profitability of agriculture and related industries. The following topics were discussed: i) significance of fungal genomics in understanding toxin synthesis; ii) the role of genomics in establishing biodiversity; iii) use of information on genetics and biodiversity to develop systems for rapid detection of toxigenic fungi; and iv) analytical methods for rapid detection of mycotoxins. The conference attracted 120 participants from 30 countries in Africa, Europe, Asia, North America, South America and Australia. Participants included scientists, heads of institutions, and trade and health specialists from government and non-government organizations and the private sector. A program booklet (114 pages) containing abstracts of 37 oral presentations and 40 posters was published and distributed to participants.

MycoGlobe sponsored two training activities. One was the 2006 *Fusarium* Laboratory workshop, which was held at the University of Bari, Faculty of Biotechnology, June 4-9, 2006 with the participation of 39 students and researchers. The workshop was taught by seven international *Fusarium* experts. Participants were introduced to standard morphological, genetic and molecular biological techniques used to identify and characterize strains of *Fusarium*. This workshop was similar to those held elsewhere since 2000 (USA 2000, 2001, 2003, 2005, 2007; Australia – 2002, South Africa – 2004, and Malaysia – 2008; Leslie and Summerell, 2006).

A training course also followed the Bari conference in September 2006 with the objective of upgrading skills of persons involved in detection methods for mycotoxins and toxigenic fungi. The goal of this course was to provide hands-on experience with traditional methods and new molecular and immunological systems for rapid, robust and user-friendly identification of mycotoxins and toxigenic fungi in the food chain. Participation was limited to 10 students/young researchers from developing countries (fully supported by MycoGlobe) and 10 participants from other countries.

A conference in Argentina also was sponsored in connection with the "Vth Latin American Congress on Mycotoxicology" in March 2006 and followed the general outlines of the Accra and Bari MycoGlobe conferences, but with simultaneous translation to/from English/Spanish. A total of 145 researchers and professionals from public organizations (universities, quality control laboratories) and private companies attended the conference and 16 countries were represented.

A three-day EU-USA bilateral workshop was held in New Orleans (July 2005) and the main topics were: biodiversity, genomics, and advanced methods for detection of mycotoxins and toxigenic fungi. The primary aim of the workshop was to provide a common platform and facilitate academic interactions among well-established scientists from the EU and the United States with the following objectives: i) information dissemination on (a) technical and economic problems encountered in the EU and the United States with respect to mycotoxin contamination, (b) current state of research on mycotoxins, and (c) identifying future scientific strategies for combating fungal invasion and mycotoxin contamination of foods and feeds, and for developing rapid detection methods for mycotoxins and toxigenic fungi; ii) establishing and designing collaborative research activities.

The EU-Australia bilateral workshop in Sydney, Australia (February 2006) focused on technical issues related to mycotoxins and food safety, with particular emphasis on: regulatory issues and international developments, research progress on major mycotoxins on both continents, identification of emerging problems, biodiversity and detection of toxigenic fungi, and advances in mycotoxin prevention and control.

The formation of the International Society for Mycotoxicology (www.mycotox-society.org) was discussed at the New Orleans and Accra meetings, and the Society launched at the 3rd World Mycotoxin Forum in November 2005 in Nordwijk (The Netherlands). The journal *Food Additives and Contaminants* was affiliated with the Society. The conclusions and recommendations from the two conferences should alter food safety regulation and research.

Socio-economic significance

Mycotoxin contamination is a worldwide problem of special importance in the developing countries of Africa, Asia and South America from which the EU often imports large quantities of food. A key hurdle for exporting food crops to the EU is meeting the EU regulations and standards. Thus, food safety and quality assurance programs are required to improve food production chains for economically important crops.

The MycoGlobe project has a significant impact by combining education and information dissemination programs with a training component that includes information on HACCP and risk analysis thereby enabling a whole food chain approach that eases international trade. The general objective of international cooperation activities under the VIth Framework Program is to open European research to the rest of the world. These activities represent a particular contribution of the Framework Program to this opening-up process. Success of this project will lead to: i) enhanced productivity and export potential of agriculture from tropical Africa and other third world countries in Asia and South America; ii) establishment of a global collaborative network in the field of mycotoxins and toxigenic fungi to help limit losses due to mycotoxins; iii) transfer of knowledge to developing countries; iv) more exchange of information between advanced countries; and v) improved coordination of EU food safety policies with those of other countries and international bodies.

Conclusion

The MycoGlobe project has created a network of mycotoxin specialists keen on information exchange and has resulted in submission of joint project proposals to generate a better understanding of the toxigenic fungi and the mycotoxins they produce with the overall goal of improving of food safety in domestic and international markets. Various donors and policy makers in Europe, Africa and South America have been sensitized to the importance of mycotoxins in enhancing food security and trade. Several scientists, technicians, policy makers and institutional specialists have been brought under one umbrella to help manage mycotoxins in a holistic manner. This interdisciplinary network of mycotoxin professionals should continue their interactions to improve the safety of the food supply globally beyond the life of the formal program.

The project also provides a mid-term contribution to European standards on the matter of mycotoxins and toxigenic fungi in food safety through the recommendations and deliverables of the conferences, which can be integrated into the general framework of regulations and their standardization/harmonization.

References

Blount, W.P. (1961) Turkey "X" disease. *Journal of the British Turkey Federation* 9, 52-54.

CAST. (2003) *Mycotoxins: Risks in Plant, Animal and Human Systems*. Council for Agricultural Sciences and Technology, Ames, Iowa, USA.

Leslie, J.F. and Summerell, B.A. (2006) *Fusarium* Laboratory Workshops – A recent history. *Mycotoxin Research* 22, 73-74.

Logrieco, A. Visconti, A. (2004) *An Overview on Toxigenic Fungi and Mycotoxins in Europe*. Kluwer Academic Publishers, Dordrecht, The Netherlands.

Robens, J. and Cardwell, K. (2003) The costs of mycotoxin management to the USA: Management of aflatoxins in the United States. *Journal of Toxicology: Toxin Reviews* 22, 139-152.

Shephard, G.S. (2006) Mycotoxins in the context of food risks and nutrition issues. In: Barug, D., Bhatnagar, D., van Egmond, H.P., van der Kamp, J.W., van Osenbruggen, W.A. and Visconti, A. (eds.) *The Mycotoxin Factbook: Food and Feed Topics*. Wageningen Academic Publishers, Wageningen, The Netherlands, pp. 21-36.

Wu, F. (2004) Mycotoxin risk assessment for the purpose of setting international regulatory standards. *Environmental Science and Technology* 38, 4049-4055.

2

IITA's Research-for-Development Agenda for Africa

Stanford Blade*

Abstract

Agriculture is a primary driver for economic development in Africa. African agriculture must benefit from the research that can be conducted with the full arsenal of cutting-edge scientific tools to ensure food security, minimization of risk and the creation of wealth. The International Institute of Tropical Agriculture (IITA) is committed to assisting the African agricultural sector realize its potential through a research-for-development approach that ensures that innovative technologies reach the hands of users. IITA partners with advanced research institutions, national partners, the private sector and non-governmental organizations to keep improving technologies based on the suggestions and insight of users.

Introduction

IITA has contributed to the improvement of African cropping systems productivity through the development of innovative technologies. The importance of the agricultural sector, which employs 2/3 of the population of sub-Saharan Africa, ensures that agriculture plays a key role in the continent's economic development (InterAcademy Council, 2004).

Agriculture is not just the sowing of a seed, nor the milking of a cow. It is instead a complex network of skills and expertise that begins with the conception of an idea for a specific agricultural product and is complete only when it nourishes a satisfied customer. This process can range from a farmer knowing when to plant her yam crop so that she can prepare a nutritious meal for her family following a bountiful harvest, to the investment in the infrastructure and organization needed for African pineapples to be marketed world-wide to consumers who are willing to pay a premium for quality products.

Agriculture is an information-intense industry. Producers plan their production based on economic opportunity, family responsibilities, resource limitations and their knowledge of the natural environment. Successful production requires an intricate set of knowledge-based decisions on the timing and rigor of management. Marketing decisions are influenced by local needs, regional markets, the possibility of local processing and the potential for post-harvest losses. Products that leave the farm are affected by the needs (and the opportu-

[1] Address correspondence to iita@iita.org.

Table 1. United Nations (2005) Millennium Development Goals and Targets

Goal 1	**Eradicate extreme poverty and hunger**
	Target 1. Halve, between 1990 and 2015, the proportion of people whose income is < US$1.00 per day
	Target 2. Halve, between 1990 and 2015, the proportion of people who suffer from hunger
Goal 2	**Achieve universal primary education**
	Target 3. Ensure that, by 2015, children everywhere, both boys and girls, can complete a full course of primary schooling
Goal 3	**Promote gender equality and empower women**
	Target 4. Eliminate gender disparity in primary and secondary education, preferably by 2005, and in all levels of education no later than 2015
Goal 4	**Reduce child mortality**
	Target 5. Reduce by two-thirds, between 1990 and 2015, the under-five mortality rate
Goal 5	**Improve maternal health**
	Target 6. Reduce by 75%, between 1990 and 2015, the maternal mortality rate
Goal 6	**Combat HIV/AIDS, malaria, and other diseases**
	Target 7. Have halted by 2015 and begun to reverse the spread of HIV/AIDS
	Target 8. Have halted by 2015 and begun to reverse the incidence of malaria and other major diseases
Goal 7	**Ensure environmental sustainability**
	Target 9. Integrate the principles of sustainable development into country policies and programs and reverse the loss of environmental resources
	Target 10. Halve, by 2015, the proportion of people without sustainable access to safe drinking water and basic sanitation
	Target 11. Have achieved by 2020 a significant improvement in the lives of at least 100 million slum dwellers
Goal 8	**Develop a global partnership for development**
	Target 12. Develop further an open, rule-based, predictable, nondiscriminatory trading and financial system (includes a commitment to good governance, development, and poverty reduction – both nationally and internationally)
	Target 13. Address the special needs of the Least Developed Countries (includes tariff- and quota-free access for Least Developed Countries' exports, enhanced program of debt relief for heavily indebted poor countries and cancellation of official bilateral debt, and more generous official development assistance for countries committed to poverty reduction)
	Target 14. Address the special needs of landlocked developing countries and small island developing states (through the Program of Action for the Sustainable Development of Small Island Developing States and 22nd General Assembly provisions)

Table 1 (continued). United Nations (2005) Millennium Development Goals and Targets

Goal 8 Develop a global partnership for development

Target 15. Deal comprehensively with the debt problems of developing countries through national and international measures in order to make debt sustainable in the long term

Targets listed below may be monitored separately for the least developed countries, Africa, landlocked developing countries, and small island developing states

Target 16. In cooperation with developing countries, develop and implement strategies for decent and productive work for youth

Target 17. In cooperation with pharmaceutical companies, provide access to affordable essential drugs in developing countries

Target 18. In cooperation with the private sector, make available the benefits of new technologies, especially information and communications technologies

nities they present) of transporters, consolidators, traders and the retailers who finally sell the item to a consumer. Beyond this network there are additional spheres of influence: input supply, government policy, regional trading networks and globalization. African agriculture is not simple, and should not be simplified for the sake of convenience.

The importance of African agriculture means that it must be given primary consideration if the Millennium Development Goals are to be met by 2015 (Table 1). The first goal is to reduce the number of people who live in poverty (with the specific measure of halving the number of people who live on US$ 1.00 per day or less). The interaction between agriculture and the natural resource base will directly affect the achievement of Millennium Development Goal (#7) of ensuring environmental sustainability. Successful development of new agricultural opportunities within Africa also will influence other Millennium Development Goals on nutrition, health care and education for vulnerable groups, *e.g.*, children, women and those who face debilitating diseases such as HIV/AIDS, malaria and tuberculosis.

Agriculture encompasses a multiplicity of stakeholders and systems that lead from the "field to the fork." Agricultural research has created important innovations around the world to address needs such as new cassava processing machineries to promote cassava as an industrial raw material or commercializing a bio-pesticide to combat desert locust or high-yielding maize cultivars with aflatoxin resistance. Such work also can capture diversification opportunities (Blade *et al.*, 2002) and exploit new niches that develop within the agricultural sector. IITA works with partners within Africa and beyond to enhance crop quality and productivity to impact the lives of poor people within the continent (both rural and urban). In addition, the Institute develops technologies for Africans who possess the expertise, initiative and resources to go beyond food security and to produce enough food or fiber to realize a financial profit. If African agriculture is to serve as an engine of economic development, IITA understands that it also must produce research to model how industries and enterprises succeed (as well as maintain their success).

IITA researchers have responded to the immense impact that mycotoxins have on African populations through direct health effects as well as a potential trade limitation. IITA researchers have been actively seeking new maize genotypes that are less heavily colonized by mycotoxin producing fungi, biocontrol mechanisms to reduce the impact of toxigenic strains of fungi, and post-harvest management strategies to decrease the problems associated with mycotoxigenic fungal contamination. IITA research has led to partnerships that have brought attention to this issue across West Africa (including 10 million people in Ghana, Togo and Benin).

Delivering the African agricultural development potential: The IITA vision

Although little appreciated outside the region, the combined national and international agricultural research efforts on food crops since the mid-1970s have brought very significant benefits to sub-Saharan Africa (IITA, 2000). These benefits can clearly be seen with two of the major food crops, cassava and maize. More productive varieties that are resistant to the prevailing diseases have been introduced, and the effective biological control of the cassava mealy bug has prevented large-scale famine. Without these research efforts, 25% less maize would be produced currently in sub-Saharan Africa, *i.e.*, the food requirements of 40 million people, and cassava production would be 50% less, *i.e.*, the food requirements of 65 million people. The research on these two crops alone has meant that > 100 million more Africans can be fed.

Increasing agricultural production, as well as farm productivity and profitability, in sub-Saharan Africa, where soils are infertile, rainfall is erratic, inputs are not regularly available, and most of the farmers have few technology options at their disposal, is a most challenging mission. Moreover, the local capacity to undertake the full range of agricultural research and development activities remains generally weak even in comparison to other developing regions, such as those in South America or Southeast Asia.

Increased globalization offers new opportunities and challenges for agricultural research and development, such as better access to information and new technologies. At the same time increased competitiveness requires, for example, the development of more cost-efficient production practices and well-structured marketing systems. Research should effectively explore the entire food production and distribution system and address these challenges for the benefit of agricultural development in sub-Saharan Africa.

Guiding principles for research

IITA aims to improve food systems in sub-Saharan Africa in a sustainable manner through the following:

- Adopting a "systems" orientation throughout the institute to identify and focus on the critical components of the system and put together improved elements from different disciplines. This includes:
  - Giving high priority to the sustainable management of natural resources, the preservation of biodiversity, and the strengthening of system resilience in the face of socioeconomic and environmental changes.
  - Conducting research in areas such as social science, soil and crop management, enhanced post-harvest utilization and food safety, and breeding for increased production, quality, and tolerance to various stresses.
  - Undertaking crop improvement activities on all crops as required to fulfill the eco-regional mandate, with special emphasis on cassava, maize, and soybean for Africa, and cowpea, plantain and banana, and yam at the global level.
  - Protecting and preserving what is produced by using the institute's capacity to deal

with introduced pests through biological control, and with endemic pests, including post-harvest pests, through integrated pest management.
  - Shifting emphasis to higher external input use and sustainable strategies for land-use and agricultural intensification as alternatives to slash-and-burn cultivation and other forms of subsistence agriculture, and decreasing the emphasis on low external input strategies, on which IITA has more than 30 years of research experience.
- Analyzing market opportunities for existing and novel value-added products and developing appropriate post-harvest systems, resulting in commercially viable farm and agro-industrial enterprises.
- Evaluating and developing specific policy issues that help regional bodies and national and local governments create an environment conducive to agricultural development, reduce poverty, and limit the effects of major human diseases and global warming.
- Defining and using poverty indicators to ensure that the institute's research and related activities have a positive, sustainable impact on the livelihoods and well-being of low-income people.
- Exploring fully the benefits derived from the optimal use of modern science, information technology and collaborative alliances.

Partnerships

IITA enters into partnerships at international, regional, national, and local levels, but especially forms strong partnerships with the national and regional bodies within its mandate region. The three regional agricultural research organizations, the Association for Strengthening Agricultural Research in Eastern and Central Africa (ASARECA), the West and Central African Council for Agricultural Research and Development (CORAF/WECARD), and the Southern Africa Centre for Cooperation in Agricultural Research and Training (SADC/SACCAR) currently provide strong leadership for regional research planning and coordination. Following an extensive priority-setting exercise and broad consultation with their stakeholders each organization prepared a strategic plan that together form an important basis for the development of IITA's research agenda.

IITA also collaborates closely with other CGIAR centers to increase synergy and participates in appropriate CGIAR system-wide and eco-regional programs. IITA also works closely with other CGIAR members to develop regional programs, particularly in West and Central Africa and the Great Lakes Region of Africa, and in global collaborations such as the Generation, Harvest Plus and the Sub-Saharan Challenge Programs.

The research agenda

The IITA research agenda is heavily influenced by past and future trends in sub-Saharan Africa (IITA, 2005). Agricultural land available per rural inhabitant is continuously declining and urban populations continue growing rapidly. By 2010, 60% of the economically active population in sub-Saharan Africa is projected to be employed in agriculture. At the same time there will be a rapid growth in demand for food, both in quantity and quality. Although current production practices are unsustainable, appropriate technologies can lead to very positive results in sub-Saharan Africa. The strategic plans and priorities developed by the three regional agricultural research organizations provide an initial research agenda that is further refined through the analysis of specific agricultural development challenges and research needs in the agroecological zones in which IITA is working.

Changes in the IITA research agenda since the mid-1990s include: (i) a shift in focus to encompass all of the humid forest, moist savanna, and mid-altitude agroecological zones of sub-Saharan Africa, (ii) a move from multidisciplinary crop programs based on divisional program arrangements to a project-based management system, and (iii) the evolution of an agroecological zone focus. These changes have occurred as unrestricted core funding declined.

For the next 10 years, a number of strategic issues related to agricultural research and development in sub-Saharan Africa influencing IITA's research agenda include:

- Increased agricultural productivity with emphasis on improved food quality and commercial competitiveness.
- Use of new technologies throughout the research-technology transfer development continuum to increase impact.
- A broadened research agenda with respect to crops, post-harvest technologies, income-generating opportunities, marketing, food quality, and policy options to reduce poverty.
- The prevailing need to improve soil fertility and sequester more carbon through increased use of organic and inorganic fertilizers.
- The impact of the HIV/AIDS pandemic and other major diseases on rural populations.
- The need for increased synergy and collaboration between crop and livestock production sectors.
- Optimized use of biotechnology in agricultural research and development, and effective collaboration with advanced research institutes.
- Increasing farm labor productivity and availability.
- More effective and less labor-intensive weed management practices.
- Greater use of integrated pest management methods.
- Environmental hazards and human safety issues related to the use of pesticides, particularly in peri-urban agriculture.
- Strengthened partnerships with NARS, NGOs, farmer organizations, and the private sector, including the devolution of activities to these groups.
- Impact of intellectual property rights on the availability of improved technologies.
- Continued strengthening of NARS capacity.
- Impacts of increased globalization and global climate change.

While addressing these issues, assumptions made regarding the execution of IITA's research agenda include:

- IITA will take a holistic approach, through which constraints and opportunities are assessed at the farm household and community levels to reduce poverty without destroying the resource base.
- IITA will work on the genetic improvement of its current crops and related production systems, and will add crops that provide additional income opportunities.
- Research will be conducted in representative benchmark areas, and then extrapolated to the wider agroecological zone.
- Maintain an appropriate balance between strategic and applied research.
- Focus research on small- and medium-scale farmers, with emphasis on increased productivity through intensified production systems.
- IITA and its partners will retain access to the required research disciplines.
- All activities will contribute to impact and development.

The research-for-development model

During a discussion with Dr. Marco Quinnoes, a member of Nobel Laureate Dr. Norman Borlaug's team, Dr. Quinnoes described how the group created the first "green revolution" in Mexico and the second in India (Hartmann, 2005). Dr. Quinnoes thought that there would have been **no** "green revolution" to speak of, if they had stopped at research. In Mexico in the 1950s, the extension service was dysfunctional or non-existent, similar to the situation now in some African countries. For their research to have any effect, the Borlaug team helped locals to organize, led them, trained them, set up multiplication systems, and worked with farmers to produce seed. The research-to-development model used in Mexico also was followed, out of necessity, in the second success in Asia. In India, the Borlaug team had to organize systems to move improved cultivars into the country, and worked with local groups to design systems that bypassed uncooperative research institutions, and conducted "trials" directly with farmers until the benefits were obvious even to the skeptics. Simultaneously they also worked, as IITA does today, with policy makers and legislators to convey the urgency and importance of the task.

The research community must produce needed technologies if Africa is to enjoy the full benefit of its agricultural resources. IITA works with key partners to ensure that innovative ideas result in more money in the hands of the end users.

References

Blade, S. F., Clayton G. W. and Lyon, D. (2002) Managing agricultural biodiversity: Opportunities and challenges for crop diversification in the Northern Great Plains. *Agronomy Journal* 94,173-174.

Hartmann, P. (2005) Research-for-development: My view. Centre Directors Committee Discussion Paper, Consultative Group on International Agricultural Research.

InterAcademy Council. (2004) Realizing the promise and potential of African agriculture: Science and technology strategies for improving agricultural productivity and food security in Africa. InterAcademy Council, Amsterdam, The Netherlands.

International Institute of Tropical Agriculture. (2000) IITA Strategic Plan. IITA Press, Ibadan, Nigeria.

International Institute of Tropical Agriculture. (2005) 2004 Annual Report. IITA Press, Ibadan, Nigeria.

United Nations. (2005) The Millennium Development Goals report. Department of Public Information, United Nations, New York, USA.

3

Priorities for Mycotoxin Research in Africa Identified by Using the Nominal Group Discussion Technique

Ranajit Bandyopadhyay*, John F. Leslie
and Richard A. Frederiksen

The Nominal Group technique was used as the discussion format for small group discussions during a conference in Accra sponsored by the MycoGlobe project of the European Union, the International Institute of Tropical Agriculture, INTSORMIL and the Council of Scientific and Industrial Research (Ghana). This chapter is based on these group discussions. The Nominal Group technique was originally developed for enhancing the productivity of work conferences and has its roots in social-psychological studies of decision conferences, management science studies of pooled group judgments, and social work studies on program planning (Delbecq *et al.*, 1975). The three most widespread applications of the Nominal Group technique are identifying problems, developing solutions, and establishing priorities. Recently, the Nominal Group technique has gained extensive recognition and wide application in conferences scheduled by health, social service, education, industrial and governmental organizations (Potter *et al.*, 2004). A few agriculture-related conferences also have used the Nominal Group technique (Durbin *et al.*, 1980; Mughogho, 1984; Leslie and Frederiksen, 1995).

The Nominal Group technique is an effective "face-to-face" discussion procedure for developing a large number of new or unique ideas and for increasing the total number of high-quality ideas proposed. A single train of thought is avoided in discussions that use the Nominal Group technique since the discussion format is tightly defined. Unlike many other decision-making processes, the Nominal Group technique avoids peer pressure that may keep some individuals from freely expressing their ideas thereby inhibiting their participation in making creative decisions. The Nominal Group technique allows minority opinions and ideas to be heard and integrated into the conclusions whenever valid. Finally, all participants have an equal opportunity to influence the direction of the group's decisions. This method enhances freedom for group discussion, avoids making "quick" decisions, and improves the quality of the proposals finally made. Good ideas are not lost in the progress of discussions, and adequate time is allocated for each suggestion.

The amount and quality of preparation prior to the meeting determines the efficiency and constructiveness of the Nominal Group technique. The participants need to be provided with relevant background material and an understanding of the desired outcomes. It is necessary to focus on a small number of key questions and to follow a strict pattern of procedure throughout the meeting.

Table 1. Nominal group responses by two groups to the statement, "How can health issues related to mycotoxin exposure in Africa be mitigated?"

1		2		Response
#[a]	S[b]	#	S	
9	37	8	29	Intensify multi-pronged awareness campaigns targeting relevant stakeholders to promote information dissemination on mycotoxin-related health risks and best-bet practices for their mitigation
4	15	4	11	Sensitize policy-makers on food safety issues and obtain political commitment to fund agriculture interventions to reduce mycotoxin risk
5	16	3	7	Educate farmers and consumers on good management practices to reduce mycotoxin contamination during storage and processing
3	9	4	14	Study and popularize use of enterosorption clays, and of other food supplements as interventions to reduce bio-availability of mycotoxins
4	7	3	8	Improve analytical capabilities for mycotoxin analysis in accredited labs
4	9	1	3	Develop and disseminate good agriculture practices and biocontrol methods
1	1	2	9	Invest in additional research to further demonstrate health effects (including sub-clinical effects) of mycotoxins
0[c]	0	2	10	Create an intersectorial forum on mycotoxins and health-related issues
3	7	1	2	Diversify diets by including less mycotoxin-prone food (*e.g.*, sorghum and millet) and procuring food from geographic areas with low mycotoxin risk
1	4	1	5	Develop resistant varieties using conventional and GM approaches
2	6	1	2	Promote interdisciplinary and international collaboration
2	6	1	1	Increase crop productivity and food security enabling people to access good quality food
1	3	2	4	Improve mycotoxin crisis management systems by incorporating risk-forecasting and risk-mitigation protocols
1	2	1	5	Establish and enforce regional standards to reduce mycotoxin exposure
2	5	0[c]	0	Implement a system for routine and effective surveillance for mycotoxins
1	4	1	2	Improve farmers' ability to identify contaminated grains to minimize consumption of visibly moldy foods
1	3	0	0	Vaccinate people against HBV to reduce liver damage due to aflatoxins
0	0	2	2	Identify uses of contaminated foods other than for human consumption
0	0	1	2	Train health professionals and equip labs to diagnose mycotoxin exposure
0	0	1	1	Conduct studies to estimate economic impact of production losses and human exposure to mycotoxins
1	1	0	0	Improve nutrition of people to reduce the health effect of mycotoxins
0	0	0	0	Fund farmers/consumers to conduct participatory research on mycotoxins
0	0	0	0	Use biocontrol interventions to reduce mycotoxin risks
0	0	0	0	Implement vaccination programs when mycotoxin exposure is lowest
0	0	0	0	Develop cheaper mycotoxin reduction methods for adoption by stakeholders

[a]# Number of persons ranking this response as one of the five most important.

[b]S Weighted priority score, with each member allowed to rank five topics, giving five points to the most important and one to the least.

[c]0 This topic was identified by at least one member of this group, but was not identified as one of the five most important topics by any member of the group.

Table 2. Nominal group responses by three groups to the statement, "Identify approaches you recommend to foster and integrate agriculture, nutrition and health research for mycotoxin management in Africa".

1		2		3		Response
#[a]	S[b]	#	S	#	S	
5	15	11	33	6	23	Establish nationwide and African mycotoxin task forces in association with regional and international organizations to conduct regionally harmonized activities related to agriculture and health
16	52	3	11	2	8	Encourage integration of expertise from agriculture, trade and health sectors through joint funding mechanisms from agriculture, health and finance ministries
7	24	9	19	8	24	Support awareness campaigns and share information on the impact of mycotoxins on health with the community
2	9	1	3	13	34	Develop and disseminate pre- and post-harvest and medical intervention strategies with participation of stakeholders from agriculture and health sectors
3	6	1	2	11	33	Develop and foster a forum for networking opportunities with stakeholders from agriculture, food and health sectors with interest in mycotoxin
2	4	2	6	2	6	Establish mycotoxin-testing laboratories managed by trained personnel who can serve common needs of health, agriculture and trade professionals
4	13	*[c]	*	*	*	Develop regional and national policies for mycotoxin management and promote political commitments for their implementation
*	*	4	11	*	*	Organize joint surveys by agriculture and health professionals to establish the occurrence, intensity and hazards posed by mycotoxins to aid in risk assessment
1	3	*	*	2	3	Further intensify cause and effect medical research to link impact of mycotoxin on health of people

[a]# Number of persons ranking this response as one of the five most important.
[b]S Weighted priority score, with each member allowed to rank five topics, giving five points to the most important and one to the least.
[c]* This topic not identified by any member of the group.

We took several steps to help insure the success of this discussion process. First, a book of abstracts of papers presented during the conference was provided to all participants. Second, all the participants had an understanding of the broad topics for discussion since it took place after all the papers were presented. Thus, all of the participants had a written summary of the thoughts of their fellow participants to which they could refer during the meeting. Third, the rationale and method used in Nominal Group discussions was explained to all participants, and a mock practice session preceded the smaller group discussions. Finally, the members of the groups were mixed in terms of their country of origin, their scientific background, and their area of interest. We prepared five questions that were used to focus the discussion within the nominal groups, with results of the discussion of four of these questions included here.

Table 3. Nominal group responses by four groups to the statement, "Identify institutional innovations needed for African farmers to produce and market safe food both domestically and internationally."

1		2		3		4		Response
#[a]	S[b]	#	S	#	S	#	S	
11	40	10	24	8	27	13	44	Intensify public education on and communication of food contaminants and mycotoxins, including trade issues and international requirements for export
5	13	4	13	6	19	12	31	Establish market inspection and lab services, and implement mechanisms for enforcement of regulations
3	7	7	29	1	4	1	3	Implement national action plan for practicing good agricultural practices
2	5	3	9	2	7	1	5	Improve infrastructure and create quality assurance and certification schemes to augment production for export
0[c]	0	2	7	4	10	1	4	Foster demand-driven interdisciplinary research and collaboration on food safety with international labs
0	0	2	4	*[d]	*	5	17	Create cooperatives for marketing and commodity boards
2	6	1	3	4	10	*	*	Improve infrastructure and transport system for rapid movement of domestic and export agriculture goods
2	2	0	0	2	7	4	8	Enhance capacity of people and improve research facilities for mycotoxin analysis
4	11	*	*	1	5	*	*	Improve food processing technology and infrastructure to minimize mycotoxin contamination in food products
*	*	1	1	4	16	1	1	Provide capital and economic incentives to farmers for producing high quality food
*	*	2	4	*	*	1	3	Make available consistent intervention policies for food safety in mycotoxin risk areas
1	1	1	5	*	*	*	*	Create a rapid alert reporting system of mycotoxin-related events using agro-meteorological information
*	*	*	*	1	5	*	*	Reduce poverty by diversifying employment opportunities so that people do not consume contaminated food
1	2	*	*	*	*	1	1	Develop local and regional networks of stakeholders in mycotoxin management and food safety
*	*	*	*	1	3	*	*	Create independent food safety directors in national systems
*	*	*	*	1	2	*	*	Inclusion of courses on mycotoxicology in medical, veterinary and agricultural curricula
*	*	0	0	*	*	*	*	Establish/strengthen agriculture insurance schemes for mycotoxin-contaminated crops

[a]# Number of persons ranking this response as one of the five most important.

[b]S Weighted priority score, with each member allowed to rank five topics, giving five points to the most important and one to the least.

[c]0 This topic identified by at least one member of this group, but not identified as one of the five most important topics by any member of this group.

[d]* This topic not identified by any member of this group.

Table 4. Nominal group responses by three groups to the statement, "Identify appropriate interventions and approaches for mycotoxin management in Africa".

1		2		3		Response
#[a]	[b]	#	S	#	S	
6	24	7	23	10	32	Create awareness about mycotoxins among consumers, health officials, farmers and households through education and training
3	5	1	3	7	18	Adopt food processing techniques to reduce toxins in food products
4	14	2	6	2	7	Use sustainable mycotoxin-binding adjuvants (enterosorption) and other dietary interventions mechanisms such as nutrient fortification and diet diversification
3	7	3	8	4	10	Develop capacity for mycotoxin testing nationally and regionally by training laboratory staff and creating the infrastructure for international accreditation
1	3	3	9	4	12	Use conventional resistance breeding and GM approaches to develop crop varieties tolerant to mycotoxins
1	2	2	8	4	15	Institutionalize quality assurance systems by adopting HACCP and good practices for agriculture, hygiene, storage and manufacturing
2	5	1	5	4	11	Develop better post-harvest drying, storage & processing technologies
*[c]	*	*	*		20	Assess impact of mycotoxin management on health and trade to sensitize policymakers and the general populace
1	6	6	12	*	*	Improved surveillance and regulation of mycotoxin in the food chain through appropriate harmonization, inspection procedures, legislative reforms and national action plans
0[d]	0	2	7	3	9	Develop biological control agents for sustainable management of mycotoxins
3	10	0	0	2	5	Ensure availability of mycotoxin-free food from strategic grain reserves thus enabling procurement and destruction of contaminated crops
2	6	*	*	3	9	Lobby policy makers and politicians to commit resources for improving health and income through mycotoxin management
*	*	3	6	*	*	Strengthen agricultural extension services for dissemination of mycotoxin management practices
1	5	*	*	*	*	Reduce poverty and increase income so that people do not consume mycotoxin-contaminated food
1	1	0	0	1	2	Develop early warning systems to predict & manage mycotoxin risks
*	*	1	3	*	*	Ensure adequate record keeping in hospitals and clinics for mycotoxin-related illnesses
1	2	0	0	*	*	Adopt good pest management practices at pre- & post-harvest stages
*	*	0	0	*	*	Adopt crop rotations that reduce mycotoxin accumulation
0	0	*	*	*	*	Create better coordination mechanisms between institutions and development organizations with a stake in mycotoxin management
*	*	0	0	0	0	Organize conferences, workshops and networking opportunities for the exchange of ideas and the dissemination of information

[a]# Number of persons ranking this response as one of the five most important.

[b]S Weighted priority score, with each member allowed to rank five topics, giving five points to the most important and one to the least.

[c]* This topic not identified by any member of this group.

[d]0 This topic identified by at least one member of this group, but not identified as one of the five most important topics by any member of the group.

The discussion procedure had six steps (Leslie and Frederiksen, 1995):

1. **Presentation of the problem or question** to the team in writing, together with a statement of the level and type of ideas sought.
2. **Recording of ideas** in a round-robin fashion.
3. **Serial discussion** of each idea or concept for the purpose of clarification.
4. **A preliminary vote** ranking individual ideas.
5. **Discussion of the preliminary vote** to clarify the voting.
6. **Final vote** (a repetition of step 4).

Steps 5 and 6 were optional and were deleted if not needed or if time was a constraint. This six-step process allowed each team to formulate a statement containing a list of the most important problems, which affect further research in the topic area. Findings of the teams were presented to the entire group collectively in a plenary session, with the goal that additional ideas and modifications would be solicited and recorded. The ideas were summarized into cohesive statements keeping the voting pattern intact to reflect the priorities attached to each idea that was amalgamated into individual statements (Tables 1-4). We attempted to preserve all of the ideas expressed in these discussions as they were recorded because we expect ideas that were of lesser importance now might come to dominate the field in the future.

Conclusions

The meeting participants identified several key areas. The foremost area was that of communication – with the farming and consumer constituency, with government and international aid officers, and with one another. Both farmers and consumers need to understand that there is a real or potential problem with mycotoxins in their food, and that the chronic long-term effects of exposure to low levels of mycotoxins may be either debilitating or lethal. Mitigation of the worst of these effects is possible with some relatively simple technologies that need to be better publicized. Government and international aid officials need to hear a similar message and to participate in the notification process. As the mycotoxin problem is interdisciplinary in nature, this communication will be difficult since many of these officials have a mandated area that will be touched by mycotoxins, *e.g.*, health, trade or agriculture, but will be dominated by other problems with a narrower focus. Finally communication amongst the various scientists was viewed as essential, since many of the scientists at the meeting were unaware of the breadth of research in progress on mycotoxin-related topics.

A second key area was to develop the protocols and infrastructure to remediate and/or quantify toxin problems. Some of this work needs more research and development, but much, including some dietary interventions and biocontrol strategies are essentially ready to implement. Development of testing laboratories is a matter of building and equipping the necessary laboratories and recruiting/training qualified personnel to staff them. Action in these areas is for implementation on a relatively broad scale, as the techniques needed are already available.

Finally there is a need for further research. The medical impacts of chronic, low-level exposure to mycotoxins are poorly understood for aflatoxins and not understood at all for any other mycotoxins. Differences between pathogens from tropical and temperate areas are poorly described, even though temperate zone models and hypotheses often fare poorly when applied to tropical conditions. The identification and development of technologies aimed at low-input, subsistence agriculture is a continuing challenge with unique attributes that receives, at least relatively speaking, little specific attention.

The central theme running through all of the meeting and the discussion groups was that food quality, food security, food safety and mycotoxins are all facets of a common problem that will require a multi-disciplinary, cross-sector effort to characterize and remediate. Single sector approaches will help, but any sustainable solution will go far beyond what any single sector approach can be expected to supply, since the energy derived from the synergistic interactions accompanying a cross-sector approach will be at least a part of the overall solution.

References

Delbecq, A.L., van de Ven, A.H. and Gustafson, D.H. (1975) *Group Techniques for Program Planning: A Guide to Nominal Group and Delphi Processes*. Scott, Foresman and Company, Glenview, Illinois.

Durbin, R.D., Frederiksen, R.A., Kelman, A. and Renfro, B.L. (eds.). (1980) *Proceedings of the International Conference Downy Mildew Diseases (28 November – 3 December 1979, Bellagio, Italy)*. The Rockefeller Foundation, New York.

Leslie, J.F., and Frederiksen, R.A. (eds.). (1995) *Disease Analysis through Biotechnology: Interdisciplinary Bridges to Improved Sorghum and Millet Crops*. Iowa State University Press, Ames, Iowa.

Mughogho, L.K. (ed.). (1984) *Sorghum Root and Stalk Rots: A Critical Review*. International Crops Research Institute for the Semi-Arid Tropics, Patancheru, A.P., India.

Potter, M., Gordon, S. and Hamer, P. (2004) The Nominal Group technique: A useful consensus methodology in physiotherapy research. *New Zealand Journal of Physiotherapy* 32, 126–130.

Health and Trade Issues

4

Mycotoxins: A Global Problem

Walter F. O. Marasas*, Wentzel C. A. Gelderblom, Gordon S. Shephard and Hester F. Vismer

Abstract

The five most important naturally occurring mycotoxins in human foods and animal feeds are aflatoxin, ochratoxin, deoxynivalenol, zearalenone and fumonisin. Risk assessment is used to manage the risk from mycotoxins to protect human and animal health. Conventional risk assessment has two major components, *i.e.*, exposure assessment and hazard assessment, which data are used to establish Maximum Tolerated Levels (MTLs). Most countries have established MTLs for total aflatoxins ranging from 4-20 ng/g. The US Food and Drug Administration (FDA) has proposed MTLs for total fumonisins of 4 μg/g in whole maize and 2 ng/g in maize products for human consumption. The MTLs proposed by developed countries apply to commodities that they import and to foodstuffs consumed within their borders, but not to agricultural products that they export. Thus conventional risk assessment has helped manage the risk from mycotoxins in developed countries but has not helped in developing countries that import foodstuffs (or receive food aid). The situation with fumonisins in maize is complicated further by large differences in maize consumption by different populations, *e.g.*, from ~5 g/person/day in Europe to ~500 g/person/day in rural Africa. The differences in maize intake have a marked effect on the Probable Daily Intake (PDI) of fumonisins by different populations. Subsistence farmers in Africa who consume home-grown maize have the highest maize intakes and also consume maize with the highest levels of fumonisin contamination. Conventional risk assessment has not been of value to them and leaves the people who are at the highest risk for mycotoxin exposure the least protected.

Introduction

Mycotoxins have undoubtedly presented a global problem to human and animal health since the earliest times, and this threat will only increase as the demand on the available food supply increases in response to the growth of the world population (Marasas and Nelson, 1987). If the food supply is limited, the mycotoxin hazard is exacerbated in at least two ways. First, more fungus-damaged, potentially mycotoxin-containing foodstuffs are consumed rather than discarded, and second, malnutrition enhances the susceptibility to lower levels of foodborne mycotoxins.

Natural outbreaks of mycotoxicoses occur world-wide, from the humid tropics to Siberia. Although the climate in a particular country may not favor the elaboration of a specific mycotoxin, such as aflatoxin, the problem may be imported from another country in the form of agricultural products, such as peanuts or maize.

Mycotoxins affect both animals and humans acutely as well as chronically. Acute outbreaks of mycotoxicoses are the tip of the iceberg, whereas chronic effects, such as growth stunting, immune suppression and cancer, are much more important although they may not be as evident. A problem cannot be controlled before it is recognized, and acceptance by governments of developed and developing countries that mycotoxins represent a serious health hazard in addition to serving as a trade barrier with significant economic impacts, is a matter of urgency. Mycotoxins are a global problem that requires a global solution to prevent or reduce the development of mycotoxigenic fungi, their insect vectors and the resulting mycotoxin contamination of agricultural crops in the field and in storage. The extent of the mycotoxin problem, particularly with respect to the foodborne carcinogenic mycotoxins, aflatoxin and fumonisin, risk assessment and possible solutions are discussed in this chapter.

Mycotoxigenic Fungi, Mycotoxins and Mycotoxicoses

Globally, the five most important mycotoxin-producing fungi are (Miller, 2002): *Aspergillus flavus*, *Aspergillus ochraceus*, *Penicillium verrucosum*, *Fusarium graminearum* and *Fusarium verticillioides*. The five most important mycotoxins that occur naturally in agricultural products are (Miller, 1995): aflatoxin produced by *A. flavus*; ochratoxin produced by *A. ochraceus* and *P. verrucosum*; deoxynivalenol and zearalenone produced by *F. graminearum*; and fumonisin produced by *F. verticillioides*. Human diseases that have been associated with two of these mycotoxins in foods are: acute toxic hepatitis and liver cancer with aflatoxin; and esophageal cancer and neural tube defects with fumonisin.

Aflatoxins

Aflatoxins are carcinogenic mycotoxins produced by some *Aspergillus* species in a wide range of agricultural commodities, primarily by *A. flavus* in maize and peanuts. Aflatoxin B_1 was first identified in the United Kingdom in 1960 in a shipment of peanuts from Brazil. Subsequently, aflatoxin B_1 was shown to cause outbreaks of acute hepatitis in animals and humans, to cause liver cancer in animals, and to be associated with liver cancer in humans, particularly in combination with hepatitis B virus infection in sub-Saharan Africa and Southeast Asia (Turner *et al.*, 2002). The International Agency for Research on Cancer (IARC) evaluated aflatoxin B_1 as a Group 1 carcinogen, *i.e.*, carcinogenic to humans (IARC, 1993). Maximum tolerated levels of aflatoxins in foods and feeds are regulated in most countries world-wide and commonly range from 4-20 ng/g (FAO, 2004). The danger of consuming foodstuffs contaminated with aflatoxin at levels above the regulatory limit was again demonstrated in 2004 in Kenya where 125 people died following the consumption of home-grown maize containing high levels of aflatoxin (Lewis *et al.*, 2005).

Fumonisins

Fumonisins are carcinogenic mycotoxins produced by some *Fusarium* species, primarily *F. verticillioides* growing in maize. Fumonisins were first isolated and identified in South Africa in 1988 from cultures of *Fusarium verticillioides* (= *F. moniliforme*) strain MRC 826 (Gelderblom *et al.*, 1988). During 1989/1990, broken maize kernels (screenings) from the 1989 maize crop in the United States caused widespread outbreaks of leukoencephalomalacia (LEM) in horses and pulmonary edema syndrome in pigs throughout the country. By 1990 both of these syndromes were proven to be caused by fumonisin B_1 (Marasas, 2001). Analytical methods for the detection of fumonisin B_1 and fumonisin B_2 in maize also were developed in 1990 (Shephard *et al.*, 1990). Reports followed of naturally occurring levels of the toxin in maize screenings associated with field outbreaks of leukoencephalomalacia and pulmonary edema syndrome as well as in home-grown maize in high-incidence areas of human esophageal cancer in the Transkei region of South Africa (Rheeder *et al.*, 1992) and China (Chu and Li, 1994). During 1991, fumonisin B_1 was shown to cause liver cancer in rats (Gelderblom *et al.*, 1991). The carcinogenicity of fumonisin B_1 was confirmed by the National Toxicology Program (NTP) of the United States Food and Drug Administration (FDA) in a two-year feeding study in rats and mice (Howard *et al.*, 2001). The Joint FAO/WHO Expert Committee on Food Additives (JECFA) evaluated the fumonisins and allocated a group provisional maximum tolerable daily intake (PMTDI) of 2 μg/kg body weight to fumonisin B_1, fumonisin B_2 and fumonisin B_3, alone or in combination (WHO, 2002). The IARC evaluated fumonisin B_1 as a Group 2B carcinogen, *i.e.*, possibly carcinogenic to humans (IARC, 2002). Although the role of fumonisin B_1 in esophageal cancer has not been proven, fumonisin has to be considered as a risk factor, particularly in rural populations living on a maize-based staple diet.

Fumonisin B_1 is known to inhibit folic acid transport by the folate receptor and because folic acid deficiency causes neural tube defects, some birth defects in humans may be caused by dietary exposure to fumonisin B_1 (Hendricks, 1999). Fumonisin B_1 causes cranial neural tube defects in mouse embryos and folic acid prevents fumonisin B_1-induced neural tube defects in these animals (Sadler *et al.*, 2002; Gelineau-van Waes *et al.*, 2005). The high-incidence areas of esophageal cancer in South Africa and China, where high levels of fumonisin in the maize staple diet have been reported, also are high incidence areas for neural tube defects in humans (Marasas *et al.*, 2004). The possible role of fumonisin as a cause of birth defects in high incidence areas of Africa, Asia and South America requires further investigation.

At least 28 fumonisin analogs are now known (Rheeder *et al.*, 2002) and three of these (fumonisin B_1, fumonisin B_2 and fumonisin B_3) occur naturally in maize world-wide, sometimes at very high levels of up to 330 μg/g (Shephard *et al.*, 1996). Fumonisins and aflatoxins often co-occur in maize and a synergistic interaction between fumonisin B_1 and aflatoxin B_1 is known (Gelderblom *et al.*, 2002).

The FDA has published a "Guidance for Industry" (FDA, 2001) with respect to fumonisin levels in human foods and animal feeds that the FDA considers to be adequate to protect human and animal health. These levels range from 2-4 μg/g in maize intended for human consumption and 1-50 μg/g in animal feeds. However, the risk of fumonisin contamination of maize to the consumer is determined by both maize intake and level of contamination (Table 1). In general, the highest maize consumers in rural areas also consume the most highly contaminated home-grown maize (Gelderblom *et al.*, 1996; Marasas, 1997).

Table 1. Interactive associations between fumonisin contamination (μg/g maize) and maize intake profiles (g/60 kg person/day) shown as Probable Daily Intake (PDI; μg/kg body weight/day). Provisional Maximum Tolerable Daily Intake (PMTDI) calculations are based on nephrotoxic effects as set forward by the JECFA (WHO, 2002) or hepatocarcinogenic effects as proposed by Gelderblom *et al.* (1996).

FB (μg/g)	Maize intake (g/60 kg person/day)						
	10	50	100	150	200	400	500
0.2	0	0.2	0.3	0.5	0.7	1.4	1.7
0.5	0.1	0.4	0.8	1.3	1.7	3.4	4.2
1	0.2	0.8	1.7	2.5	3.3	6.6	8.3
2	0.3	1.7	1.3	5.0	6.7	13	17
3	0.5	2.5	5.0	7.5	10	20	25
4	0.7	3.3	6.7	10	13	27	33
5	0.8	4.2	8.3	13	17	33	43
10	1.7	8.3	17	25	33	67	83
12	2	10	20	30	40	80	100

White areas: PMTDI falls within the tolerable daily intake level; lightly shaded areas: PMTDI = 0.8 μg/kg body weight/day (hepatocarcinogenicity); medium shaded areas: PMTDI between 0.8 and 2 μg/kg body weight/day (nephrotoxicity); dark shaded areas: PMTDI values above the maximum tolerable daily intake levels.

Such maize is not subject to national or international regulations based on the MTLs of aflatoxin or the guidelines proposed for fumonisin by the FDA.

Economic Impacts of Mycotoxins

Few attempts to estimate the economic costs of mycotoxins in monetary terms have been published. Lubulwa and Davies (1995) estimated the social costs of the impacts of fungi and aflatoxins in maize and peanuts in Indonesia, Philippines and Thailand during 1991 to be Aus.$ 477 million. The economic models used to make this estimate included the evaluation of product spoilage effects, human health effects with respect to disability and premature death due to aflatoxin-related primary liver cancer and livestock health effects due to reduced feed efficiency and increased mortality. The estimate, however, did not include the costs associated with immune suppression and growth stunting (see Chapters 5 and 6) or those from the loss of export markets for the contaminated commodities.

The effects of regulating mycotoxin levels on trade in agricultural products have been investigated by the World Bank (Otsuki *et al.*, 2001*a,b*). The implementation of a new European Union (EU) aflatoxin standard which is lower (4 ng/g aflatoxin B_1) than the internationally accepted *Codex Alimentarius* standard would reduce health risks by 2.3 deaths per billion people per year, but with a reduction of 64% in the export of cereals and peanuts from Africa to Europe at a cost of US$ 670 million (Otsuki *et al.*, 2001*b*). In a subsequent World Bank study by Jaffee and Henson (2004) these findings were challenged because the estimated "cost" of US$ 670 million had been misinterpreted as actual losses of trade rather than an estimate from an econometric simulation. Jaffee and Henson (2004) concluded that

EU imports from Africa would increase due to the more stringent aflatoxin standards, whereas some competing countries, *e.g.* Turkey, incurred more rejections. Similarly, Wu (2004*a*) stated that the developing countries most likely to experience large losses from the tighter mycotoxin standards are not sub-Saharan African nations, but China and Argentina. Among developed countries the United States would experience the heaviest economic losses. The three largest maize exporting countries are the United States, China and Argentina. Wu (2004*a*) calculated that if the current FDA guideline of 2 μg/g fumonisin were adopted internationally, then the export losses to each of these three countries would range from US$ 20-40 million annually with a total loss amongst the three countries of US$ 100 million. A fumonisin standard of 0.5 μg/g would increase the maize export losses to the United States to US$ 170 million, to China to US$ 60 million, and to Argentina to US$ 70 million, for a total of US$ 300 million.

The potential annual cost of contamination of food and feed crops in the United States with three mycotoxins (aflatoxin, fumonisin and deoxynivalenol) is estimated to range from US$ 418 million to US$ 1.66 billion, with a mean estimated cost of US$ 946 million (CAST, 2003). In addition, the costs of mycotoxin management, including research and monitoring, are estimated at between US$ 500 million and US$ 1.5 billion (Robens and Cardwell, 2003).

Risk Assessment and Regulation of Mycotoxins

Conventional risk assessment of mycotoxins has two major components, *i.e.*, exposure assessment and hazard assessment (Gelderblom *et al.*, 1996; Marasas, 1997). Exposure is calculated from food intake and naturally occurring levels of a mycotoxin and expressed as the Probable Daily Intake (PDI). Hazard is calculated from toxicological studies in experimental animals and is expressed as the Tolerable Daily Intake (TDI). The PDI and TDI data are used to assess the risk of a mycotoxin and establish MTLs.

Risk assessment of fumonisins to human health has been performed (IARC, 2002; WHO, 2002) and MTLs proposed ranging from 100-200 ng/g (Gelderblom *et al.*, 1996, Marasas, 1997), to 2-4 μg/g (FDA, 2001). It remains to be seen if and when the fumonisin levels proposed in the FDA Guidance for Industry (FDA, 2001) will be implemented. Viljoen and Marasas (2003) supported the fumonisin levels proposed by the FDA and pointed out that lower MTLs could seriously limit the food supply and affect the entire grain chain from producer to consumer.

Conventional risk assessment and MTLs do not apply to subsistence farmers in Africa who consume the largest amounts of maize containing the highest levels of fumonisins due to the interaction between maize intake and fumonisin contamination (Table 1). The intake profiles for the best quality maize with the lowest fumonisin contamination levels provide PDI values well below the PMTDI of 2.0 μg/kg bw/day proposed by the JECFA (WHO, 2002). This scenario reflects the typical situation in developed countries where maize consumption is low and mycotoxin contamination of foodstuffs is strictly regulated. In contrast, a completely different situation prevails in developing countries, particularly in rural and subsistence farming communities, where home-grown maize is the major dietary staple. High fumonisin contamination levels together with high maize consumption patterns (400-500 g/person/day) result in PDI values well above (10-50 fold higher) the PMTDI. The risk of developing disease due to fumonisin intake is further increased when considering the maize consumption patterns in children. Detailed maize intake profiles in different population groups, particularly in Southern and Eastern Africa where maize is the staple diet, are required to accurately assess the risk of fumonisins to human health. The implementation of MTLs based on conventional risk assessment in developed countries to protect the health of the lowest maize consumers may make the situation worse as food security problems will lead the highest consumers in the producing countries to consume the contaminated maize rejected by the importing countries. A similar rationale applies to MTLs for aflatoxin in peanuts.

People in rural areas of developing countries, who are at the highest risk from mycotoxins in staple foods, particularly subsistence farmers, are completely unprotected by mycotoxin regulations. Moreover, in developing countries in Africa and elsewhere, food safety is an issue that frequently must be balanced against issues of food security (Shephard, 2003). Given the choice between starvation and consuming foods containing mycotoxins at levels higher than the prescribed MTLs, most people in developing countries would probably eat the foodstuffs that would be rejected by developed countries. Thus, people who are at the highest risk, also have the most urgent need for solutions other than regulation for the mycotoxin problem.

Possible Solutions

The ultimate solution to the global mycotoxin problem is not regulation, but reduction of fungal infection and mycotoxin levels in crop plants (Marasas and Nelson, 1987; WHO, 2000). Attempts to achieve this goal by conventional plant breeding have not been very successful for various reasons including the lack of major single genes and difficulties in selecting appropriate germplasm due to time-consuming and expensive mycotoxin analyses (Gressel *et al.*, 2004; Munkvold, 2003). Although several sources of resistance to *A. flavus* infection and/or aflatoxin production in maize have been identified, the levels of genetic resistance are not sufficient to prevent the development of unacceptable aflatoxin levels. The same problem also occurs with the polygenic sources of resistance to *F. verticillioides* and fumonisin levels in maize (Munkvold, 2003). Molecular markers are being used increasingly to facilitate selection and to combine resistance genes from different sources in order to develop varieties with high yields and low mycotoxin levels, but potentially commercial lines have yet to be identified.

The most promising approach for innovative solutions is biotechnology. Genetic engineering approaches are the most attractive methods now under development, and the future of fumonisin reduction may lie in the hands of biotechnologists. The potential of transgenic resistance to mycotoxigenic fungi and/or their mycotoxins as biotechnology solutions for the global mycotoxin problem is receiving intensive international attention. The following strategies that might be used to reduce fumonisins in maize were reviewed by Duvick (2001).

Reducing infection by the mycotoxigenic fungus

Several antifungal compounds in plants are potential candidates for genetic engineering to alter maize genotypes for resistance to mycotoxigenic fungi.

Inserting genes capable of degrading the mycotoxin

Progress has been made with this strategy of *in planta* detoxification of fumonisins in maize. Duvick (2001) reported that two species of saprophytic fungi (*Exophiala spinifera* and *Rhinocladiella atrovirens*) from moldy maize ears can utilize FB_1 as their sole carbon source. These fungi produce enzymes capable of hydrolyzing and further metabolizing fumonisins by oxidative deamination. The genes coding the specific enzymes that carry out the detoxification steps have been cloned and the effects of the expression of these genes in transgenic maize on fumonisin levels are currently being evaluated in the United States (Duvick, 2001).

Interfering with mycotoxin biosynthesis

An α-amylase inhibitor has been identified in the legume *Lablab purpureus* that inhibits aflatoxin biosynthesis (Munkvold, 2003). This gene is a candidate for expression in genetically modified crops to reduce aflatoxin contamination. Genes that regulate fumonisin production by *F. verticillioides* have been identified (Proctor *et al.*, 2003; Brown *et al.*, 2005) and this information has the potential to be used in transgenic maize to disrupt fumonisin biosynthesis.

Inserting genes for insect resistance

There is a close association between insects, *e.g.*, the European corn borer (*Ostrinia nubilalis*), and the infection of maize by *F. verticillioides*, so transgenic maize hybrids carrying genes encoding insecticidal proteins from *Bacillus thuringiensis* (*Bt*) are a potential solution to the fumonisin problem. *Bt* hybrids, which are resistant to the European corn borer and have correspondingly less *F. verticillioides* ear rot, had significantly lower fumonisin levels than did conventional hybrids grown in Iowa, USA (Munkvold *et al.*, 1999). Similar results were subsequently reported for *Bt* hybrids elsewhere in the world where fumonisin contamination of maize is associated with insect damage such as France, Italy and Spain (Munkvold, 2003). *Bt* hybrids represented approximately 25% of the field maize planted in the United States and the estimated annual saving to farmers due to reduced mycotoxin (fumonisin and deoxynivalenol) levels alone, is US$ 17 million (Wu *et al.*, 2004). Whether currently available *Bt* hybrids also contain significantly less fumonisin than non-*Bt* hybrids in countries where other species of corn borers predominate, *e.g.*, *Busseola fusca* in West Africa (Cardwell *et al.*, 2000) and South Africa (Flett and van Rensburg, 1992), remains unknown. In South Africa, significant reductions in fumonisin levels have been found in some *Bt* cultivars in some seasons in some locations (Vismer *et al.*, 2005). If additional *Bt* genes are deployed to control a broader range of maize insects, then the reduction in fumonisin levels may be improved further.

Conclusions

The EU continues to make mycotoxin standards more stringent by lowering the MTL in imported agricultural products on the one hand, while prohibiting the importation of GM crops on the other. This impasse is a serious obstacle to the implementation of biotechnology solutions to the mycotoxin problems of exporting countries. The reasons why decision makers at both the government and individual consumer levels have not endorsed *Bt* maize and other GM crops are complex (Wu, 2004*b*). The potential benefits of *Bt* maize, *e.g.*, increased yield, decreased use of pesticides and reduced mycotoxin levels, should be emphasized in educational programs to improve public understanding of biotechnology by providing accurate and balanced information.

During the Workshop on Mycotoxins in Food in Africa (Cardwell, 1996; Cardwell and Miller, 1996) held in Cotonou, Benin, 6-10 November 1995 it was resolved by an international, multidisciplinary team of scientists that:

Recognizing the increasing importance of food grains in sub-Saharan Africa;

Realizing that heavy losses are caused by mycotoxins at all levels of production, storage, processing, and utilization; and being

Concerned that mycotoxins are having a direct negative impact on human and animal health, and on trade; and being

Aware that African governments are fully committed to the promotion of food security and safety, and to the improvement of public health and quality of life for their citizenry

Be it Resolved that The Pan African Mycotoxins Initiative Committee:

Does Hereby Reiterate that appropriate measures must be taken to reduce the grain losses caused by mycotoxins to internationally accepted standards, and to increase production of good quality grains; and

Advocates that resources be mobilized by African governments and the international community for support to mycotoxin research and intervention initiatives.

We strongly recommended that not only governments of African countries, but that governments of all developed and developing countries support the search for, as well as the development and implementation of, solutions for the global mycotoxin problem.

Acknowledgements

We thank all of our colleagues, past and present, at the PROMEC Unit of the South African Medical Research Council (MRC) for their invaluable contributions to the global knowledge of mycotoxins, and the MRC for continuous funding from 1975 to the present.

References

Brown, D.W., Cheung, F., Proctor, R.H., Butchko, R.A., Zheng, L., Lee, Y., Utterback, T., Smith, S., Feldblyum, T., Glenn, A.E., Plattner, R.D., Kendra, D.F., Town, C.D. and Whitelaw, C.A. (2005). Comparative analysis of 87,000 expressed sequence tags from the fumonisin-producing fungus *Fusarium verticillioides*. *Fungal Genetics and Biology* 42, 848-861.

Cardwell, K.F. (ed). 1996. *Proceedings of the Workshop on Mycotoxins in Foods in Africa* (November 6-10, 1995, Cotonou, Benin). IITA, Ibadan, Nigeria

Cardwell, K., and Miller, J.D. (1996) Mycotoxins in foods in Africa. *Natural Toxins* 4, 103-107.

Cardwell, K.F., Kling J.G., Maziya-Dixon, B. and Bosque-Pérez, N.A. (2000) Interactions between *Fusarium verticillioides, Aspergillus flavus* and insect infestation in four maize genotypes in lowland Africa. *Phytopathology* 90, 276-284.

CAST. (2003) *Mycotoxins: Risks in Plant and Animal Systems.* Task Force Rep No 38. Council for Agricultural Science and Technology (CAST), Ames, Iowa, USA.

Chu, F.S. and Li, G.Y. (1994) Simultaneous occurrence of fumonisin B_1 and other mycotoxins in moldy corn collected from the People's Republic of China in regions with high incidences of esophageal cancer. *Applied and Environmental Microbiology* 60, 847-852.

Duvick, J. (2001) Prospects for reducing fumonisin contamination of maize through genetic modification. *Environmental Health Perspectives* 109 (Suppl. 2), 337-342.

FAO (Food and Agriculture Organization). (2004) *Worldwide Regulations for Mycotoxins in Foods and Feeds in 2003.* FAO Food and Nutrition Paper 81. Rome, Italy.

FDA (Food and Drug Administration). (2001) Guidance for industry: Fumonisin levels in human foods and animal feeds; Availability. *Federal Register* November 9, 2001; 66 (218), 56688-56689. See also FDA – CFSAN background papers on Final Guidance (http://www.cfsan.fda.gov~/dms/fumongu2.html) and in support of fumonisin levels in foods (http://www.cfsan.fda.gov~/dms/fumonbg3.html) and feeds (http://www.cfsan.fda.gov/dms/fumonbg4.html).

Flett, B.C. and van Rensburg, J.B.J. (1992) Effect of *Busseola fusca* on the incidence of maize ear rot caused by *Fusarium moniliforme* and *Stenocarpella maydis. South African Journal of Plant and Soil* 9, 177-179.

Gelderblom W.C.A., Jaskiewicz, K., Marasas, W.F.O., Thiel, P.G., Horak, M.J., Vleggaar, R., and Kriek, N.P.J. (1988) Fumonisins - Novel mycotoxins with cancer promoting activity produced by *Fusarium moniliforme. Applied and Environmental Microbiology* 54, 1806-1811.

Gelderblom, W.C.A., Kriek, N.P.J., Marasas, W.F.O., and Thiel, P.G. (1991) Toxicity and carcinogenicity of the *Fusarium moniliforme* metabolite, fumonisin B_1, in rats. *Carcinogenesis* 12, 1247-1251.

Gelderblom, W.C.A., Snyman, S.D., Abel, S., Lebepe-Mazur, S., Smuts, C.M., van der Westhuizen, L., Marasas, W.F.O., Victor, T.C., Knasmüller, S., and Huber, W. (1996) Hepatotoxicity and carcinogenicity of the fumonisins in rats. A review regarding mechanistic implications for establishing risk in humans. *Advances in Experimental Medicine and Biology* 392, 279-296.

Gelderblom, W.C.A., Marasas, W.F.O., Lebepe-Mazur, S., Swanevelder, S., Vessey, C.J., de la M. Hall, P. (2002) Interaction of fumonisin B_1 and aflatoxin B_1 in a short-term carcinogenesis model in rat liver. *Toxicology* 171, 161-173.

Gelineau-van Waes, J., Starr, L., Maddox, J., Alleman, F., Voss, K.A., Wilberding, J. and Riley, R.T. (2005) Maternal fumonisin exposure and risk for neural tube defects: Mechanisms in an *in vivo* mouse model. *Birth Defects Res (Part A): Clinical and Molecular Teratology* 73, 487-497.

Gressel, J., Hanafi, A., Head, G., Marasas, W., Obilana, B., Ochanda, J., Souissi, T., and Tzotzos, G. (2004) Major heretofore intractable biotic constraints to African food security that may be amenable to novel biotechnological solutions. *Crop Protection* 23, 661-689.

Hendricks, K. (1999) Fumonisins and neural tube defects in south Texas. *Epidemiology* 10, 198-200.

Howard, P.C., Eppley, R.M., Stack, M.E., Warbritton, A., Voss, K.A., Lorentzen, R.J., Kovach, R.M., Bucci, T.J. (2001) Fumonisin B_1 carcinogenicity in a two-year feeding study using F344 rats and $B6C3F_1$ mice. *Environmental Health Perspectives* 109 (Suppl. 2), 277-282.

IARC (International Agency for Research on Cancer). (1993) Aflatoxins. *IARC Monographs on the Evaluation of Carcinogenic Risks to Humans: Some Naturally Occurring Substances: Food Items and Constituents, Heterocyclic Aromatic Amines and Mycotoxins*. 56, 245-395.

IARC. (2002) Fumonisin B_1. *IARC Monographs on the Evaluation of Carcinogenic Risks to Humans: Some Traditional Medicines, Some Mycotoxins, Naphthalene and Styrene* 82, 301-366.

Jaffee, S. and Henson, S. (2004) Standards and agro-food exports from developing countries: Rebalancing the debate. *World Bank Policy Research Working Paper* 3348, 1-43.

Lewis, L., Onsongo, M., Njapau, H., Schurz-Rogers, H., Luber, G., Kieszak, S., Nyamongo, J., Backer, L., Dahiye, A.M., Misore, A., DeCock, K., and Rubin, C. (2005) Aflatoxin contamination of commercial maize products during an outbreak of acute aflatoxicosis in eastern and central Kenya. *Environmental Health Perspectives* 113, 1763-1767.

Lubulwa, A.S.G., and Davis, J.S. (1995) Estimating the social costs of the impacts of fungi and aflatoxins in maize and peanuts. *Proceedings of the 6th International Working Conference on Stored Product Protection*, pp. 1017-1042.
Marasas, W.F.O. (1997) Risk assessment of fumonisins produced by *Fusarium moniliforme* in corn. *Cereal Research Communications* 25, 399-406.
Marasas, W.F.O. (2001) Discovery and occurrence of the fumonisins. A historical perspective. *Environmental Health Perspectives* 109 (Suppl. 2), 239-243.
Marasas, W.F.O. and Nelson, P.E. (1987) *Mycotoxicology. Introduction to the Mycology, Plant Pathology, Chemistry, Toxicology and Pathology of Naturally Occurring Mycotoxicoses in Animals and Man.* The Pennsylvania State University Press, University Park, Pennsylvania.
Marasas, W.F.O., Riley, R.L., Hendricks, K.A., Stevens, V.L., Sadler, T.W., Gelineau-van Waes, J., Missmer, S.A., Cabrera, J., Torres, O., Gelderblom, W.C.A., Allegood, J., Martinez, C., Maddox, J., Miller, J.D., Starr, L., Sullards, M.C., Roman, A.V., Voss, K.A., Wang, E., and Merrill, A., Jr. (2004) Fumonisins disrupt sphingolipid metabolism, folate transport, and neural tube development in embryo culture and *in vivo*: A potential risk factor for human neural tube defects among populations consuming fumonisin-contaminated maize. *Journal of Nutrition* 134, 711-716.
Miller, J.D. (1995) Fungi and mycotoxins in grain: Implication for stored product research. *Journal of Stored Products Research* 31, 1-16.
Miller, J.D. (2002) Aspects of the ecology of *Fusarium* toxins in cereals. *Advances in Experimental Medicine and Biology* 504, 19-27.
Munkvold, G.P. (2003) Cultural and genetic approaches to managing mycotoxins in maize. *Annual Review of Phytopathology* 41, 99-116.
Munkvold, G.P., Hellmich, R.L. and Rice L.G. (1999) Comparison of fumonisin concentrations in kernels of transgenic *Bt* maize hybrids and nontransgenic hybrids. *Plant Disease* 83, 130-138.
Otsuki, T., Wilson, J.S. and Sewadeh, M. (2001*a*) What price precaution? European harmonisation of aflatoxin regulations and African groundnut exports. *European Review of Agricultural Economics* 28, 263-283.
Otsuki, T., Wilson, J.S. and Sewadeh, M. (2001*b*) Saving two in a billion: Quantifying the trade effect of European food safety standards on African exports. *Food Policy* 26, 493-514.
Proctor, R.H., Brown, D.W., Plattner, R.D. and Desjardins, A.E. (2003) Co-expression of 15 contiguous genes delineates a fumonisin biosynthetic cluster in *Gibberella moniliformis*. *Fungal Genetics and Biology* 38, 237-249.
Rheeder, J.P., Marasas, W.F.O., Thiel, P.G., Sydenham, E.W., Shephard, G.S., van Schalkwyk, D.J. (1992) *Fusarium moniliforme* in corn in relation to human esophageal cancer in Transkei. *Phytopathology* 82, 353-357.
Rheeder, J.P., Marasas, W.F.O. and Vismer, H.F. (2002) Production of fumonisin analogs by *Fusarium* species. *Applied and Environmental Microbiology* 68, 2101-2105.
Robens, J. and Cardwell, K. (2003) The costs of mycotoxin management to the USA: Management of aflatoxins in the United States. *Journal of Toxicology – Toxin Reviews* 22, 139-152.
Sadler, T.W., Merrill, A.H., Stevens, V.L., Sullards, M.C., Wang, E. and Wang, P. (2002) Prevention of fumonisin B_1-induced neural tube defects by folic acid. *Teratology* 66, 169-176.
Shephard, G.S. (2003) Aflatoxin and food safety: Recent African perspectives. *Journal of Toxicology – Toxin Reviews* 22, 267-286.
Shephard, G.S., Sydenham, E.W., Thiel, P.G., Gelderblom, W.C.A. (1990) Quantitative determination of fumonisins B_1 and B_2 by high-performance liquid chromatography with fluorescence detection. *Journal of Liquid Chromatography* 13, 2077-2087.
Shephard, G.S., Thiel, P.G., Stockenström, S. and Sydenham, E.W. (1996) Worldwide survey of fumonisin contamination of corn and corn-based products. *JAOAC International* 79, 671-687.
Turner, P.C., Sylla, A., Diallo, M.S., Castegnaro, J.J., Hall, A.J. and Wild, C.P. (2002) The role of aflatoxins and hepatis viruses in the etiopathogenesis of hepatocellular carcinoma: A basis for primary prevention in Guinea-Conakry, West Africa. *Journal of Gastroenterology and Hepatology* 17, 5441-5448.

Viljoen, J.H. and Marasas, W.F.O. (2003) A review of proposed maximum tolerated levels for fumonisins in maize and maize products. *Advances in Stored Product Protection. Proceedings of the 8th International Working Conference on Stored Product Protection*, pp 448-455.

Vismer, H.F., Rheeder, J.P., van der Westhuizen, L., Imrie, G., Gatyeni, P.M., Thomas, D., Shephard, G.S., Marasas, W.F.O. and Flett, B.C. (2005) Effect of *Bt* corn hybrids on insect damage, incidence of fumonisin-producing *Fusarium* species and fumonisin levels in South Africa (Abstract). *Phytopathology* 95, 588.

WHO (World Health Organization). (2000) Fumonisin B_1. *Environmental Health Criteria* 219, 1-150.

WHO. (2002) Evaluation of Certain Mycotoxins in Food. *Fifty-sixth Report of the Joint FAO/WHO Expert Committee on Food Additives (JECFA). WHO Technical Report* 906, 16-27.

Wu, F. (2004*a*) Mycotoxin risk assessment for the purpose of setting international regulatory standards. *Environmental Science and Technology* 38, 4049-4055.

Wu, F. (2004*b*) Explaining public resistance to genetically modified corn: An analysis of the distribution of benefits and risks. *Risk Analysis* 24, 715-726.

Wu, F., Miller, J.D. and Casman, E.A. (2004) The economic impact of *Bt* corn resulting from mycotoxin reduction. *Journal of Toxicology – Toxin Reviews* 23, 397-424.

5

Modulation of the Human Immune System by Aflatoxin

Pauline E. Jolly*, Yi Jiang, William O. Ellis, Jia Sheng-Wang, Evans Afriyie-Gyawu, Timothy D. Phillips and Jonathan H. Williams

Abstract

This chapter provides a brief review of the literature on immunomodulation by aflatoxin and other mycotoxins, examines immunomodulation by mycotoxins that co-occur in food, reviews possible mechanisms through which mycotoxins exert their effects, and considers future research in the field. Studies conducted in several species of animals and in animal and human cell cultures have shown that aflatoxins act as immunomodulators, primarily as immunosuppressors, of various aspects of cell-mediated immunity and phagocytic cell function. The effect of aflatoxin on humoral immunity in animals is less consistent. Recently, two studies have been published that show association between aflatoxin B_1 albumin adduct levels in blood and antibody and cellular immune status of humans chronically exposed to aflatoxin in their diet. Several other immunotoxic mycotoxins co-occur with aflatoxin in foods, and are likely to have an additive, if not synergistic, effect on the immune system. Aflatoxin seems to exert its immunomodulatory effects by modulating cytokine production either at the mRNA or protein level, but, the mechanisms of immune modulation by aflatoxin and other mycotoxins remain to be clearly determined.

Introduction

Aflatoxins are a group of secondary metabolites produced by strains of some *Aspergillus* species, mainly *A. flavus* and *A. parasiticus* (Gourama and Bullerman, 1995). These compounds are potent carcinogens found in crops such as maize, groundnuts, legumes and other grains that form the staple diet in many developing countries (Cardwell, 2001). Among the aflatoxin chemotypes (B_1, B_2, G_1 and G_2), B_1 is the most common and toxic form (Park *et al.*, 2002). Aflatoxin M_1 is a hydroxylated metabolite of aflatoxin B_1 found in the milk and urine of humans or other mammals that consume a diet contaminated with aflatoxin B_1

(Neal *et al.*, 1998). Chronic exposure of humans to low levels of aflatoxins occurs more commonly than acute toxic exposure. Chronic aflatoxin exposure is associated with the development of hepatocellular carcinoma in humans (Bosch and Munoz, 1988; Ozturk, 1991; IARC, 1996), especially in those infected with hepatitis B virus (Ross *et al.*, 1992; Qian *et al.*, 1994; Wang *et al.*, 1996, 2001). Thus, chronic aflatoxin exposure is a major and significant public health problem.

Aflatoxins are less well known for their role in immune suppression, although numerous studies have been published since the late 1960s (Gallikeev *et al.*, 1968; Pier and Heddleston, 1970; Savel *et al.*, 1970) on the effects of aflatoxin on the immune system of animals (*in vivo* studies) and on animal and human immune cells *in vitro*. These studies have been reviewed extensively (Pier, 1986; Richard, 1991; Bondy and Pestka, 2000; Oswald *et al.*, 2005). However, data on the immunotoxic effects of aflatoxins in humans are limited and only two reports have been made in recent years on the effects of aflatoxins on humans who are chronically exposed to it in the diet (Turner *et al.*, 2003; Jiang *et al.*, 2005).

Immunomodulation by aflatoxin in animals and animal & human cell cultures

The cells of the immune system are continually proliferating and differentiating and are vulnerable to the immunomodulatory, primarily immunosuppressive, effects of mycotoxins. Previous studies on the effect of aflatoxin on the immune system in animals and in cell cultures examined cell mediated and antibody responses, NK cell activity, macrophage phagocytic function, and infectivity and host-resistance challenges. In most of these studies, cellular immunity was examined with methods such as delayed type hypersensitivity (Reddy and Sharma, 1989), expression of regulatory cytokines such as interleukin 2 (IL-2) production by spleenocytes (Hatori *et al.*, 1991; Dugyala and Sharma, 1996), graft versus host response, leukocyte migration and lymphoblastogenesis using [^{3}H]-thymidine (Ghosh *et al.*, 1991; Kadian *et al.*, 1988). The effect of aflatoxin on the humoral immune system was usually investigated by examining antibody responses to sheep red blood cells and the results have been inconsistent (van Heugten *et al.*, 1994). Studies conducted on the immunotoxic effect of aflatoxin up to the year 2001 have shown that exposure to aflatoxin decreased T or B lymphocyte activity (Reddy *et al.*, 1987; Richard *et al.*, 1978), impaired macrophage/neutrophil effector functions (Neldon-Ortiz, 1991; Cusumano *et al.*, 1996; Silvotti *et al.*, 1997; Moon *et al.*, 1999), modified synthesis of inflammatory cytokines (Jakab *et al.*, 1994; Moon, 1999), suppressed NK cell-mediated cytolysis (Reddy and Sharma, 1989), decreased resistance to infectious diseases (Hamilton and Harris, 1971; Edds *et al.*, 1973; Boonchuvit and Hamilton, 1975; Wyatt *et al.*, 1975; Cysewki *et al.*, 1978; Joens *et al.*, 1981; Pier, 1986), induced reactivation of chronic infection (Venturini *et al.*, 1996; Kubena *et al.*, 2001), decreased immunity to vaccination (Gabal and Azzam, 1998; Gabal and Dimitri, 1998) and impaired immune function in developing animals (Pier *et al.*, 1984; Silvotti *et al.*, 1997). Studies of interleukin-1 (IL-1) production by peritoneal macrophages of rats given a single injection of 1 μg/g aflatoxin B_1 showed increase in IL-1 (Cukrova *et al.*, 1992). However, aflatoxin B_1 did not have an effect on IL-1 production by bovine macrophages (Walsh *et al.*, 1991) suggesting that aflatoxin may have an immunomodulatory rather than simply an immunosuppressive role.

One study in animals published after 2001 that merits special review is that by Marin *et al.* (2002) on the effect of aflatoxin on cellular and humoral immunity in weanling piglets. In this study, several parameters of immune function were examined in relation to levels of aflatoxin that are normally allowed in animal feed (low-dose = 140 ppb and high-dose = 280 ppb). There was a biphasic change in the total number of white blood cells depending on whether aflatoxin was fed at the low or high dose. Low-dose aflatoxin decreased the total number of white blood cells while high-dose aflatoxin increased the number of white blood cells. Marin *et al.* (2002) found an increase in γ–globulin concentration in serum and a non-significant reduction in antibody levels in piglets fed 280 ppb aflatoxin and immunized with *Mycoplasma agalacticae*. Other investigators have found significant changes in γ-globulin levels in swine fed high levels of aflatoxin (Annau *et al.*, 1964; Cysewki *et al.*, 1978; Miller *et al.*, 1981) and in other species [discussed in Marin *et al.* (2002)]. Marin *et al.* (2002) also reported a significant decrease in mRNA expression of interleukin-1β (IL-1β) and a slight non-significant decrease in mRNA of tumor necrosis factor-α (TNF-α). There was an increase in cytokine mRNA expression of the anti-inflammatory cytokine interleukin-10 (IL-10), which might account for down regulation of the inflammatory cytokines IL-1β and TNF-α. The authors found no effect on the total number of leukocytes or on expression of either Th1 (IL-2) or Th2 (interleukin 4; IL-4) cytokines.

Immunomodulation by aflatoxin in humans exposed to dietary aflatoxin

Only two studies (Turner *et al.*, 2003; Jiang *et al.*, 2005) are available on the association between aflatoxin levels and immune status/function in humans chronically exposed to aflatoxins in their diets. In these studies, aflatoxin B_1-albumin adducts in blood was used as a biomarker of aflatoxin exposure. The aflatoxin B_1 assay is a sensitive method that measures aflatoxin that is covalently bound to albumin in peripheral blood (Nyathi *et al.*, 1987) and reflects aflatoxin exposure in the previous 2 to 3 months. Turner *et al.* (2003) investigated cell mediated and antibody responses in Gambian children exposed to aflatoxin in their diets. They found markedly reduced secretory immunoglobulin A (sIgA) levels in the saliva of children with detectable aflatoxin B_1-albumin adducts compared to those without. Cell-mediated immunity was evaluated with the cell-mediated immunity multitest (Marcel Merieux, Lyon, France) in which test antigens (tetanus, diphtheria, streptococcus, tuberculin, candida, tricophyton, proteus, and glycerin as a control) were applied to the skin and a response (induration of 2 mm or greater) read 48 hours after application. The investigators found no association between cell-mediated immune responses to the test antigens and aflatoxin B_1-albumin adducts.

Jiang *et al.* (2005) used three-color flow cytometric techniques to investigate the cellular immune status in relation to aflatoxin levels of adult Ghanaians chronically exposed to aflatoxin in their diets. They measured aflatoxin B_1-albumin adduct levels in participants and examined the potential role of aflatoxin in modifying the distribution and function of peripheral blood leukocyte subsets (CD3, CD4, CD4CD69, CD8, CD19, CD19CD69, CD14, CD56), lymphocyte proliferation of CD4+ T cells; and cytokine production by CD4+ and CD8+ T cells and CD3-CD56+ (NK) cells. The proportion of, and cytokine secretion by, the different cellular subsets indicate the strength of specific immune responses carried out by each subset of cells and the overall strength of the immune response. All participants in the study had detectable aflatoxin B_1-albumin adduct levels (mean $\pm$ standard deviation = 0.99 ± 0.40 pmol/mg albumin; range = 0.33-2.27 and median = 0.90 pmol/mg albumin); on average, participants consumed at least 10 μg of aflatoxin B_1 daily in their diet. For analysis, the median aflatoxin B_1-albumin adduct level was used to separate participants into high aflatoxin B_1 or low aflatoxin B_1 groups.

The participants with high aflatoxin B_1 levels had non-significantly lower percentages of CD3+, CD8+, CD19+, CD14+, mature cytotoxic T cells, CD3-CD56+CD16- NK cells, perforin+ NK cells and monocyte phagocytoses than those with low aflatoxin B_1 levels. Although there were no significant differences in the percentages of CD3+ and CD19+ cells between the low and high aflatoxin B_1 groups, participants in the low aflatoxin B_1 group had significantly higher percentages of activated T and B cells (CD3+CD69+ and CD19+CD69+) than did participants in the high aflatoxin B_1 group. This report is the first of the association of aflatoxin B_1 albumin adducts with decreased level of activated T and B cells and is a significant finding since activation of T and B cells results in proliferation and amplification of immune responses that allow the immune system to fight infectious agents and produce effective antibody responses to vaccines. The decreased number of activated T and B cells in those with high aflatoxin B_1 levels indicate that aflatoxin may decrease expression of the CD69 activation molecule and that the cells may not mount appropriate and effective immune responses. These results also indicate that aflatoxin may change the percentages of subsets of T lymphocytes and affect CMI without affecting the overall T-cell population.

No differences in the frequency of interferon-gamma (IFN-γ) or IL-4 expressing CD4+ or CD8+ T cells were found in the study by Jiang *et al.* (2005), similar to findings from studies conducted in piglets, rats and mice exposed to aflatoxin in their diets (Dugyala and Sharma, 1996; Watzl *et al.*, 1999; Marin *et al.*, 2002). The investigators also found no difference in proliferation of CD4+ T cells in relation to aflatoxin B_1 levels. Studies in animals have shown that aflatoxin decreased CD4+ T cell proliferation (Raisuddin *et al.*, 1993; van Heugten *et al.*, 1994; Harvey *et al.*, 1995) but an *in vitro* study using human cells found that aflatoxin up to a concentration of 10 μg/ml had no effect on CD4+ T cell proliferation (Meky *et al.*, 2001). The concentration of aflatoxin used in the *in vitro* experiments with human cells or present in humans in the study by Jiang *et al.* (2005) may not be high enough to affect CD4+ T cell proliferation. The T cell responses observed in the animal studies could be induced by the type and amount of exposure to aflatoxin, the dosing schedule and the animal species used.

It is important to investigate cytokine expression by effector CD8+ T cells since these cells function in killing infected cells and preventing the spread of infectious pathogens in the body. Perforin and granzymes are two of the cytokines used by effector-type CD8+ cells in cell killing. Jiang *et al.* (2005) found significantly lower levels of perforin and granzyme-A expressing CD8+cells in participants with high aflatoxin B_1 levels suggesting that CD8+T cell function (and consequently the cellular response to infectious agents) is impaired in those with high aflatoxin B_1 levels. No other studies have been reported in the literature on cytokine expression by effector CD8+ cells in relation to aflatoxin or any other mycotoxin. Thus, this area warrants further investigation.

The study by Jiang *et al.* (2005) also found a non-significantly lower percentage of $CD56^{bright}$ NK cells (cells that express high levels of CD56 and CD16 negative) in people with high aflatoxin B_1 levels. Since NK cells lyse target cells and provide early regulatory cytokines such as IFN-γ to macrophages and other antigen presenting cells (Cooper *et al.*,

2001), further research needs to be conducted to determine whether aflatoxin affects NK cell subsets, and thereby, interferes with efficient control of infected and transformed cells.

Macrophages represent the first line of defense against infectious agents and through secretion of cytokines regulate T-lymphocyte activity. Jiang *et al.* (2005) found no difference in the percentage of monocytes in peripheral blood of study participants in relation to aflatoxin B_1 levels. This finding is similar to that of Marin *et al.* (2002) who found no effect of aflatoxin on the number of monocytes in peripheral blood of weanling piglets. Jiang *et al.* (2005) also found a non-significant decrease in macrophage phagocytic rate in those with high aflatoxin B_1 levels. Several animal studies have shown that aflatoxins suppressed macrophage phagocytic activity (Mohapatra and Roberts, 1985; Sorenson *et al.*, 1986; Cusamano *et al.*, 1990; Neldon-Ortiz, 1991; Neldon-Ortiz *et al.*, 1992; Moon *et al.*, 1999). An *in vitro* study of the effect of aflatoxin on human monocytes showed impairment in phagocytic and microbiocidal activity by aflatoxin B_1 at doses as low as 0.1 pg/ml (Cusamano *et al.*, 1996). In contrast, a study by Silviotti et al. (1997) with monocyte-derived macrophages from sows fed aflatoxin in their diets found that the phagocytic ability of the macrophages was not compromised although the cells failed to efficiently produce superoxide anions after oxidative burst stimulation. Data from Moon *et al.* (1999) on murine macrophages suggest that aflatoxin B_1 can inhibit the production of nitrous oxide, superoxide anion, hydrogen peroxide, TNF-α, IL-1 and IL-6, all of which are associated with macrophage dysfunction. Dugyala and Sharma (1996) also reported suppression of IL-1 and IL-6 secretion by murine macrophages treated with aflatoxin B_1 and Kurtz and Czuprynski (1992) reported suppression of IL-1 by aflatoxin in bovine mononuclear phagocytes.

Mechanism of the immunomodulatory effects of mycotoxins

The molecular mechanism(s) of immunomodulation by aflatoxin and other mycotoxins remains to be clearly determined. However, available evidence suggests immunosuppression by several mycotoxins is due to inhibition of DNA, RNA and protein synthesis through many different mechanisms. T cells seem to be more sensitive to the effect of aflatoxin than B cells. Many studies (some are discussed in the previous section) show that aflatoxin exerts its immunomodulatory effects by modulating cytokine production (Rossano *et al.*, 1999; Liu *et al.*, 2002). Rossano *et al.* (1999) investigated the release and genetic expression of IL-1α, IL-6 and TNF-α by human monocytes incubated with low concentrations (0.01-1.0 pg/ml) of aflatoxin B_1 and activated with bacterial lipopolysaccharide. At a concentration of 0.05 pg/ml aflatoxin B_1 decreased each of the three cytokines and completely blocked transcription of their mRNAs. Since the mRNA of β-actin was not effected, the authors concluded that aflatoxin B_1 exerts its effect on cytokine production through selective inhibition of specific mRNAs, without affecting protein synthesis in general. They propose that aflatoxin B_1 may inactivate some kinases involved in the activation of genes that code for cytokines.

A study by Dugyala and Sharma (1996) showed that aflatoxin B_1 markedly increased mRNA levels of major cytokines produced by macrophages but suppressed their corresponding protein levels. Corresponding doses of aflatoxin B_1 slightly decreased mRNA and protein levels of cytokines produced by lymphocytes. Their data suggest that aflatoxin B_1 preferentially affects macrophage functions and differentially affects transcription and translation in macrophages. Inhibition of DNA, RNA and protein synthesis by several different mechanisms seem to be responsible for the immunosuppressive effects of many other mycotoxins. Taranu *et al.* (2005) showed that fumonisin B_1 increased IFN-γ production by swine Peripheral Blood Mononuclear Cells (PBMCs) at both the mRNA and protein levels.

The data of Marin *et al.* (2002) and Jiang *et al.* (2005) support previous conclusions (Pier, 1986) that total T cell populations of peripheral blood lymphocytes are not affected by oral consumption of aflatoxin. However, the alterations in the different lymphocyte subsets and their cytokine secretions found by Jiang *et al.* (2005) and the significant decrease in secretory IgA found by Turner *et al.* (2003) indicate impairment in cellular and humoral immunity in people (adults or children) chronically exposed to aflatoxin in their diets. These individuals had accumulated high levels of aflatoxin B_1-albumin adducts in their blood, and based on the immune results are expected to have decreased resistance to infection and cancer (supported by the association between aflatoxin with hepatocellular carcinoma), decreased immune responses to vaccines, and increased susceptibility to reactivation of chronic infections.

Immunotoxicity of other mycotoxins

Several other mycotoxins, beside aflatoxin, are present in the food crops consumed by people chronically exposed to aflatoxin (Cardwell, 2001). Fusarium mycotoxins, including fumonisins produced primarily by *F. verticillioides,* and deoxynivalenol (DON, sometimes termed vomitoxin) produced primarily by *F. graminearum,* are contaminants of food in areas of the world where people are exposed to aflatoxin. These mycotoxins occur worldwide on maize, wheat and other cereal grains (Bullerman, 1996). *Aspergillus* and *Fusarium* are two of the most abundant fungi found in grains (Hill *et al.*, 1984). Fumonisin B_1 is a major contaminant of maize and maize products worldwide (Nelson *et al.*, 1992). There have been several reports on the co-occurrence of aflatoxin and fumonisins and aflatoxin and other mycotoxins such as trichothecenes and ochratoxin in maize and other food products (Chamberlain *et al.*, 1993; Chu and Li, 1994; Kpodo *et al.*, 2000; Park *et al.*, 2002).

Fumonisin B_1 stimulates T-cell proliferation and induces nitric oxide production by rat spleenic macrophages (Dombrink-Kurtzman *et al.*, 2000). Also, macrophages from mice treated with fumonisin B_1 showed increased TNF-α production (Dugyala *et al.*, 1998). The phagocytic activity of swine alveolar macrophages was suppressed to 55% and 36% of control levels by fumonisin B_1 (50 ng/ml fumonisin B_1) and aflatoxin B_1 (100 ng/ml aflatoxin B_1) respectively. Fumonisin B_1 was toxic at lower concentrations than aflatoxin B_1 (Liu *et al.*, 2002). A dramatic decrease in mRNA levels of IL-1β and TNF-α was observed in alveolar macrophages incubated with 2 and 10 μg/ml fumonisin B_1 for 24 hours, while aflatoxin B_1 did not alter IL-1β and TNF-α mRNA expression (Liu *et al.*, 2002). Humoral immune responses in mice or rats were not affected by fumonisin B_1 (Martinova and Merrill, 1995; Tryphonas *et al.*, 1997).

In 2005, Taranu *et al.* (2005) reported that fumonisin B_1 altered the cytokine profile (IL-4, IFN-γ) and decreased antibody titer to a vaccine in pigs. In pig PBMC *in vitro*, they found that fumonisin B_1 decreased IL-4 and increased IFN-γ synthesis at both the mRNA and protein levels. In a brief *in vivo* study, they found that fumonisin B_1 altered cytokine balance in mesenteric lymph nodes and spleen of weanling piglets exposed to 1.5 mg fumonisin B_1/kg body weight for 7 days in a manner similar to that observed in the *in vitro* results. *In vivo* exposure of weanling piglets for a prolonged period (28 days) to 8 mg fumonisin B_1/kg in feed resulted in a significant decrease in the expression of IL-4 mRNA by porcine whole blood cells. Specific antibody titer after vaccination against *Mycoplasma agalactiae* was diminished but there was no effect of fumonisin B_1 on serum concentrations of IgG, IgA and IgM.

Fumonisin B_1 is structurally analogous with sphingosine and sphinganine and interferes with sphingolipid metabolism causing accumulation of free sphingoid bases (Merrill *et al.*, 2001). These free sphingoid bases inhibit lymphocyte growth, and Th2 lymphocytes are more sensitive to their action than Th1 lymphocytes (Tokura *et al.*, 1996). Thus, Tokura *et al.* (1996) hypothesize that fumonisin B_1 may selectively act on Th2 cells decreasing the synthesis of IL-4. Fumonisin B_1 also modulates glycolipid synthesis (Merrill *et al.*, 2001) but how modulation of glycolipid synthesis by fumonisin B_1 affects lymphocyte function has not been determined. Since Th2 cytokines function in the development of immune responses that lead to antibody production, the authors postulate that decrease of IL-4 by fumonisin B_1 resulted in decrease of the specific antibody response to the *M. agalactiae* vaccine.

Data from Martinova and Merrill (1995) suggest that fumonisin-induced immunomodulation may result from changes in the expression of certain cell surface receptors involved in immune communication in mice such as CD3 and sphingomyelin. However, the immunomodulation also may result from changes in cytokine secretion such as TNF-α (Dugyala *et al.*, 1998). Further study of the role of fumonisin B_1 on sphingolipids in immune cell signaling is needed to provide information on the immunomodulatory effects of fumonisin B_1.

The trichothecenes (produced by *Fusarium* spp. and other species of fungi) present in cereal grains also are immunotoxins (Bondy and Pestka, 2000). Trichothecenes at high doses can cause leukocyte apoptosis since they target actively dividing cells in the bone marrow, lymph node, spleen and thymus. Low dose trichothecenes seem to promote expression of a diverse array of cytokines that can either up-regulate or down-regulate immune functions (Bondy and Pestka, 2000). Immunosuppression by trichothecenes may be explained by their ability to bind ribosomes and inhibit protein synthesis (Bamburg, 1983).

Ochratoxin A is another immunotoxic mycotoxin that has multiple effects on immune cells (Richard, 1991; Bondy and Pestka, 2000). Harvey *et al.* (1992) showed that ochratoxin A suppressed cell-mediated and phagocytic cell responses in pigs. Lea *et al.* (1989) reported that ochratoxin A abrogated the ability of purified human lymphocytes to respond to activating stimuli *in vitro*. They found that expression of both IL-2 and IL-2 receptors was severely affected in cells exposed to ochratoxin A. Antibody production also was inhibited in purified B lymphocytes. This effect was not due only to the blocking of T helper cell function but also to the inability to respond to polyclonal activation after treatment with ochratoxin A. Therefore, the authors concluded that ochratoxin A interferes with essential processes in cell metabolism regardless of the subpopulation of lymphocytes.

Berek *et al.* (2001) tested the effects of eight important fusarium mycotoxins (deoxynivalenol, 3-acetyldeoxynivalenol, fusarenon-X, T-2 toxin, zearalenone, α-zearalenol, β-zearalenol, and nivalenol) on T and B cell proliferation, antibody-dependent cellular cytotoxicity (ADCC) and NK cell activity in healthy monocyte-free human PBMCs. Concentrations of mycotoxins similar to those found in normal human peripheral blood (0.2-1800 ng/ml) were used. T-2 toxin, fusarenon-X, nivalenol and 3-acetyldeoxynivalenol had the highest immunosuppressive effects and depressed T and B lymphocyte proliferation in a dose-dependent manner. The other five mycotoxins had significantly less or no effect. Deoxynivalenol, T-2 toxin and nivalenol each significantly inhibited the ADCC reaction with the effect of T-2 toxin being dose-dependent. These three mycotoxins also inhibited NK cell activity in a dose-dependent manner with significant inhibition at higher concentrations.

The combined effect of several mycotoxins as immunotoxins, each with a potentially different mechanism of toxicity, is unknown. However, a mixture of mycotoxins should have at least additive, if not synergistic, effects (Miller and Wilson, 1994).

Future studies on immune effects of aflatoxin and other mycotoxins

Currently, there is a scarcity of data on the immunosuppressive effects of mycotoxins in humans. Future research needs to be conducted to: i) establish the association between mycotoxin exposure in humans and modulation of various aspects of immune function; ii) determine the mode of action of the different mycotoxins and their interaction; and iii) evaluate immunomodulatory interactions between mycotoxins, nutritional factors and infectious agents, especially those that are immunosuppressive, such as HIV/AIDS.

The study by Jiang *et al.* (2005) was relatively small and exploratory in nature. The results suggest that additional larger investigations are needed on the association between aflatoxin levels and immune status/function. Thus, large randomized follow-up studies to assess the association between aflatoxin levels, immune status, susceptibility to infectious diseases and failure to produce an effective immune response to vaccines need to be conducted. The availability of appropriate biomarkers for aflatoxin makes these studies possible. However, appropriate biomarkers also need to be developed for many other mycotoxins. Molecular and advanced immunologic methods used in recent mycotoxin research, mostly in animal and in *in vitro* studies, now need to be applied in human studies.

The proportion of, and secretion of major cytokines by, subsets of lymphocytes, macrophages or other leukocytes can be determined by using immunoassays and more advanced flow cytometric methods such as five color staining. Specific mechanisms of immunomodulation and synergistic effects of commonly occurring mycotoxins also need to be investigated. A major area for investigation is the immunomodulatory interactions between mycotoxins and infectious agents that are immunosuppressive, *e.g.*, HIV, and between mycotoxins and immune modulators, *e.g.*, nutritional factors. The dietary interactions in immunosuppression by aflatoxin are not known, but a less than fully functioning immune system due to mycotoxin suppression certainly could enable many secondary disease conditions.

References

Annau, E., Corner, A.H., Magwood, S.E. and Jericho, K. (1964) Electrophoretic and chemical studies on sera of swine following the feeding of toxic groundnut meal. *Canadian Journal of Comparative Medicine and Veterinary Science* 28, 267-270.

Bamburg, J.R. (1983) Biological and biochemical actions of trichothecene mycotoxins. *Progress in Molecular and Subcellular Biology* 8, 41-110.

Berek, L., Petri, I.B., Mesterhazy, A., Teren, J. and Molnar, J. (2001) Effects of mycotoxins on human immune functions *in vitro*. *Toxicology in Vitro* 15, 25-30.

Bondy, G.S. and Pestka, J.J. (2000) Immunomodulation by fungal toxins. *Journal of Toxicology and Environmental Health, Part B* 3, 109-143.

Boonchuvit, B. and Hamilton, P.B. (1975) Interaction of aflatoxin and paratyphoid infections in broiler chickens. *Poultry Science* 54, 1567-1573.

Bosch, F. X. and Munoz, N. (1988) Prospects for epidemiological studies on hepatocellular cancer as model for assessing viral chemical interactions. In: Bartsch H., Hemminki, K. and O'Neill, I.K. (eds.) *Methods for Detecting DNA Damaging Agents in Humans: Applications in Cancer Epidemiology and Prevention, IARC Scientific Publication No. 89*, International Agency for Research on Cancer, Lyon, France, pp. 427-438.

Bullerman, L.B. (1996) Occurrence of *Fusarium* and fumonisins on food grains and in foods. *Advances in Experimental Medicine and Biology* 392, 27-38.

Cardwell, K.F. (2001) Mycotoxin contamination of foods in Africa: Anti-nutritional factors. *Food Nutrition Bulletin* 21, 488-492.

Chamberlain, W.J., Bacon, C.W., Norred, W. P. and Voss, K. A. (1993) Levels of fumonisin B_1 in corn naturally contaminated with aflatoxins. *Food and Chemical Toxicology* 12, 995-998.

Chu, F.S. and Li, G.Y. (1994) Simultaneous occurrence of fumonisin B_1 and other mycotoxins in moldy corn collected from the People's Republic of China in regions with high incidences of esophageal cancer. *Applied and Environmental Microbiology* 60, 847-852.

Cooper, M.S., Fehniger, T.A., Turner, S.C., Chen K.S., Ghaheri, B.A., Ghayur, T., Carson, W.E. and Caligiuri, M.A. (2001) Human natural killer cells: A unique innate immunoregulatory role for the $CD56^{bright}$ subset. *Blood* 97, 3146-3151.

Cukrova, V., Kurita,N. and Asao, M. (1992) An early effect of aflatoxin B_1 administered *in vivo* on the growth of bone marrow CFU-GM and the production of some cytokines in rats. *Mycopathologia* 120, 113-119.

Cusamano, V., Costa,G.B. and Seminara, S. (1990) Effect of aflatoxins on rat peritoneal macrophages. *Applied and Environmental Microbiology* 56, 3482-3484.

Cusumano, V., Rossano, F., Merendino, R.A., Arena, A., Costa, G.B., Mancuso, G., Baroni, A. and Losi, E. (1996) Immunobiological activities of mould products: functional impairment of human monocytes exposed to aflatoxin B_1. *Research Microbiology* 147, 385-391.

Cysewki, S.J., Wood, R.L., Pier, A.C. and Baetz, A.L. (1978) Effects of aflatoxin on the development of acquired immunity to swine erysipelas. *American Journal of Veterinary Research* 39, 445-448.

Dombrink-Kurtzman, M.A., Gomez-Flores, R. and Weber, R.J. (2000) Activation of rat splenic macrophage and lymphocyte functions by fumonisin B_1. *Immunopharmacology* 49, 401-409.

Dugyala, R.R. and Sharma, R.P. (1996) The effect of aflatoxin B_1 on cytokine mRNA and corresponding protein levels in peritoneal macrophages and splenic lymphocytes. *International Journal of Immunopharmacology* 18, 599-608.

Dugyala, R.R., Sharma, R.P., Tsunda, M. and Riley, R.T. (1998) Tumor necrosis factor-alpha as a contributor in fumonisin B_1 toxicity. *Journal of Pharmacology and Experimental Therapeutics* 285, 317-324.

Edds, G.T., Nair, K.P.C. and Simpson, C.F. (1973) Effect of aflatoxin B_1 on resistance in poultry against cecal coccidiosis and Marek's disease. *American Journal of Veterinary Research* 34, 819-826.

Gabal, M.A. and Azzam, A.H. (1998) Interaction of aflatoxin in the feed and immunization against selected infectious diseases in poultry. II. Effect on one-day-old layer chicks simultaneously vaccinated against Newcastle disease, infectious bronchitis and infectious bursal disease. *Avian Pathology* 27, 290-295.

Gabal, M.A. and Dimitri, R.A. (1998) Humoral immunosuppressant activity of aflatoxin ingestion in rabbits measured by response to *Mycobacterium bovis* antigens using enzyme-linked immunosorbent assay and serum protein electrophoresis. *Mycoses* 41, 303-308.

Gallikeev, K.H.L., Raipov, O.R. and Manyasheva, R.A. (1968) Effect on aflatoxin on dynamics of antibody formation (Tr). *Biulleten' eksperimental'noĭ biologii i meditsiny (USSR)* 65, 88-90.

Ghosh, R.C., Chauhan, H.V.S. and Jah, G.J. (1991) Suppression of cell-mediated immunity by purified aflatoxin B_1 in broiler chicks. *Veterinary Immunology and Immunopathology* 28, 165-172.

Gourama, H. and Bullerman, L.B. (1995) *Aspergillus flavus* and *Aspergillus parasiticus*, aflatoxigenic fungi of concern in foods and feed – a review. *Journal of Food Protection* 58, 1395-1404.

Hamilton, P.B. and Harris, J.R. (1971) Interaction of aflatoxicosis with *Candida albicans* infections and other stresses in chickens. *Poultry Science* 50, 906-912.

Harvey, R.B., Edrington, T.S., Kubena, L.F., Elissalde, M.H. and Rottinghaus, G.E. (1995) Influence of aflatoxin and fumonisin B_1-containing culture material on growing barrows. *American Journal of Veterinary Research* 56, 1668-1672.

Harvey, R.B., Elissalde, M.H., Kubena, L.F., Weaver, E.A., Corrier, D.E. and Clement, B.A. (1992) Immunotoxicity of ochratoxin A to growing gilts. *American Journal of Veterinary Research* 53, 1966-1970.

Hatori, Y., Sharma, R.P. and Warren, R.P. (1991) Resistance of C57B1/6 mice to immunosuppressive effects of aflatoxin B_1 and relationship with neuroendocrine mechanisms. *Immunopharmacology* 22, 127-136.

Hill, R.A., Wilson, D.M., Burg, W.R. and Shotwell, O.L. (1984) Viable fungi in corn dust. *Applied and Environmental Microbiology* 47, 84-87.

IARC. (1996) Some naturally occurring substances: food items and constituents, heterocyclic aromatic amines and mycotoxins. *IARC Monographs on the Evaluation of Carcinogenic Risks to Humans, vol. 56*. International Agency for Research on Cancer, Lyon, France.

Jakab, G.J., Hmieleski, R.R., Zarba, A., Hemenway, D.R. and Groopman, J.D. (1994) Respiratory aflatoxicosis: suppression of pulmonary and systemic host defenses in rats and mice. *Toxicology and Applied Pharmacology* 125, 198-205.

Jiang, Y., Jolly, P.E., Ellis, W.O., Wang, J.S., Phillips, T.D. and Williams, J.H. (2005) Aflatoxin B_1 albumin adduct levels and cellular immune status in Ghanaians. *International Immunology* 17, 807-814.

Joens, L.A., Pier, A.C. and Cutlip, R.C. (1981) Effects of aflatoxin consumption on the clinical course of swine dysentery. *American Journal of Veterinary Research* 42, 1170-1172.

Kadian, S.K., Monga, D.P. and Goel, M.C. (1988) Effect of aflatoxin B_1 on the delayed type hypersensitivity and phagocytic activity of reticuloendothelial system in chickens. *Mycopathologia* 104, 33-36.

Kpodo, K., Thrane, U. and Hald, B. (2000) Fusaria and fumonisins in maize from Ghana and their co-occurrence with aflatoxins. *International Journal of Food Microbiology* 61, 147-157.

Kubena, L.F., Bailey, R.H., Byrd, J.A., Young, C.R., Corrier, D.E., Stanker, L.H. and Rottinhaus, G.E. (2001) Cecal volatile fatty acids and broiler chick susceptibility to *Salmonella typhimurium* colonization as affected by aflatoxins and T-2 toxin. *Poultry Science* 80, 411-417.

Kurtz, R.S. and Czuprynski, C.J. (1992) Effects of aflatoxin B_1 on *in vitro* production of interleukin-1 by bovine mononuclear phagocytes. *Veterinary Immunology and Immunopathology* 34, 149-158.

Lea, T., Setién,K. and Stormer, F.C. (1989) Mechanism of ochratoxin A-induced immunosuppression. *Mycopathologia* 107, 153-159.

Liu, B.H., Yu, F.Y., Chan, M.H. and Yang, Y.L. (2002) The effects of mycotoxins, fumonisin B_1 and aflatoxin B_1, on primary swine alveolar macrophages. *Toxicology and Applied Pharmacology* 180, 197-204.

Marin, D.E., Taranu, I., Bunaciu, R.P., Pascale, F., Tudor, D.S., Avram, N., Sarca, M., Cureu, I., Criste, R.D., Suta, V. and Oswald, I.P. (2002) Changes in performance, blood parameters, humoral and cellular immune responses in weanling piglets exposed to low doses of aflatoxin. *Journal of Animal Science* 80, 1250-1257.

Martinova, E.A., and Merrill, A.H. Jr. (1995) Fumonisin B_1 alters sphingolipid metabolism and immune function in BALB/c mice: Immunological responses to fumonisin B_1. *Mycopathologia* 130, 163-170.

Meky, F.A., Hardie, L.J., Evans, S.W. and Wild, C.P. (2001) Deoxynivalenol-induced immunomodulation of human lymphocyte proliferation and cytokine production. *Food Chemistry and Toxicology* 39, 827-836.

Merrill, A.H. Jr., Sullards, M.C., Wang, E., Voss, K.A. and Riley, R.T. (2001) Sphingolipid metabolism: Roles in signal transduction and disruption by fumonisins. *Environmental Health Perspectives* 109, 283-289.

Miller, D.M., Stuart, B.P. and Crowell, W.A. (1981) Experimental aflatoxicosis in swine: Morphological and clinical pathological results. *Canadian Journal of Comparative Medicine* 45, 343-351.

Miller, J.D. and Wilson, D.M. (1994) Veterinary diseases related to aflatoxins. In: Eaton, D. and Groopman, J. (eds.) *The Toxicology of Aflatoxin: Human Health, Veterinary, and Agricultural Significance*. Academic Press, Chicago, pp. 347-367.

Mohapatra, N.K. and Roberts, J.F. (1985) *In vitro* effect of aflatoxin B_1 on rat liver macrophages (Kupffer cells). *Toxicology Letters* 29, 177-181.
Moon, E.Y. (1999) Inhibition of various functions in murine peritoneal macrophages by aflatoxin B_1 exposure *in vivo*. *International Journal of Immunopharmacology* 21, 47-58.
Moon, E.Y., Rhee, D.K. and Pyo, S. (1999) *In vitro* suppressive effect of aflatoxin on murine peritoneal macrophage functions. *Toxicology* 133, 171-179.
Neal, G.E., Eaton, D.L., Judah, D.J. and Verma, A. (1998) Metabolism and toxicity of aflatoxins M_1 and B_1 in human-derived *in vitro* systems. *Toxicology and Applied Pharmacology* 151, 152-158.
Neldon-Ortiz, D.L. (1991) Direct and microsomal activated aflatoxin B_1 exposure and its effects on turkey peritoneal macrophage functions *in vitro*. *Toxicology and Applied Pharmacology* 109, 432-442.
Neldon-Ortiz, D.L. and Qureshi, M.A. (1992) Effect of aflatoxin B_1 embryonic exposure on mononuclear phagocytic cell functions. *Developmental and Comparative Immunology* 16, 187-196.
Nelson, P.E., Plattner, R.D., Schackelford, D.D. and Desjardins, A.E. (1992) Fumonisin B_1 production by species other than *Fusarium moniliforme* in section *Liseola* and some related species. *Applied and Environmental Microbiology* 58, 984-989.
Nyathi, C.B., Mutiro, C.F., Hasler, J.A. and Chetsanga, C.J. (1987) A survey of urinary aflatoxin in Zimbabwe. *International Journal of Epidemiology* 6, 516-519.
Oswald, I.P., Marin, D.E., Bouhet, S., Pinton, P., Taranu, I. and Accensi, F. (2005) Immunotoxicological risk of mycotoxins for domestic animals. *Food Additives and Contaminants* 22, 354-360.
Ozturk, M. (1991) p53 mutation in hepatocellular carcinoma after aflatoxin exposure. *Lancet* 338, 1356-1359.
Park, J.W., Kim, E.K., Shon, D.H. and Kim, Y.B. (2002) Natural co-occurrence of aflatoxin B_1, fumonisin B_1, and ochratoxin A in barley and maize foods from Korea. *Food Additives and Contaminants* 19, 1073-1080.
Pier, A.C. (1986) Immunologic changes associated with mycotoxicoses. 13. Immunomodulation in aflatoxicosis In: Richard, J.L. and Thurston, J.R. (eds.) *Diagnosis of Mycotoxicosis,* Martinus Nijhoff, Boston, pp. 143-148.
Pier, A.C. and Heddleston, K.L. (1970) The effect of aflatoxin on immunity in turkeys. I. Impairment of actively acquired resistance to bacterial challenge. *Avian Diseases* 14, 797-809.
Pier, A.C., McLoughlin, M.E., Richard, J.L., Baetz, A.L. and Dahlgren, R.R. (1984) *In utero* transfer of aflatoxin and selected effects on neonatal pigs. In: Lacey, J. (ed.) *Trichothecenes and Other Mycotoxins*. John Wiley and Sons, New York, pp 495-506.
Qian, G.S., Ross, R.K., Yu, M.C., Yuan, J.M., Gao, Y.T., Henderson, B.E., Wogan, G.N. and Groopman, J.D. (1994) A follow-up study of urinary markers of aflatoxin exposure and liver cancer risk in Shanghai, People's Republic of China. *Cancer Epidemiology, Biomarkers & Prevention* 3, 3-10.
Raisuddin, S., Singh, K.P., Zaidi, S.I., Paul, B.N. and Ray, P.K. (1993) Immunosuppressive effects of aflatoxin in growing rats. *Mycopathologia* 124, 189-194.
Reddy, R.V. and Sharma, R.P. (1989) Effects of aflatoxin B_1 on murine lymphocytic functions. *Toxicology* 54, 31-44.
Reddy, R.V., Taylor, M.J. and Sharma, R.P. (1987) Studies of immune function of CD-1 mice exposed to aflatoxin B_1. *Toxicology* 43, 123-132.
Richard, J.L. (1991) Mycotoxins as immunomodulators in animal systems. In: Bray, G.A. and D. H. Ryan, D.H (eds.) *Mycotoxins, Cancer and Health.* State University Press, Baton Rouge, Louisiana, pp. 197-220.
Richard, J., Thurston, J.R. and Pier, A.C. (1978) Effects of mycotoxins on immunity. In: Rosenberg, P. (ed.) *Toxins: Animal, Plant and Microbial*, Pergamon Press, New York, pp. 801-817.
Ross, R.K., Yuan, J.M., Yu, M.C., Wogan, G.N., Qian, G.S., Tu, J.T., Groopman, J.D., Gao, Y.T. and Henderson, B.E. (1992) Urinary aflatoxin biomarkers and risk of hepatocellular carcinoma. *Lancet* 339, 943-946.
Rossano, F., de Luna, L.O., Buommino, E., Cusumano, V., Losi, E. and Catania, M.R. (1999) Secondary metabolites of *Aspergillus* exert immunobiological effects on human monocytes. *Research Microbiology* 150, 13-19.

Savel, H., Forsyth, B., Schaeffer, W. and Cardella, T. (1970) Effect of aflatoxin B_1 upon phytohemaglutinin-transformed human lymphocytes. *Proceedings of the Society for Experimental Biology and Medicine* 134, 1112-1115.

Silvotti, L., Petterino, C., Bonomi, A. and Cabassi, E. (1997) Immunotoxicological effects on piglets of feeding sows diets containing aflatoxins. *Veterinary Record* 141, 469-472.

Sorenson, W.G., Gerberik, G.F., Lewis, D.M. and Castranova, V. (1986) Toxicity of mycotoxins for the rat pulmonary macrophages in vitro. *Environmental Health Perspectives* 66, 45-53.

Taranu, I., Marin, D.E., Bouhet, S., Pascale, F., Bailly, J. D., Miller, J.D., Pinton, P. and Oswald, I.P. (2005) Mycotoxin fumonisin B_1 alters the cytokine profile and decreases the vaccinal antibody titer in pigs. *Toxicological Sciences* 84, 301-307.

Tokura, Y., Wakita, H., Nishimura, K., Furukawa, M. and Tolleson, W.H. (1996) Th2 suppressor cells are more susceptible to sphingosine than Th1 cells in murine contact photosensitivity. *Journal of Investigative Dermatology* 107, 34-40.

Tryphonas, H., Bondy, G., Miller, J.D., Lacroix, F., Hodgen, M., McGuire, P., Fernie, S., Miller, D. and Hayward, S. (1997) Effects of fumonisin B_1 on the immune system of Sprague-Dawley rats following a 14-day oral (gavage) exposure. *Fundamental and Applied Toxicology* 39, 53-59.

Turner, P.C., Moore, S.E., Hall, A.J., Prentice, A.M. and Wild, C.P. (2003) Modification of immune function through exposure to dietary aflatoxin in Gambian children. *Environmental Health Perspectives* 111, 217-220.

van Heugten, E., Spears, J.W., Coffey, M.T., Kegley, E.B. and Qureshi, M.A. (1994) The effect of methionine and aflatoxin on immune function in weanling pigs. *Journal of Animal Science* 72, 658-664.

Venturini, M.C., Quiroga, M.A., Risso, M.A., di Lorenzo, C., Omata, Y., Venturini, L. and Godoy, H. (1996) Mycotoxin T-2 and aflatoxin B_1 as immunosuppressants in mice chronically infected with *Toxoplasma gondii*. *Journal of Comparative Pathology* 115, 229-237.

Walsh, C.J., Bodine, A.M. and Scout, T.R. (1991) Co-mitogenic assay for assessing the effects of aflatoxin B_1 on interleukin-1 production in bovine macrophages. *Drug Development Research* 24, 157-166.

Wang, L.Y., Hatch, M., Chen, C.J., Levin, B., You, S.L., Lu, S.N., Wu, M.H., Wu, W.P., Wang, L.W., Wang, Q., Huang, G.T., Yang, P.M., Lee, H.S. and Santella, R.M. (1996) Aflatoxin exposure and risk of hepatocellular carcinoma in Taiwan. *International Journal of Cancer* 67, 620-625.

Wang, J.S., Huang, T., Su, J., Liang, F., Wei, Z., Liang, Y., Luo, H., Kuang, S.Y., Qian, G.S., Sun, G., He, X., Kensler, T.W. and Groopman, J.D. (2001) Hepatocellular carcinoma and aflatoxin exposure in Zhuqing Village, Fusui County, People's Republic of China. *Cancer Epidemiology, Biomarkers & Prevention* 10, 143-146.

Watzl, B., Neudecker, C., Hansch, G.M., Rechkemmer, G. and Pool-Zovel, B.L. (1999) Short-term moderate aflatoxin exposure has only minor effects on the gut-associated lymphoid tissue of Brown Norway rats. *Toxicology* 138, 93-102.

Wyatt, R.D., Ruff, M.D. and Page, R.K. (1975) Interaction of aflatoxin with *Eimeria tenella* infection and monensin in young broiler chickens. *Avian Diseases* 19, 730-740.

6

Aflatoxin Exposure and Impaired Child Growth in West Africa: An Unexplored International Public Health Burden?

Yun Yun Gong, Paul C. Turner, Andrew J. Hall and Christopher P. Wild*

Abstract

Aflatoxins are common contaminants of staple foods in sub-Saharan Africa. These toxins are human hepatocarcinogens, especially in combination with chronic infection with hepatitis B virus. Exposure to aflatoxins begins early in life and recent studies in West Africa demonstrated an association between exposure and slowed growth, particularly stunting, in young children. The underlying mechanisms for these latter effects are unknown but may include impairment of immunity, altered intestinal integrity and increased susceptibility to infections. Simple post-harvest intervention strategies were successful in reducing aflatoxin exposure in a subsistence farm setting, providing a rationale for prevention of aflatoxin-related disease(s).

Introduction

Approximately 40% of the 11 million deaths in children younger than 5 years of age worldwide occur in sub-Saharan Africa (Black *et al.*, 2003). Both child morbidity and mortality are exceptionally high in the region, with 175 deaths per 1000 live births compared to 6 per 1000 in industrialized nations, a 29-fold difference. The major terminal causes of death in these children are diarrhea, pneumonia, malaria, and neonatal disorders (Bryce *et al.*, 2005). These problems are associated with poverty, including insufficient and poor quality food and a high prevalence of malnutrition. Indeed under-nutrition and slowed growth have been identified as an underlying cause of ~50% of the deaths linked to infectious diseases in sub-Saharan African children (Black *et al.*, 2003). Notably, slowed growth and micronutrient deficiencies are associated with decreased immune and non-immune host defenses that increase susceptibility to the above infectious diseases (Caulfield *et al.*, 2004*a*,*b*).

Mycotoxins are metabolites of fungi that contaminate up to 25% of the human food supply. The aflatoxins, including aflatoxins B_1, B_2, G_1 and G_2, are a potent family of mycotoxins produced by *Aspergillus* spp. that occur on cereals, oilseeds (including peanuts) and tree nuts in parts of the world where the tropical climate facilitates fungal growth. Whilst climatic conditions do not restrict the occurrence of aflatoxin to developing countries, in practice high-technology agricultural practices and the presence and enforcement of government regulatory limits in developed countries translate to exposures that typically are low and sporadic for these latter populations. In contrast, there is little emphasis on agricultural practice to combat aflatoxins in developing countries, and the few regulatory controls are rarely enforced. In addition, it is difficult to envisage such controls being effective when 60-85% of the population are subsistence farmers whose trade often is limited to local markets, where most regulations are unenforceable. Thus, in reality, dietary aflatoxin exposure is a continuous daily problem for the world's poor.

Aflatoxins are potent hepatocarcinogens in animals and epidemiological evidence in human populations, particularly from prospective cohort studies, has led to aflatoxin being classified by the International Agency for Research on Cancer (IARC) as a Class 1 human carcinogen (IARC, 2002). The risk of hepatocellular carcinoma is particularly elevated in individuals with chronic hepatitis B virus infection who also are exposed to aflatoxins (IARC, 2002). Hepatocellular carcinoma is the most common cancer in many parts of sub-Saharan Africa, parts of southeast Asia, and China, with up to 10% of all adult male deaths attributed to this disease.

In addition to hepatocellular carcinoma, aflatoxins are associated with occasional outbreaks of acute aflatoxicosis leading to death shortly after exposure (see Hall and Wild, 1994). In 2004, the consumption of heavily aflatoxin-contaminated maize in Kenya was linked to more than a hundred deaths (Azziz-Baumgartner *et al.*, 2005). In this incident the levels of aflatoxins in maize frequently exceeded 5000 ng/g. Levels > 100 ng/g are common in foods in parts of Africa whilst permitted levels for foods for human consumption in Europe and North America, for example, are generally < 20 ng/g (van Egmond and Jonker, 2004).

Aflatoxin exposure is a result of both the level of contamination in a given commodity and the quantity of the commodity that is consumed. Thus in some areas of the world aflatoxin levels in foods might be relatively high but exposures modest because of a varied diet. However, in sub-Saharan Africa, similar levels of food contamination will translate to a much higher exposure. Dietary staples in the region include two that are highly susceptible to contamination by aflatoxins, namely peanuts and maize. In many countries and regions these staples are consumed daily for the majority of the year and may constitute >50% of the diet. The long-term storage of both crops in hot, humid conditions results in fungal proliferation and increased post-harvest toxin contamination. The result is that aflatoxin exposure in parts of West Africa is ubiquitous and at high levels throughout the life of the people living there (Wild and Turner, 2002).

The problem of aflatoxin contamination must be considered in the context of food sufficiency. Aflatoxin contamination is frequently superimposed on severe problems of food production, storage and availability. For example, one-third to one-half of the people in West Africa live in poverty with a *per capita* daily energy intake of < 2,500 Kcal (UNEP GEO, 2000). There is marked seasonal variation in food availability, resulting in the so-called "hungry period" where nutrition is inadequate. Consequently, even clearly moldy food may be eaten, as occurred in the recent aflatoxicosis cases in Kenya. This interaction between food safety and food security is crucial when addressing the problem of aflatoxins.

Childhood aflatoxin exposure patterns in West Africa

Aflatoxin exposure has been difficult to characterize in the past, partly because the heterogeneous nature of contamination makes representative sampling of foodstuffs difficult (Whitaker, 2006). This inability to accurately measure individual exposure has in turn hampered efforts to understand the health effects of aflatoxins. Overcoming the difficulties inherent to these types of assessment has been the driving force for the development of aflatoxin exposure biomarkers (Groopman and Kensler, 1999; Wild and Turner, 2002). Aflatoxins require bio-activation to a reactive aflatoxin 8,9-epoxide in order to exert their toxic effects and this metabolic activation is effected in humans by enzymes of the cytochrome P450 family, specifically CYP3A4, CYP3A5, CYP3A7 and CYP1A2 (Guengerich *et al.*, 1998; IARC, 2002). Once activated to the epoxide, aflatoxin can bind to and damage cellular targets such as DNA and proteins. One consequence is that aflatoxin binds covalently to albumin and this aflatoxin-albumin (AF-alb) adduct can be measured in the peripheral blood as a useful biomarker of exposure over the 2-3 months prior to sampling (Wild and Turner, 2002). The ability to measure exposure using AF-alb has permitted a number of investigations of aflatoxin exposure and disease in the developing world. This chapter focuses on the consequences of aflatoxins for child health.

Aflatoxin exposure begins early in life in West Africa. The lipophilic nature of aflatoxins means that they can cross the placental barrier, and aflatoxin-albumin adducts have been reported in Gambian cord blood samples (Wild *et al.*, 1991). This exposure continues in infancy. In The Gambia and Guinea, for example, children as young as three years of age have similar high prevalence (> 90%) and levels of adduct as observed in adults from the same populations (Allen *et al.*, 1992; Wild *et al.*, 1993, 2000; Diallo *et al.*, 1995; Turner *et al.*,

Table 1. Aflatoxin-albumin adducts in children by country.

COUNTRY (no. subjects)	Year	AF-Alb (pg/mg) Mean[a] (range)	Frequency of positive samples (%)	Age group (years)	Reference
The Gambia (391)	1988	57 (nd-720)	83	3-8	Allen *et al.*, 1992
The Gambia (444)	1990-1991	41 (nd-459)	100	3-4	Turner *et al.*, 2000
The Gambia (466)	1998-1999	24 (nd-456)	93	6-9	Turner *et al.*, 2003
Benin & Togo (479)	2000	33 (nd-1064)	99	1-5	Gong *et al.*, 2002
Benin[b] (200)	2001	[1]37 (nd-688) [2]39 (nd-744) [3]88 (5-1568)	98 99.5 100	1-3	Gong *et al.*, 2004
Guinea (124)	2002	9.2 (nd-66)	96	2-5	Turner *et al.*, 2005*b*

nd: non-detected.
[a]Geometric mean.
[b]Data from an eight month longitudinal study with 200 children aged 1-3 years at recruitment, tested for AF-alb in [1]February, [2]June and [3]October of 2001.

2000, 2003, 2005*a,b*). These levels result from early weaning and introduction to family foods. The common family foods, *e.g.*, maize, are contaminated with aflatoxins. Aflatoxin-albumin adducts can be detected in sera of Beninese children aged less than one year of age (Gong *et al.*, 2003). Various studies of children from West Africa have found a uniformly high occurrence of exposure across countries and over time (Table 1). Comparisons across countries, however, should be interpreted cautiously because the data are drawn from different studies whose designs are not directly comparable. Nevertheless, the overall observations leave no doubt that children are chronically exposed to high levels of aflatoxins in this part of the world.

The pattern of aflatoxin exposure in early infancy is dynamic because the infant will initially consume exclusively breast milk and then be slowly introduced to weaning foods followed by family foods. The weaning food, typically a thin cereal gruel made from maize, sorghum or millet, is introduced early, sometimes by two months of age. The nature of the weaning food, the relative quantities compared to breast milk and the duration of weaning before the introduction of family foods will affect the amount of aflatoxin exposure during this potentially critical developmental period. Factors affecting the weaning process also vary and may depend on seasonal variation in availability of different foods, sibling number, local cultural practices, and maternal or infant health. Despite the variation in individual circumstances, the general pattern is that as weaning foods slowly replace breast milk the level of aflatoxin exposure increases markedly (Gong *et al.*, 2002, 2003, 2004).

The period of breast-feeding is generally associated with lower levels of aflatoxin-albumin in a child's blood (Turner *et al.*, 2007) because the mother's metabolism limits transfer of dietary aflatoxins into the milk. Little is known about how the components of breast milk might interfere with the absorption of aflatoxins from milk, a point worthy of additional research. However, there may be some toxic impact of aflatoxin metabolites even during this period of life, because the hydroxylated metabolite, aflatoxin M_1, can be transferred to breast milk (IARC, 2002). This metabolite is less carcinogenic than the parent aflatoxin B_1 found in food but may still possess cytotoxic activity.

Whilst aflatoxins have long been investigated in the etiology of HCC, there has been remarkably little attention paid to the other potential adverse effects on health despite the widespread exposure. Given the growing belief that early life exposures influence health and disease later in life (Barker, 2002), aflatoxin exposure in children should receive more attention. This point was highlighted by a recent World Health Organization/Center for Disease Control, USA report on aflatoxins (Strosnider *et al.*, 2006).

Childhood exposure and impaired growth

Studies in different animal species indicate that aflatoxin exposure can severely affect growth and development (Dersjant-Li *et al.*, 2003). In the veterinary field the adverse effects on growth of poultry, swine and other species are the primary concern due to the association of aflatoxin contamination of feed with lower efficiency of food use and reduced feed intake. In a recent review it was estimated that for each ng/g increase in aflatoxin in the diet, growth rate would be reduced by 16% in swine and 5% in broilers (Dersjant-Li *et al.*, 2003). Aflatoxin B_1 impairs development and increases mortality in chicken embryos (Edrington *et al.*, 1995) and results in reduced body weight in broilers (Smith *et al.*, 1992). Marin *et al.* (2002) reported a reduction in weight gain in piglets fed 140 or 280 μg aflatox-

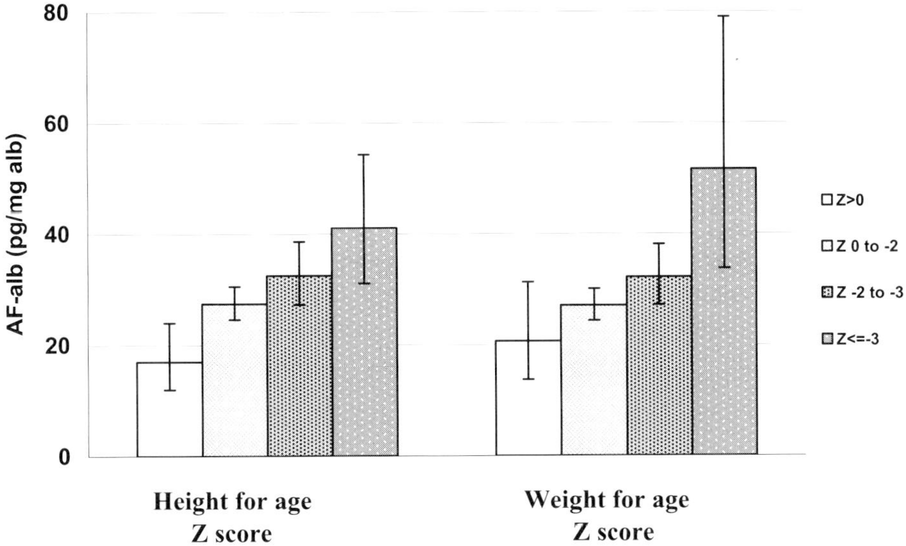

Figure 1. Aflatoxin-albumin adduct level and slowed growth in children (Gong *et al.*, 2002). Concentrations of aflatoxin-albumin adduct categorized into four groups for height for age and weight for age *z* scores based on the WHO classification of malnutrition (*z* score < -2) and severe malnutrition (< -3). Note: a *z* score is a measure of the distance, in standard deviations, of a measured value from the mean. Geometric mean adduct concentrations are shown, with 95% confidence intervals, adjusted for weaning status, agro-ecological zone, and socioeconomic status.

in B_1/kg body weight for 4 weeks (Marin *et al.,* 2002). These and other studies provide convincing evidence that aflatoxins impair animal growth and development. However, until recently these effects of aflatoxin have not been considered in relation to human exposure.

Given the widespread nature of aflatoxin exposure in African children, it is essential to evaluate the potential effects of exposure on child health, including growth and susceptibility to diseases. Some early studies suggested a link between aflatoxin exposure and Kwashiorkor (Hendrickse *et al.,* 1982), but the weakness in study design makes interpretation of these data difficult (Hall and Wild, 1994). In particular, aflatoxin exposure was assessed as the presence of free urinary aflatoxins, a short-term marker of exposure. Levels of urinary aflatoxins could also be affected by the disease, because aflatoxin metabolism and excretion may be altered by severe protein energy malnutrition. More recently we have applied the AF-alb biomarker to explore the association between aflatoxin and child growth in Benin and Togo in West Africa. This work has been published elsewhere (Gong *et al.,* 2002, 2003, 2004) and is summarized briefly below.

In a cross-sectional study with 480 children (aged 1-5 years) recruited randomly from sixteen villages in four geographic zones of Benin and Togo, aflatoxin-albumin adducts were detected in 99% of the samples, with a geometric mean of 33 pg/mg albumin (range 5-1060 pg/mg) representing some of the highest AF-alb levels we have observed. Means of weight for age *Z*-score (WAZ), height for age *Z*-score (HAZ) and weight for height *Z*-score (WHZ) in these children were all below zero with a prevalence of stunting (HAZ < -2) of 33% and of underweight (WAZ< -2) of 29% (Gong *et al.*, 2002). A striking inverse asso-

Table 2. Height increase relative to aflatoxin-albumin level over an 8 month period (after Gong *et al.*, 2004).

Aflatoxin Exposure Group	Height increase (cm)	
	Unadjusted	Adjusted[a]
Lower quartile	4.9 (4.5,5.3)[b]	5.9 (5.2,6.6)
Mid-lower quartile	4.4 (4.1,4.7)	5.3 (4.8,5.9)
Mid-upper quartile	4.1 (3.8,4.5)	4.8 (4.4,5.2)
Upper quartile	4.1 (3.8,4.5)	4.2 (3.9,4.6)

[a]Data are adjusted for age, height and weaning status at February and for mothers' socioeconomic status and village.
[b]Mean (confidence interval).

ciation was found between AF-alb and growth, with children that were stunted (HAZ) or underweight (WAZ) having 30-40% higher mean AF-alb levels compared to children with normal Z-scores. In a categorical analysis the association with AF-alb was again significant, with clear dose-response relationships with HAZ and WAZ (Fig. 1; Gong *et al.*, 2002, 2003).

To better understand this novel finding, an 8-month longitudinal study was conducted in some of the same areas in Benin. Two hundred children aged 16-37 months were recruited from four villages, two previously characterized with high and two with low aflatoxin exposure (50 children per village). AF-alb, anthropometry, information on weaning foods, weaning practices, food consumption and various demographic data were measured at recruitment (February) and at two subsequent time points (June and October). Two possible dietary confounding factors, vitamin A and zinc, were taken into consideration when analyzing the association between aflatoxin exposure and growth. Consistent with the earlier cross-sectional study there was a strong negative correlation ($p < 0.0001$) between AF-alb and height increase over the 8-month follow-up; the highest quartile of AF-alb was associated with a mean 1.7 cm reduction in height increase over eight months compared to the lowest quartile (Table 2; Gong *et al.*, 2004). This study further supports the hypothesis that aflatoxin exposure is associated with growth impairment, in particular stunting, in young children.

Despite these observations, the mechanism(s) by which aflatoxin may affect child growth are unknown. Possibilities include immune suppression, altered growth factor expression or intestinal toxicity (Williams *et al.,* 2004; Bouhet and Oswald, 2005). In relation to the possibility of a direct effect of aflatoxins on the intestinal epithelium, the CYP3A enzymes that bio-activate aflatoxin B_1 are expressed in human intestinal epithelium (Paine *et al.,* 2006). In this context genetic polymorphisms in aflatoxin-metabolizing enzymes (Sim *et al.,* 2005; Wojnowski *et al.*, 2004) could be significant modifiers of the relationship between aflatoxin exposure and child growth.

Immune suppression, altered growth factor expression or intestinal toxicity could each contribute at different times to growth faltering during the dynamic changes in nutrition early in life. As the pattern of nutrition changes, with a shift from breast milk to solid food, the child is exposed to dietary contaminants such as bacteria, viruses and environmental chemicals (including mycotoxins) that could affect immunity and the integrity of the intestinal tract. In parallel studies in the same areas of West Africa, Lunn and colleagues (Lunn *et al.*, 1991; Lunn, 2000) have reported a marked change in intestinal membrane permeability ("leakiness") in children of similar ages as those in our studies, and this "leakiness" was associated with an intestinal enteropathy and subsequent growth faltering. This disorder is characterized by villous shortening, crypt hyperplasia and lymphocyte infiltration (Lunn, 2000; Campbell *et al.*, 2002, 2003*a*,*b*). These children also exhibit decreased absorption of sugars (Northrop-Clewes *et al.*, 1997), probably as a consequence of the decrease in intestinal surface area, and elevated inflammatory markers (Campbell *et al.*, 2003*a*,*b*).

One hypothesis is that recurrent exposure to infectious agents and mycotoxins, due to introduction of unhygienic food, damages the intestinal epithelium thus contributing to the observations of Lunn *et al.* (Lunn *et al.*, 1991; Lunn, 2000, 2002; Campbell *et al.*, 2004). In Gambian children, chronic inflammatory enteropathy could explain ~40% of the slowed growth in the first 18 months of life (Lunn *et al.*, 1991; Lunn, 2000). Nutritional supplementation had only a modest effect on child growth during this period (Prentice, 1993), which is consistent with a problem in uptake of nutrients, rather than simple inadequacy of nutrition. We observed altered permeability in aflatoxin-treated, differentiated human intestinal cell monolayers, consistent with the potential to compromise intestinal integrity *in vivo* (Gratz *et al.*, 2007). Potential targets for aflatoxin toxicity are the intestinal tight junction proteins which are closely related to permeability function (McLaughlin *et al.*, 2004) or proteins involved in nutrient uptake (Thomson *et al.*, 2003).

In addition to susceptibility to infections (through compromised intestinal barrier function) and impaired nutrient uptake (through reduced epithelial absorption capacity), slowed growth in infancy may be linked to an altered immune response (Finch and Crimmins, 2004). Immune suppression by aflatoxins is a common finding in animals (Williams *et al.*, 2004). However, evidence of the immunosuppressive action of aflatoxins in humans is limited. In *in vitro* studies, extremely low doses (0.05-0.1 pg/ml) of aflatoxin B_1 decreased phagocytosis and the microbicidal activity (against *Candida albicans*) of human monocytes (Cusumano *et al.*, 1996) as well as to decreasing the secretion of IL-1, IL-6 and TNFα (Rossano *et al.*, 1999).

We have observed reduced salivary sIgA, a vital component of mucosal immunity, in aflatoxin exposed children (Turner *et al.*, 2003). In Ghana, aflatoxin exposure was associated with alterations in a number of parameters of cellular immunity in adults (Jiang *et al.*, 2005). The effect of aflatoxin on reducing levels of sIgA is of particular interest given the high levels of aflatoxin exposure and frequency of infectious insult in West African populations. In saliva, breast milk, tears, and mucus of the bronchial, genito-urinary and digestive tracts, sIgA binds to bacterial and viral surface antigens and provides an important component of the mucosal barrier. Thus, impairment of mucosal immunity could further contribute to decreased intestinal resistance to bacteria, and to increased inflammation of the intestinal epithelium. Mycotoxins are recognized as potentially influencing barrier functions (see Bouhet and Oswald, 2005). Indeed in Gambian children both enteropathy and growth retardation were associated with increased translocation of antigenic macromolecules from the gut lumen, with subsequent systemic immune stimulation (Campbell *et al.*, 2003, 2004).

In addition to sIgA and its contribution to intestinal barrier function, it is possible that the immunosuppressive action of aflatoxins is exerted in other ways. For example, suppression of cell-mediated immunity (CMI) is a common observation with, most notably, an impairment of delayed type hypersensitivity at low doses in different species (Pier and McLoughlin, 1985). Aflatoxin exposure also increases susceptibility to bacterial and parasitic infections and has an adverse effect on acquired immunity as evidenced following experimental challenge with infectious agents after vaccination (Denning, 1987). The immune system of the newborn also may be altered by maternal exposure to aflatoxin in a variety of species including pigs, rats, chickens and rainbow trout (Williams *et al.*, 2004).

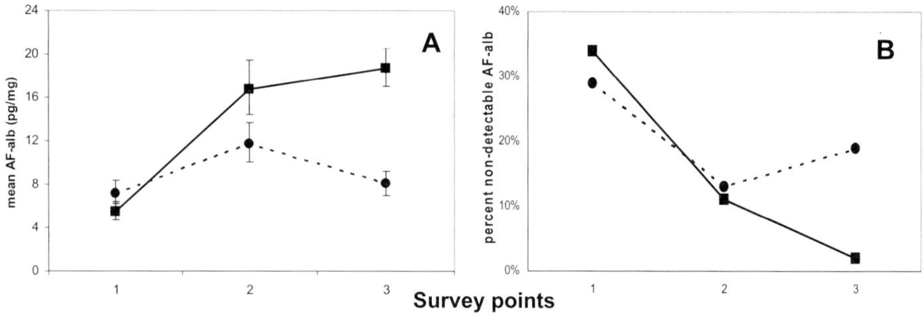

Figure 2. (A) Aflatoxin-albumin levels and (B) percentage of individuals with non-detectable aflatoxin-albumin in the intervention and control groups from each of three surveys (Turner *et al.*, 2005*a*). The three survey points were (1) – at peanut harvest, (2) – 3 months after harvest, and (3) – 5 months after harvest. - - ● - - intervention; —■— control.

In summary, slowed growth in West African children occurs with the introduction of solid foods when there is an initial high exposure to aflatoxin. The mechanisms of action are currently unknown although the possibility of a compromised intestinal integrity, through altered barrier function as a consequence of endothelial cell toxicity or immune suppression, is a valid hypothesis that merits further research.

Intervention strategies against aflatoxins

Aflatoxins are clearly of public health importance, although at present the full-scale of their impact is difficult to estimate. Nevertheless, the effects on child growth, hepatocellular carcinoma and the incidence of aflatoxicosis suggest reducing exposure is a valid public health goal (Strosnider *et al.*, 2006). In addition, reducing contamination of foods with aflatoxins would have benefits both in reducing food spoilage in communities where food often is scarce and in making the foodstuffs more desirable in world markets (Wu, 2004).

A number of strategies for reducing aflatoxin exposure or the biological effects once ingested have been discussed (Wild and Hall, 2000). These include pre-harvest interventions (for example, better irrigation, use of pesticides, fungicides, biocontrol or genetic modification to improve crop resistance), post-harvest measures (for example, better drying and storage) or chemoprevention. In our own studies in West Africa we have focused on post-harvest measures because much of the contamination occurs at this stage and post-harvest approaches are simple and cheap compared to many other strategies. In particular, we consider methods that can be implemented at the subsistence farm level rather than the larger-farm scale approaches, *e.g.*, Hazard Analysis and Critical Control Point (HACCP), whose strategies may not be readily applicable to small-scale farmers.

In a community intervention study in Guinea (Turner *et al.*, 2005a), we implemented a number of measures to reduce fungal growth and toxin production post-harvest on the peanut crop; peanuts are the main dietary staple in the lower Kindia region of Guinea where the study was conducted. None of the introduced practices were new to the local farmers but they typically are used sporadically and without any training. The implemented intervention package comprised: drying the crop thoroughly in the sun; drying peanuts on mats rather than the ground; hand sorting to discard visibly moldy kernels or nuts before storage; use of natural fiber sacks for storage, rather than plastic bags that promote humidity; placing storage sacks on wooden pallets to avoid dampness; and using insecticide to reduce insect damage to the crop.

The intervention was conducted over a five month period in sixteen villages: eight intervention villages and eight control. The eight control villages followed their usual post-harvest practices. AF-alb was measured in 600 subjects at three time points. In the control villages the mean AF-alb levels increased post-harvest, whereas in the intervention villages the levels after 5 months of storage were similar to that immediately post-harvest. Mean AF-alb levels at this time were 60% lower in intervention villages than in the control villages ($p < 0.001$; Fig. 2A). The number of subjects with non-detectable levels of adduct at the time of harvest was ~30%. This value decreased to 2% five months later in control villages, but to only 20% in the intervention group over the same time ($p < 0.001$; Fig. 2B). The mean level of aflatoxin B_1 in peanuts in household stores in intervention and control villages mirrored the pattern seen for AF-alb levels. The effectiveness of this intervention suggests that significant reductions in exposure to this potent environmental toxin can be achieved by using low technology approaches at the subsistence farm level in sub-Saharan Africa.

This work provides evidence that aflatoxin exposure is preventable on a community basis with low cost input. The use of the aflatoxin-albumin biomarker enabled the impact of aflatoxin reduction measures on human exposure to be assessed, as opposed to simply demonstrating lower levels in food commodities in field trials. We are now conducting a similar intervention to assess the impact of reduced aflatoxin exposure on child growth and immunity in villages in Guinea. This information will provide an indication of the benefits of such strategies, not just in reducing aflatoxin exposure but also in terms of reduced morbidity in these populations.

Conclusions

Over a 20-year period of research in West Africa we have demonstrated, by using biomarkers, the ubiquitous nature of high level exposure to aflatoxins throughout life. One cannot help but speculate as to what the reaction would be if continued exposure to high levels of a potent carcinogen was to be left unaddressed in Europe or North America. Indeed it is ironic that aflatoxins have evoked the most interest recently in the context of the slim risk of their use as chemical weapons by the late Saddam Hussein (http://www.publications.parliament.uk/pa/cm200203/cmselect/cmfaff/813/813we31.htm) or the relatively limited numbers of acute aflatoxicosis outbreaks, such as that in Kenya, rather than in terms of the major public health problems of hepatocellular carcinoma and, potentially, of child morbidity and mortality (Table 3). The scale of the former consequences of aflatoxin exposure is dwarfed by the impact of these toxins if they truly impair growth and immunity in children and thereby make a significant contribution towards the 4.5 million deaths annually of children aged five years and younger in sub-Saharan Africa. The development of affordable and acceptable intervention strategies, such as that demonstrated for peanuts in Guinea, should be a high priority together with further etiologic studies of aflatoxins and child health. These parallel approaches collectively provide the foundation and rationale for effective prevention strategies in affected countries.

Table 3. The wrong emphasis on aflatoxin and human disease.

Health Effect	Possible number of deaths	Relative Attention
Chemical weapon	0 (?)	Very high
Acute aflatoxicosis	100's	High
Hepatocellular carcinoma	10,000's	Medium
Growth impairment and immunosuppression	100,000's (?)	Low/None

Acknowledgments

We thank our collaborators in many countries for their studies of aflatoxin and human health. This project was supported in part by the NIEHS (USA), grant ES06052.

References

Allen, S.J., Wild, C.P., Wheeler, J.G., Riley, E.M., Montesano, R., Bennett, S., Whittle, H.C. and Hall, A.J. (1992) Aflatoxin exposure, malaria and hepatitis B infection in rural Gambian children. *Transactions of the Royal Society of Tropical Medicine and Hygiene* 86, 426-430.

Azziz-Baumgartner, E., Lindblade, K., Gieseker, K., Rogers, H.S., Kieszak, S., Njapau, H., Schleicher, R., McCoy, L.F., Misore, A., DeCock, K., Rubin, C. and Slutsker, L. (2005) Case-control study of an acute aflatoxicosis outbreak, Kenya. *Environmental Health Perspectives* 113, 1779-1783.

Barker, D.J. (2002) Fetal programming of coronary heart disease. *Trends in Endocrinology and Metabolism* 13, 364-368.

Black, R.E., Morris, S.S. and Bryce, J. (2003) Where and why are 10 million children dying every year? *The Lancet* 361, 2226-2234.

Bouhet, S. and Oswald, I.P. (2005) The effects of mycotoxins, fungal food contaminants, on the intestinal epithelial cell-derived innate immune response. *Veterinary Immunology and Immunopathology* 108, 199-209.

Bryce, J., Boschi-Pinto, C., Shibuya, K. and Black, R.E. (2005) WHO estimates of the causes of death in children. *The Lancet* 365, 1147-1152.

Campbell, D.I., Lunn, P.G. and Elia, M. (2002) Age-related association of small intestinal mucosal enteropathy with nutritional status in rural Gambian children. *British Journal of Nutrition* 88, 499-505.

Campbell, D.I., Elia, M. and Lunn, P.G. (2003*a*) Growth faltering in rural Gambian infants is associated with impaired small intestinal barrier function, leading to endotoxemia and systemic inflammation. *Journal of Nutrition* 133, 1332-1338.

Campbell, D.I., Murch, S.H., Elia, M., Sullivan, P.B., Sanyang, M.S., Jobarteh, B. and Lunn, P.G. (2003*b*) Chronic T cell-mediated enteropathy in rural West African children: Relationship with nutritional status and small bowel function. *Pediatric Research* 54, 306-311.

Campbell, D.I., McPhail, G., Lunn, P.G., Elia, M. and Jeffries, D.J. (2004) Intestinal inflammation measured by fecal neopterin in Gambian children with enteropathy: Association with growth failure, *Giardia lamblia*, and intestinal permeability. *Journal of Pediatric Gastroenterology and Nutrition* 39, 153-157.

Caulfield, L.E., de Onis, M., Blössner, M. and Black, R.E. (2004*a*) Undernutrition as an underlying cause of child deaths associated with diarrhea, pneumonia, malaria and measles. *American Journal of Clinical Nutrition* 80, 193-198.

Caulfield, L.E., Richard, S.A. and Black, R.E. (2004*b*) Undernutrition as an underlying cause of malaria morbidity and mortality in children less than five years old. *American Journal of Tropical Medicine and Hygiene* 71 (suppl. 2), 55-63.

Cusumano. V., Rossano, F., Merendino, R.A., Arena, A., Costa, G.B., Mancuso, G., Baroni, A. and Losi, E. (1996) Immunobiological activities of mould products: Functional impairment of human monocytes exposed to aflatoxin B_1. *Review of Microbiology* 147, 385-391.

Denning, D.W. (1987) Aflatoxin and human disease. *Adverse Drug Reaction and Acute Poisoning Review* 6, 175-209.

Dersjant-Li, Y., Verstegen, M.W.A. and Gerrits, W.J.J. (2003) The impact of low concentrations of aflatoxin, deoxynivalenol or fumonisin in diets on growing pigs and poultry. *Nutrition Research Reviews* 16, 223-239.

Diallo, M.S., Sylla, A., Sidibé, K., Sylla, B.S., Trepo, C.R. and Wild, C.P. (1995) Prevalence of exposure to aflatoxin and hepatitis B and C viruses in Guinea, West Africa. *Natural Toxins* 3, 6-9.

Edrington, T.S., Harvey, R.B. and Kubena, L.F. (1995) Toxic effects of aflatoxin B_1 and ochratoxin A, alone and in combination, on chicken embryos. *Bulletin of Environmental Contamination and Toxicology* 54, 331-336.

Finch, C.E. and Crimmins, E.M. (2004) Inflammatory exposure and historical changes in human lifespans. *Science* 305, 1736-1739.

Gong, Y.Y., Cardwell, K., Hounsa, A., Egal, S., Turner, P.C., Hall, A.J. and Wild, C.P. (2002) Dietary aflatoxin exposure and impaired growth in young children from Benin and Togo, West Africa: Cross sectional study. *British Medical Journal* 325, 20-21.

Gong, Y.Y., Egal, S., Hounsa, A., Turner, P.C., Hall, A.J., Cardwell, K.F. and Wild, C.P. (2003) Determinants of aflatoxin exposure in young children from Benin and Togo, West Africa: The critical role of weaning. *International Journal of Epidemiology* 32, 556-562.

Gong, Y.Y., Hounsa, A., Egal, S., Turner, P.C., Sutcliffe, A.E., Hall, A.J., Cardwell, K. and Wild, C.P. (2004) Post-weaning exposure to aflatoxin results in impaired child growth: A longitudinal study in Benin, West Africa. *Environmental Health Perspectives* 112, 1334-1338.

Gratz, S., Wu, Q.K., El-Nezami, H., Juvonen, R.O., Mykkänen, H. and Turner, P.C. (2007) Alteration of intestinal transport, metabolism and toxicity of aflatoxin B_1 by *Lactobacillus rhamnosus* strain GG *in vitro* in Caco-2 cells. *Applied and Environmental Microbiology* 73, 3958-3964.

Groopman, J.D. and Kensler, T.W. (1999) The light at the end of the tunnel for chemical-specific biomarker: Daylight or headlight? *Carcinogenesis* 20, 1-11.

Guengerich, F.P., Johnson, W.W., Shimada, T., Ueng, Y.F., Yamazaki, H. and Langouet, S. (1998) Activation and detoxification of aflatoxin B_1. *Mutation Research – Fundamental and Molecular Mechanisms of Mutagenesis* 402, 121-128.

Hall, A.J. and Wild, C.P. (1994) Epidemiology of aflatoxin-related disease. In: Eaton, D.A. and Groopman, J.D. (eds.) *The Toxicology of Aflatoxins: Human Health, Veterinary and Agricultural Significance*. Academic Press, San Diego, California. pp. 233-258.

Hendrickse, R.G.J.B., Coulter, S.M., Lamplugh, S.B., Macfarlane, S.B.J., Williams, T.E., Omer, M.I.A. and Suliman, G.I. (1982) Aflatoxins and Kwashiorkor: A study in Sudanese children. *British Medical Journal* 285, 843-846.

IARC Working Group on the Evaluation of Carcinogenic Risks to Humans. (2002) Some traditional herbal medicines, some mycotoxins, naphthalene and styrene. *IARC Monographs on the Evaluation of Carcinogenic Risks to Humans* 82, 1-556.

Jiang, Y., Jolly, P.E., Ellis, W.O., Wang, J.S., Phillips, T.D. and Williams, J.H. (2005) Aflatoxin B_1 albumin adduct levels and cellular immune status in Ghanaians. *International Immunology* 17, 807-814.

Lunn, P.G. (2000) The impact of infection and nutrition on gut function and growth in childhood. *Proceedings of the Nutrition Society* 59, 147-154.

Lunn, P.G. (2002) Growth retardation and stunting of children in developing countries - Invited commentary. *British Journal of Nutrition* 88, 109-110.

Lunn, P.G., Northro-Clewes, C.A. and Downes, R.M. (1991) Intestinal permeability, mucosal injury, and growth faltering in Gambian infants. *The Lancet* 338, 907-910.

Marin, D.E., Taranu, I., Bunaciu, R.P., Pascale, F., Tudor, D.S., Avram, N., Sarca, M., Cureu, I., Criste, R.D., Suta, V. and Oswald, I.P. (2002) Changes in performance, blood parameters, humoral and cellular immune responses in weanling piglets exposed to low doses of aflatoxin. *Journal of Animal Science* 80, 1250-1257.

McLaughlin, J., Padfield, P.J., Burt, J.P.H. and O'Neill, C.A. (2004) Ochratoxin A increases permeability through tight junctions by removal of specific claudin isoforms. *American Journal of Physiology Cell Physiology* 287, C1412-C1417.

Northrop-Clewes, C.A., Lunn, P.G. and Downes, R.M. (1997) Lactose maldigestion in breast-feeding Gambian infants. *Journal of Pediatric Gastroenterology and Nutrition* 24, 257-263.

Paine, M.F., Hart, H.L., Ludington, S.S., Haining, R.L., Rettie, A.E. and Zeldin, D.C. (2006) The human intestinal cytochrome P450 "pie". *Drug Metabolism and Disposition* 34, 880-886.

Pier, A.C. and McLoughlin, M.E. (1985) Mycotoxic suppression of immunity. In: Lacey, J. (ed.) *Tricothecenes and Other Mycotoxins*. John Wiley, New York, pp. 507-519.

Prentice, A. (1993) Nutrient requirements for growth, pregnancy and lactation: The Keneba experience. *South African Journal of Clinical Nutrition* 6, 33-38.

Rossano, F., DeLuna, L.O., Buommino, E., Cusumano, V., Losi, E. and Catania, M.R. (1999) Secondary metabolites of *Aspergillus* exert immunobiological effects on human monocytes. *Research in Microbiology* 150, 13-19.

Sim, S.C., Edwards, R.J., Boobis, A.R. and Ingelman-Sundberg, M. (2005) *CYP3A7* protein expression is high in a fraction of adult human livers and partially associated with the *CYP3A7*1C* allele. *Pharmacogenetics and Genomics* 15, 625-631.

Smith, E.E., Kubena, L.F., Braithwaite, R.B., Harvey, R.B., Phillips, T.D. and Reine, A.H. (1992) Toxicological evaluation of aflatoxin and cyclopiazonic acid in broiler chickens. *Poultry Science* 71, 1136–1144.

Strosnider, H., Azziz-Baumgartner, E., Banziger, M., Bhat, R.V., Breiman, R., Brune, M.N., DeCock, K., Dilley, A., Groopman, J., Hell, K., Henry, S.H., Jeffers, D., Jolly, C., Jolly, P., Kibata, G. N., Lewis, L., Liu, X., Luber, G., McCoy, L., Mensah, P., Miraglia, M., Misore, A., Njapau, H., Ong, C-N., Onsongo, M.T.K., Page, S.W., Park, D., Patel, M., Phillips, T., Pineiro, M., Pronczuk, J., Schurz Rogers, H., Rubin, C., Sabino, M., Schaafsma, A., Shephard , G., Stroka, J., Wild, C.P., Williams, J.T. and Wilson, D. (2006) Public health strategies for reducing aflatoxin exposure in developing countries: a workgroup report. *Environmental Health Perspectives* 114, 1898–1903.

Thomson, A.B.R., Drozdowski, L., Iordache, C., Thomson, B.K.A., Vermeire, S., Clandinin, M.T. and Wild, G. (2003) Small bowel review - Normal physiology, part 1. *Digestive Diseases and Sciences* 48, 1546-1564.

Turner, P.C., Mendy, M., Whittle, H., Fortuin, M., Hall, A.J. and Wild, C.P. (2000) Hepatitis B infection and aflatoxin biomarker levels in Gambian children. *Tropical Medicine and International Health* 5, 837-841.

Turner, P.C., Moore, S.E. Hall, A.J., Prentice, A.M. and Wild, C.P. (2003) Modification of immune function through exposure to dietary aflatoxin in Gambian children. *Environmental Health Perspectives* 111, 217-220.

Turner, P.C., Sylla, A., Gong, Y.Y., Diallo, M.S., Sutcliffe, A.E., Hall, A.J. and Wild, C.P. (2005*a*) Reduction in exposure to carcinogenic aflatoxins by postharvest intervention measures in West Africa: a community-based intervention study. *The Lancet* 365, 1950-1956.

Turner, P.C., Sylla, A., Kuang, S-Y., Marchant, C.L., Diallo, M.S., Hall, A.J., Groopman, J.D. and Wild, C.P. (2005*b*) Absence of *TP53* codon 249 mutations in Guinean infants with high aflatoxin exposure. *Cancer Epidemiology, Biomarkers & Prevention* 14, 2053-2055.

Turner P.C., Collinson A.C., Cheung Y.B., Gong Y.Y., Hall A.J., Prentice A.M. and Wild C.P. (2007) Aflatoxin exposure *in utero* causes growth faltering in Gambian infants. *International Journal of Epidemiology* 36, 1119-1125.

UNEP GEO-2000. http://www.unep.org/Geo2000/english/i56b.htm

Van Egmond, H.P. and Jonker, M.A. (2004) Worldwide regulations on aflatoxins – The situation in 2002. *Journal of Toxicology-Toxin Review* 23, 273-293.

Whitaker, T.B. (2006) Sampling foods for mycotoxins. *Food Additives and Contaminants* 23, 50-61.

Wild, C.P. and Hall, A.J. (2000) Primary prevention of hepatocellular carcinoma in developing countries. *Mutation Research* 462, 381-393.

Wild, C.P. and Turner, P.C. (2002) The toxicology of aflatoxins as a basis for public health decisions. *Mutagenesis* 17, 471-481.

Wild, C.P., Fortuin, M., Donato, F., Whittle, H.C., Hall, A.J., Wolf, C.R. and Montesano, R. (1993) Aflatoxin, liver enzymes and hepatitis B virus infection in Gambian children. *Cancer Epidemiology, Biomarkers & Prevention* 2, 555-561.

Wild, C.P., Rasheed, F.N., Jawla, M.F. Hall, A.J., Jansen, L.A. and Montesano, R. (1991) *In utero* exposure to aflatoxin in West Africa. *The Lancet* 337, 1602.

Wild, C.P., Yin, F., Turner, P.C., Chemin, I., Chapot, B., Mendy, M., Whittle, H., Kirk, G.D. and Hall, A.J. (2000) Environmental and genetic determinants of aflatoxin-albumin adducts in The Gambia. *International Journal of Cancer* 86, 1-7.

Williams, J.H., Phillips, T.D., Jolly, P.E., Stiles, J.K., Jolly, C.M. and Aggarwal, D. (2004) Human aflatoxicosis in developing countries: a review of toxicology, exposure, potential health consequences, and interventions. *American Journal of Clinical Nutrition* 80, 1106-1122.

Wojnowski, L., Turner, P.C., Pedersen, B., Hustert, E., Brockmoller, J., Whittle, H.C., Kirk, G. and Wild, C.P. (2004) Increased levels of aflatoxins-albumin adducts are associated with CYP3A5 polymorphisms in The Gambia, West Africa. *Pharmacogenetics* 14, 691-700.

Wu, F. (2004) Mycotoxin risk assessment for the purpose of setting international regulatory standards. *Environmental Science and Technology* 38, 4049-4055.

7

Economic Impact of Aflatoxin Contamination in Sub-Saharan Africa

Ousmane Coulibaly*, Kerstin Hell, Ranajit Bandyopadhyay, Said Hounkponou and John F. Leslie

Abstract

Globalization, including new international trade standards and regulations, has placed significant constraints on competitiveness in and access to international markets for developing countries, especially those in Sub-Saharan Africa. Agricultural commodities from these regions must overcome problems ranging from lower productivity and product quality to higher per unit transportation costs, and lower capacity to manage product flow from suppliers to end-users (value chain management). The competitiveness of African commodities also is impeded by poor policy and institutional environments that result in high transaction costs. One factor reducing African agricultural commodity competitiveness for export is aflatoxin contamination. Aflatoxin lowers product quality and discounts export values, which may lead to significant economic losses for the countries and the agents in commodity value chains. Losses from rejected export shipments and lower prices due to poor quality may exceed 100% if the product is destroyed and the exporter is paying for the shipping. Negative impacts on human health and household include mortality, loss of productivity and reduced income due to lower productive capacity, and related health costs. In this context, the cost of complying with food safety and agricultural health standards has been a major source of concern in the international development community and for African economies, but without increased food quality, neither competitiveness nor more revenue from exports will result. We address the problem of access of African commodities to international markets by recommending an increase in public awareness of the costs and ill effects incurred due to mycotoxins, and the diffusion of aflatoxin control technology and related capacity building to improve food quality in Sub-Saharan Africa. We also recommend strengthening the capacity for conducting impact assessments and collecting the data needed to make optimal decisions amongst possible aflatoxin control measures.

Introduction

International trade of agricultural commodities, especially of high-value food products, has expanded substantially over the last decades. For example Kenya's horticultural exports have grown by > 6% per year since the mid-1970s to become the country's third largest source of foreign exchange (IFPRI, 2004). The explanation is that the world trade environment is no longer dominated by bulk non-processed products but is diversifying to include more processed high-value products, which doubled as a share of world agricultural trade between 1980 and 1998 (IFPRI, 2004). Production and policy incentives, lower per unit transport costs, and advances in agricultural and post-harvest technologies also have increased agricultural exports. However, processed products face stringent requirements for norms and standards, mainly sanitary and phytosanitary standards and other trade barriers, which may limit access to international markets.

The complicated trade environment and complying with norms and standards are major challenges for African countries. Standards for aflatoxin contamination in agricultural commodities threaten the competitiveness and the profitability of African exports. The allowable contamination level varies based on commodity, geographic destination and intended use (Williams *et al.*, 2004). Standards for human foods in most countries, for example, range between 4 and 20 ng/g, although a few have limits as low as 1 ng/g and a few others permit more than 30 ng/g (FAO, 2004). Meeting these standards and paying the cost of compliance are prerequisites for access to international markets.

In this chapter we analyze the environment facing Sub-Saharan Africa as an exporter of high-value food and agricultural products to the European Union, the United States and other developed countries. We focus on the problems associated with aflatoxin contamination in terms of incidence and economic consequences in developing countries.

Constraints to competitiveness of African commodities

Many constraints impede the competitiveness of African commodities for exports and access to international markets. These include: (i) unfavorable or discriminatory trading terms, (ii) delays in the logistics chain, (iii) low quality data and information on the constraints and opportunities to value chains, and (iv) non-conducive policy environments with high transaction costs including bureaucracy in import and export licensing and border crossing procedures. The capacity for high productivity and good quality products along the value chains from input supply to exports is low and is reflected in low productivity, high scrap rates in industry, and large post-harvest losses in agriculture, sometimes as much as 30-50% of the crop. Trade transaction processes such as licenses and quality assurance systems are completed slowly and erratically resulting in unpredictable delivery and long delivery times.

International norms and standards

Standards are a legitimate means of facilitating production and exchange. In 1961, *Codex Alimentarius* was developed as a single international reference point for food safety and quality. Similarly, international standards developed by the International Standards Organization (ISO) provide a basis for the choices of norms recognized in foreign markets.

The application of international norms and standards by countries has been controversial. This controversy is linked to the abuse of standards as non-tariff barriers to international trade and protection of domestic industries and markets. As noted by Zarrilli (1999), the abuse of standard setting can take the form of: (i) unjustified differential requirements in different markets, (ii) unnecessary expensive or time-consuming tests, and (iii) duplicative conformity assessment procedures.

The World Trade Organization (WTO) developed two new international agreements to address these concerns and regulate international trade. These are: (i) the WTO Agreement on the Application of Sanitary and Phytosanitary Measures (SPS Agreement); and (ii) the WTO Agreement on Technical Barriers to Trade (TBT Agreement). These agreements are designed to minimize discriminatory and adverse effects of food regulations (Boakye-Yiadom, 2003). According to the Global Trade Network (GTN), standards affect trade in at least three ways: (i) They help to achieve social or public goals, *e.g.*, health or safety standards may be established to regulate the production, sales, or importation of a good as a public health policy. (ii) They may be used to clearly define products, thus facilitating compatibility and usability. Finally, (iii) standard setting may be employed as a disguised tool for unfairly restricting trade. The SPS and TBT Agreements aim to regulate the first two effects, while minimizing the third (Boakye-Yiadom, 2003).

Increasingly stringent food safety and agricultural health standards in industrialized countries create major challenges for continued developing country success in international markets for high-value food products, such as fruits, vegetables, fish, meat, nuts and spices. Yet such standards have had a positive role, by providing the catalyst and incentives for the modernization of export supply and regulatory systems and for the adoption of safer and more sustainable production and processing practices (World Bank, 2005).

Impact of new standards on international market access by African countries

The cost of complying with food safety and agricultural health standards has been a major source of concern in the international development community and among developing countries. Many worry that sanitary and phytosanitary standards will work increasingly to the disadvantage of developing countries that lack the administrative, technical, and other capacities to comply with new or more stringent requirements. However, the available evidence indicates that, in many instances, these challenges are manageable and that the compliance costs are a worthwhile investment, especially relative to the value of exports and associated benefits (World Bank, 2005).

For example, Mbaye (2004) estimated that an added capital investment of FCFA 2 billion (US$ 1 = FCFA 490) for ammoniation to detoxify aflatoxin plus recurrent costs of approximately 15% in the Senegalese peanut oil-mill subsector would yield higher quality products that would attract a 30% price differential relative to non-detoxified products. Exportable quantities of peanut cake would increase from 25,000 tons for the non-ammoniated product to 210,000 tons of an ammoniated detoxified product. After deducting the investment costs to detoxify aflatoxins, the increased quantity of peanut cake sales and the price differential provide an estimated benefit of FCFA 138 billion. Mbaye (2004) also estimated that the adoption of simple management practices to reduce aflatoxin contamination in confectionary peanuts in Senegal would accrue a benefit of FCFA 22 billion. Therefore, investments in meeting quality standards for aflatoxins would have a multiplier effect on Senegal's peanut exports.

African countries need to conform to emerging norms and standards. Only occasionally do sanitary and phytosanitary standards pose an absolute barrier to international market access and then usually in relation to animal diseases and plant pests. Costs of compliance with sanitary and phytosanitary regulations may reduce imports of foreign commodities (Hooker and Caswell, 1999). Otsuki *et al.* (2001*a,b*) argue that the European Union standards are unnecessarily stringent given the estimated risk reduction that would result. They used an econometric model to estimate that the impact on African countries of these standards would be a loss of approximately US$ 400 million, while use of the *Codex Alimentarius* standards would result in an increase of approximately US$ 250 million. These results were widely misinterpreted to mean that the European Union regulations would cost African economies some US$ 650 million in direct trade (Annan, 2001). Actual losses appear to be approximately US$ 40 million, with much larger losses occurring in the United States, China and Argentina than in all of the Sub-Saharan African countries combined (Wu, 2004). The major focus relates to the value of exports before and after the adoption of a standard and the lessons that stakeholders take from this example (Jaffe and Henson, 2004).

Aflatoxin and food quality

Types of aflatoxin

Hundreds of fungal species from > 12 fungal genera produce > 300 identified mycotoxins which can pose a threat to the health of at least some mammalian species. Of these mycotoxins, the aflatoxins are the greatest known concern because they are both highly toxic and potentially carcinogenic. Aflatoxin is a naturally occurring mycotoxin produced by three fungal species: *Aspergillus flavus*, *Aspergillus parasiticus* and *Aspergillus nominus*, of which *A. flavus* is the most common. These fungi occur principally in soil and decaying vegetation. There are four main types of aflatoxin: B_1, B_2, G_1, G_2, and two additional minor ones, M_1 and M_2, usually associated with milk. Aflatoxin B_1 is the most common member of this family of mycotoxins and has extremely high carcinogenic potency. All countries with mycotoxin regulations in 2003 have at least regulatory limits for aflatoxin B_1 or for the sum of aflatoxins B_1, B_2, G_1 and G_2 in foods and/or feeds (FAO, 2004). It is unlikely that commodities will contain aflatoxins B_2, G_1 or G_2 and not aflatoxin B_1 (Yabe and Nakajima, 2004), and the sum of aflatoxins B_2, G_1 and G_2 generally is less than that of aflatoxin B_1 alone.

Aflatoxin contamination and food quality

Agricultural crops can be contaminated by aflatoxin during production, storage, processing and transportation when temperature and humidity conditions are suitable. Mold growth and aflatoxin production are favored by warm temperatures and high humidity. Aflatoxin production requires high humidity and poor storage conditions often are conducive to the growth of *Aspergillus* (Hell *et al.*, 2000). Aflatoxin contaminates a wide range of agricultural products, including cereals, dried fruits, nuts, coffee beans and oilseeds, which are the agricultural backbone of most developing economy's exports (Bankole and Adebanjo, 2003). Total aflatoxin (B_1, B_2, G_1 and G_2) content in ng/g of product can give an indication of product quality and can be used as a threshold for separating high, medium and low quality produce. This grading is used for pricing either with a premium (high quality) or a discount (poor quality). The risk for spoilage is a function of factors including: the variety of crop, the time and method of harvest and storage, the storage temperature, the moisture content and the drying method prior to storage.

Adoption and diffusion of aflatoxin control methods

A number of approaches have been developed and tested for technical and economic feasibility for the management of aflatoxin contamination, but no single approach has been fully successful. Control measures include awareness and information campaigns aimed at producers and consumers on the health risks and economic losses associated with aflatoxin contaminated diets and products (James *et al.*, 2007). Farmer field forums also may be organized for building capacity in the use of cost effective and simple techniques like early harvesting, rapid drying, sorting, sanitation, use of improved storage structures, the use of botanicals and low risk pesticide in storage fumigation, biological control, the use of resistant varieties, fungicides and preservatives against fungal growth, and detoxification of mycotoxin contaminated grains. Most of these technologies are cost effective, do not require large capital investments and can yield high returns in controlling aflatoxin contamination (Mbaye, 2004). There is a need for locally-developed methods for adaptability and cost effectiveness. Simple transfer of aflatoxin control techniques from developed countries to Sub-Saharan African countries may not work because of the differences in food systems and infrastructure (Williams *et al.*, 2004).

Socio-economic impacts of aflatoxin contamination

Impact on nutrition, health and environment

Aflatoxins are among the main food contaminants with significant negative impact on health, food and nutritional security and incomes at the household, community and national levels. Food contaminated with aflatoxins may result in fatal aflatoxicosis and chronic mutagenic and carcinogenic effects with long latency periods. Aflatoxins also are associated with exacerbation of the protein malnutrition syndrome Kwashiorkor in human children (Ramjee *et al.*, 1992). Estimating the human health effects of aflatoxins in terms of primary liver cancer, requires data on human exposure to aflatoxins. In developing countries, many individuals are not only malnourished but also are exposed chronically to high levels of this mycotoxin in their diet (Cardwell, 2001). The health risks decrease labor productivity, while increasing health costs and overall income losses due to opportunity costs linked to lost days of work (Lubulwa and Davis, 1994).

Aflatoxins also are amongst the most potent mutagenic and carcinogenic substances known for animals (Henry *et al.*, 2001). Symptoms of severe aflatoxicosis include hemorrhagic necrosis of the liver, bile duct proliferation, edema, and lethargy (Anonymous, 2005). The impact of fungal toxins on animals extends beyond their obvious effects in producing death in the wide variety of animals that are likely to consume mycotoxin-contaminated grains or feeds. Aflatoxins are proven carcinogens, immunotoxins and causes of growth retardation in domesticated animals (Williams *et al.*, 2004).

Economic impacts of aflatoxin

Four potential impacts of aflatoxin have been identified (Lubulwa and Davis, 1994): (i) deterioration of the food and nutritional quality of agricultural products with an accompanying reduction in sensory characteristics, *e.g.*, taste, odor, texture and color, (ii) health-related productivity losses due to mutagenic and carcinogenic effects on humans who consume aflatoxin-contaminated food over an extended period of time, (iii) loss of income from livestock resulting from feeding aflatoxin-contaminated feedstuffs, *e.g.*, higher mortality rates and lower feed to weight conversion ratios for chickens, ducks, egg-layers, and pigs, and (iv) losses of export markets and related economic gains due to regulations that restrict international trade of aflatoxin-contaminated grain. Aflatoxin contamination in agricultural commodities can result in considerable microeconomic losses, *e.g.*, at the farm, household and community levels, as well as macro-economic losses, *e.g.* at the country or regional level, especially in Sub-Saharan Africa. Losses from rejected shipments and lower prices for poor quality grain can devastate a developing country's export markets (IFPRI, 2003).

The economic impact of aflatoxin on livestock production includes mortality and reductions in productivity, weight gain, feed efficiency, fertility, and the ability to resist disease. A decrease in either the quantity or the quality of meat, milk and eggs produced may result in significant financial losses. For example, in Indonesia, the Philippines and Thailand, 5% of the maize and peanuts produced are discarded due to fungal contamination. The annual costs of contamination due to aflatoxin and other molds in these countries in terms of product spoilage, human health effects, and losses in the poultry and pork sectors was estimated to be US$ 367 million (Cardwell, 2001).

Human health impacts of chronic aflatoxin exposure are just now being evaluated (Jolly *et al.*, Chapter 5; Gong *et al.*, Chapter 6). The interaction of aflatoxins with Hepatitis B and Hepatitis C to increase the risk of hepatocellular carcinoma is well known. The risk aflatoxin imposes on children who fail to thrive when their diet is highly contaminated with the toxin appears to be large enough to account for a significant proportion of the deaths of children under age 5 in Sub-Saharan Africa. For older children and adults, immune system suppression appears to occur and may reduce the effectiveness of vaccines and lower innate resistance to seemingly unrelated diseases. These losses in mental and physical development, the quality and quantity of output from a worker, and the added costs to already overburdened health care systems will be both economically and socially significant but have yet to be quantified or examined in any detail.

Aflatoxin is a major constraint affecting exports and foreign exchange for developing countries including Sub-Saharan Africa. FAO estimates that 25% of the world's food crops are affected by mycotoxins each year (Scholthof, 2004). Income losses due to aflatoxin contamination cost US producers more than US$ 100 million per year, on average, including US$ 26 million paid to peanut farmers alone (US$ 69.34/ha). These losses will be larger and more important for Sub-Saharan Africa due to the favorable environmental conditions for the growth of mycotoxigenic fungi and the lack of adequate storage infrastructure. The economic losses are compounded by lowered productivity, reduced prices, *i.e.*, poor quality discount, reduced feed efficiency, and greater disease incidence as a result of immune system suppression. Many countries have regulations on aflatoxin levels that restrict international trade in food and feeds contaminated with unacceptable levels of aflatoxin contamination (FAO, 2004). Member states of the African Groundnut Council (Gambia, Mali, Niger, Nigeria, Senegal, and Sudan) have estimated that the annual cost of implementing a program to reduce aflatoxin contamination is approximately US$ 7.5 million (IFPRI, 2003). Economic returns from preventing aflatoxin contamination through better production, harvesting, and storage practices under these conditions can be quite high.

Data for estimating the economic impact of aflatoxin

Estimating the economic impact of aflatoxins requires good data sets and expertise in the use of various economic impact assessment models. Both of these prerequisites generally are missing in Sub-Saharan Africa. Data are needed on the effects and related costs of aflatoxin contamination on human health. Cost elements include mortality, *i.e.*, the cost of productive capacity lost with premature death and the cost of morbidity, *i.e.*, losses resulting from productivity loss, hospitalization and the costs of health care services both public and private. Finally, there is the intangible cost of pain, suffering, anxiety, and reduction in the quality of life (Lubulwa and Davis, 1994). The data on the economic impact of aflatoxin on livestock should include income losses due to mortality as well as those due to reductions in productivity, weight gain, and the yield of meat, milk and eggs, as well as those due to feed use inefficiency and increased susceptibility to disease.

Even trade data are difficult to obtain. The trade data used in the gravity model by Otsuki *et al.* (2001*a,b*) to establish the baseline put African exports to the European Union (in 1998) at US$ 472 million for dried fruit and nuts and US$ 298 million for cereals, with the bulk of this trade occurring with France. These figures seem implausible, especially for cereals, given Africa's lack of competitiveness in this sector relative to Europe, and statistics from the United Nations COMTRADE database show that European imports from Africa were not this large. In 1998, for example, Africa exported to Europe approximately US$ 104 million of dried fruit, US$ 45 million of peanuts, US$ 27 million for other edible nuts, and < US$ 14 million for cereals and cereal products. Thus, the baseline against which economic impacts should be defined is at best fuzzy and often lacking (Jaffe and Henson, 2004).

Likely near-term impacts

Africa is not the only region to lose export markets or to be paid a lesser price due to aflatoxin concerns. The United States, India, China, Argentina and some Southeast Asian countries are encountering similar difficulties (Dohlman, 2003; Wu, 2006). African losses are most likely to occur for peanuts, since little maize is exported from Africa other than from South Africa (Wu, 2004). The major economic losses in both maize and peanut trade will accrue to China, the United States and Argentina.

Lowering allowable aflatoxin contamination levels to 10 ng/g in the developed countries will have a minimal and epidemiologically undetectable benefit of saving two lives per billion people, as neither the hepatitis B nor the hepatitis C viruses are widely distributed in most of these countries (Henry *et al.*, 1999). A further reduction from 10 ng/g to 2 ng/g is unlikely to have any additional health benefit while clearly imposing additional economic costs. Application of such standards in developing countries could have significant health, and presumably economic, benefits as many of those living in such countries consume a steady diet composed primarily of maize (Shephard *et al.*, 2007), some of which is heavily contaminated. Such consumption often is of necessity rather than by choice due to the need to sell the least contaminated material for cash and of the lack of food security in many developing countries. The current lowering of aflatoxin contamination limits by the European Union could exacerbate this problem by encouraging the consumption of more contaminated grain by those who can least afford the additional exposure to the toxin, and by discouraging farmers from these countries from even attempting to grow crops suitable for sale in international markets (Wu, 2004). Note that a reduction of the allowable aflatoxin contamination in peanuts from 20 ng/g to a 4 ng/g harmonized standard would lead to rejection of 86% of the peanuts now being exported worldwide for excessive aflatoxin contamination (Wu, 2006).

The cost of managing mycotoxins and doing the requisite testing also must not be overlooked. In the United States, the estimated cost for testing of all mycotoxins, not just aflatoxins, is between US$ 500 million and US$ 1.5 billion per year (Robens and Cardwell, 2003). If the number of commodities evaluated and the number of mycotoxins analyzed for is reduced, then it is possible to establish effective mycotoxin testing laboratories in Africa (Waliyar *et al.*, Chapter 31). To be economically justifiable, such laboratories usually need to be regional in nature since the associated costs often would weigh heavily on the economy of many developing countries. Thus, the establishment of standards that are based on clear health benefits and the development of both infrastructure and regulatory frameworks that enable these benefits to be realized remain high priority issues in determining the economic impact of mycotoxins on the safety of various agricultural commodities.

Conclusions

The economic losses associated with aflatoxin contamination are difficult to assess in a consistent and uniform way. The lack of information on the health costs and other economic losses from mycotoxin induced human illness is partly due to the difficulty of establishing cause-and-effect relationships between aflatoxins and the chronic diseases they are suspected of causing. For Sub-Saharan Africa, capacity building for economic impact assessment is urgent for both impact assessment and trade analysis. Baseline data do not exist in most of the countries, so comparisons often have no basis for normalization. Research for development efforts should focus on the potential for improved health, enhanced trade, increased income and enhanced welfare of farmers and consumers.

References

Annan, K. (2001) *Third United Nations Conference on the Least Developed Countries*. www.globalpolicy.org/socecon/un/unctad/2001/anna0514.htm.

Anonymous. (2005) Aflatoxicosis: Health implications. *East African Medical Journal* 82, 273-274.

Bankole, S.A. and Adebanjo, A. (2003) Mycotoxins in food in West Africa: Current situation and possibilities of controlling it. *African Journal of Biotechnology* 2, 254-263.

Boakye-Yiadom, L. (2003) *An Economic Surplus Evaluation of Aflatoxin-reducing Research: A Case Study of Senegal's Confectionery Groundnut Sector*. M.Sc. Thesis, Department of Agricultural and Applied Economics, Virginia Polytechnic Institute and State University, Blacksburg.

Cardwell, K. F. (2001) Mycotoxin contamination of foods in Africa: Anti-nutritional factors. *Food and Nutrition Bulletin* 21, 488-492.

Dohlman, E. (2003) Mycotoxin hazards and regulations: Impacts on food and animal feed crop trade. In Buzby, J.C (ed.) *International Trade and Food Safety: Economic Theory and Case Studies*. USDA Agricultural Economic Report No. 828. United States Department of Agriculture, Washington, D.C., pp. 97-108.

FAO. (2004) *Worldwide Regulations for Mycotoxins in Food and Feed in 2003 – FAO Food and Nutrition Paper 81*. FAO, Rome, Italy.

Hell, K., Sétamou, M., Cardwell, K.F. and Poehling, H.-M. (2000) The influence of storage practices on aflatoxin contamination in maize in four agroecological zones in Benin, West Africa. *Journal of Stored Products Research* 36, 365-382.

Henry, S.H., Bolger, P.M. and Troxell, T.C. (2001) *The Costs of Mycotoxin Management to the World: Regulatory Standards, Risk, and Appropriate Public Health Strategies.* Center for Food Safety and Applied Nutrition, U.S. Food and Drug Administration, Washington, D.C.

Henry, S.H., Bosch, F.X., Troxell, T.C. and Bolger, P.M. (1999) Reducing liver cancer – Global control of aflatoxin. *Science* 286, 2453-2454.

Hooker, N.H. and Caswell, J. (1999) A framework for evaluating non-tariff barriers to trade related in sanitary and phytosanitary regulation. *Journal of Agricultural Economics* 2, 234-246.

IFPRI. (2003) African agriculture: Past performance and future imperatives. Background paper No. 2 presented at the conference *"Successes in African Agriculture: Building for the Future"*. Pretoria, South Africa, 1-3 December 2003. IFPRI, Washington, D.C.

IFPRI. (2004) *Building on Successes in African Agriculture*. 2020 Focus 12, Brief 7. IFPRI, Washington, D.C.

Jaffe, S. and Henson S. (2004) *Standards and Agro-Food Exports from Developing Countries: Rebalancing the Debate*. World Bank Policy Research Working Paper 3348, World Bank, Washington D.C.

James, B., Adda, C., Cardwell, K., Annang, D., Hell, K., Korie, S., Edorh, M., Gbeassor, F., Nagatey, K. and Houenou, G. (2007) Public information campaign on aflatoxin contamination of maize grains in market stores in Benin, Ghana and Togo. *Food Additives and Contaminants* 24, 1283-1291.

Lubulwa, A.S.G. and Davis, J.S. (1994) Estimating the social costs of the impacts of fungi and aflatoxins. In Highley, E., Wright, E.J., Banks, H.J. and Champ, B.R. (eds.) *Stored Products protection, Proceedings of the 6th International Working Conference in Stored-product Protection, 17-23 April 1994, Canberra, Australia*. CAB International, Wallingford, UK.

Mbaye, A.A. (2004) *Sanitary and Phytosanitary Requirements and Developing-Country Agro-Food Exports: An Assessment of the Senegalese Groundnut Subsector.* World Bank, Washington D.C.

Otsuki, T., Wilson, J.S. and Sewadeh, M. (2001*a*) *A Case Study of Food Safety Standards and African Exports*. World Bank Policy Research working paper N° 2563, World Bank, Washington, D.C.

Otsuki, T., Wilson, J.S., and Sewadeh, M. (2001*b*) What price precaution? European harmonisation of aflatoxin regulations and African groundnut exports. *European Review of Agricultural Economics* 28: 263-283.

Ramjee, G., Berjak, P., Adhikari, M. and Dutton, M.F. (1992) Aflatoxins and kwashiorkor in Durban, South Africa. *Annals of Tropical Pediatrics* 12, 241-247.

Robens, J. and Cardwell, K. (2003) The costs of mycotoxin management to the USA: Management of aflatoxins in the United States. *Journal of Toxicology: Toxin Reviews* 22, 139-152.

Scholthof, K.-B.G. (2004) One foot in the furrow: Linkages between agriculture, plant pathology and public health. *Annual Review of Public Health* 24, 153-174.

Shephard, G.S., Marasas, W.F.O., Burger, H.-M., Somdyala, N.I.M., Rheeder, J.P., van der Westhuizen, L., Gatyeni, P. and van Schalkwyk, D.J. (2007) Exposure assessment of fumonisins in the former Transkei region of South Africa. *Food Additives and Contaminants* 24, 621-629.

Williams, J.H., Phillips, T.D., Jolly, P., Styles, J.K., Jolly, C.M. and Aggarwal, D. (2004) Human aflatoxicosis in developing countries: A review of toxicology, exposure, potential health consequences, and interventions. *American Journal of Clinical Nutrition* 80, 1106-1122.

World Bank. (2005) *Food Safety and Agricultural Health Standards: Challenges and Opportunities for Developing Country Exports*. World Bank, Washington, D.C.

Wu, F. (2004) Mycotoxin risk assessment for the purpose of setting international regulatory standards. *Environmental Science and Technology* 38, 4049-4055.

Wu, F. (2006) Economic impact of fumonisin and aflatoxin regulations on global corn and peanut markets. In: Barug, D., Bhatnagar, D., van Egmond, H.P., van der Kamp, J.W., van Osenbruggen, W.A., and Visconti, A. (eds.) *The Mycotoxin Factbook: Food and Feed Topics*. Wageningen Academic Publishers, Wageningen, The Netherlands, pp. 83-93.

Yabe, K. and Nakajima, H. (2004) Enzyme reactions and genes in aflatoxin biosynthesis. *Applied Microbiology and Biotechnology* 64, 745-755.

Zarrilli, S. (1999) *WTO Sanitary and Phytosanitary Agreement: Issues for Developing Countries*. Working paper. South Center, Geneva, Switzerland.

8

European Union Legislation on Mycotoxins in Food and Feed: Overview of the Decision-making Process and Recent and Future Developments

Frans Verstraete*

Abstract

This chapter provides a comprehensive overview of the European Union (EU) regulatory framework for feed and food safety legislation and the relevant decision-making process with particular emphasis on mycotoxins. Principles and objectives of the "General Food Law" of the EU are reported and details are provided on the European Food Safety Authority (EFSA), the Rapid Alert System for Food and Feed (RASFF) and the Standing Committee on the Food Chain and Animal Health. The regulatory framework and procedures for setting regulatory limits for contaminants in feed and food are described and relevant information on maximum levels, sampling and analysis, including specific provisions for mycotoxins (updated as of July 2007) is provided. The maximum levels for aflatoxins, ochratoxin A, patulin and major *Fusarium* toxins (deoxynivalenol, zearalenone, fumonisins, T-2 and HT-2 toxins) in food are given as well as the maximum levels for aflatoxin B_1 and ergot alkaloids in feed and guideline levels for ochratoxin A and *Fusarium* toxins in feed. The principle of a regular updating of the provisions in relation to new developments in technology and science is highlighted.

Regulatory framework for EU feed and food safety legislation

Regulation (EC) 178/2002 of the European Parliament and of the Council of 28 January 2002 lays down the general principles and requirements of food law, establishes the European Food Safety Authority and establishes procedures in matters of food safety (1). This Regulation, more commonly known as the "General Food Law", provides the general principles of food law that must be followed when measures to ensure feed and food safety are taken.

General principles and objectives of food law

General food law applies to all stages of the production, processing and distribution of food and also of feed produced for, or fed to, food-producing animals. Food law pursues one or more

 Mycotoxins: Detection Methods, Management, Public Health and Agricultural Trade (eds. J. F. Leslie *et al.*).

general objectives including: (i) a high level of protection of human health, (ii) protection of consumers' interests, and, where appropriate, (iii) protection of animal health and welfare, plant health and the environment. Food law provides for the free movement within the European Union of feed and food manufactured or marketed according to the general principles and requirements of food law. When international standards exist, or their completion is imminent, they must be taken into consideration in the development of food law, except where such standards would be an ineffective or inappropriate means for the fulfilling the objectives of food law.

To achieve the general objective of a high level of protection of human health, the General Food Law specifies that EU food legislation is based on risk analysis, except when such analyses are not appropriate to the circumstances or the nature of the measure, *e.g.*, labeling. Risk assessment is to be based on the available scientific evidence and to be undertaken in an independent, objective and transparent manner. Risk management shall take into account the results of risk assessment, other factors relevant to the matter under consideration and the precautionary principle where appropriate. Under the precautionary principle, EU Member States and the European Commission, *i.e.*, the "Commission", may take appropriate provisional risk-management measures when an assessment points to the likelihood of harmful health effects and there is a lack of scientific certainty.

Transparent public consultation is required, either directly or through representative bodies, during the preparation, evaluation and revision of the food law. When a food or feed product is deemed to constitute a risk, then the authorities must inform the general public of the nature of the risk to human or to animal health.

General requirements

Food may not be placed on the market if it is unsafe, *i.e.*, if it is harmful to health and/or unfit for consumption. In determining whether a food is unsafe, multiple factors are considered, including: the normal conditions of use, the information provided to the consumer, the likely immediate or delayed effects on health, cumulative toxic effects, if any, and, where appropriate, particular health sensitivities of specific categories of consumers. If food that is unsafe forms part of a batch, lot or consignment, then the entire quantity is presumed to be unsafe. Feed may not be placed on the market or given to any food-producing animal if it is unsafe. Feed is deemed unsafe if it has an adverse effect on human or animal health. Again, the entire quantity of a batch, lot or consignment is considered unsafe if any part of it fails to satisfy the requirements.

Across all stages of the food production chain, business operators must ensure that food and feed satisfies the requirements of food law and that those requirements are adhered to. The Member States enforce the law, ensuring that operators comply with it and apply appropriate measures and penalties for infringements. The traceability of food, feed, food-producing animals and all substances incorporated into foodstuffs must be established at all stages of production, processing and distribution. Business operators must implement appropriate systems and procedures to enable this tracing. If an operator considers a food or feed product that has been imported, produced, processed, manufactured or distributed to be harmful to human or animal health, then steps must be taken immediately to withdraw the product from the market and to notify competent authorities. If a product may have reached consumers, then the operator must inform them and recall the product(s) already supplied.

EFSA, RASFF and Standing Committee

EFSA

Regulation (EC) 178/2002 established the European Food Safety Authority (EFSA). The Authority provides scientific advice and scientific and technical support for the Community's legislation and policies in all fields with a direct or indirect impact on food and feed safety. EFSA provides independent information on all matters within these fields and communicates on risks. The scientific opinions provided by the Authority serve as the scientific basis for drafting and adopting Community measures related to feed and food safety and to other matters relating to animal health and welfare and plant health.

RASFF

Recent food crises have shown that procedures relating to food safety need improvement. Thus, the scope of the Rapid Alert System for Food and Feed (RASFF) has been broadened to cover animal feed, to identify emergency measures and to manage crises. The rapid alert system now covers all foodstuffs and animal feeds. The network involves the Member States, with the Commission in a management capacity, and EFSA as a member of the network. Member States notify the Commission of problems and the Commission uses RASFF to transmit the information throughout the network. Member States notify the Commission of:

- Measures that restrict the sale of food or feed.
- Measures that force the withdrawal or recall of food or feed.
- Measures involving professional operators that prevent or control the use of food or feed.
- Rejection of a batch or consignment of food or feed by a competent authority at a border post of the European Union.

Standing Committee

Regulation (EC) 178/2002 establishes a Standing Committee on the Food Chain and Animal Health, *i.e.*, the "Committee". The Committee is composed of representatives of the Member States and is chaired by a representative of the Commission. The Committee's mandate covers the entire food supply chain, ranging from animal health issues on the farm to products that arrive on the consumer's table. It can target risks to health wherever they arise in the food production system. The Committee is organized into the following sections:

- General food law.
- Biological safety of the food chain.
- Toxicological safety of the food chain.
- Controls and import conditions.
- Animal nutrition.
- Genetically modified food and feed and environmental risk.
- Animal health and animal welfare.
- Phytopharmaceuticals.

The "Toxicological safety of food chain" section is responsible for measures related to mycotoxins in foods and the "Animal nutrition" section is responsible for measures related to mycotoxins in feed.

The Committee is a regulatory committee. The Commission may adopt implementing measures only if they obtain a favorable opinion from the Committee that is given by a qualified majority of the Member States. Failing that, the proposed measure is referred to the Council, which takes a decision by a qualified majority. If the Council fails to reach a decision, then the Commission adopts the implementing measure unless the Council opposes it by a qualified majority. Qualified majority voting is based on the weighted votes of the Member States, with larger countries having more votes than smaller ones: Germany (29), France (29), United Kingdom (29), Italy (29), Spain (27), Poland (27), Romania (14), Netherlands (13), Greece (12), Portugal (12), Belgium (12), Czech Republic (12), Hungary (12), Sweden (10), Austria (10), Bulgaria (10), Denmark (7), Slovakia (7), Finland (7), Ireland (7), Lithuania (7), Latvia (4), Slovenia (4), Estonia (4), Cyprus (4), Luxembourg (4), Malta (3). In the present EU with 27 Member States, a proposed measure requires at least 255 of the 345 possible votes from a majority of the Member States (at least 14 delegations) in favor of the proposed measure, before it can be adopted by the Commission or the Council.

Emergencies

If food or feed originating in the EU or imported from a third country is likely to constitute a serious risk to human health, animal health or the environment, and the risk cannot be controlled by measures taken by the Member State(s) concerned, then the Commission is to adopt immediately by the Committee procedure, either on its own initiative or at the request of a Member State, one or more of the following measures, depending on the gravity of the situation:

- For products of EU origin, suspend the sale and/or use of the product in question, impose special conditions on the marketing of the product in question such as requiring a certificate of compliance based upon the results of sampling and analysis of the product, and adopt any other appropriate interim measure(s).
- For products imported from outside the EU, suspend imports, impose special conditions on the import such as the requirement of certifying the imported goods and adoption of any other appropriate interim measure(s).

In emergencies, the Commission may act unilaterally to provisionally adopt the necessary measures, after consulting the Member State(s) concerned and informing the other Member States. In such cases, the Committee must confirm, amend, revoke or extend the provisional measures within no more than 10 working days. If a Member State officially informs the Commission of the need for emergency measures and the Commission does not act, then the Member State may take interim protective measures. If such actions are taken, then the Member State taking them must immediately inform the other Member States and the Commission. Within 10 working days, the Commission must refer the matter to the Committee for a decision on extending, amending or revoking the national interim protective measures.

Regulatory framework for contaminants in food

Council Regulation (EEC) 315/93 of 8 February 1993 specifies Community procedures for contaminants in food (2) and is the basic regulation governing the measures on contaminants in food. A contaminant is defined as any substance unintentionally added to food and present therein in the form of a residue from production, manufacture, processing, preparation, treatment, packing, packaging, transport or storage or as a result of environmental contamination. The Regulation does not apply to contaminants covered by more specific EU-legislation. The Regulation provides:

- That food containing a contaminant at an unacceptable level from the public health viewpoint, and in particular at a toxicological level, shall not be sold.
- That contaminant levels shall be as low as can reasonably be achieved following good practices at all stages of production and distribution.
- That, when necessary for protecting public health, maximum levels shall be established for specific contaminants. These limits may include a reference to the sampling and analytical methods to be used for official control.
- For an obligatory consultation with the Scientific Panel on "Contaminants in the Food Chain" of EFSA before provisions that affect public health are adopted.

The Regulation also provides that Member States may not prohibit trade in foods that comply with this Regulation for any reason related to the aspects covered by the provisions of the Regulation. A safeguard clause also is included that enables Member States to temporarily suspend or restrict marketing on their territory of any foodstuff suspected of containing contaminants that would endanger human health.

Procedure for setting regulatory limits for contaminants in food

Regulatory limits

The scientific risk assessment is an assessment of the risks related to the presence of a contaminant in foodstuffs for human health and is the basis for any and all measures to be taken. If the contaminant is not a genotoxic carcinogen, then a health-based guidance value is derived. Human exposure (average and 95th percentile) is assessed in relation to this health-based guidance value. Particular attention is paid to vulnerable groups within the population and to high-level consumers. The exposure assessment also identifies the foods/food groups that contribute significantly to exposure. The occurrence data are obtained for these foods and food groups following the application of good practices and used to determine the appropriate maximum level to be set for the protection of consumer health. Maximum levels are set at a strict level that is reasonably achievable when good agricultural, fishery and manufacturing practices are followed and that takes into account the risk related to the consumption of the food.

If the contaminant(s) is a genotoxic carcinogen or if current exposure of the population or of vulnerable groups in the population is close to or greater than the tolerable daily intake, then the maximum levels are set at a level that is "as low as reasonably achievable" (ALARA). These approaches ensure that food business operators apply measures that prevent and reduce contamination as much as possible to protect the public health. To protect infants and young children, a vulnerable group, the maximum levels are set as the lowest achievable through the strict selection of raw materials used to manufacture these foods. This strict selection of raw materials also is appropriate for the production of some specific foodstuffs, *e.g.*, bran, for direct consumption.

Sampling and analysis

Adequate sampling is critically important for estimating the average level in a lot when the contaminants are heterogeneously distributed throughout the lot, as is common for mycotoxins, and is an essential component in the development of any maximum level (see Miraglia *et al.*, Chapter 16). In EU legislation, the maximum level for a contaminant is always tied to sampling protocols and requirements for analytical methods. EU policy is that the sampling procedure must be practical and must minimize the consumer's risk without rendering trade impossible. In terms of methods of analysis, no methods are specified in EU legislation but performance criteria that the methods must meet are defined.

The application of Regulation (EC) No 882/2004 of the European Parliament and of the Council of 29 April 2004 on official controls ensures the verification of compliance with feed and food law, animal health and animal welfare rules (3). Commission Regulation (EC) No 401/2006 of 23 February 2006 defines the methods of sampling and analysis for the official control of the levels of mycotoxins in foodstuffs (4) and replaces four former Commission Directives on sampling and analytical methods for the official control of mycotoxins in foodstuffs, in particular for aflatoxins (5), ochratoxin A (6), patulin (7) and *Fusarium* toxins (8). This Regulation did not substantially modify the previous sampling procedures and requirements as regards the analytical methods, but did specify the sampling procedure by category of foodstuffs instead of by mycotoxin. The collection of the sampling provisions and the performance criteria for the analytical methods to be used for the official control of all mycotoxins into a single legal text significantly improved the clearness and the applicability of the legal provisions.

In addition, a guidance document for competent authorities for the control of EU legislation on aflatoxins was prepared. The guidance document provides additional practical information but this information is subordinate to the Regulation's provisions. A guidance document for the practical sampling of very large lots of cereals, *e.g.*, large static lots stored in warehouses or in silos is currently being developed and will be available on the DG Health and Consumer website of the European Commission when it is finalized.

Regulatory framework for contaminants in feed

Directive 2002/32/EC of the European Parliament and of the Council of 7 May 2002 on undesirable substances in animal feed governs the measures on undesirable substances in feed. An "undesirable substance" is any substance or product, with the exception of pathogenic agents, that are present in and/or on a product intended for animal feed that presents a potential danger to animal or human health or to the environment or could adversely affect livestock production. The Regulation does not apply to contaminants or undesirable substances covered by more specific EU-legislation. The Regulation provides:

- That products intended for animal feed may enter the Community and be marketed and used in the Community only if they are sound, genuine and of merchantable quality and therefore do not present a danger to human health, animal health or to the environment, and do not adversely affect livestock production.
- That maximum levels and action levels can be set for undesirable substances in all products intended for animal feed. Feed containing an undesirable substance above the "maximum level" cannot be marketed and/or used for animal feeding. If a feed contains an undesirable substance at a level greater than the "action level", then Member States, in cooperation with the economic operator(s) concerned, must investigate to identify the source(s) of the substance(s). They also must inform the Commission of the outcome of

these investigations and the measures taken as regards the sources of contamination to reduce the level of the substances or eliminate them.

- That products intended for animal feeding containing levels of an undesirable substance that exceed the maximum level may not be mixed for dilution purposes with the same, or other, products intended for animal feed.
- That detoxification is allowed by chemical treatment. Member States shall ensure that measures are taken to guarantee the correct application of these detoxification processes and to guarantee the conformity of the detoxified products intended for animal feed with the provisions of the Directive.
- For an obligatory consultation with the Scientific Panel on Contaminants in the Food Chain of the EFSA before provisions with effects on public or animal health or on the environment are adopted.

The Directive, furthermore, prevents Member States from prohibiting trade in feeds that comply with this Directive for any reason related to the aspects covered by the provisions of the Directive. A safeguard clause also is included that allows Member States to temporarily suspend or restrict sales in their territory of any feedstuff suspected of containing contaminants that endanger human or animal health or the environment.

Procedure for setting regulatory limits for contaminants in feed

The scientific risk assessment is an assessment of the risks related to the presence of a contaminant in feed for animals or to human health and is the basis for the measures to be taken. In this scientific assessment the toxic exposure level is determined above which harmful animal health effects can occur in different animal species. Animal species are neither universally nor uniformly sensitive to particular toxins, so the toxic exposure level can differ significantly from one animal species to another. To assess the risks of the presence of a contaminant in feed for human health, the carryover of the undesirable substance from feed into food of animal origin, *e.g.*, meat, edible offal, milk, eggs, *etc.*, is assessed, where possible quantitatively. The contribution of the presence of the contaminant in food of animal origin to the total human exposure also is assessed.

The animal exposure assessment also enables the identification of the feeds/feed groups that contribute significantly to the exposure. An appropriate maximum level to protect animal and public health is determined from the occurrence data in these feeds and feed groups following the application of good agricultural and management practices. Maximum levels are set at a strict level that is reasonably achievable when following good agricultural, fishery and manufacturing practices and taking into account the risk related to the consumption of the feed or the corresponding animal product. Such an approach ensures that feed business operators apply measures to prevent and reduce the contamination as much as possible to protect both animal and human health.

Provisions regarding mycotoxins in food

Commission Regulation (EC) No 1881/2006 of 19 December 2006 sets maximum levels for some contaminants in foods (11) and establishes maximum levels for the following mycotoxins: aflatoxins, ochratoxin A, patulin and the *Fusarium* toxins (zearalenone, fumonisins, deoxynivalenol, T-2 and HT-2 toxins). This Regulation replaces Commission Regulation (EC) No 466/2001 of 8 March 2001 which had been amended 18 times.

General provisions

Regulation (EC) 1881/2006 provides that:

- Foods containing a contaminant at a level exceeding the maximum level established shall not be sold.
- The maximum level applies to the edible part.
- For dried, diluted, processed or compound foods for which no specific maximum level has been established, a concentration/dilution factor, or relative proportion factor is to be applied. This factor has to be provided and justified by the food business operator. If the food business operator does not provide the necessary concentration or dilution factor, then a competent authority shall itself define that factor, based on the available information and with the objective of maximum protection of human health.
- The maximum level also applies to products used as food ingredients.
- Mixing contaminated and non-contaminated consignments and detoxification of mycotoxins by chemical treatment are prohibited.
- In the absence of a clear indication of the intended use on the label of each individual bag, box, *etc.*, or on the original accompanying document of a batch of groundnuts and derived products or cereals, the maximum levels established for these products for human consumption shall be applicable.

Maximum levels for aflatoxins

Aflatoxins are mycotoxins produced by certain species of *Aspergillus*, usually at high temperatures and humidity levels. Aflatoxins may be present in a large number of foods, *e.g.*, nuts, groundnuts, dried fruit, cereals (especially maize) and spices. The European Scientific Committee for Food expressed in its opinion of 23 September 1994 (12) that aflatoxins are genotoxic carcinogens. For such substances there is no threshold below which no harmful effect is observed, and, therefore, no tolerable daily intake can be set. Current scientific and technical knowledge and improvements in production and storage techniques do not suffice to prevent aflatoxin biosynthesis and it is not technically possible to completely eliminate aflatoxins in food. Thus, the maximum levels are set as low as reasonably achievable (ALARA) (Table 1).

The aflatoxin group is a series of related compounds with varying toxicity, incidence and levels in food. Aflatoxin B_1 is the most toxic member of the aflatoxin group. For safety reasons, there are limits to both the total aflatoxin (aflatoxins B_1, B_2, G_1 and G_2) content of food and the aflatoxin B_1 content alone. Aflatoxin M_1 is a metabolic product of aflatoxin B_1 that occurs in milk and milk products from animals that have consumed contaminated feed. Aflatoxin M_1 also is a genotoxic carcinogen. Therefore, its presence is minimized in milk and milk products intended for human consumption to levels that are reasonably achievable. For aflatoxin M_1 in foods for infants and young children, the current maximum level may be reduced further in light of developments in analytical procedures.

Sorting or other physical treatments enable the reduction of the aflatoxin content of groundnuts, nuts, dried fruits and maize. To minimize the effects on trade, higher aflatoxin contents are permitted for products that are not intended for direct human consumption or that are ingredients

Table 1. Maximum levels (ng/g) for aflatoxins in foods.

	Food(s)	Maximum aflatoxin level		
		B_1	$B_1 + B_2 + G_1 + G_2$	M_1
1	Groundnuts to be sorted or otherwise physically treated before human consumption or use as an ingredient in foods	8[1]	15[1]	-
2	Nuts to be sorted or otherwise physically treated before human consumption or use as an ingredient in foods	5[1]	10[1]	-
3	Groundnuts, nuts and processed products thereof, intended for direct human consumption or for use as an ingredient in foods	2[1]	4[1]	-
4	Dried fruit to be sorted or otherwise physically treated before human consumption or use as an ingredient in foods	5	10	-
5	Dried fruit and processed products thereof, intended for direct human consumption or use as an ingredient in foods	2	4	-
6	All cereals and all products derived from cereals, including processed cereal products, with the exception of foods listed in 7, 10 and 12	2	4	-
7	Maize to be sorted or otherwise physically treated before human consumption or use as an ingredient in foods	5	10	-
8	Raw milk, heat-treated milk and milk for the manufacture of milk-based products	-	-	0.05
9	Following species of spices: *Capsicum* spp. (dried fruits thereof, whole or ground, including chilies, chili powder, cayenne and paprika), *Piper* spp. (fruits thereof, including white and black pepper), *Myristica fragrans* (nutmeg), *Zingiber officinale* (ginger) and *Curcuma longa* (turmeric)	5	10	-
10	Processed cereal-based foods and baby foods for infants and young children[2]	0.1	-	-
11	Infant formulae and follow-on formulae, including infant milk and follow-on milk[3]	-	-	0.025
12	Dietary foods for special medical purposes intended specifically for infants[4]	0.1	-	0.025

[1]The maximum levels refer to the edible portion of groundnuts and nuts. If groundnuts and nuts "in the shell" are analyzed, then all of the aflatoxin contamination is assumed to be on the edible portion.
[2]The maximum level in the dry matter.
[3]The maximum level is in the product ready to use, *i.e.*, marketed as such or after reconstitution as instructed by the manufacturer.
[4]The maximum level is in the product ready to use, *i.e.*, marketed as such or reconstituted as instructed by the manufacturer, for milk and milk products. For other products the maximum level is in the dry matter.

in foods. In these cases, the maximum aflatoxins levels have been determined by taking into consideration the possible effects of various treatments for groundnuts, nuts, dried fruit and maize and the need to comply after treatment with the maximum limits fixed for these products intended for direct human consumption or for use as an ingredient in a food.

Maximum levels for ochratoxin A

Ochratoxin A is a mycotoxin produced by several species of fungi including *Penicillium* and *Aspergillus*. It occurs naturally in a wide variety of plant products. The European Scientific Committee for Food adopted a scientific opinion on 17 September 1998 regarding ochratoxin A. An assessment of the dietary intake of ochratoxin A by the population of the Community (14) was performed (SCOOP) in the framework of Council Directive 1993/5/EEC of 25 February 1993 on assistance to the European Commission and cooperation by the Member States in the scientific examination of questions relating to food (15). The EFSA has, on request from the European Commission, adopted an updated scientific opinion relating to ochratoxin A in food on 4 April 2006 (16), taking into account new scientific information and derived a tolerable weekly intake (TWI) of 120 ng/kg body weight (bw).

The main contributors to ochratoxin A exposure are cereals and cereal products. Wine, coffee and beer also are significant contributors to human ochratoxin A exposure. Dried vine fruits and grape juice also contribute a significant portion of the ochratoxin A exposure of some vulnerable consumer groups, *e.g.*, children. Ochratoxin A also has been detected in dried fruits other than dried vine fruits, cocoa and cocoa products, spices, liquorice and in some products of animal origin, particularly pig kidneys.

Maximum levels of ochratoxin A have been set for cereals, cereal products, dried vine fruits, roasted coffee, wine, grape juice and foods for infants and young children, which are as low as reasonably achievable (Table 2). The level of ochratoxin A in beer is dependent on the ochratoxin A level in the malt, a cereal product and for which a maximum level has been established. The appropriateness of setting a maximum level for ochratoxin A in foods such as dried fruits other than dried vine fruits, cocoa and cocoa products, spices, meat products, green coffee, beer and liquorice, as well as a review of the existing maximum levels (Table 2), in particular for ochratoxin A in dried vine fruits and grape juice, is being considered in light of a recent EFSA scientific opinion. These technical discussions led to following provisional conclusions (17):

- To keep the current maximum levels unchanged.
- For food commodities in which ochratoxin A has been observed and for which no EU maximum level has yet been established, to consider setting a maximum level for food commodities that are significant contributors to the exposure of ochratoxin A (for the entire population, for vulnerable groups in the population, or for a significant part of the population), or for foods that are not necessarily a significant contributor to the exposure of ochratoxin A but there is evidence of very high levels of ochratoxin A in these commodities. In the latter case, the setting of a maximum level could be appropriate to keep these highly contaminated commodities from entering the food chain.
- To continue monitoring ochratoxin A in foods for which no maximum levels are set. If high levels of ochratoxin A are found frequently then these findings will be brought to the attention of the Commission and other Member States and discussions on the appropriateness of setting a maximum level for ochratoxin A in these commodities initiated in light of the new findings.

Table 2. Maximum levels (ng/g) for ochratoxin A in foods.

	Food(s)	Maximum level
1	Unprocessed cereals	5
2	All products derived from unprocessed cereals, including processed cereal products and cereals intended for direct human consumption except for foods listed in 9 and 10	3
3	Dried vine fruits (currants, raisins and sultanas)	10
4	Roasted coffee beans and ground roasted coffee, excluding soluble (instant) coffee	5
5	Soluble (instant) coffee	10
6	Wine, including sparkling wines, but excluding liqueur wines and wines with an alcoholic strength of ≥ 15% by volume, and fruit wines	2[1]
7	Aromatized wine, aromatized wine-based drinks and aromatized wine-product cocktails	2[1]
8	Grape juice, concentrated grape juice as reconstituted, grape nectar, grape must and concentrated grape must as reconstituted, intended for direct human consumption	2[1]
9	Processed cereal-based foods and baby foods for infants and young children[2]	0.5
10	Dietary foods for special medical purposes intended specifically for infants[3]	0.5
11	Green coffee, dried fruit other than dried vine fruit, beer, cocoa and cocoa products, liqueur wines, meat products, spices and liquorice	-

[1]The maximum level applies to products produced from the 2005 harvest onwards.
[2]The maximum level refers to the dry matter.
[3]The maximum level refers in the case of milk and milk products, to the products ready for use (marketed as such or reconstituted as instructed by the manufacturer) and in the case of products other than milk and milk products, to the dry matter.

Maximum levels for patulin

Patulin is a mycotoxin produced by fungi belonging to several genera, including *Penicillium* spp., *Aspergillus* spp. and *Byssochlamys* spp. Although patulin can occur in many moldy fruits, cereals and other foods, the major sources of patulin exposure are apple products. The European Scientific Committee on Food endorsed in its meeting on 8 March 2000 a tolerable daily intake (TDI) of 0.4 μg/kg bw for patulin (18).

In 2001, a SCOOP-task "Assessment of the dietary intake of patulin by the population of EU Member States" in the framework of Directive 1993/5/EEC was performed (19). Based on this assessment, the average exposure is well below the TDI. When considering specific groups of consumers, especially small children, and assuming worst case scenarios, the exposure to patulin is more significant but is still below the TDI.

Maximum levels of patulin have been set for some foods to protect consumers from unacceptable contamination (Table 3). These maximum levels should be reviewed and, if necessary, reduced taking into account progress in scientific and technological knowledge and the implementation of Commission Recommendation 2003/598/EC of 11 August 2003 on the prevention and reduction of patulin contamination in apple juice and apple juice ingredients in other beverages (20).

Table 3. Maximum levels (ng/g) for patulin in foods.

	Food(s)	Maximum level
1	Fruit juices, concentrated fruit juices as reconstituted and fruit nectars	50
2	Spirit drinks, cider and other fermented drinks derived from apples or containing apple juice	50
3	Solid apple products, including apple compote, apple puree intended for direct consumption except for foods listed in 4 and 5	25
4	Apple juice and solid apple products, including apple compote and apple puree, for infants and young children and labeled and sold as such[1]	10
5	Baby foods other than processed cereal-based foods for infants and young children[1]	10

[1]The maximum level refers to the products ready to use (marketed as such or after reconstitution as instructed by the manufacturer).

Maximum levels for *Fusarium* toxins

A number of *Fusarium* fungi produce different mycotoxins in the trichothecene class including deoxynivalenol, nivalenol, T-2 toxin and HT-2 toxin, as well as some other chemically unrelated toxins, *e.g.*, zearalenone and fumonisins. The *Fusarium* fungi commonly are found on cereals grown in the temperate and tropical regions of the Americas, Europe and Asia. Several of the toxin-producing *Fusarium* species can produce, to at least some degree, two or more of these toxins.

The European Scientific Committee for Food has adopted several scientific opinions on these toxins, evaluating deoxynivalenol in December 1999 (21) establishing a TDI of 1 µg/kg bw, zearalenone in June 2000 (22) establishing a temporary TDI of 0.2 µg/kg bw, fumonisins in October 2000 (23), updated in April 2003 (24), establishing a TDI of 2 µg/kg bw, nivalenol in October 2000 (25) establishing a temporary TDI of 0.7 µg/kg bw, T-2 and HT-2 toxins in May 2001 (26) establishing a combined temporary TDI of 60 ng/kg bw, and the trichothecenes as group in February 2002 (27).

In the framework of Directive 93/5/EEC the SCOOP-task "Collection of occurrence data on *Fusarium* toxins in food and assessment of dietary intake by the population of EU Member States" was performed and finalized in September 2003 (28). Based on the scientific opinions and the assessment of the dietary intake, maximum levels have been set (Table 4) for deoxynivalenol, zearalenone and fumonisins. With respect to fumonisins, analyses of grain from recent harvests indicate that maize and maize products can be very highly contaminated by fumonisins and measures should be taken to prevent highly contaminated maize and maize products from entering the food chain.

Intake estimates indicate that T-2 and HT-2 toxins can be of concern for public health. The development of a reliable and sensitive detection method, collection of more occurrence data and further investigation/research into the factors involved in determining the presence of T-2 and HT-2 toxins in cereals and cereal products, in particular in oats and oat products, is needed and of high priority.

It is not necessary, due to co-occurrence, to consider specific measures for 3-acetyl deoxynivalenol, 15-acetyl deoxynivalenol and fumonisin B_3, as limits on deoxynivalenol and fumonisins B_1 and B_2 also would protect the human population from unacceptable exposure to these three toxins. A similar argument applies to nivalenol, which to a certain degree co-occurs with deoxynivalenol. Furthermore, human exposure to nivalenol is estimated to be significantly below the temporary TDI. With respect to the other trichothecenes considered in the SCOOP-task, *e.g.*, 3-acetyldeoxynivalenol, 15-acetyldeoxynivalenol, fusarenon-X, T2-triol, diacetoxyscirpenol, neosolaniol, monoacetoxyscirpenol and verrucarol, the limited information available indicates that they do not occur widely and that the levels found generally are low.

Climatic conditions during plant growth, particularly at flowering, have a major influence on *Fusarium* toxin content. However, good agricultural practices, whereby risk factors are minimized, can prevent or reduce contamination by *Fusarium* fungi. Commission Recommendation 2006/583/EC of 17 August 2006 on the prevention and reduction of *Fusarium* toxins in cereals and cereal products (29) contains general principles for the prevention and reduction of *Fusarium* toxin contamination (zearalenone, fumonisins and trichothecenes) in cereals to be implemented by the development of national codes of practice based on these principles.

Maximum levels of *Fusarium* toxins are set for unprocessed cereals sold for first stage processing. Cleaning, sorting and drying procedures are not considered first-stage processing insofar as no physical action is exerted on the grain kernel itself. Scouring is considered first-stage processing. The degree to which *Fusarium* toxins in unprocessed cereals are removed by cleaning and processing may vary. Thus, to have enforceable legislation, maximum levels also have been set for final consumer cereal products and for major food ingredients derived from cereals. Maximum levels were established in 2005 for *Fusarium* toxins in cereals and cereal products, including maize and maize products. For maize, all of the factors involved in the formation of *Fusarium* toxins, particularly zearalenone and fumonisins B_1 and B_2, are not known precisely. The maximum levels in maize and maize products were implemented beginning 1 July 2007 for deoxynivalenol and zearalenone, and on 1 October 2007 for fumonisins B_1 and B_2.

Data for the 2005 and 2006 harvests indicate that higher levels of zearalenone and fumonisins, and to a lesser extent deoxynivalenol, were observed in European maize than in the 2003 and 2004 harvests. These differences appear to be linked to weather conditions. The maximum levels for zearalenone and fumonisins were exceeded for maize under some weather conditions, even when applying the recommended prevention measures to the greatest extent possible. Therefore, the maximum levels for deoxynivalenol, zearalenone and fumonisins B_1 and B_2 in maize and maize products were reevaluated to avoid disrupting the markets while still protecting the public's health. The Standing Committee on the Food Chain expressed on 20 July 2007 (30) a favorable opinion by unanimity on the modified maximum levels (Table 4). The Commission adopted on 28 September 2007 these measures by Commission Regulation (EC) No 1126/2007 as an amendment to Regulation (EC) 1881/2006.

Given the low contamination levels of *Fusarium* toxins found in rice, no maximum levels were set for rice or for rice products.

Discussions of possible maximum levels for T-2 and HT-2 toxins in cereals and cereal products were initiated in autumn of 2007 and should be finalized by 1 July 2008.

Provisions regarding mycotoxins in feeds

Directive 2002/32/EC of the European Parliament and of the Council of 7 May 2002 on undesirable substances in animal feed (10) establishes maximum levels for aflatoxin B_1 and rye ergot.

Table 4. Maximum levels (ng/g) allowed for *Fusarium* toxins in foods.

	Toxin(s)/Food(s)	Maximum level
1	**Deoxynivalenol**[1]	
1.1	Unprocessed cereals[2] other than durum wheat, oats and maize	1250
1.2	Unprocessed durum wheat and oats[2]	1750
1.3	Unprocessed maize[2], except for unprocessed maize for wet milling[3]	1750[4]
1.4	Cereals, cereal flour, bran and germ as end products sold for direct human consumption, except for foods listed in 1.7, 1.8 and 1.9	750
1.5	Pasta (dry)[5]	750
1.6	Bread (including small bakery wares), pastries, biscuits, cereal snacks and breakfast cereals	500
1.7	Processed cereal-based foods and baby foods for infants and young children[6]	200
1.8	Milling fractions of maize with particle sizes > 500 µm falling within CN code 1103 13 or 1103 20 40 and other maize milling products with particle size > 500 µm not used for direct human consumption falling within CN code 1904 1010	750[4]
1.9	Milling fractions of maize with particle sizes ≤ 500 µm falling within CN code 1102 20 and other maize milling products with particle sizes ≤ 500 µm not used for direct human consumption falling within CN code 1904 1010	1250[4]
2	**Zearalenone**[1]	
2.1	Unprocessed cereals[2] other than maize	100
2.2	Unprocessed maize[2], except for unprocessed maize for wet milling[3]	350[4]
2.3	Cereals intended for direct human consumption, cereal flour, bran and germ as end products sold for direct human consumption, except for foods listed in 2.6, 2.7, 2.8, 2.9 and 2.10	75
2.4	Refined maize oil	400[4]
2.5	Bread (including small bakery wares), pastries, biscuits, cereal snacks and breakfast cereals, excluding maize snacks and maize-based breakfast cereals	50
2.6	Maize intended for direct human consumption, including maize-based snacks and maize-based breakfast cereals	100[4]
2.7	Processed cereal-based foods (excluding processed maize-based foods) and baby foods for infants and young children[6]	20
2.8	Processed maize-based foods for infants and young children[6]	20[4]
2.9	Milling fractions of maize with particle sizes > 500 µm falling within CN code 1103 13 or 1103 20 40 and other maize milling products with particle size > 500 µm not used for direct human consumption falling within CN code 1904 1010	200[4]
2.10	Milling fractions of maize with particle sizes ≤ 500 µm falling within CN code 1102 20 and other maize milling products with particle size ≤ 500 µm not used for direct human consumption falling within CN code 1904 1010	300[4]

Table 4 (continued). Maximum levels (ng/g) allowed for *Fusarium* toxins in foods.

	Toxin(s)/Food(s)	Maximum level
3	**Fumonisins (B_1 + B_2)**	
3.1	Unprocessed maize[2], except for unprocessed maize for wet milling[3]	4000[4]
3.2	Maize intended for direct human consumption, maize based foods for direct human consumption, except for foods listed in 3.3 and 3.4	1000[4]
3.3	Maize based breakfast cereals and maize-based snacks	800[4]
3.4	Processed maize-based foods and baby foods for infants and young children[6]	200[4]
3.5	Milling fractions of maize with particle sizes > 500 µm falling within CN code 1103 13 or 1103 20 40 and other maize milling products with particle sizes > 500 µm not used for direct human consumption falling within CN code 1904 1010	1400[4]
3.6	Milling fractions of maize with particle sizes ≤ 500 µm falling within CN code 1102 20 and other maize milling products with particle sizes ≤ 500 µm not used for direct human consumption falling within CN code 1904 1010	2000[4]
4	**T-2 and HT-2 toxins[1] (T-2 + HT-2)**	
4.1	Unprocessed cereals[2] and cereal products	-

[1]For the purpose of the application of maximum levels for deoxynivalenol, zearalenone, T-2 and HT-2 toxins established in points 1 and 2, rice is not included in "cereals" and rice products are not included in "cereal products".

[2]The maximum level applies to unprocessed cereals placed on the market for first-stage processing. "First-stage processing" shall mean any physical or thermal treatment, other than drying, of or on the grain. Cleaning, sorting and drying procedures are not considered to be "first-stage processing" insofar no physical action is exerted on the grain kernel itself and the whole grain remains intact after cleaning and sorting. In integrated production and processing systems, the maximum level applies to the unprocessed cereals in case they are intended for first-stage processing.

[3]The exemption applies only for maize for which it is evident e.g. through labeling, destination, that it is intended for use in a wet milling process only (starch production).

[4]Maximum level applies from 1 October 2007.

[5]Pasta (dry) means pasta with a water content of approximately 12%.

[6]The maximum level refers to the dry matter.

General provisions

The maximum levels are relative to a feed with a moisture content of 12%.

Maximum levels for aflatoxin B_1

The Scientific Panel on Contaminants in the Food Chain of the EFSA expressed on 3 February 2004 an opinion relating to aflatoxin B_1 as an undesirable substance in animal feeds (31). The EFSA Panel concluded that the carry-over of aflatoxins in feedstuffs to milk from animals exposed to maximally contaminated, albeit in compliance with the levels set for feed materials, could result in the milk obtained from high-yielding dairy cows and other milk-producing animals including small ruminants, buffalo and camels, containing aflatoxin M_1 at levels exceeding the present statutory limits. Surveys conducted by various EU

Member States, however, revealed a very low incidence of milk samples that did not comply with the current EU regulations. Given the present agricultural practices and the possibility that aflatoxin B_1 is present in staple feeds grown in Europe, the monitoring activity of aflatoxin M_1 in milk should be intensified and expanded to consumable milk from animal species other than dairy cows.

Based on the conclusions of the EFSA Panel, the Committee agreed on the following conclusions at its meeting on 29 April 2004 (32):

- To continue and possible intensify monitoring aflatoxin B_1 in feed materials, both imported and produced domestically. Aflatoxin B_1 content in compound feeds intended for dairy animals other than dairy cows also should be controlled.
- For food controls, the control of aflatoxin M_1 in milk should be focused on milk at the farm level and the monitoring of milk should be expanded to milk and milk products from dairy animals other than dairy cows.
- As the Scientific Panel concluded that the current maximum levels of aflatoxin B_1 in animal feeds provided adequate protection from adverse health effects in target animal species and seemed to successfully prevent undesirable concentrations of aflatoxin M_1 in milk, there was no need to modify the current maximum levels for aflatoxin B_1 (Table 5) in the Annex to Directive 2002/32/EC on undesirable substances in animal feed.

Maximum level for rye ergot

A maximum level of 1000 μg/g of rye ergot (*Claviceps purpurea)* sclerotia was established by Directive 2002/32/EC for all feeds containing unground cereals. The Scientific Panel on Contaminants in the Food Chain of EFSA adopted an opinion on request from the Commission related to ergot as undesirable substance in animal feed on 19 April 2005 (33).

The term ergot refers to fungal structures produced by *Claviceps* species that develop in place of kernels on grain ears or seeds on grass heads, which are visible as large discolored sclerotia. These sclerotia contain different classes of alkaloids, the most prominent being ergometrine, ergotamine, ergosine, ergocristine, ergocryptine and ergocornine and their related -inines. The amount and toxin pattern produced vary by fungal strain and depend on the host plant and the geographic region. At present, the degree of variation in ergot alkaloid pattern in relation to fungal species, its geographic distribution, and its relation to the host plant, *e.g.*, the alkaloid pattern in rye ergot is different from that in grass ergot, is not known. More data are needed to identify all of the major factors responsible for the variation in ergot alkaloid patterns in individual plant species.

Ergot alkaloids (ergolines) have toxic effects in all animal species. Data on the sensitivity of agricultural animal species towards ergot alkaloids are incomplete and do not suffice for the establishment of tolerance levels for individual ergot alkaloids or mixtures thereof. The available data indicate that adverse effects may occur in agricultural animals, particularly in pigs, after intake of feed contaminated with ergot at levels close to the current maximum levels. Physical determination of the rye ergot contamination rate of feeds often is inaccurate, as the size and the weight of the sclerotia may vary considerably. Moreover, this physical determination is impossible to make in processed feeds. Hence, in addition to control by physical methods, control by chemical analysis of potentially contaminated feed materials also has been suggested, as various chromatographic methods are available to detect ergot alkaloids in feeds. The methods are limited, however, to only a relatively few ergot alkaloids.

Table 5. Maximum levels (ng/g) for aflatoxin B_1 in feeds with a moisture content of 12%.

Product(s) intended for animal feed	Maximum level
All feed materials	20
Complete feeds:	
Cattle, sheep and goats except for:	20
Dairy animals	5
Calves and lambs	10
Pigs and poultry (except young animals)	20
Other complete feeds	10
Complementary feeds:	
Cattle, sheep and goats, except for dairy animals, calves and lambs	20
Pigs and poultry (except young animals)	20
Other complementary feeds	5

At present, the data on the toxicological properties of individual ergot alkaloids are too limited to confirm that the ergot alkaloids that can be measured by chromatographic analytical methods are the most relevant toxic compounds. Data on the toxicity of individual ergot alkaloids are scarce, as animals, under field conditions, are exposed to complex mixtures of varying composition of ergot alkaloids depending on the fungal strain, the host plant and/or environmental factors. These possibilities imply that at present neither the total alkaloid content, nor any single alkaloid can be recommended as reliable indicator of the potential adverse effects on livestock associated with the ingestion of ergot-contaminated feeds. The limited data on tissue distribution and residual concentrations in edible tissues, milk and eggs are insufficient to estimate carry-over rates. There is, however, no evidence that ergot alkaloids accumulate in edible tissues. Levels of alkaloids measured in animal tissues so far indicate that these tissues are unlikely to be an important source of human exposure to ergot alkaloids. More data are needed on the presence of these ergot alkaloids, not only in unground cereals but also in processed cereals and compound feeds, to obtain reliable data on the ergot alkaloid patterns occurring in feeds and to correlate the presence of ergot alkaloids to the amount of sclerotia present. This monitoring should focus on the six most common ergot alkaloids, *i.e.*, ergometrine, ergotamine, ergosine, ergocristine, ergocryptine and ergocornine.

Based on the available data, a provisional relationship between the amount of sclerotia and the level of these individual ergot alkaloids can be established. In particular, the levels of individual ergot alkaloids corresponding to a content of 1 g of rye ergot sclerotia are 600 μg for ergocristine, 300 μg for ergotamine and 100 μg for ergocryptine, ergometrine, ergosine and ergocornine. This relationship includes a high level of uncertainty/variation and more reliable data are required to establish these relationships with certainty.

Provisions for ochratoxin A, deoxynivalenol, zearalenone, fumonisin B_1 + B_2 and T-2 and HT-2 toxins in feeds

At the request of the Commission the EFSA adopted opinions on the mycotoxins deoxynivalenol on 2 June 2004 (34), zearalenone on 28 July 2004 (35), ochratoxin A on 22 September 2004 (36), and fumonisins on 22 June 2005 (37).

Deoxynivalenol

Deoxynivalenol has toxic effects in humans and all other animal species investigated thus far. Species sensitivity varies considerably, but pigs are generally recognized as the most sensitive animal species. The initial adverse effect observed after deoxynivalenol exposure is reduced feed intake. At higher toxin concentrations vomiting and feed refusal occur. These effects lead to reduced body weight gain, particularly in growing animals. Deoxynivalenol also reduces immune response. The lowest reported levels of deoxynivalenol in feeds with a negative effect on feed intake for pigs range from 0.35 to 0.9 μg/g. With respect to other animal species, healthy ruminants tolerate several mg of deoxynivalenol per kg of feed. Poultry are less sensitive than pigs to the effects of deoxynivalenol on feed intake and weight gain, but the data do not suffice to estimate a no effect level. Other animal species, including rabbits, horses, cats and dogs, all seem to have a higher tolerance to deoxynivalenol than do pigs.

Deoxynivalenol is metabolized rapidly in animals and carry over to edible tissues, milk and eggs is very low. Thus, animal-derived foods contribute marginally, if at all, to total human exposure to deoxynivalenol.

Zearalenone

The most prominent effects of zearalenone result from its interaction with estrogen receptors resulting in apparent hyper-estrogenism, including reduced fertility. Pigs generally are considered the animal species most sensitive to zearalenone, with boars less sensitive than sows. Limited experimental studies indicate that sheep are the next most sensitive to the adverse effects of zearalenone, whilst cattle are relatively less sensitive. Poultry (chickens and turkeys) are the least sensitive to the hormonal effects of zearalenone. No reliable data are available for other species such as rabbits, horses, cats and dogs.

Zearalenone deposition in meat and other edible tissues is limited, and the transmission rate into milk and eggs is low. Thus, animal-derived foods are expected to contribute only marginally to the total human exposure to zearalenone, with the bulk of the exposure expected from cereals and grain products.

Ochratoxin A

Pigs are considered the most sensitive farm animal species to the nephrotoxicity of ochratoxin A. No no-observed-effect-level (NOEL) can be established, but based on effects on renal (diagnostic) enzyme levels and kidney function, a dietary concentration of 0.2 μg/g is considered the lowest-observed-effect-level (LOEL). Chickens also are a sensitive species, and ochratoxin A is assumed to be the most important cause of poultry nephropathy. Ruminants are less sensitive to ochratoxin A than are monogastric species. This result is consistent with data indicating that prior to absorption, significant microbial degradation of ochratoxin A to the less toxic ochratoxin α occurs in the rumen. Herbivores such as horses, rabbits and related species that rely on caecal rather than ruminal fermentation may absorb intact ochratoxin A in the small intestine and, therefore, are more likely to be sensitive to this toxin than are ruminants. Other monogastric animal species including dogs, cats and fish are expected to be sensitive to renal toxicity and immunosuppressive effects, as these effects have been observed in all species tested so far.

Ochratoxin A is retained in blood serum and may accumulate in blood, liver and kidney while significantly lower residual levels are found in muscle tissue, fat and milk. Carry-over into eggs occurs with high exposures under experimental conditions. Exposure assessments indicate that food of animal origin contributes only to a minor extent, on average 3% and in populations with dietary preferences < 10%, to human dietary exposure to ochratoxin A.

Fumonisins

Intoxications associated with the occurrence of fumonisins in animal feeds comprise distinct syndromes such as ELEM (equine leukoencephalomalacia) and PPE (porcine pulmonary edema). Fumonisins exhibit toxic effects in all animal species evaluated thus far. Susceptibility varies considerably amongst species, with pigs, rabbits, horses and other *Equidae* being the most sensitive. Relative to other animal species, adult ruminants are significantly less sensitive than are calves. For broiler chickens the LOEL level is ~2 mg/kg bw/day. Data from ducks, ducklings and turkeys provide no evidence that these species are more sensitive than chickens. Adult ruminants are not sensitive and have a low responsiveness to fumonisins.

Available data on the carry-over of fumonisins from animal feeds into edible tissues, including milk and eggs, indicate that carry-over is limited and that products of animal origin do not contribute substantially to human exposure.

T-2 and HT-2 toxins

Data on the presence of T-2 and HT-2 toxins in products intended for animal feeding are very limited. There also is an urgent need to develop and validate a sensitive analytical method. However, there are indications that the presence of T-2 and HT-2 toxins in products intended for animal feed could be of concern. Therefore, the development of a sensitive analytical method, collection of more occurrence data and more investigations/research into the factors involved in the presence of T-2 and HT-2 toxins in cereals and cereal products, in particular in oats and oat products, is needed.

Based on current scientific knowledge, the lack of reliable data on T-2 and HT-2 toxins, and the large year-to-year variation that occurs in these mycotoxins, the recommended first step is to collect more data on these mycotoxins in different feed materials and feeds.

Recommendations for deoxynivalenol, zearalenone, fumonisins, ochratoxin A and T-2 and HT-2 toxins

Commission Recommendation 2006/576/EC of 17 August 2006 on the presence of deoxynivalenol, zearalenone, ochratoxin A, T-2 and HT-2 toxins and fumonisins in products for animal feeding (38) provides recommendations for monitoring the presence of these mycotoxins in feed. The Recommendation is based on Article 211 of the Treaty of the European Community that enables the Commission to formulate recommendations on matters dealt

Table 6. Guidance values (µg/g relative to a feed with 12% moisture content) for deoxynivalenol, zearalenone, ochratoxin A and fumonisins in animal feeds.

Mycotoxin	Product intended for animal feed	Guidance value
Deoxynivalenol	Feed materials[1]:	
	Cereals and cereal products[2] except for maize by-products	8
	Maize by-products	12
	Complementary and complete feeds except for:	5
	Pigs	0.9
	Calves (< 4 months), lambs and kids	2
Zearalenone	Feed material[1]:	
	Cereals and cereal products[2] except for maize by-products	2
	Maize by-products	3
	Complementary and complete feeds for:	
	Piglets and gilts	0.1
	Sows and fattening pigs	0.25
	Calves, dairy cattle, sheep (including lambs) and goats (including kids)	0.5
Ochratoxin A	Feed material[1]:	
	Cereals and cereal products[2]	0.25
	Complementary and complete feeds for:	
	Pigs	0.05
	Poultry	0.1
Fumonisins B_1+B_2	Feed material[1]:	
	Maize and maize by-products[3]	60
	Complementary and complete feeds for:	
	Pigs, horses (*Equidae*), rabbits and pet animals	5
	Fish	10
	Poultry, calves (< 4 months), lambs and kids	20
	Adult ruminants (> 4 months) and mink	50

[1]Particular attention must be paid to cereals and cereal products fed directly to animals so that their use in a daily ration does not lead to the animal being exposed to a higher level of these mycotoxins than the corresponding levels of exposure when only a complete feed is used as a daily ration.
[2]The term "Cereals and cereal products" also includes cereal forages and roughages.
[3]The term "Maize and maize by-products" also includes maize forages and roughages.

with in the Treaty if the Commission considers such recommendations necessary to ensure the proper functioning and development of the common market.

To provide orientation to the Member States on the acceptability of cereals, cereal products and compound feeds, and to avoid disparities in the values accepted by the competent authorities of the different Member States and the consequent risk of distortion of competition, guidance values are recommended (Table 6). In applying these guidance values, Member States should bear in mind that the guidance values for cereals and cereal products were determined for the most tolerant animal species and are therefore to be considered as upper guidance values.

For feeds for more sensitive animals, Member States should ensure that lower guidance values for cereals and cereal products are applied by feed manufacturers as necessary to enable compliance with the guidance values determined for compound feeds for these animal species. Feed business operators should incorporate into their Hazard Analysis and Critical Control Points (HACCP) system (39) the guidance values to determine the critical limits at critical control points which separate acceptability from unacceptability, for the prevention, elimination or reduction of identified hazards.

An assessment of the approach provided for by this Recommendation is expected by 2009 in particular to assess its contribution towards protecting animal health. The monitoring data obtained as a result of this Recommendation also will enable a better understanding of the year-to-year variance and the presence of these mycotoxins in a wide range of animal by-products used for animal feed, which is of importance for taking any further legislative measures, if necessary.

Conclusions

In this chapter, a comprehensive overview of the current provisions of European Union legislation as regards mycotoxins in feed and food has been provided. These provisions are updated regularly taking into account new developments in technology and science. As regards the provisions in food, in particular on aflatoxins, discussions are in progress in the context of *Codex Alimentarius* that might result in changes to the current provisions.

References

1. *Official Journal of the European Communities*, L 31, 1.2.2002, pp. 1-24.
2. *Official Journal of the European Communities,* L 37, 13.2.1993, pp. 1-3.
3. *Official Journal of the European Union* L 165, 30.04.2004, pp. 1-141. Corrigendum published in *Official Journal of the European Union* L191, 28.5.2004, pp. 1-52.
4. *Official Journal of the European Union* L 70, 9.03.2006, pp. 12-34.
5. *Official Journal of the European Communities* L 201, 17.7.1998, p. 93. Directive as last amended by Directive (EC) 2004/43/EC of 13 April 2004 (*Official Journal of the European Union* L 113, 20.4.2004, p. 14).
6. *Official Journal of the European Communities* L75, 16.3.2002, p. 38. Directive as last amended by Directive (EC) 2005/5/EC of 26 January 2005 (*Official Journal of the European Union* L 27, 29.1.2005, p. 38).
7. *Official Journal of the European Union* L203, 12.8.2003, p. 40.
8. *Official Journal of the European Union* L143, 7.6.2005, p. 18.
9. Guidance document for competent authorities for the control of compliance with EU legislation on aflatoxins. http://ec.europa.eu/food/food/chemicalsafety/contaminants/comm_dec_2006_504 guidance_en.pdf.
10. *Official Journal of the European Communities* L 140, 30.5.2002, p. 10. Directive as last amended by Directive 2006/13/EC (*Official Journal of the European Union* L32, 4.2.2006, p. 44.
11. *Official Journal of the European Communities* L 364, 20.12.2006, p. 5.
12. Reports of the Scientific Committee for Food, 35th series, Opinion of the Scientific Committee for Food on aflatoxins, ochratoxin A and patulin, pp. 45-50, http://ec.europa.eu/food/fs/sc/scf/reports/scf_reports_35.pdf.

13. Opinion of the Scientific Committee on Food on Ochratoxin A (expressed on 17 September 1998) http://ec.europa.eu/food/fs/sc/scf/out14_en.html.
14. Reports on tasks for scientific co-operation, Task 3.2.7 "Assessment of dietary intake of Ochratoxin A by the population of EU Member States, 153 pp. http://ec.europa.eu/food/ food/chemicalsafety/contaminants/task_3-2-7_en.pdf.
15. *Official Journal of the European Communities* L 52, 4.3.1993, p. 18.
16. Opinion of the Scientific Panel on contaminants in the Food Chain of the EFSA on a request from the Commission related to ochratoxin A in food. http://www.efsa.europa.eu/etc/media lib/efsa/science/contam/contam_opinions/1521.Par.0001.File.dat/contam_op_ej365_ochratoxin_a_food_en1.pdf.
17. Summary record of the Standing Committee on the Food Chain and Animal Health held in Brussels on 9 February 2007 "Section Toxicological Safety of the Food Chain", agenda item 2.1. http://ec.europa.eu/food/committees/regulatory/scfcah/toxic/summary09022007_en.pdf.
18. Minutes of the 120th Meeting of the Scientific Committee on Food held on 8-9 March 2000 in Brussels, Minute statement on patulin. http://ec.europa.eu/food/fs/sc/scf/out55_en.pdf.
19. Reports on tasks for scientific co-operation, Task 3.2.8 "Assessment of dietary intake of Patulin by the population of EU Member States," 138 pp. http://ec.europa.eu/food/food/chemicalsafety/contaminants/3.2.8_en.pdf.
20. *Official Journal of the European Union* L 203, 12.8.2003, p. 34.
21. Opinion of the Scientific Committee on Food on *Fusarium* toxins Part 1: Deoxynivalenol (DON), (expressed on 2 December 1999) http://ec.europa.eu/food/fs/sc/scf/out44_en.pdf.
22. Opinion of the Scientific Committee on Food on *Fusarium*-toxins Part 2: Zearalenone (ZEA), (expressed on 22 June 2000) http://ec.europa.eu/food/fs/sc/scf/out65_en.pdf.
23. Opinion of the Scientific Committee on Food on *Fusarium*-toxins Part 3: Fumonisin B_1 (FB_1) (expressed on 17 October 2000) http://ec.europa.eu/food/fs/sc/scf/out73_en.pdf.
24. Updated opinion of the Scientific Committee on Food on Fumonisin B_1, B_2 and B_3 (expressed on 4 April 2003) http://ec.europa.eu/food/fs/sc/scf/out185_en.pdf.
25. Opinion of the Scientific Committee on Food on *Fusarium*-toxins Part 4: Nivalenol (expressed on 19 October 2000), http://ec.europa.eu/food/fs/sc/scf/out74_en.pdf.
26. Opinion of the Scientific Committee on Food on *Fusarium*-toxins Part 5: T-2 toxin and HT-2 toxin (adopted on 30 May 2001) http://ec.europa.eu/food/fs/sc/scf/out88_en.pdf.
27. Opinion of the Scientific Committee on Food on *Fusarium*-toxins Part 6: Group evaluation of T-2 toxin, HT-2toxin, nivalenol and deoxynivalenol. (adopted on 26 February 2002) http://ec.europa.eu/food/fs/sc/scf/out123_en.pdf.
28. Reports on tasks for scientific co-operation, Task 3.2.10 "Collection of occurrence data of Fusarium toxins in food and assessment of dietary intake by the population of EU Member States", 606 pp. http://ec.europa.eu/food/fs/scoop/task3210.pdf.
29. *Official Journal of the European Union* OJ L 234, 29.8.2006, p. 35.
30. Summary record of the Standing Committee on the Food Chain and Animal Health held in Brussels on 20 July 2007 "Section Toxicological Safety of the Food Chain", agenda item 1. http://ec.europa.eu/food/committees/regulatory/scfcah/toxic/summary20072007_en.pdf.
31. Opinion of the Scientific Panel on contaminants in the Food Chain of the EFSA on a request from the Commission related to Aflatoxin B_1 as an undesirable substance in animal feed. http://www.efsa.europa.eu/etc/medialib/efsa/science/contam/contam_opinions/294.Par.0001.File.dat/opinion_contam_02_en_final1.pdf.
32. Summary minutes of the meeting of the Standing Committee on the Food Chain and Animal Health, Animal Nutrition section, Brussels, 29 April 2004. http://ec.europa.eu/food/committees/regulatory/scfcah/animalnutrition/summary20_en.pdf
33. Opinion of the Scientific Panel on contaminants in the Food Chain of the EFSA on a request from the Commission related to ergot as undesirable substance in feed. http://www.efsa.europa.eu/etc/media lib/efsa/science/contam/contam_opinions/941.Par.0001.File.dat/contam_op_ej225_ergot_en1.pdf.

34. Opinion of the Scientific Panel on contaminants in the Food Chain of the European Food Safety Authority (EFSA) on a request from the Commission related to deoxynivalenol as undesirable substance in animal feed, adopted on 2 June 2004. http://www.efsa.europa.eu/etc/medialib/efsa/science/contam/contam_opinions/478.Par.0005.File.dat/opinion05_contam_ej73_deoxynivalenol_v2_en1.pdf.
35. Opinion of the Scientific Panel on contaminants in the Food Chain of the European Food Safety Authority (EFSA) on a request from the Commission related to zearalenone as undesirable substance in animal feed, adopted on 28 July 2004. http://www.efsa.europa.eu/etc/medialib/efsa/ science/contam/contam_opinions/527.Par.0004.File.dat/opinion_contam06_ej89_zearalenone_v3_en1.pdf.
36. Opinion of the Scientific Panel on contaminants in the Food Chain of the European Food Safety Authority (EFSA) on a request from the Commission related to ochratoxin A as undesirable substance in animal feed, adopted on 22 September 2004. http://www.efsa.europa.eu/etc/medialib/efsa/science/contam/contam_opinions/645.Par.0001.File.dat/opinion_contam09_ej101_ochratoxina_en1.pdf.
37. Opinion of the Scientific Panel on contaminants in the Food Chain of the European Food Safety Authority (EFSA) on a request from the Commission related to fumonisins as undesirable substance in animal feed, adopted on 22 June 2005. http://www.efsa.europa.eu/etc/medialib/efsa/science/contam/contam_opinions/1037.Par.0001.File.dat/contam_op_ej235_fumonisins_en1.pdf.
38. *Official Journal of the European Union* L 229, 23.8.2006, pp.7-9.
39. Regulation (EC) No 183/2005 of the European Parliament and of the Council of 12 January 2005 laying down requirements for feed hygiene. *Official Journal of the European Union* L 35, 8.2.2005, p. 1.

Mycotoxin Contamination and Toxigenic Fungi in Africa and the Mediterranean Basin

9

Mycotoxin Contamination in Foods in West and Central Africa

Kafui A. Kpodo* and Samuel A. Bankole

Abstract

Mycotoxins, natural toxic metabolites of fungi produced under favorable conditions on a wide range of foods and feeds are a world-wide problem. The five most important groups of mycotoxins are aflatoxins, fumonisins, ochratoxin A, zearalenone and deoxynivalenol. The weather conditions in most West and Central African countries are favorable for the growth of toxigenic fungi and mycotoxin production. We review mycotoxin contamination of foods from these two sub-regions. Aflatoxins have been detected in maize, peanuts and their products, sorghum, millet and rice, among others, from eight West African countries. Fumonisins were reported in maize and sorghum from six countries, whilst ochratoxin A was detected in foods from Ghana, Nigeria and Sierra Leone. Zearalenone was reported in maize from Cameroon and Nigeria with only one report of deoxynivalenol available. Limited data are available on mycotoxin contamination of foods in the Central African countries, suggesting an urgent need for increased mycotoxin research activities and surveillance.

Introduction

Mycotoxins are a group of toxic chemical compounds produced by strains of some fungal species when they grow under favorable conditions on a wide range of foods and feeds (CAST, 2003). There are three main genera of fungi that produce mycotoxins: *Aspergillus*, *Fusarium*, and *Penicillium*. Of the mycotoxins, five types are of major agricultural and human health significance: (i) aflatoxins, (ii) fumonisins, (iii) ochratoxin A, (iv) zearalenone and (v) the trichothecenes, *e.g.*, T-2, diacetoxyscirpenol, deoxynivalenol and nivalenol. Mycotoxins generally are of concern in human health, food safety and trade because of their acute and chronic effects on humans and domesticated animals. The presence of excessive mycotoxins can cause grain shipments to be rejected by importing countries resulting in a loss in consumer confidence in the importing country and severe economic losses for the exporting country.

Mycotoxins affect several agricultural products, including cereals, oilseeds, pulses, root crops, dried fruits, and coffee beans which form the agricultural economic backbone of most developing countries in Africa. Contamination of agricultural products occurs as a result of infection by toxigenic fungi under favorable environmental conditions in the field and may occur at various stages in the food chain, *e.g.*, preharvest, during harvest, drying, storage and/or processing. Whether fungi will grow and produce toxins depends on the environmental conditions and the specific temperature and water activity requirements of the particular fungus (Marín *et al.*, 2001, 2004).

Mycotoxin contamination is a world-wide problem and is not confined to any one geographical area or country. Countries in West Africa include Benin, Burkina Faso, Cameroon, Cape Verde, Côte d'Ivoire, Equatorial Guinea, Gabon, The Gambia, Ghana, Guinea, Guinea Bissau, Liberia, Mali, Niger, Nigeria, Senegal, Sierra Leone, and Togo. Central Africa consists of Burundi, Central African Republic, Chad, Congo Brazzaville, Democratic Republic of Congo (DRC) and Rwanda. These countries lie within latitudes 3° 30' south and 17° 00' north of the Equator and longitudes 32° 00' east and 17° 00' west of Greenwich. Countries within West and Central Africa contain a wide variation in climatic conditions ranging from almost temperate with frost and possible snow in parts of Rwanda through tropical hot and humid conditions in most of West Africa to desert and arid areas in the Northern parts of Mali and Niger. Generally, the conditions of temperature and humidity found in most of these countries are favorable for the growth of toxigenic fungi and mycotoxin production.

In this chapter, we summarize the health effects of each major group of mycotoxins and present data on the levels of these mycotoxins in foods in the West and Central African countries. Evidence of human exposure to mycotoxins also is presented through data on mycotoxin levels in body fluids.

Aflatoxins

Aflatoxins are produced by species of *Aspergillus flavus* and *Aspergillus parasiticus* in various commodities including maize, rice, barley, wheat, sorghum, peanuts and copra. *Aspergillus flavus* produces only aflatoxins B_1 and B_2 whereas *A. parasiticus* produces aflatoxins B_1, B_2, G_1 and G_2. The toxicological effects of aflatoxins are dose-dependent. At high doses they are lethal if consumed, causing liver, myocardial and kidney tissue damage. At sub-lethal doses aflatoxins cause chronic toxicity, *e.g.* liver cirrhosis, and at low-level exposure, they are potent human hepatocellular carcinogens (Wild and Turner, 2002). Aflatoxins also are mutagenic and teratogenic and can depress cell-mediated immunity (Williams *et al.*, 2004). Recently these toxins have been implicated in the stunting of children in Benin and Togo, where growth was reduced in children eating foods highly contaminated with aflatoxins introduced at weaning (Gong *et al.*, 2002). Aflatoxin B_1 acts synergistically with hepatitis B virus infection to increase its carcinogenic potency thirty-fold in hepatitis B surface antigen positive populations. The International Agency for Research on Cancer (IARC) has evaluated aflatoxin B_1 as a Group 1 carcinogen producing liver cancer in humans (IARC, 1993).

Benin

Nationwide studies for aflatoxins in stored maize in Benin have been conducted (Setamou *et al.*, 1997). The study involved the sampling of 80 and 60 maize fields in 1994 and 1995, respectively, to monitor *Aspergillus* infection and aflatoxin contamination of preharvest maize found *A. flavus* the most prevalent of the *Aspergillus* species isolated. Aflatoxins were detected in at least 30% of the maize fields sampled. There was a trend towards higher aflatoxin accumulation per percentage of *A. flavus* infection from the south to the north of the country. Damage by the ear borer (*Mussidia nigrivenella*) increased aflatoxin accumulation in maize and its incidence may be related to aflatoxin contamination.

Other studies of fungal infection and mycotoxins in maize stores of small scale farmers from different agroecological zones of Benin showed aflatoxin contamination in 25% of the maize stores in the southern zones with a mean of 100 ng/g after six to eight months of storage. In the northernmost zone, the Sudan Savanna, 56% of the maize stores were contaminated with aflatoxins at a mean level of 220 ng/g after six months of storage (Hell *et al.*, 2000). Nearly 40% of 744 maize samples from the southern Guinea, Sudan Savanna and the forest mosaic zones of Benin were positive for aflatoxins with an average of 105 ng/g (Hell *et al.*, 2000).

Apart from maize, dried yam chips used for the preparation of "amala", a staple food in Benin, also were contaminated with aflatoxins. Ninety-eight percent of the samples contained aflatoxins at levels ranging from 2.2 to 220 ng/g with a mean value of 14 ng/g (Bassa *et al.*, 2001). Another study found that 23% of 107 samples contained > 15 ng/g total aflatoxins and that 80% contained total aflatoxins above the EU standard of 4 ng/g (Mestres *et al.*, 2004).

Gambia

Aflatoxin levels were measured in a variety of cooked foods including maize, rice, millet, peanut sauces, and leaf sauces (Hudson *et al.*, 1992). The highest levels of contamination were found in the peanut sauces. Eighteen of twenty samples were positive for aflatoxins at 19-940 ng/g and a mean of 160 ng/g. Nine of 10 maize samples contained low levels of aflatoxins ranging from 2-35 ng/g with a mean of 9.7 ng/g. All nine millet samples analyzed contained aflatoxins (1-27 ng/g) as did 2/8 sorghum samples at levels of 2 and 16 ng/g. Fourteen of 20 rice samples contained aflatoxins (2-19 ng/g) and three leaf sauces contained aflatoxins at levels of 21, 26 and 34 ng/g (Hudson *et al.*, 1992).

In another study conducted in the Gambia (Wild *et al.*, 1992), 87 leaf sauces, 22 flour sauces and 47 peanut sauces were evaluated for aflatoxins; 64%, 59%, and 87% of these sauces, respectively, contained detectable aflatoxins. The highest level of aflatoxins was 770 ng/g, which was detected in a leaf sauce that contained raw peanuts. Fifteen samples of boiled rice in this study did not contain aflatoxins.

Ghana

In Ghana, aflatoxin studies have been conducted on maize, peanuts, and their products (Kpodo and Halm, 1990; Kpodo, 1995, 1997, 2001; Awuah and Kpodo, 1996; Kpodo *et al.*, 1996). Studies on aflatoxin contamination of maize stored in some silos and warehouses in parts of Ghana revealed that all the samples contained aflatoxins with 80% of the samples containing aflatoxin levels > 30 ng/g (Kpodo and Halm, 1990). Further studies on the occurrence of mycotoxins in fermented maize dough and "Ga kenkey" (fermented dough boiled for three hours) from markets and processing sites in Accra also were contaminated with aflatoxins (Kpodo *et al.*, 1996). Thirty-one of 32 fermented maize dough samples contained aflatoxins at levels up to 310 ng/g. Fifteen of 16 Ga kenkey samples from four production sites in Accra were contaminated with aflatoxins at levels up to 200 ng/g (Kpodo *et al.*, 1996).

In a more recent study (Kpodo, 2001), 58/75 Ga kenkey samples again were contaminated with aflatoxins at levels up to 200 ng/g and 84/128 maize kernel samples from markets and maize processing sites throughout the country also contained aflatoxins at levels up to 2,000 ng/g. In this same study, aflatoxins and fumonisins were found to co-occur in 42/90 maize kernel samples and in 41/75 kenkey samples (Kpodo, 2001).

Peanuts, purchased in markets in and around Accra contained 3-220 ng/g aflatoxins (Mintah and Hunter, 1979). In a nation-wide survey of aflatoxin contamination in stored peanuts in Ghana covering 21 markets in all 10 regions of Ghana, total aflatoxin levels ranging from 5.7 to 22,000 ng/g were identified in damaged kernels. In the same study, aflatoxins were not detected in 50% of visibly undamaged kernels tested and were present at low levels (0.1-12 ng/g) in the remaining undamaged kernels (Awuah and Kpodo, 1996). Peanut samples from the 1994 crop season in six locations in the southern parts of Ghana contained aflatoxins at levels ranging from 12-110 ng/g (Kpodo, 1995). In another study, 100 peanut paste samples purchased from selected major markets in all 10 regions of Ghana were screened. Eighty-six samples contained aflatoxins at varying levels with 65 samples containing total aflatoxin levels > 30 ng/g. The highest total aflatoxin level recorded in this study was 3,300 ng/g (Kpodo, 1997).

Apart from maize, peanuts and their products, other miscellaneous commodities have been screened for aflatoxins in Ghana (Kpodo, 2005). Aflatoxins were detected in 3/10 sorghum samples at levels of 7.5, 8.0, and 81 ng/g, 5/8 soybean meal samples at levels up to 36 ng/g, all four cassava flour samples at levels ranging from 4-2100 ng/g, 3/4 cashew paste samples with levels up to 370 ng/g. Relatively lower levels were detected in rice (< 2 ng/g), cocoa cake (< 8 ng/g) and "Agushie" (< 15 ng/g).

Côte d'Ivoire

In Côte d'Ivoire some aflatoxin studies have been conducted on maize and peanuts (Pollet *et al.*, 1989; Wyers *et al.*, 1991). Aflatoxin at up to 360 ng/g were detected in maize analyzed in a 2-year survey of peanut stocks from 434 farmers. Aflatoxin levels > 250 ng/g were detected in 7.9% of the samples with 4.4% containing > 1,000 ng/g and 73% having > 10 ng/g. In the same study, 13/72 peanut samples from markets contained > 50 ng/g aflatoxins.

Niger

In Niger, 25 peanut lines including germplasm, advanced *A. flavus* resistant breeding lines, and cultivars from West Africa, were tested at three sites – Sadore, Bengou and Maradi – during the 1989, 1990, and 1991 rainy seasons (Waliyar and Hassan, 1993). Seeds from these sites were tested for *A. flavus* contamination, aflatoxin content and the average proportion of contaminated seeds. Aflatoxin content of the seeds ranged from 1-750 ng/g and 5 to 37% of the seeds in a seed lot were contaminated. Total aflatoxin contamination in a sample generally increased with the percentage of contaminated seeds in the sample.

Nigeria

Several studies have been conducted in Nigeria on aflatoxins in foods. A two-year countrywide survey was initiated in Nigeria in 1980 to determine storage losses of food crops at the market level (Opadokun and Ikeorah, 1983). During the study, 145 maize samples from markets in Kano and Plateau States of Nigeria were assayed for moisture and aflatoxin content.

Over 20% (30 samples) had detectable aflatoxin B_1 (> 5 ng/g). Twenty-one samples had aflatoxin B_1 levels > 30 ng/g and one sample contained 1,300 ng/g of aflatoxin B_1. Various human foods and feeds also were screened for aflatoxins and 9/22 maize samples were positive for aflatoxins with a mean level of 21 ng/g and a range of 20-110 ng/g (Atawodi *et al.*, 1994).

Moldy maize samples from Plateau State contained aflatoxins, zearalenone and ochratoxin A. An aflatoxin level of 960 ng/g was obtained for a sample from Langtang (Gbodi *et al.*, 1986). Of 48 samples of maize-based gruels used as weaning food for children admitted to the Wesley Guild Hospital in Ilesha, 12 were positive for aflatoxins with levels up to 20 ng/g (Oyelami *et al.*, 1996).

In other studies, 80% of maize samples from different locations in southeastern Nigeria were positive for aflatoxin B_1 (Aja-Nwachukwu and Emejuaiwe, 1994). In southwestern Nigeria, 45% of raw maize from farms, 80% of maize cakes and 12% of maize rolls were contaminated with aflatoxins at mean levels of 200, 233, and 55 ng/g, respectively (Adebajo *et al.*, 1994).

The agroecological zones in which maize was grown affected the amount of aflatoxin present, as did the storage method. The highest level of contamination, 3,100 ng/g, was obtained from the Humid Forest zone while the lowest aflatoxin content, 670 ng/g, was obtained from the Southern Guinea Savanna zone. No aflatoxins were detected in maize samples from the Northern Guinea Savanna zone (Udoh, 1997). Storage of maize in bags and "rhumbu" (traditional clay stores) was correlated with lower aflatoxin levels in the Sudan Savanna region of Nigeria (Udoh, 1997). Similar regional studies have been conducted on other commodities and products from the savanna and forest regions (Nwokolo and Okonkwo, 1978). High risk foods (> 200 ng/g of aflatoxin B_1) included peanuts, dried fish, guinea corn (sorghum) and millet. Maize, rice, beans, and crude palm-oil contained 30-200 ng/g of aflatoxin B_1. Low risk foods (< 30 ng/g) included cereal "acha" ("findi"), some cassava products, yams and refined vegetable oils. Confirmation of peanuts as a food at high risk for mycotoxin contamination was obtained from studies conducted by several researchers (Akano and Atanda, 1990; Atawodi *et al.*, 1994). Eighty-seven percent of peanut cake ("Kulikuli") samples purchased from markets in Ibadan contained aflatoxin B_1. Levels ranged from 20-460 ng/g. In other studies (Atawodi *et al.*, 1994), levels as high as 1,900 ng/g were observed in peanut cake samples whilst 64% of 106 roasted peanut samples from retail outlets in Nigeria contained aflatoxin B_1 with a mean value of 25 ng/g.

Aflatoxins were detected in 18/100 samples of foods – "gari", yam flour, cassava flour, melon, onion, rice, plantain, red pepper and eggs – from Benin City (Ibeh *et al.*, 1991). Aflatoxins also were detected in five samples of yam flour, four samples of cassava flour, three samples of "gari", two samples each of beans and melon, and one sample of rice. Extremely high concentrations of aflatoxins were recorded in yam flour (4,000-7,600 ng/g). Cassava flour also contained high levels of aflatoxins ranging from 3,500-5,400 ng/g. Pepper, onion, plantain and eggs did not contain detectable amounts of aflatoxins.

In another study, aflatoxins were detected in melon seed and tiger nut (*Cyperus esculentus*) samples. Aflatoxin B_1 was detected in 32% of melon seed samples collected from Nigerian markets, with the means of 14 and 11 ng/g, respectively, for samples from the forest and savanna zones (Bankole *et al.*, 2004). The presence of aflatoxins in 35% of the tiger nut samples obtained from different parts of Nigeria at levels ranging from 10-120 ng/g also has been reported (Bankole and Eseigbe, 1996). Fifty-four percent of dried yam chips for sale in various parts of western Nigeria were contaminated with aflatoxin B_1 at levels of 4-190 ng/g (mean 23 ng/g), and 32% with aflatoxin B_2 (2-55 ng/g). Five percent of the samples contained aflatoxin G_1 (4-18 ng/g) with two samples testing positive for aflatoxin G_2 (Bankole and Adebanjo, 2003).

Aflatoxins have been detected in Nigerian indigenous beverages (Alozie *et al.*, 1980; Okoye and Ekpenyong, 1984). In the Jos metropolis aflatoxin B_1 contamination was common in the traditionally brewed millet-based beers "pito" and "burukutu" (Okoye and Ekpenyong, 1984). Seventeen of 20 "pito" samples contained aflatoxin at levels ranging from 16-140 ng/g while 15/20 "burukutu" samples contained aflatoxins at levels ranging from 1.7 to 140 ng/g. In this study the millet raw material was not analyzed for aflatoxins.

In another survey (Alozie *et al.*, 1980), 16 Nigerian indigenous beverages and foodstuffs, all eight beverage samples from Ugbowo in Benin City were contaminated with aflatoxins. The beverages were "burukutu", "pito", "emu aran" (fermentable sap of the *Raphia* palm; *R. vinifera* and *R. raphia*), and "ogoro" (fermentable sap of immature shoots of the oil palm, *Elais guinensis*). Aflatoxin levels ranged from 83 ng/g in "emu aran" to 260 ng/g in "burukutu". Foodstuffs evaluated in this study also were purchased from markets in Benin City and included "garï" (Cassava *farina*), "ogbono" (*Irvingia gabunesis*), "egusi" meal (*Cucumeropsis edulis*), "ogili-ugba" (prepared from castor bean, *Riccinus communis*), "dawadawa" (prepared from locust bean, *Parkia filicoden*), "ewedu" (*Coconus seratus*) and "shoko yokoto" (*Ceropsia* sp.). All the foodstuffs except "dawadawa", "ewedu", and "shoko yokoto" contained aflatoxins (Alozie *et al.*, 1980).

Senegal

In Senegal high levels of aflatoxins were found in peanut cake and unrefined oil, with > 90% of market samples of unrefined oil contaminated with > 300 ng/g of aflatoxins (Clavel, 1995). In peanut oil foods in Kaolack and Diourbel, > 85% of the samples were contaminated with aflatoxin B_1 at a mean of 40 ng/g (Ndiaye *et al.*, 1999). The mean level of total aflatoxin in peanut oil from small-scale units in two regions of Senegal ranged from 57-82 ng/g (Diop *et al.*, 2000).

Sierra Leone

There are few studies of aflatoxin contamination in Sierra Leone. Twenty samples of Bonga, a smoke-dried fish, obtained from homes and markets in Njala were contaminated with four *Aspergillus* spp. – *A. flavus*, *A. ochraceus*, *A. tamarii* and *A. niger*. Varying amounts of aflatoxins B_1, G_1 and G_2 were detected in the moldy fish (Jonsyn, 1992). Aflatoxins also were reported in fermented sesame seeds (Jonsyn, 1990).

Central African countries

No reports were available of aflatoxins in foods in any Central African country.

Fumonisins

Fumonisins are mycotoxins produced by *Fusarium verticillioides*, formerly termed *F. moniliforme* (Seifert *et al.*, 2003), and *Fusarium proliferatum* which are among the fungi most commonly associated with maize. These toxins normally are synthesized before harvest or during the early stages of drying, and, except under unusual conditions, do not increase during grain storage (Fandohan *et al.*, 2005). Since the discovery of fumonisins in 1988 (Gelderblom *et al.*, 1988), these toxins have been found as natural contaminants of maize and maize-based foods and feeds in many parts of the world

Fumonisins have been associated with a high incidence of esophageal cancer in certain areas of the Transkei region of the Eastern Cape Province in South Africa (van Rensburg, 1985; Rheeder *et al.*, 1992) and high levels of fumonisin levels have been found in maize from this and other parts of the world known to have unusually high rates of esophageal cancer such as the Linxian county region in China (Yang, 1980), northeastern Iran (Kmet and Mahboubi, 1972; Shephard *et al.*, 2000), northeastern Italy (Franceschi *et al.*, 1990). Fumonisins also may be a risk factor in primary liver cancer and could act synergistically with aflatoxins, microcystins and/or deoxynivalenol (Ueno *et al.*, 1996).

Whether fumonisin B_1 is a carcinogen in human beings has not been established; however, based on toxicological evidence, the International Agency for Research on Cancer (IARC) has designated fumonisin B_1 to be possibly carcinogenic to humans (class 2B carcinogen) (IARC, 2002). The development of neural tube defects in rural populations known to consume contaminated maize has been related to the inhibition of uptake of folic acid by fumonisins and, although not yet confirmed, fumonisins are believed to play a critical role in the disruption of folate membrane transport (WHO, 2002).

Limited surveys of good-quality maize from hybrids grown in Benin in 1992 found that 82% of the samples contained fumonisins at levels up to 3,300 ng/g. The mean fumonisin level for the positive samples was 700 ng/g (Doko *et al.*, 1995). In a more recent study in Benin, widespread fumonisin occurrence in maize also was reported. Samples from the 1999-2000 crop year contained fumonisins at up to 12,000 ng/g. For the 2000-2001 crop year, fumonisin levels were as high as 6,700 ng/g, and as high as 6,100 ng/g for the 2002-2003 crop year (Fandohan *et al.*, 2005).

Maize from Burkina Faso, Ghana, Nigeria and Cameroon also are contaminated with fumonisins (Kpodo *et al.*, 2000; Bankole *et al.*, 2003; Nikiema *et al.*, 2004). In Burkina Faso, 72 market samples of maize contained fumonisins at levels ranging from 130-16,000 ng/g, as did 52 samples obtained from farms with fumonisins present at between 10 and 3,100 ng/g (Nikiema *et al.*, 2004). In Ghana, 14 maize samples contained between 70 and 4,200 ng/g of fumonisins with eight samples co-contaminated with aflatoxins (Kpodo *et al.*, 2000). In another study from Ghana, 68/75 maize samples from different sites contained fumonisins at levels ranging from 11-2,500 ng/g, while 55/75 kenkey samples contained fumonisins at levels of 15-1,000 ng/g (Kpodo, 2001). In one study from Nigeria, 55/108 maize samples were positive for fumonisin B_1 with levels ranging from 65 to 1,800 ng/g with a mean value of 390 ng/g for the positive samples (Bankole *et al.*, 2003). Studies in the Cameroon (Ngoko *et al.*, 2001) identified fumonisins in 13/15 maize samples at levels ranging from 300-26,000 ng/g.

The only report of fumonisin contamination of foods in Central Africa was a study conducted in Burundi on 50 food samples by Munimbazi and Bullerman (1996). Fumonisin B_1 was detected in six maize samples at levels ranging from 12-75 ng/g. In the same study, fumonisin B_1 was detected in one sorghum meal sample.

Ochratoxin A

The ochratoxins (ochratoxin A and ochratoxin B) are produced by *Aspergillus ochraceus*, *Penicillium verrucosum* and *Aspergillus carbonarius* (Frisvad and Thrane, 2000). In warm climates, such as those found in West and Central Africa, ochratoxin production is more commonly associated with *A. ochraceus* than it is with *P. verrucosum*, which often produces ochratoxin A in temperate climates (Sweeney and Dobson, 1998). Ochratoxin A is the major metabolite found as a natural contaminant of cereal grains such as maize, barley, wheat, oats and rye (Shotwell *et al.*, 1971, 1976). *Aspergillus carbonarius* grows at high temperatures and is more often associated with fruits, especially grapes.

Ochratoxin A is the more toxic of the two derivatives and is nephrotoxic, immunosuppressive, carcinogenic and teratogenic in all experimental animals tested (WHO, 1990). It has been classified as a Group 2B compound, *i.e.*, possibly carcinogenic to humans (IARC, 1987). The toxin has been associated with endemic nephropathy which is a fatal renal disease found among rural populations in Croatia, Bosnia, Herzegovina, Serbia, Bulgaria and Romania.

Limited data exist on the occurrence of ochratoxin A in foods in Africa. In Ghana, studies of fermented maize dough found ochratoxin A in 5/20 samples at levels < 6.4 ng/g (Kpodo *et al.*, 1996). In Nigeria, ochratoxin A was detected in moldy maize samples from the Plateau State and the highest level recorded was 150 ng/g (Gbodi *et al.*, 1986). Four of 48 maize-based gruel samples used as weaning food for children admitted to the Wesley Guild Hospital in Ilesha, Nigeria contained ochratoxin A at low levels (Oyelami *et al.*, 1996). Another study in Nigeria also detected the toxin in Tiger nuts (Adebajo, 1993). In Sierra Leone, ochratoxin A was detected in fermented sesame seeds ("ogiri") and moldy fish samples (Jonsyn, 1988).

Deoxynivalenol

Deoxynivalenol is a trichothecene mycotoxin associated primarily with *Fusarium graminearum* (*Gibberella zeae*) and *Fusarium culmorum*. *Fusarium graminearum* is the more common species and may occur in warmer climates. Deoxynivalenol has been found in grains such as wheat, barley, oats, rye and maize but it has not received much attention in Africa. Consumption of grain contaminated with deoxynivalenol has been associated with outbreaks of acute disease involving nausea, vomiting, gastrointestinal upset, dizziness, diarrhea and headache in Asia (WHO, 2002).

Only one report of deoxynivalenol in foods is available from these African regions. In the Cameroon, deoxynivalenol was present in 12/15 maize samples at levels ranging from < 100-1,300 ng/g (Ngoko *et al.*, 2001).

Zearalenone

Zearalenone is produced mainly by *F. graminearum* and *F. culmorum* and may occur in most cereals, including maize, wheat, barley, oats and rye. Zearalenone has estrogenic properties in various animal species that include infertility, vulval edema, and mammary hypertrophy in females (Peraica *et al.*, 1999). Little information is available regarding the effects of zearalenone in humans, although in Puerto Rico, precocious sexual development in young children was attributed to food contaminated with zearalenone (Saenz de Rodriguez *et al.*, 1985).

Zearalenone has been reported in maize and maize products from several African countries including Botswana, Egypt, Lesotho, Swaziland, South Africa and Zambia. However from the West and Central African sub-regions, the only reports available are from Nigeria and Cameroon. In Nigeria, the toxin was detected in moldy maize samples at levels of 960 ng/g (Gbodi *et al.*, 1986). In a study of maize beer in Nigeria, 28/46 samples were contaminated with zearalenone at levels ranging from 13-200 ng/g with a mean of the positive samples being 82 ng/g (Okoye, 1986). In Cameroon, 12/15 maize samples contained zearalenone at levels up to 1,100 ng/g (Ngoko *et al.*, 2001).

Other mycotoxins in foods in West and Central African countries

In addition to the five groups of mycotoxins already discussed, other mycotoxins have been detected in some foods in some countries in West Africa. These mycotoxins include sterigmatocystin, patulin, cyclopiazonic acid, penicillic acid and tenuazonic acid from dried cassava chips in Ghana (Waering *et al.*, 2001). Ten of 49 samples contained sterigmatocystin (170-1,700 ng/g); four samples each contained either patulin (550-850 ng/g) or cyclopiazonic acid (80-720 ng/g); five contained penicillic acid (60-230 ng/g); and three contained tenuazonic acid (20-340 ng/g). In other studies, citrinin was detected in fermented sesame seeds ("ogiri") in Sierra Leone (Jonsyn, 1988) and at levels up to 580 ng/g in 20/20 fermented maize dough samples from Ghana (Kpodo *et al.*, 1996).

Mycotoxins in humans living in West and Central African countries

Further evidence of exposure to mycotoxins by sections of the populations in West African countries has been obtained through the detection of mycotoxins in breast-milk and other body fluids of people in Burkina Faso, The Gambia, Ghana, Guinea, Nigeria and Sierra Leone.

Studies in The Gambia have reported aflatoxins in sera from umbilical cord, maternal venous blood and breast-milk (Wild *et al.*, 1991; Zarba *et al.*, 1992; Miele *et al.*, 1996). Thirty-two of 35 individuals were positive for aflatoxin-albumin adducts (Miele *et al.*, 1996). In eastern Gambia, samples of sera from umbilical cord and maternal venous blood from 30 pregnant women were assayed for the aflatoxin-albumin adducts. Twenty-nine venous blood and 21 cord sera were positive for aflatoxin-albumin adducts (Wild *et al.*, 1991). Aflatoxin M_1 was identified in breast-milk of five subjects by using a preparative monoclonal antibody immunoaffinity column/HPLC method. Aflatoxin G_1 was found in 3/5 women (Zarba *et al.*, 1992). In a third study in The Gambia (Turner *et al.*, 2000), sera from 444 children aged 3-4 years was analyzed for aflatoxin-albumin adducts. Adduct levels ranged from 2.2 to 460 pg AF-lysine eq/mg albumin with a strong association between high AF-albumin adduct levels and acute HBV infection.

In Accra, Ghana, aflatoxins were detected in 32% of the breast milk and in 31% of the cord blood samples (Maxwell *et al.*, 1989). Earlier studies of the pathological effects of aflatoxins found that aflatoxins were present in liver samples from children who died of kwashiorkor (Apeagyei *et al.*, 1986). Further studies of toxin levels in serum, urine and fecal specimens of a group of 40 apparently healthy Ghanaian adults found that aflatoxins G_1 and B_1 and aflatoxins Q_1 and M_1, both metabolites of aflatoxin B_1, were present in one or more of the specimens from 35% of the subjects. Twenty-six of the 40 subjects had only aflatoxin G_1, in their specimens (Ankrah *et al.*, 1994).

Only one study was reported in Guinea. Of 75 blood serum samples from men living in Kindia (lower Guinea) analyzed for aflatoxin bound to serum albumin, > 90% of the samples contained detectably high levels for adults. The highest level was equivalent to 360 pg aflatoxin-lysine per mg of albumin. Eleven of these patients also were positive for hepatitis B and eight were positive for hepatitis C (Diallo and Wild, 1995).

Of 161 human urine samples collected in Lagos, Nigeria, aflatoxin B_{2a} was detected in 33% of the samples, B_1 in 3.1%, M_1 in 8.7%, G_1 in 9.9% and L in 9.3% of the samples (Bean *et al.*, 1989). Aflatoxin G_1 was present at the highest mean concentration of 12 ng/100 ml urine. In another study of human serum involving residents of Nigeria, Nepal and the United Kingdom, aflatoxins were detected in 76% of the Nigerian residents, from all of the Nepalese residents, and from none of the UK residents (Wilkinson *et al.*, 1989).

In Sierra Leone, 54 urine samples from children < 5 years of age (Jonsyn, 1999) were all contaminated with aflatoxins and 24% and 20%, respectively, also were contaminated with ochratoxin A and ochratoxin B. For serum, 94% of the samples contained aflatoxins, 33% contained ochratoxin A and 23% contained ochratoxin B. Ninety-four percent of the stool specimens also contained aflatoxins.

Aflatoxin regulations

Several International and Regional groups as well as individual countries have established standards for aflatoxins and other mycotoxins in some cases. The European Union has established separate maximum permissible limits for aflatoxins in food products for direct consumption and those to be sorted or physically treated before consumption. For example, the maximum permissible limit for maize and maize products intended for direct consumption is 2 ng/g of aflatoxin B_1 and 4 ng/g for total aflatoxins. Maize and maize products to be sorted or physically treated before consumption have a maximum limit of 5 ng/g of aflatoxin B_1 and 10 ng/g for total aflatoxins. The Food and Drug Administration (FDA) of the United States also has set an action level of 20 ng/g for all human foods except milk, which is lower.

Based on the data presented above, some foods in the West and Central African countries contain mycotoxins, and certainly aflatoxins at levels well above the internationally recommended maximum limits. Only eight African countries had regulatory limits for aflatoxins in foods in a 1994/1995 worldwide survey of mycotoxin regulations (FAO, 1997), with the number increasing to 15 in a 2003 study (FAO, 2004). Countries such as The Gambia and Benin have taken steps to address aflatoxin contamination problems, but even in countries with regulations, food that does not move through formal marketing channels, *e.g.*, almost all food sold in local markets, is effectively unregulated.

Conclusions

Portions of the populations in some West and Central African countries are consuming foods containing unacceptable levels of aflatoxins and other mycotoxins. Data on mycotoxin contamination of foods are lacking for some West African countries and most countries in Central Africa. Mycotoxin testing laboratories with trained personnel must be made available to initiate studies in these countries. Intensified mycotoxin surveillance and monitoring programs are needed in the remaining countries in both subregions. Finally, there is an urgent need for extensive mycotoxin awareness creation and education programs as well as the adoption of good agricultural practices inclusive of proper storage and handling of foods in countries of the two subregions.

References

Adebajo, L.O. (1993) Survey of aflatoxins and ochratoxin A in stored tubers of *Cyperus esculentus*. *Mycopathologia* 124, 41-46.

Adebajo, L.O., Idowu, O. and Adesanya, A.M. (1994) Mycoflora and mycotoxin production in Nigerian corn and corn-based snacks. *Mycopathologia* 126, 183-192.

Aja-Nwachukwu, J. and Emejuaiwe, S.O. (1994) Aflatoxin producing fungi associated with Nigerian maize. *Environmental Toxicology and Water Quality* 9, 17-23.

Akano, D.A. and Atanda, O.O. (1990) The present level of aflatoxin in Nigerian groundnut cake (Kulikuli). *Letters in Applied Microbiology* 10, 187-189.

Alozie, T.C., Rotimi, C.N. and Oyibo, B.B. (1980) Production of aflatoxin by *Aspergillus flavus* (UBMI) in some Nigerian indigenous beverages and foodstuffs. *Mycopathologia* 70, 125-128.

Ankrah N.-A., Rikimaru, T. and Ekuban, F.A. (1994) Observations on aflatoxins and the liver status of Ghanaian subjects. *East African Medical Journal* 71, 739-741.

Apeagyei, F., Lamplugh, S.M., Hendrickse, R.G., Affram, K. and Lucas, S. (1986) Aflatoxins in the livers of children with kwashiorkor in Ghana. *Tropical and Geographical Medicine* 38, 273-276.

Atawodi, S.E., Atiku, A.A. and Lamorde, A.G. (1994) Aflatoxin contamination of Nigerian foods and feeding stuffs. *Food and Chemical Toxicology* 32, 61-63.

Awuah, R.T. and Kpodo, K.A. (1996) High incidence of *Aspergillus flavus* and aflatoxins in stored groundnut in Ghana and the use of a microbial assay to assess the inhibitory effects of plant extracts on aflatoxin synthesis. *Mycopathologia* 134, 109-114.

Bankole, S.A. and Adebanjo, A. (2003) Aflatoxin contamination of dried yam chips marketed in Nigeria. *Tropical Science* 43, 201-203.

Bankole, S.A. and Eseigbe, D.A. (1996) Occurrence of mycoflora and aflatoxins in marketed tigernut. *Crop Research* 11, 219-223.

Bankole, S.A., Mabekoje, O.O. and Enikuomehin, O.A. (2003) *Fusarium moniliforme* and fumonisin B_1 in stored maize from Ogun State, Nigeria. *Tropical Science* 43, 76-79.

Bankole, S.A., Ogunsanwo, B.M. and Mabekoje, O.O. (2004) Natural occurrence of moulds and aflatoxins in melon seeds from markets in Nigeria. *Food and Chemical Toxicology* 44, 1209-1214.

Bassa, S., Mestres, C., Hell, K. Vernia, P. and Cardwell, K. (2001) First report of aflatoxin in dried yam chips in Benin. *Plant Disease* 85, 1032.

Bean, T.A., Yourtee, D.M., Akanda, B. and Ogunlewe, J. (1989) Aflatoxin metabolites in the urine of Nigerians: Comparison of chromatographic methods. *Toxin Reviews* 8, 43-52.

CAST. (2003). *Mycotoxins: Risks in Plant, Animal and Human Systems, Task Force Report No. 139*. CAST, Ames, Iowa, USA.

Clavel, D. (1995) Present status of research on the groundnut aflatoxin problem in Senegal. *Arachide Infos* 6, 5.

Diallo, M.S. and Wild, C.P. (1995) The occurrence of exposure to aflatoxins and to hepatitis B and C viruses in Guinea. *Proceedings of the Workshop on Mycotoxins in Foods in Africa (6-10 November 1995, Cotonou, Benin)*, p. 27.

Diop, Y., Ndiaye, B., Diouf, A., Fall, M., Thiaw, C., Thiam, A., Barry, O., Ciss, M. and Ba, D. (2000) Artisanal peanut oil contamination by aflatoxins in Senegal. *Annales Pharmaceutiques Françaises* 58, 470-474.

Doko, M.B., Rapior, S., Visconti, A. and Schjoth, J.E. (1995) Incidence and levels of fumonisin contamination in maize genotypes grown in Europe and Africa. *Journal of Agricultural and Food Chemistry* 43, 429-434.

FAO. (2004) *Worldwide Regulations for Mycotoxins in Food and Feed in 2003 – FAO Food and Nutrition Paper 81*. FAO, Rome, Italy.

Fandohan, P., Gnonlonfin, B., Hell, K., Marasas, W.F.O. and Wingfield, M.J. (2005) Natural occurrence of *Fusarium* and subsequent fumonisin contamination in preharvest and stored maize in Benin, West Africa. *International Journal of Food Microbiology* 99, 173-183.
Franceschi, S., Bidoli, E., Baron, A.E. and la Vecchia, C. (1990) Maize and risks of cancers of the oral cavity, pharynx and esophagus in Northern Italy. *Journal of the National Cancer Institute* 82, 1407-1411.
Frisvad, J.C. and Thrane, U. (2000) Mycotoxin production by common filamentous fungi. In: Samson, R.A., Hoekstra, E.S., Frisvad, J.C. and Filtenborg, O. (eds.), *Introduction to Food and Airborne Fungi*. Centraalbureau voor Schimmelcultures, Utrecht, The Netherlands, pp. 321-331.
Gbodi, T.A., Nwude, N., Aliu, Y.O. and Ikediobi, C.O. (1986) The mycoflora and some mycotoxins found in maize (*Zea mays*) in the Plateau State of Nigeria. *Veterinary and Human Toxicology* 28, 1-5.
Gelderblom, W.C.A., Jaskiewicz, K., Marasas, W.F.O., Thiel, P.G., Horak, R.M., Vleggaar, R. and Kriek, N.P.J. (1988) Fumonisins – Novel mycotoxins with cancer-promoting activity produced by *Fusarium moniliforme*. *Applied and Environmental Microbiology* 54, 1806–1811.
Gong, Y.Y., Cardwell, K., Hounsa, A. Egal, S., Turner, P.C., Hall, A.J. and Wild, C.P. (2002) Dietary aflatoxin exposure and impaired growth in young children from Benin and Togo: cross sectional study. *British Medical Journal* 325, 20-21.
Hell K., Cardwell, K.F., Setamou, M. and Poehling, H.M. (2000) The influence of storage practices on aflatoxin contamination in maize in four agro-ecological zones of Benin, West Africa. *Journal of Stored Products Research* 36, 365-382.
Hudson, G.J., Wild, C.P., Zarba, A. and Groopman, J.D. (1992) Aflatoxins isolated by immunoaffinity chromatography from foods consumed in The Gambia, West Africa. *Natural Toxins* 1, 100-105.
Ibeh, I.N., Uraih, N. and Ogonor, J.I. (1991) Dietary exposure to aflatoxin in Benin City, Nigeria: A possible public health concern. *International Journal of Food Microbiology* 14, 171-174.
IARC (International Agency for Research on Cancer). (1987) *Overall evaluations of carcinogenicity: An updating of IARC Monographs volumes 1 to 42.* Report of an IARC Expert Committee, Lyon, (IARC Monographs on the Evaluation of Carcinogenic Risks in Humans, Supplement 7).
IARC (International Agency for Research on Cancer). (1993) Aflatoxins. *IARC Monographs on the Evaluation of Carcinogenic Risks to Humans: Some Naturally Occurring Substances: Food Items and Constituents, Heterocyclic Aromatic Amines and Mycotoxins* 56, 245-395.
IARC. (2002) Fumonisin B_1. *IARC Monographs on the Evaluation of Carcinogenic Risks to Humans: Some Traditional Medicines, Some Mycotoxins, Naphthalene and Styrene* 82, 301-366.
Jonsyn, F.E. (1988) Seedborne fungi of sesame (*Sesamum indicum* L) in Sierra Leone and their potential aflatoxin/mycotoxin production. *Mycopathologia* 104, 123-127.
Jonsyn, F.E. (1990) Mycological contamination of ogiri (fermented sesame seeds) during production in Sierra Leone. *Mycopathologia* 110, 113-117.
Jonsyn, F.E. (1992) Mycotoxic flora and mycotoxins in smoke-dried fish from Sierra Leone. *Nahrung* 36, 485-489.
Jonsyn, F.E. (1999) Intake of aflatoxins and ochratoxins by infants in Sierra Leone: Possible effects on the general health of these children. *Journal of Nutritional and Environmental Medicine* 9, 15-22.
Kmet, J., and Mahboubi, E. (1972) Esophageal cancer in the Caspian Littoral of Iran: Initial studies. *Science* 175, 846-853.
Kpodo, K.A. (1995) Present status of research on the aflatoxin problem in groundnut in Ghana. *Arachide Infos* 6, 5.
Kpodo, K.A. (1997) Final Report of the Food Research Institute (FRI)/Texas A & M Peanut Collaborative Research Support Program (Peanut CRSP) Project titled "Mycotoxin Management in Peanut by Prevention of Contamination and Monitoring".
Kpodo, K.A. (2001) *Fusaria and Fumonisins in Maize and Fermented Maize Products in Ghana.* Ph.D. Thesis. University of Ghana, Legon, Ghana.
Kpodo, K.A. (2005) Incidence and levels of mycotoxin contamination of agricultural commodities in Ghana. *Book of Abstracts, Regional Workshop on Mycotoxins organised by the National Agency for Food and Drug Administration and Control (NAFDAC), Nigeria, 7th – 11th February 2005*, p. 16.
Kpodo, K.A. and Halm, M. (1990) *Fungal and Aflatoxin Contamination of Maize Stored in Silos and Warehouses in Ghana.* Food Research Institute Project Report. Food Research Institute, Accra, Ghana.

Kpodo, K.A., Sørensen, A.K. and Jakobsen, M. (1996) The occurrence of mycotoxins in fermented maize products. *Food Chemistry* 56, 147-153.

Kpodo, K.A., Thrane, U. and Hald, B. (2000) *Fusaria* and fumonisins in maize from Ghana and their co-occurrence with aflatoxins. *International Journal of Food Microbiology* 61, 147-157.

Marín, S., Albareda, X., Ramos, A.J. and Sanchis, V. (2001) Impact of environment and interactions of *Fusarium verticillioides* and *Fusarium proliferatum* with *Aspergillus parasiticus* on fumonisin B_1 and aflatoxins on maize grain. *Journal of the Science of Food and Agriculture* 11, 1060-1068.

Marín, S., Magan, N., Ramos, A.J. and Sanchis, V. (2004) Fumonisin-producing strains of *Fusarium*: A review of their ecophysiology. *Journal of Food Protection* 67, 1792-1805.

Maxwell, S.M., Apeagyei, F., de Vries, H.R., Mwanmut, D.D. and Hendrickse, R.G. (1989) Aflatoxins in breast milk, neonatal cord and sera of pregnant women. *Journal of Toxicology – Toxin Reviews* 8, 19-29.

Mestres, C., Bassa, S., Fagbohoun, E., Nago, M., Hell, K., Vernia, P., Champiat, D., Hounhouigan, J. and Cardwell, K. (2004) Yam chip food sub-sector: hazardous practices and presence of aflatoxins in Benin. *Journal of Stored Products Research* 40, 575-585.

Miele, M., Donato, F., Hall, A.J., Whittle, H., Chapot, B., Bonatti, S., de Ferrari, M., Artuso, M., Gallerina, E., Abbondandolo, A., Montesano, R. and Wild, C. (1996) Aflatoxin exposure and cytogenetic alterations in individuals from the Gambia, West Africa. *Mutation Research* 349, 209-217.

Mintah, S. and Hunter, R.B. (1979) The incidence of aflatoxin found in groundnuts (*Arachis hypogea* L.) purchased from markets in and around Accra, Ghana. *Peanut Science* 5:13-16.

Munimbazi, C. and Bullerman, L.B. (1996) Molds and mycotoxins in foods from Burundi. *Journal of Food Protection* 59, 869-875.

Ndiaye, B., Diop, Y.M., Diouf, A., Fall, M., Thiaw, C., Thiam, A., Barry, O., Ciss, M. and Ba, D. (1999) Measurement and levels of aflatoxins in small-scale pressed peanut oil prepared in the Diourbel and Kaolack regions of Senegal. *Dakar Medicine* 44, 202-205.

Ngoko, Z., Marasas, W.F.O., Rheeder, J.P., Shephard, G.S., Wingfield, M.J. and Cardwell, K.F. (2001) Fungal infection and mycotoxin contamination of maize in the humid forest and western highlands of Cameroon. *Phytoparasitica* 29, 352-360.

Nikiema, P.N., Worrillow, L., Traore, A.S., Wild, C.P. and Turner, P.C. (2004) Fumonisin contamination of maize in Burkina Faso. *Food Additives and Contaminants* 21, 865-870.

Nwokolo, C. and Okonkwo, P. (1978) Aflatoxin load of common food in savanna and forest regions of Nigeria. *Transactions of the Royal Society of Tropical Medicine and Hygiene* 72, 329-332.

Okoye, Z.S.C. (1986) Zearalenone in native cereal beer brewed in Jos metropolis of Nigeria. *Journal of Food Safety* 7, 233-239.

Okoye, Z.S.C. and Ekpenyong, K.I. (1984) Aflatoxin B_1 in native millet beer brewed in Jos suburbs. *Transactions of the Royal Society of Tropical Medicine and Hygiene* 78, 417-418.

Opadokun, J.S., and Ikeorah, J.N.. (1983) Moisture and aflatoxin contents of market grain samples in Kano and Plateau states of Nigeria. *Nigerian Stored Products Research Institute Technical Report* No. 3, 35-41.

Oyelami, O.A., Maxwell, S.M. and Adeoba, E. (1996) Aflatoxins and ochratoxin A in the weaning food of Nigerian children. *Annals of Tropical Paediatrics* 16, 137-140.

Peraica, M., Radić, B., Lucić, A. and Pavlović, M. (1999) Toxic effect of mycotoxins in humans. *Bulletin of the World Health Organization (The International Journal of Public Health)* 77, 754-766.

Pollet, A., Declert, C., Wiegandt, W., Harkema, J. and de Lisdonk, E. (1989) Traditional groundnut storage and aflatoxin problems in Côte d'Ivoire: Ecological approaches. *Proceedings of the International Workshop on Aflatoxin Contamination of Groundnut (ICRISAT Center, India, 6th – 9th October, 1987)*, ICRISAT, Patancheru, India. pp. 263-268.

Rheeder, J.P., Marasas, W.F.O., Thiel, P.G., Sydenham, E.W., Shepard, G.S. and van Schalkwyk, D.J. (1992) *Fusarium moniliforme* and fumonisins in corn in relation to human esophageal cancer in Transkei. *Phytopathology* 82, 353-357.

Saenz de Rodriguez ,C.A., Bongiovanni, A.M. and Conde de Borrego, L. (1985) An epidemic of precocious development in Puerto Rican children. *Journal of Pediatrics* 107, 393-396.

Seifert, K.A., T. Aoki, T., Baayen, R.P., Brayford, D., Burgess, L.W., Chulze, S., Gams, W., Geiser, D., de Gruyter, J., Leslie, J.F., Logrieco, A., Marasas, W.F.O., Nirenberg, H.I., O'Donnell, K.,

Rheeder, J.P., Samuels, G.J., Summerell, B.A., Thrane, U. and Waalwijk, C. (2003) The name *Fusarium moniliforme* should no longer be used. *Mycological Research* 107, 643-644.
Setamou, M., Cardwell, K.F., Schulthess, F. and Hell, K. (1997) *Aspergillus flavus* infection and aflatoxin contamination of preharvest maize in Benin. *Plant Disease* 81, 1323-1327.
Shephard, G.S., Marasas, W.F.O., Leggott, N.L., Yazdanpanah, H., Rahimian, H. and Safavi, N. (2000) Natural occurrence of fumonisins in corn from Iran. *Journal of Agricultural and Food Chemistry* 48, 1860-1864.
Shotwell, O.L., Hesseltine, C.W., Vandegraft, E.E. and Goulden, M.L. (1971) Survey of corn from different regions for aflatoxins, ochratoxin, and zearalenone. *Cereal Science Today* 16, 266-268.
Shotwell, O.L., Goulden, M.L. and Hesseltine, C.W. (1976) Survey of US wheat for ochratoxin and aflatoxin. *Journal of the AOAC* 59, 122-124.
Sweeney, M.J. and Dobson, A.D.W. (1998) Mycotoxin production by *Aspergillus*, *Fusarium*, and *Pencillium* species. *International Journal of Food Microbiology* 43,141-158.
Turner, P.C., Mendy, M., Whittle, H., Fortuin, M., Hall, A.J. and Wild, C.P. (2000) Hepatitis B infection and aflatoxin biomarker levels in Gambian children. *Tropical Medicine and International Health* 5, 837-841.
Udoh, J. (1997) *Production, Harvest and Storage Practices of Maize in Agroecological Zones of Nigeria as They Affect Aflatoxin Content of the Grains*. Ph.D Thesis, University of Ibadan, Nigeria.
Ueno, Y., Nagata, S., Tsutsumi, T., Hasega, A. Watanabe, F. and Park, H. (1996) Detection of microcystins, a blue-green algal hepatotoxin, in drinking water sampled in Haimen and Fusui, endemic areas of primary liver cancer in China, by highly sensitive immunoassay. *Carcinogenesis* 17, 1317-1321.
van Rensburg, S.J. (1985) Recent studies on the aetiology of oesophageal cancer. *South African Cancer Bulletin* 29, 22-31.
Waering, P.W., Westby, A., Gibbs, J.A., Allotey, L.T. and Halm, M. (2001) Consumer preferences and fungal and mycotoxin contamination of dried cassava products from Ghana. *International Journal of Food Science and Technology* 36, 1-10.
Waliyar, F. and Hassan, H. (1993) Aflatoxin contamination of groundnut in Niger. In Waliyar, F., Ntare, B.R. and Williams, J.H. (eds.) *Summary Proceedings of the Third ICRISAT Regional Groundnut Meeting for West Africa (14 -17 September 1992, Ouagadougou, Burkina Faso)*. ICRISAT, Patancheru, India. , pp. 54-55.
WHO. (1990) *Selected Mycotoxins: Ochratoxins, Trichothecenes, Ergot*. Report of an Expert Committee, Geneva, Environmental Health Criteria No. 105. WHO, New York.
WHO. (2002) *Evaluation of certain mycotoxins in foods* (Fifty-sixth report of the Joint FAO/WHO Expert Committee on Food Additives) Technical Report Series, No. 906. WHO, New York.
Wild, C.P. and Turner, P.C. (2002) The toxicology of aflatoxins as a basis for public health decisions. *Mutagenesis* 17, 471-481.
Wild, C.P., Rasheed, F.N., Jawla, M.F.B., Hall, A.J., Jansen, L.A.M. and Montesano, R. (1991) *In-utero* exposure to aflatoxins in West Africa. *Lancet* 337, 1602.
Wild, C.P., Hudson, G.J., Sabbioni, G., Chapot, B., Hall, A., Wogan, G.N., Whittle, H., Montesano, R. and Groopman, J.D. (1992) Dietary intake of aflatoxins and the level of albumin-bound aflatoxin in peripheral blood in The Gambia, West Africa. *Cancer Epidemiology Biomarkers and Prevention* 1, 229-234.
Wilkinson, A.P., Denning, D.W. and Morgan, M.R.A. (1989) Immunoassay of aflatoxin in food and human tissue. *Toxin Reviews* 8, 69-79.
Williams, J.H., Phillips, T.D., Jolly, P., Styles, J.K., Jolly, C.M. and Aggarwal, D. (2004) Human aflatoxicosis in developing countries: a review of toxicology, exposure, potential health consequences, and interventions. *American Journal of Clinical Nutrition* 80, 1106-1122.
Wyers, M., Mobio, M.G., Schricke, E. and Guetta, A.N. (1991) Recherche dans trios élevages industriels de poules en Côte-d'Ivoire des lesions de l'intoxication par l'aflatoxine B_1. *Revue d'élevage et de Médecine Vétérinaire des pays Tropicaux* 44, 15-21.
Yang, C.S. (1980) Research on esophageal cancer in China: a review. Cancer *Research* 40, 2633-2644.
Zarba, A., Wild, C.P., Hall, A.J., Montesano, R., Hudson, G.J. and Groopman, J.D. (1992) Aflatoxin M_1 in human breast milk from the Gambia, West Africa, quantified by combined monoclonal antibody immunoaffinity chromatography and HPLC. *Carcinogenesis* 13, 891-894.

10

Mycotoxin Contamination in Food Systems in Eastern and Southern Africa

Bupe A. Siame* and Ireen N. Nawa

Abstract

Mycotoxin contamination of food systems is a major problem in many countries in Eastern and Southern Africa. Most diets in the region are maize based, but cassava, sorghum, millet and peanuts also are important. These food systems are susceptible to aflatoxins, fumonisin B_1, zearalenone and deoxynivalenol contamination. Aflatoxins have been found in foods from Botswana, Ethiopia, Kenya, Lesotho, Malawi, South Africa, Swaziland, Tanzania, Uganda, Zambia, and Zimbabwe. Maize and maize products from the region also are almost always contaminated by fumonisin B_1. Other mycotoxins, such as zearalenone and deoxynivalenol also have been reported in various food commodities and beer.

Introduction

Maize, sorghum and cassava comprise the major components of the human diet in Eastern and Southern Africa. Based on data from the U.N. Food and Agriculture Organization (FAO), the daily maize intake is > 200 g/person/day in Eastern and Southern African countries. The highest daily intakes of 469, 419, and 331 g/person/day have been reported in Malawi, Zambia, and Zimbabwe respectively (FAO, 1992). Other crops such as wheat, rice, millet, and peanuts also are eaten, but to a lesser extent. These crops are susceptible to fungal infestation, with *Fusarium*, *Penicillium* and *Aspergillus* being the fungal genera found most commonly on these crops. These fungi can attack the crop while it is still in the field or during storage. *Fusarium* spp. usually are problematic in the field, whereas *Penicillium* spp. and *Aspergillus* spp. usually are more problematic during storage (Miller, 1995). Apart from general food spoilage, fungi growing on foods may produce mycotoxins. Although many mycotoxins are produced by fungi, aflatoxins, fumonisin B_1, deoxynivalenol, and zearalenone are the most commonly reported mycotoxins from countries in Eastern and Southern Africa.

Occurrence and toxicity of aflatoxins

In Eastern and Southern Africa, as in most developing countries, aflatoxins are the most important mycotoxins from the point of view of occurrence, toxicity and economy. The aflatoxins are produced primarily by strains of *Aspergillus flavus* and *Aspergillus parasiticus*. The four major aflatoxins commonly reported in foods are aflatoxin B_1, B_2, G_1, and G_2. In grains, aflatoxins are primarily a problem in maize although other grains such as sorghum, millet, and wheat also may be contaminated (Table 1). Other products such as peanuts, peanut butter, figs, dried fruits, cottonseed, spices, milk, and opaque beer also may be contaminated. However, it is not so much the type of food contaminated, as it is the amount of food ingested and its level of contamination that may lead to toxicity.

In Southern Africa, aflatoxins are not a major contaminant of maize and levels usually are < 20 ng/g. Several surveys of commercial maize in South Africa have found very low incidence of aflatoxins in maize (Viljoen *et al.*, 1993). Similar low levels of aflatoxins in commercial maize in Botswana and Swaziland have been reported (Table 1), although higher levels have been found in some maize products from Zambia (Lovelace and Nyathi, 1977). Peanut and peanut butter samples from Southern Africa usually are more heavily contaminated with aflatoxins than maize, and levels > 20 ng/g are common (Table 1). In an analysis of peanuts and peanut butter for sale in Botswana supermarkets, the total aflatoxin concentration ranged from 3-64 ng/g (Mphande *et al.*, 2004). In South Africa, a recent survey of peanut butter used in a national feeding scheme for school children found that 42% of the samples were contaminated with aflatoxins at levels as high as 470 ng/g (National Monitoring Programme, 2004).

Aflatoxins also have been reported in food samples from Eastern Africa at extremely high levels (Table 1). For example, aflatoxins levels of 1,700 ng/g were reported in cassava from Uganda and levels as high as 46,000 ng/g were reported in maize from Kenya (Lewis *et al.*, 2005). Aflatoxin levels ranging between 100 and 525 ng/g also have been reported in red pepper and mixed legumes from Ethiopia (Fufa and Urga, 1996).

Aflatoxins are lethal when consumed in large doses and fatalities have been reported in Uganda and Kenya. In Uganda, a 15-year old boy died two days after hospitalization with a subsequent examination of foodstuffs in the boy's home finding moldy cassava containing aflatoxins at up to 1,700 ng/g (Alpert *et al.*, 1971). An outbreak of aflatoxicosis was reported in 1982 in Kenya where 12 people died after consuming maize samples contaminated with aflatoxin B_1 at levels ranging between 3,200 and 12,000 ng/g (Ngindu *et al.*, 1982). In a more recent outbreak of aflatoxicosis in Kenya, over 100 people died after consuming maize contaminated with aflatoxins at up to 46,000 ng/g (Lewis *et al.*, 2005).

Aflatoxins are potent carcinogens and mutagens. Naturally occurring mixtures of aflatoxins are classified as Class 1 human carcinogens (IARC, 1993). There is a correlation between the incidence of liver cancer in humans in some areas of Africa and dietary exposure to aflatoxins. Studies in Kenya, Mozambique, Swaziland, and South Africa have found that aflatoxin levels in the diet and the incidence of primary liver cancer are correlated (Groopman *et al.*, 1988). Aflatoxin consumption also has been implicated in some infant diseases such as kwashiorkor, a form of protein malnutrition (Hendrickse, 1984), and protein-deficient diets may increase aflatoxin toxicity (Hendrickse *et al.*, 1982).

The aflatoxin problem is not only a health risk but also can result in monetary losses to farmers when contaminated produce is rejected by export markets (Wu, 2004). The marketability of the contaminated produce, particularly in international trade also is considerably reduced due to stringent limits set by importing countries. In many developing countries, there is little, if any, regulation of aflatoxin contamination in foods at the local level, hence the population is always at risk. The *Codex Alimentarius* Commission, supported by FAO and WHO, set a standard of 15 ng/g total aflatoxins in unprocessed peanuts. However, in 1998, the European Union (EU) set a tolerance level of 2 ng/g for aflatoxin B_1 and 4 ng/g for total aflatoxins in nuts and cereals meant for human consumption (Dimanche, 2001). To take advantage of the growing export market, farmers must use good agricultural practices and innovative methods, *e.g.*, forming farmers' association and using low-cost technology

Table 1. Occurrence of aflatoxins (as aflatoxin B_1) in food commodities reported from several countries from Eastern and Southern Africa.

Country	Commodity	Conc. Range (ng/g)	Reference
Botswana	Maize	Not Detected	Siame *et al.*, 1998
	Opaque beer	Not Detected	Nkwe *et al.*, 2005
	Peanut butter	12-330	Mphande *et al.*, 2004
	Peanuts	3-64	Mphande *et al.*, 2004
Burundi	Sorghum	Not Detected	Munimbazi and Bullerman, 1996
Ethiopia	Red pepper	250-530	Fufa and Urga, 1996
	Shiro (legume)	100-500	Fufa and Urga, 1996
Kenya	Maize	500-12,000	Ngindu *et al.*, 1982
	Maize	1-46,000	Lewis *et al.*, 2005
	Maize meal	0.4-20	Muriuki and Siboe, 1995
Mozambique	Peanuts	1.2-5.5	Conzane *et al.*, 2002
South Africa	Maize	< 5	Viljoen *et al.*, 1993
	Opaque beer	0-400 µg/l	Odhav and Naicker, 2002
	Peanut butter	10-470	National Monitoring Program, 2004
	Peanuts	10-510	National Monitoring Program, 2004
Swaziland	Maize meal	1-16	Peers *et al.*, 1987
	Peanuts	12-130	Martin and Gilman, 1976
Tanzania	Fish	7-19	Mugula and Lyimo, 1992
	Opaque beer	10-50 µg/l	Nikander *et al.*, 1991
Uganda	Cassava	100-1,700	Serck-Hanssen, 1970
	Maize	1-1,000	Kaaya and Warren, 2005
Zambia	Opaque Beer	120-170 µg/l	Lovelace and Nyathi, 1977

to determine aflatoxin levels in their products; Waliyar *et al.*, Chapter 31) to reduce aflatoxin contamination to a level that meets export standards.

Occurrence and toxicity of fumonisins

The fumonisins are a group of toxic metabolites produced mainly by the fungus *Fusarium verticillioides*. *Fusarium verticillioides* is the most frequently isolated fungal species from virtually all maize kernels, including those that look healthy and are intended for human consumption (Alberts *et al.*, 1990). Furthermore, fumonisins have been detected in maize and maize based foods marketed in many countries worldwide including several countries in Eastern and Southern Africa (Table 2). The presence of toxigenic *Fusarium* spp. with a prevalence of *F. verticillioides* also has been reported from millet and sorghum from South Africa, Botswana, Lesotho, Zimbabwe, and Kenya (Onyike *et al.*, 1991; Onyike and Nelson, 1993). Whether all of these strains are fumonisin producers, however, needs further investigation given the recognition that *Fusarium moniliforme*, the species to which many of these strains were identified, is a species complex. The *F. moniliforme* name should no longer be used (Seifert *et al.*, 2003) and the identity of many of the strains, especially those not from maize, may not be *F. verticillioides*, the species that usually dominates on maize and may not produce any significant amount of fumonisin (Leslie *et al.*, 2005).

Table 2. Occurrence of fumonisin B_1 in cereals and cereal-based food commodities reported in several countries from Eastern and Southern Africa.

Country	Commodity	Conc. Range (ng/g)	Reference
Botswana	Maize	0-350	Doko *et al.*, 1996
	Maize meal	20-1,300	Siame *et al.*, 1998
	Sorghum meal	20-60	Siame *et al.*, 1998
	Opaque beer	46-1,300 µg/l	Nkwe *et al.*, 2005
Burundi	Maize	12,000-75,000	Munimbazi and Bullerman, 1996
Kenya	Maize	100-12,000	Kedera *et al.*, 1999
Malawi	Maize	20-140	Doko *et al.*, 1996
Mozambique	Maize	340-400	Doko *et al.*, 1996
South Africa	Maize	0-7,900	Rheeder *et al.*, 1992
	Maize meal	0-2,800	Sydenham *et al.*, 1991
	Moldy maize	110-120,000	Sydenham *et al.*, 1990
Tanzania	Maize	25-225	Doko *et al.*, 1996
Uganda	Maize	845	Doko *et al.*, 1996
Zambia	Maize	70-1,200	Doko *et al.*, 1996
Zimbabwe	Maize	55-2,700	Doko *et al.*, 1996

Although several fumonisins have been reported to occur naturally, fumonisin B_1 is the major one found in maize and maize-based foods. Most work on the occurrence of fumonisins in Africa has been done in South Africa, where the fumonisins were first isolated and characterized (Gelderblom *et al.*, 1988). However, there are several reports of fumonisin contamination of maize and other food products from several countries in Southern and Eastern Africa (Table 2). In a limited survey of cereal and cereal-based food commodities from Eastern and Southern Africa, fumonisins were detected in 93% of the maize samples at levels ranging from 20-2,700 µg/g (Doko *et al.*, 1996).

Fumonisins have cancer-promoting activity in rats and may play an important role in human carcinogenesis. In the Transkei region of South Africa, where maize is a staple human food, consumption of fumonisin-contaminated maize is linked to esophageal cancer (Sydenham *et al.*, 1990). In maize being used primarily to brew beer, the fumonisin B_1 levels could exceed 100,000 ng/g. Fumonisins also have been associated with neural tube defects and other birth defects among populations consuming fumonisin-contaminated maize (Marasas *et al.*, 2004). Maize consumed as such or in its different processed forms remains a staple food for people in Eastern and Southern Africa and the prevalence of fumonisins in these commodities should be a major concern for these consumers.

Occurrence and toxicity of deoxynivalenol

Trichothecene mycotoxins (Desjardins, 2006) are produced by several genera of fungi, with the most important naturally occurring ones being those produced by *Fusarium* spp. The trichothecenes of major concern are deoxynivalenol, nivalenol, and T-2 toxin. In Africa, most work on trichothecenes has been with deoxynivalenol. Deoxynivalenol is produced by several species of *Fusarium* including *Fusarium graminearum*. This toxin has been reported in samples from Kenya, South Africa and Zambia (Table 3).

Deoxynivalenol is a potent antifeedant and induces vomiting in pigs at very low concentrations; it also is immunosuppressive (Moss, 2000). General signs of trichothecene toxicity in experimental animals include weight loss, decreased feed conversion, feed refusal,

Table 3. Occurrence of deoxynivalenol in cereals and cereal-based food commodities reported in several countries from Eastern and Southern Africa.

Country	Commodity	Conc. Range (ng/g)	Reference
Botswana	Maize meal	Not Detected	Doko *et al.*, 1996
	Sorghum meal	Not Detected	Doko *et al.*, 1996
Kenya	Lager beer	1.6-6.4 µg/l	Mbugua and Gathumbi, 2004
South Africa	Wheat (under grade)	3,800-14,000	Sydenham *et al.*, 1991
	Moldy maize	ND-16,000	Sydenham *et al.*, 1991
	Maize	0-820	Thiel *et al.*, 1982
Southern Africa	Maize	1,000-4,000	Marasas *et al.*, 1979
Zambia	Maize	1000	Siame and Lovelace, 1989
	Moldy maize	500-16,000	Siame and Lovelace, 1989
Zimbabwe	Maize	Not Detected	Doko *et al.*, 1996

vomiting, severe dermatitis, hemorrhage, abortion and in some cases death. However, there are no reported cases of human toxicity due to deoxynivalenol in Eastern or Southern Africa.

Occurrence and toxicity of zearalenone

Zearalenone also is produced by *Fusarium* spp., most notably *F. graminearum*. This mycotoxin has been detected in food commodities from several in Eastern and Southern Africa (Table 4). Concentrations of up to 10 µg/g have been reported in samples originating from the Transkei region of South Africa (Marasas *et al.*, 1979).

The major effects of zearalenone are estrogenic and mostly affect the urogenital system. In experimental animals the effects are similar to those produced by excessive steroidal estrogen intake, a condition known as hyperestrogenism (Nelson *et al.*, 1994; Desjardins, 2006). This condition can lead to the development of female characteristics in prepubescent male pigs. Zearalenone has been implicated in some incidents of precocious puberty changes in children (Saenz de Rodriguez, 1984; Kuiper-Goodman, 1987).

Conclusions

Mycotoxins can have adverse effects on humans that consume contaminated foods and there is constant need for surveillance of foods for the level of mycotoxins present. Lack of enforceable regulations of maximum tolerance levels (MTL) also is a problem in many countries in Eastern and Southern Africa. Only Kenya, Malawi, Mozambique, Tanzania, South Africa, and Zimbabwe have official regulations on aflatoxins in foods (FAO, 2004). Most of the countries in the region need to develop regulations on acceptable levels of mycotoxins, particularly aflatoxins, in human food. Information on the regulation of other mycotoxins is lacking and most probably does not exist in most countries in the region. The mycotoxin issue in most of these countries needs to be viewed in the context of food safety, health and agricultural issues (Shephard, 2004). The establishment of mycotoxin regula-

Table 4. Occurrence of zearalenone in cereals and cereal-based food commodities reported in several countries from Eastern and Southern Africa.

Country	Commodity	Conc. Range (ng/g)	Reference
Botswana	Sorghum meal	Not Detected - 100	Doko *et al.*, 1996
	Maize meal	Not Detected - 40	Siame *et al.*, 1998
	Sorghum malt	100-2,200	Nkwe *et al.*, 2005
	Opaque beer	20-200 μg/l	Nkwe *et al.*, 2005
Kenya	Maize	40-80	Doko *et al.*, 1996
	Maize meal	2,500-5,000	Muriuki and Siboe, 1995
	Lager beer	0-10 μg/l	Mbugua and Gathumbi, 2004
Lesotho	Maize and sorghum malt	300-2,000	Martin and Keen, 1978
South Africa	Maize	1,500-10,000	Marasas *et al.*, 1979
	Opaque beer	3-430 μg/l	Odhav and Naicker, 2002
Tanzania	Maize	25-230	Doko *et al.*, 1996
Uganda	Maize	Not Detected	Doko *et al.*, 1996
Zambia	Moldy maize	80-6,000	Siame and Lovelace, 1989
	Maize meal	50-600	Siame and Lovelace, 1989
	Opaque beer	90-4,600 μg/l	Lovelace and Nyathi, 1977
Zimbabwe	Maize	0-1,100	Doko *et al.*, 1996

tions will have limited effects in terms of health protection because farmers in most of these countries grow crops only at a subsistence level, *i.e.,* for their own consumption. When crops do enter official marketing channels, there is little enforcement of the existing regulations. Exposure assessment information also is lacking in many African countries. As regulations alone will not offer health protection in many of these countries, there is a need for simple technologies to reduce fungal and mycotoxin contamination of food. Governments in Southern and Eastern Africa also need to be more proactive in protecting their citizens against mycotoxin contaminated foods.

References

Alberts, J.F., Gelderblom, W.C.A., Thiel, P.G., Marasas, W.F.O., van Schalkwyk, D.J. and Behrend, Y. (1990) Effects of temperature and incubation period on production of fumonisin B_1 by *Fusarium moniliforme. Applied and Environmental Microbiology* 56, 1729-1733.

Alpert, M.E., Hutt, M.S.R., Wogan, G.N. and Davison, C.S. (1971) Association between aflatoxin content of food and hepatoma frequency in Uganda. *Cancer* 28, 253-260.

Conzane, R.S., Stenzel, W.R. and Kroh, L.W. (2002) Detection and determination of aflatoxins B_1, B_2, G_1 and G_2 in peanuts from Mozambique using HPLC. *Deutsche Lebensmittel-Rundschhau* 98, 289-295.

Desjardins, A.E. (2006) Fusarium *mycotoxins: Chemistry, Genetics and Biology*. APS Press, St. Paul, Minnesota.

Dimanche, P. (2001) Groundnut exporters in Southern countries penalized by new standards on aflatoxins imposed by the European Union. *OCL-Oleagineux, Corps Gras, Lipides* 8, 237-238.

Doko, M.B., Canet, C., Brown, N., Sydenham, E.W., Mpuchane, S. and Siame, B.A. (1996) Natural co-occurrence of fumonisins and zearalenone in cereals and cereal-based foods from Eastern and Southern Africa. *Journal of Agricultural and Food Chemistry* 44, 3240-3243.

FAO. (1992) *Maize in Human Nutrition. FAO Food and Nutrition Series No. 25.* FAO, Rome, Italy.
FAO. (2004) *Worldwide Regulations for Mycotoxins in Food and Feed in 2003. FAO Food and Nutrition Paper 81.* FAO, Rome, Italy.
Fufa, H. and Urga, K. (1996) Screening of aflatoxin in Shiro and ground red pepper in Addis Ababa. *Ethiopian Medical Journal* 34, 243-249.
Gelderblom, W.C.A., Jaskiewicz, K., Marasas, W.F.O., Thiel, P.G., Horak, R.M., Vleggaar, R. and Kriek, N.P.J. (1988) Fumonisins – Novel mycotoxins with cancer-promoting activity produced by *Fusarium moniliforme. Applied and Environmental Microbiology* 54, 1806–1811.
Gelderblom, W.C.A., Kriek, N.P.J., Marasas, W.F.O. and Thiel, P.G. (1991) Toxicity and carcinogenicity of *Fusarium moniliforme* metabolite, fumonisin B_1 in rats. *Carcinogenesis* 12, 1247-1251.
Groopman, J.D., Cain, L.G. and Kensler, T.W. (1988) Aflatoxin exposure in human populations: Measurements and relationship to cancer. *CRC Critical Reviews of Toxicology* 19, 113-145.
Hendrickse, R.G. (1984) The influence of aflatoxins on child health in the tropics with particular reference to kwashiorkor. *Transactions of the Royal Society of Tropical Medicine and Hygiene* 78, 427-435.
Hendrickse, R.G., Coulter, J.B.S. and Lamplugh, S.M. (1982) Aflatoxin and kwashiorkor: A study in Sudanese children. *British Medical Journal* 285, 843-846.
IARC (International Agency for Research on Cancer). (1993) Aflatoxins. *IARC Monographs on the Evaluation of Carcinogenic Risks to Humans: Some Naturally Occurring Substances: Food Items and Constituents, Heterocyclic Aromatic Amines and Mycotoxins.* 56, 245-395.
Kaaya, N. A. and Warren, H.L. (2005) A review of past and present research on aflatoxin in Uganda. *African Journal of Food, Agriculture and Nutritional Development* 5, 1-18.
Kedera, C.J., Platter, R.D. and A. E. Desjardins, A.E. (1999) Incidence of *Fusarium* spp. and levels of fumonisin B_1 in maize in western Kenya. *Applied and Environmental Microbiology* 65, 41-44.
Kuiper-Goodman, T., Scott, P.M. and Watanabe, H. (1987) Risk assessment of the mycotoxin zearalenone. *Regulatory Toxicology and Phamarcology* 7, 253-306.
Leslie, J.F., Zeller, K.A., Lamprecht, S.C., Rheeder, J.P. and Marasas, W.F.O. (2005) Toxicity, pathogenicity and genetic differentiation of five species of *Fusarium* from sorghum and millet. *Phytopathology* 95, 275-283.
Lewis, L., Onsongo, M., Njapau, H., Schurz-Rogers, H., Luber, G., Kieszak, S., Nyamongo, J., Backer, L., Dahiye, A.M., Misore, A., DeCock, K., Rubin, C. and the Kenya Aflatoxicosis Investigation Group. (2005) Aflatoxin contamination of commercial maize products during an outbreak of acute aflatoxicosis in Eastern and Central Kenya. *Environmental Health Perspectives* 113, 1763-1767.
Lovelace, C.E.A. and Nyathi, C.B. (1977) Estimation of the fungal toxins zearalenone and aflatoxin contaminating opaque maize beer in Zambia. *Journal of the Science of Food & Agriculture* 28, 288-292.
Marasas, W.F.O., Riley, R.T., Hendricks, K.A., Stevens, V.L., Sadler, T.W., Gelineau-van Waes, J., Missmer, S.A., Cabrera, J., Torres, O., Gelderblom, W.C.A., Allegood, J., Martínez, C., Maddox, J., Miller, J.D., Starr, L.M., Sullards, C., Roman, A.V., Voss, K.A., Wang, E. and Merrill, A.H., Jr. (2004) Fumonisins disrupt sphingolipid metabolism, folate transport, and neural tube development in embryo culture and *in vivo*: A potential risk factor for human neural tube defects among populations consuming fumonisin-contaminated maize. *Journal of Nutrition* 134, 711-716.
Marasas, W.F.O., van Rensburg, S.J. and Mirocha, C.J. (1979) Incidence of *Fusarium* species and the mycotoxins, deoxynivalenol and zearalenone, in corn produced in esophageal cancer areas in Transkei. *Journal of Agricultural and Food Chemistry* 27, 1108-1112.
Martin, P.M.D. and Gilman, G.A. (1976) *A Consideration of the Mycotoxin Hypothesis with a Special Reference to the Mycoflora of Maize, Sorghum, Wheat and Groundnuts.* Tropical Products Institute, London, U.K.
Martin, P.M. and Keen, P. (1978) The occurrence of zearalenone in raw and fermented products from Swaziland and Lesotho. *Sabouraudia* 16, 15-22.
Mbugua, S.K. and Gathumbi, J.K. (2004) The contamination of Kenyan lager beers with *Fusarium* mycotoxins. *Journal of the Institute of Brewing* 110, 227-229.

Miller, J.D. (1995) Fungi and mycotoxins in grain: Implications for stored product research. *Journal of Stored Products Research* 31, 1-16.

Moss, M.O. (2000) Toxigenic fungi and mycotoxins. In: Lund, B.M.., Baird-Parker, T.C. and Gould, G.W. (eds.), *The Microbiological Safety and Quality of Food*, Vol. 2. Aspen, Maryland, USA, pp. 1490-1517.

Mphande, F.A., Siame, B.A. and Taylor, J.E. (2004) Fungi, aflatoxins and cyclopiazonic acid associated with peanut retailing in Botswana. *Journal of Food Protection* 67, 96-102.

Mugula, J.K. and Lyimo, M.H. (1992) Microbiological quality of traditional market cured fish in Tanzania. *Journal of Food Safety* 13, 33-41.

Munimbazi, C. and Bullerman, L.B. (1996) Molds and mycotoxins in foods from Burundi. *Journal of Food Protection* 59, 869-875.

Muriuki, G.K. and Siboe, G.M. (1995) Maize flour contaminated with toxigenic fungi and mycotoxins in Kenya. *African Journal of Health Science* 2, 236-241.

National Monitoring Programme. (2004) Aflatoxins in groundnuts and peanut products. Results of Survey, July 2003- March 2004. Government of the Republic of South Africa, Department of Health, http://www.doh.gov.za/department/foodcontrol/docs/nmp.html.

Nelson, P.E., Dignani, M.C. and Anaaissie, E.J. (1994) Taxonomy, biology, and clinical aspects of *Fusarium* species. *Clinical Microbiology Reviews* 7, 479-504.

Ngindu, A., Johnson, B.K., Kenya, P.R., Ngira, J.A., Ocheng, D.M., Nandwa, H., Omondi, T.N., Jansen, A.J., Ngare, W., Kaviti, J.N., Gatei, D., Siongok, T.A. (1982) Outbreak of acute hepatitis caused by aflatoxins poisoning in Kenya. *Lancet* 319, 1346-1348.

Nikander, P., T. Seppälä, T., Kilonzo, G.P., Huttunen, P., Saarinen, L., Kilima, E. and Pitkänen, T. (1991) Ingredients and contaminants of traditional alcoholic beverages in Tanzania. *Transactions of the Royal Society of Tropical Medicine and Hygiene* 85, 133-135.

Nkwe, D.O., Taylor, J.E. and Siame, B.A. (2005) Fungi, aflatoxins, fumonisin B_1 and zearalenone contaminating sorghum-based traditional malt, wort and beer in Botswana. *Mycopathologia* 160, 177-186.

Odhav, O. and Naicker, V. (2002) Mycotoxins in South African traditionally brewed beer. *Food Additives and Contaminants* 19, 55-61.

Onyike, N.B.N. and Nelson, P.E. (1993) The distribution of *Fusarium* species in soils planted to millet and sorghum in Lesotho, Nigeria, and Zimbabwe. *Mycopathologia* 121, 105-114.

Onyike, N.B.N., Marasas, W.F.O. and Nelson, P.E. (1991) *Fusarium* species associated with millet grain from Nigeria, Lesotho, and Zimbabwe. *Mycologia* 83, 708-712.

Peers, F., Bosch, X., Kaldor, J. Linsell, A. and Pluijment, M. (1987) Aflatoxin exposure, hepatitis B virus infection and liver cancer in Swaziland. *International Journal of Cancer* 39, 545-553.

Rheeder, J.P., Marasas, W.F.O., Thiel, P.G., Sydenham, E.W., Shephard, G.S. and van Schalkwyk, D.J. (1992) *Fusarium moniliforme* and fumonisins in corn in relation to human esophageal cancer in Transkei. *Phytopathology* 82, 353-357

Saenz de Rodriguez, C.A. (1984) Environmental hormone contamination in Puerto Rico. *New England Journal of Medicine* 310, 1741-1742.

Seifert, K.A., T. Aoki, T., Baayen, R.P., Brayford, D., Burgess, L.W., Chulze, S., Gams, W., Geiser, D., de Gruyter, J., Leslie, J.F., Logrieco, A., Marasas, W.F.O., Nirenberg, H.I., O'Donnell, K., Rheeder, J.P., Samuels, G.J., Summerell, B.A., Thrane, U. and Waalwijk, C. (2003) The name *Fusarium moniliforme* should no longer be used. *Mycological Research* 107, 643-644.

Serck-Hanssen, A. (1970) Aflatoxin induced fatal hepatitis? A case report from Uganda. *Archives of Environmental Health* 20, 729-731.

Shephard, G.S. (2004) Mycotoxins worldwide: Current issues in Africa. *In* D. Barug, D., van Egmond, H.P., López-Garciá, R., van Osenbruggen, W.A. and Visconti, A. (eds.), *Meeting the Mycotoxin Menace*. Wageningen Academic Publishers, The Netherlands, pp. 81-88.

Siame, B.A. and Lovelace, C.E.A. (1989) Natural occurrence of zearalenone and trichothecene toxins in maize-based animal feeds in Zambia. *Journal of the Science of Food & Agriculture* 49, 25-35.

Siame, B.A., Mpuchane, S.F., Gashe, B.A., Allotey, J. and Teffera, G. (1998) Occurrence of aflatoxins fumonisin B_1, and zearalenone in foods and feeds in Botswana. *Journal of Food Protection* 61, 1670-1673.

Sydenham, E.W., Shephard, G.S., Thiel, P.G., Marasas, W.F.O. and Stockenstrom, S. (1991) Fumonisin contamination of commercial corn-based human foodstuffs. *Journal of Agricultural and Food Chemistry* 39, 2014-2018.

Sydenham, E.W., Thiel, P.G. Marasas, W.F.O., Shephard, G.S., van Schalkwyk, D.J. and Koch, K.R. (1990) Natural occurrence of some *Fusarium* mycotoxins in corn from low and high esophageal cancer prevalent areas of the Transkei, Southern Africa. *Journal of Agricultural and Food Chemistry* 38, 1900-1903.

Thiel, P.G., Marasas, W.F.O. and Meyer, C.J. (1982) Natural occurrence of *Fusarium* toxins in maize from Transkei. *Proceedings of the Fifth International IUPAC Symposium on Mycotoxins and Phycotoxins (1-3 September, Vienna, Austria)*, 126-129.

Thiel, P.G., Marasas, W.F.O., Sydenham, E.W., Shephard, G.S. and Gelderblom, W.C.A. (1992) The implications of naturally occurring levels of fumonisins in corn for human and animal health. *Mycopathologia* 117, 3-9.

Viljoen, J.H., Marasas, W.F.O. and Thiel, P.G.. (1993) Fungal infection and contamination of commercial maize. In: Taylor, J.R.N., Randall, P.G. and Viljoen, J.H. (eds.), *Cereal Science and Technology: Impact on changing Africa.* CSIR, Pretoria, South Africa, pp. 837-853.

Wu, F. (2004) Mycotoxin risk assessment for the purpose of setting international regulatory standards. *Environmental Science and Technology* 38, 4049-4055.

11

The 2004 and 2005 Aflatoxin Tragedies in Kenya – A Case Study

Mwancha N. Okioma*

Abstract

Between April and September 2004, one of the largest known aflatoxicosis outbreaks occurred in Kenya, with 317 reported cases and 125 deaths. A less severe outbreak occurred in 2005, resulting in 16 deaths. Food samples collected from households in the affected areas contained high levels of aflatoxin B_1 (20 to > 1,000 ng/g), suggesting that the outbreak was caused by acute aflatoxin poisoning. The outbreak resulted from aflatoxin contamination of locally grown maize that was stored under damp conditions.

Introduction

The deleterious effects of mycotoxins were reported as early as 857 AD and very serious epidemics of ergotism occurred from the late 900s to the 1800s, but most Kenyans are neither fully aware of nor appreciative of the magnitude of the dangers that can result from consuming mycotoxin-contaminated food. This ignorance remains even though Kenya has experienced two outbreaks of aflatoxins food poisoning between April 2004 and September 2005, in which 125 and 16 people died, respectively (Nyikal *et al.*, 2004; Azziz-Baumgartner *et al.*, 2005; Lewis *et al.*, 2005). The villages in the affected districts are remote and cases were reported from across a large geographic area. The limited medical facilities in this region probably means that some patients did not reach health facilities for diagnosis and treatment before they died and that the true magnitude of these outbreaks was probably considerably greater and the death toll much higher than has been documented.

The 2004 and 2005 cases were not the first aflatoxicosis outbreaks to be reported in Kenya or in that part of the country. Kenya has previously had sporadic outbreaks of aflatoxicosis, but the earlier outbreaks were more limited in the area and the number of people affected much smaller. The last major reported outbreak of aflatoxicosis in Kenya occurred in 1981 (Ngindu *et al.*, 1982) in which a total of 20 people died. Yet to this day, maize grain harvested from the fields routinely is sorted into clean grain, which is consumed by the family and sold for between 16 and 22 Kenyan shillings per kilo, while moldy and discolored grain is fed to livestock or sold at a price that is about half that for the clean grain. The price per kilogram of both clean and discolored moldy grain depends on the season. During prolonged drought periods, the price of both grades of maize can more than double. Depending on the poverty level of the household, during a severe famine moldy grain may be mixed with clean grain and cooked for consumption by the family.

During the 2004 and 2005 outbreaks, the area involved was large; ~40,000 sq. km covering six districts in the Eastern province (Machakos, Kitui, Mwingi, Makueni, Embu and Mbeere) and one district (Thika) in the Central Province of Kenya, although > 87% of the cases were reported from only four districts (Machakos, Kitui, Thika and Makueni). Among the questions that need answers are:

- What was the extent of the problem? *i.e.*, how widespread is aflatoxin contamination of maize in the distribution market channels?
- What was the source of the contaminated maize? Was it home grown? Was it sourced from other parts of the country? Was it imported?
- What were the risk factors associated with the outbreak?
- What types of toxins and fungal species were involved in the outbreak?

A series of studies were undertaken to answer these questions. The descriptive epidemiology investigations conducted in May 2004 resulted in a hypothesis relating aflatoxin content to the methods of harvesting, preparing and storing the maize grain. Based on this hypothesis two other concurrent studies – an assessment of market maize and case control studies were undertaken (Lewis *et al.*, 2005).

The 2004 and 2005 outbreaks affected mainly young adults, some livestock and domesticated animals. The patients were aged between 1.3 years and 80 years with a mean age of 22.5 years. However, in some instances all of the adult members of one household were affected. There were cases of recurrence in some patients who were discharged back into the community. These recurrent cases indicated that contaminated maize remained present in the local communities.

Challenges and Response

Challenges

The initial epidemiological investigations found that the outbreak posed several challenges:

- There was an upsurge of patients with acute hepatitis of unknown etiology presenting at poorly equipped local hospitals with few, if any, adequately trained personnel.
- There was an apparent widespread contamination of maize grain in the local maize distribution system with more than 55% of maize samples from markets having levels greater than the regulatory limit of 20 ng/g. About 7% of the samples were exceedingly highly contaminated (> 1,000 ng/g; Lewis *et al.*, 2005).
- In addition to the contamination problems, there was a general shortage of maize forcing starving families to consume overtly contaminated maize. Farmers usually sort and sell clean maize with what remains being consumed in the household.
- The community did not as yet associate the high number of deaths with the maize they were consuming. Awareness campaigns were needed to educate the public on the dangers associated with the consumption of contaminated maize.
- The affected area has rough terrain with a poor road network that generates distribution and other logistical problems. Thus, in the affected areas, households will consume more of the locally grown contaminated maize, as what is sold in the local market and consumed may be of poor quality as well.

Responses

The immediate efforts in rapid response focused on:

- Providing support to the local health facilities to strengthen their capacity to manage the large number of admitted cases. This support needs to be followed up with routine training of personnel to familiarize them with the symptoms of aflatoxin poisoning.
- Education of the public on the causes of the prevailing ill health and death.
- Sampling and analysis of grain stocks and the destruction of condemned contaminated grain.
- Replacement of condemned and destroyed grain. This intervention presents enormous logistical problems ranging from aflatoxin testing of a large number of samples, inadequate stocks of replacement grain, to informing vulnerable groups who often live in areas with a poor communication network of the problem, and distributing the available replacement grain over a minimal and rough road network.
- Due the general shortage of maize, some of the grain destined for the affected households was diverted to other, less needy, households.

Maize and its distribution systems

The community in the affected area consists mainly of peasant farmers who practice mixed subsistence farming with some livestock rearing. In the affected areas, as in the rest of the country, maize is the staple food and is the main crop grown in the area. Local maize often is harvested in February before the onset of the rainy season. At harvest most of the farmers store maize for household consumption and sell the rest for cash to meet household needs. When household stocks are exhausted the farmers buy the maize back from the grain traders. The maize is distributed and sold through a network of rural markets. The grain traders often pool small lots from many farmers, and may mix this pool of locally grown maize with maize from outside the region. There are no records of maize sources or movement available at this level of the food chain.

Maize standards and assessment of the maize in the local markets

In the United States and many other parts of the world, aflatoxin concentrations are limited to 20 ng/g (FAO, 2004). Although no formal limits are enforced in Kenya, the 20 ng/g level also has been adopted by Kenyan authorities as the regulatory limit for aflatoxin contamination in grain destined for human consumption and animal feed. An assessment of the maize in the local market system found widespread aflatoxin contamination. Over 55% of the maize had aflatoxin levels > 20 ng/g, and 7% of the maize had levels > 1000 ng/g. These high levels in the marketplace samples were an indication of the significant aflatoxin risk exposure of the population in the affected areas. The aflatoxin levels in the grain in the markets mirrored the geographic distribution of the aflatoxicosis cases.

Case Control Study

The objective of the case control study was to identify the risk factors for developing acute aflatoxicosis, and to quantify aflatoxin exposure in the affected population by measuring the aflatoxin level in maize and the level of aflatoxin adducts in serum. Little is known about risk factors associated with outbreaks of aflatoxicosis in this region of Africa or what protective factors may prevent exposure and/or mitigate disease. In addition, few studies have measured aflatoxin levels in biological samples from clinical cases, because unbound aflatoxin remains in circulation for only 13-90 minutes following exposure (Krishnamachari *et al.*, 1975). Between 1% and 4% of the ingested aflatoxin, however, binds irreversibly with proteins, primarily albumin, to form adducts.

Findings

- In February 2004, the onset of rains in the affected areas was earlier than expected, which forced the farmers to harvest and store maize grain with a high moisture content. Due to the severe shortage of maize, the grain was not stored in a granary (to prevent theft), but instead was stored on the floor of the huts where the families live. The roofs of the huts are thatched with grass. During the rains, some roofs leak which increases the moisture content of the grain on the floor even further. If the moisture content of the grain is > 12%, then the conditions are conducive for growth of the mold that produces aflatoxin.
- The early rains in 2004 were followed by hot humid conditions in March and subsequent months. These conditions formed an ideal environment for the growth of the aflatoxin-producing fungi and the subsequent contamination of the grain with aflatoxins.
- The consumption of contaminated maize was the primary risk for developing aflatoxicosis. Maize taken from households where affected individuals lived had higher aflatoxin contamination levels than did maize obtained from other households where no one was affected by aflatoxicosis.
- Food collected from households in the affected areas contained high levels of aflatoxins, suggesting that the aflatoxicosis outbreak was caused by aflatoxin poisoning.
- Affected individuals had higher serum levels of the aflatoxin B_1-lysine adducts than did unaffected control individuals (1.2 ng/ml vs. 0.15 ng/ml). Individuals who died of aflatoxicosis had higher aflatoxin B_1-lysine adduct levels than did individuals who survived (3.2 ng/mg vs. 0.5 ng/mg) although this difference was not statistically significant (Azziz-Baumgartner *et al.*, 2005). These levels of aflatoxin adducts are the highest ever reported.
- Levels of aflatoxin B_1-albumin adducts in apparently unaffected individuals were consistent with levels previously documented in studies that measured the level of aflatoxin B_1-albumin adducts in African populations with a high incidence of liver cancer (Peers and Linsell, 1973).

Conclusions

The 2004 and 2005 outbreaks of aflatoxicosis are among the largest, in both the size of population affected and the duration, to occur in Kenya and the world. Serious issues need to be addressed quickly if future outbreaks are to be averted. These issues include:

- Continued community education in food storage and increased awareness campaigns through local media and civic education channels.
- Active systematic food safety surveillance in designated and centralized maize marketing points and distribution channels.
- Food security should be viewed not only as the amount of food available, but also should include a measure of the quality of the food that is available.
- Studies of local methods of food storage and preservation with the goal of recommending the most culturally appropriate methods for storing and drying maize.
- Implementation and dissemination of simple pre- and postharvest technologies on a wider scale for the management of aflatoxin contamination in maize.

Acknowledgements

I thank the Ministry of Health, Kenya, World Health Organization (WHO), and Center for Disease Control and Prevention (CDC) for the investigation of this outbreak, and the National Public Health Laboratories for the analysis of the samples. I also thank the International Institute of Tropical Agriculture (IITA) and the African Agricultural Technology Foundation, Nairobi, Kenya, for support to attend this conference, and the National Assembly of Kenya for assistance in preparing the manuscript and for permission to travel to Accra to present it.

References

Azziz-Baumgartner, E., Lindblade, K., Gieseker, K., Rogers, H.S., Kieszak, S., Njapau, H., Schleicher, R., McCoy, L.F., Misore, A., de Cock, K., Rubin, C. and Slutsker, L. (2005) Case-control study of an acute aflatoxicosis outbreak, Kenya. *Environmental Health Perspectives* 113, 1779-1783.

FAO. (2004) *Worldwide Regulations for Mycotoxins in Food and Feed in 2003. FAO Food and Nutrition Paper 81.* FAO, Rome, Italy.

Krishnamachari, K.A., Nagarajan, V., Ramesh, V.B. and Tilak, T.B.G. (1975) Investigations into an outbreak of hepatitis in parts of Western India. *Indian Journal of Medical Research* 63, 1036-1049.

Lewis, L., Onsongo, M., Njapau, H., Schurz-Rogers, H., Luber, G., Kieszak, S., Nyamongo, J., Backer, L., Dahiye, A.M., Misore, A., DeCock, K., Rubin, C. and the Kenya Aflatoxicosis Investigation Group. (2005) Aflatoxin contamination of commercial maize products during an outbreak of acute aflatoxicosis in Eastern and Central Kenya. *Environmental Health Perspectives* 113, 1763-1767.

Ngindu, A., Johnson, B.K., Kenya, P.R., Ngira, J.A., Ocheng, D.M., Nandwa, H., Omondi, T.N., Jansen, A.J., Ngare, W., Kaviti, J.N., Gatei, D. and Siongok, T.A. (1982) Outbreak of acute hepatitis caused by aflatoxin poisoning in Kenya. *Lancet* 319, 1346-1348.

Nyikal, J., Misore, A., Nzioka, C., Njuguna, C., Muchiri, E., Njau, J., Maingi, S., Njoroge, J., Mutiso, J., Onteri, J., Langat, A., Kilei, I.K., Nyamongo, J., Ogana, G., Muture, B., Tukei, P., Onyango, C., Ochieng, W., Tetteh, C., Likimani, S., Nguku, P., Galgalo, T., Kibet, S., Manya, A., Dahiye, A., Mwihia, J., Mugoya, I., Onsongo, J., Ngindu, A., de Cock, K.M., Lindblade, K., Slutsker, L., Amornkul, P., Rosen, D., Feiken, D., Thomas, T., Mensah, P., Eseko, N., Nejjar, A., Onsongo, M., Kessel F., Njapau, H., Park, D.L., Lewis, L., Luber, G., Rogers, H., Backer, L., Rubin, C., Gieseker, K.E., Azziz-Baumgartner, E., Chege, W. and Bowen, A. (2004) Outbreak of aflatoxin poisoning – Eastern and Central Provinces, Kenya, January-July 2004. *Morbidity and Mortality Weekly Report* 53, 790-793.

Peers, F.G. and Linsell, C.A. (1973) Dietary aflatoxins and liver cancer – a population based study in Kenya. *British Journal of Cancer* 27, 473-484.

12

Mycotoxin Problems in Nuts and Dried Fruits from the Mediterranean Basin

Guner Özay* and Hayrettin Ozer

Abstract

The occurrence of toxigenic molds and mycotoxin formation can be a problem in the Mediterranean basin. Nuts and dried fruits, the major commodities produced and traded in the basin, may be contaminated with aflatoxins and ochratoxin A. The nuts most at risk for aflatoxin formation are peanuts, pistachios, hazelnuts, walnuts and pine nuts. The major Mediterranean dried fruits susceptible to aflatoxin contamination are figs, raisins and red peppers while ochratoxin A has been identified in figs and raisins. In the EU Mediterranean countries detailed, harmonized regulations exist for most mycotoxins. Some countries apply different limits for aflatoxins (range 1-15 ng/g for aflatoxin B_1) in certain products and there is a need for more specific and harmonized regulations. To control mycotoxin contamination, precautionary measures taken during the pre- and postharvest periods: GAP, GHP, GSP, GMP and HACCP, sharing information, and a collective effort by all interested parties is needed.

Introduction

Mediterranean countries produce high volumes of fruits and vegetables and are the main suppliers of these fresh foods to Europe. Major exporting countries include Turkey, Spain, Italy, Tunisia, Morocco, Cyprus, Egypt, Israel and Algeria (Cano and Sanchez, 2004). About 40% of the vegetable production consists of tomatoes, cabbages, watermelons, onions and carrots. The basin also produces diverse fruits such as apples, pears, peaches, apricots, citrus, grapes, plums, cherries, figs and dates. Figs, raisins, apricots and dates are major dried fruits produced in this region. In terms of nut production, hazelnuts, almonds, pistachios, chestnuts, and peanuts are all major crops. The Mediterranean area also produces some products specific to the region, *e.g.*, olives, grapes (table and wine), raisins, citrus, figs and dates.

Occurrence of toxigenic molds and mycotoxin formation can be a problem in the Mediterranean basin, as in other parts of the world. The most important mycotoxins occurring in Mediterranean crops are aflatoxins (B_1, B_2, G_1 and G_2) and ochratoxin A. Aflatoxin contamination of dried fruits, *e.g.*, figs, and nuts, *e.g.*, pistachios, hazelnuts and peanuts, is widespread in the southern and eastern parts of the Basin while ochratoxin A occurs in dried vine fruits. In the northern basin (European Mediterranean countries) trichothecenes, fumonisins (in cereals), and ochratoxin A (in grapes) are the most commonly occurring mycotoxins.

Present situation

Mycotoxins are a problem in the Mediterranean basin as a health hazard to the population as well as a trade issue for the export of local produce. The type and level of mycotoxins vary by country and in some cases by geographic location within a country. The results of existing studies are best summarized by product.

Dried figs and other dried fruits

This fruit is a good substrate for fungal invasion and aflatoxin formation. Figs are one of the top Turkish export commodities, so intensive studies have been carried out on the occurrence of aflatoxins in dried figs. Of 284 fig samples from different storage and processing stages, 4% were contaminated with aflatoxins amongst the poorer grade figs (Boyacıoğlu and Gönül, 1990). In another study, 103 fig samples from various locations and different stages of drying and processing were examined, with 29% containing aflatoxins (maximum 63 ng/g) and 3% containing ochratoxin A (maximum 8.3 ng/g). Aflatoxin formation begins while the fruit is on the tree, as the moisture and a_w values of firm, ripened and shriveled figs are suitable for mycotoxin formation (Özay and Alperden, 1991). Harvesting the fruit by hand, treating it with different solutions, and using solar drying systems all effectively reduced the contamination level (Özay *et al.*, 1995). A correlation between BGYF (Bright Greenish Yellow Fluorescence) under UV light and aflatoxin B_1 was found for figs (Steiner *et al.*, 1988) and selection of BGYF fruits on the processing line is used widely in Turkey to reduce the aflatoxin level in fig lots (Düzbastılar and Arıkan, 1992; Düzbastılar and Arasıler, 1992). In Egypt, among various dried fruits surveyed for mycotoxin contamination, apricots (50-110 ng/g), figs (60-120 ng/g), and plums (210-280 ng/g), all could be found with high levels of aflatoxins while raisins were free of aflatoxins (Zohri and Abdelgawad, 1993). Three of 25 Tunisian date samples were contaminated with aflatoxin at the first stage of maturation, but no aflatoxin was detected at the final, edible mature stage of the fruit (Shenasi *et al.*, 2002).

Grapes

Fungal growth occurs optimally at a_w between 0.95 and 0.99 and a temperature of 20-36°C, which corresponds with optimal conditions for ochratoxin production (Bellí *et al.*, 2005). In a survey in Italy during 1999-2000, ochratoxin was found in both grapes and wine (Battilani and Pietri, 2002). As a result of a survey (1998-2000) in Greece, dried vine fruits (currants and sultanas) were found contaminated with ochratoxin A. Sultanas were less contaminated ($n = 27$, median 0.6 ng/g) than were currants ($n = 54$, median 1.3 ng/g). Seasonal variation occurred in the contamination levels (Stefanaki *et al.*, 2003). The most comprehensive study of ochratoxin A in grapes and wine has been the subject of the EU project WINEOCHRA-RISK (Battilani, Chapter 21). This project identified the key risk elements for ochratoxin A in grape and wine, and developed preventive and corrective actions for the occurrence of ochratoxin producing molds and ochratoxin formation by using HACCP principles.

Nuts

Another group of mostly susceptible commodities, nuts, have been subjected to many studies. A survey was made of the occurrence of mycotoxins and mycotoxin production poten-

Table 1. Aflatoxin sampling plan designs (CX/FAC 05/37/23, December 2004).

Authority	Sample size (kg)	Product	Maximum aflatoxin level (ng/g)
European Union	30	Raw shelled peanuts destined for further processing	15 – total OR 8 – B_1
European Union	30	Tree nuts destined for further processing	10 – total OR 5 – B_1
European Union	10	Samples of consumer-ready peanuts and tree nuts	4 – total OR 2 – B_1
Codex Alimentarius	20	Raw shelled peanuts	15 – total
U.S. Department of Agriculture	22	Raw shelled peanuts destined for further processing	15 – total
U.S. Pistachio Industry	10	Pistachio nuts	15 – total

tial by molds isolated from nuts (almonds, peanuts, hazelnut and pistachios). Aflatoxin was identified in a sample from almonds (95 ng/g aflatoxin B_1 and 15 ng/g aflatoxin B_2) and peanuts (< 10 ng/g) in Spain (Jimenez *et al.*, 1991). Similar results were obtained in Tunisia (Boutrif *et al.*, 1977*b*). Pistachio nuts are the commodity most at risk for aflatoxin formation from the Mediterranean Basin. Twenty-six of 50 pine nut samples were contaminated with up to 2,000 ng/g aflatoxin B_1. Traditional pudding made from pine nuts contained 80% of the aflatoxin B_1 originally present (Boutrif *et al.*, 1977*a*). In another study of hazelnuts and walnuts, 90% of the hazelnuts were aflatoxin positive (25-175 ng/g) and 75% of the walnut samples contained aflatoxin in the range of 15-25 ng/g. One sample also was contaminated with zearalenone (Abdelhafez and Saber, 1993).

Turkey is the world's main producer of hazelnuts with a 75% share of the total world production and 70-75% of the total world's export. Sometimes Turkey encounters technical barriers for hazelnut exports due to excessive levels of aflatoxin. A wide scope project on this topic began in 2002 whose main objective was to minimize the risk of aflatoxin formation in hazelnuts during maturation, harvesting, drying and storage. Training hazelnut growers, traders and processors also was included in the project. Field studies were conducted at 72 sites representative of the hazelnut growing region near the Black Sea in Turkey. Alternative drying techniques also were used to shorten the drying period. Toxigenic mold strains were identified and their aflatoxin production ability evaluated.

Since aflatoxin is a serious problem in terms of consumer health and economic losses, both mycotoxin management efforts and studies of analytical methods have continued (Whitaker, 2003). Aflatoxin limits vary in regulations for different countries. In the United States a guidance maximum level for total aflatoxins of 20 ng/g for nuts intended for human consumption was set by the Food and Drug Administration (FDA), whereas the corresponding limit set by the European Commission is 4 ng/g. Sampling and analytical procedures for detecting and quantifying aflatoxin is very important in relation to the acceptance and rejection of shipments. Since high levels of aflatoxin may be present in a very small portion of nuts in a lot, sampling is a critical procedure and sample size usually is the key issue (Table 1). The sampling step is the largest source of error and relatively large samples are required to reduce the error associated with the aflatoxin test procedure (and thus reduce the exporters' and importers' risks) to acceptable levels (Whitaker, 2003).

We conducted a study of the sampling of hazelnuts for aflatoxin analysis to determine the total error associated with testing commercial shelled hazelnuts for aflatoxin; particularly

Table 2. Regulations on mycotoxins in nuts and dried fruits in the Mediterranean countries (after FAO, 2004).

Country	Commodity	Aflatoxins (ng/g)	Ochratoxin A (ng/g)
European Union	Nuts and dried fruits (direct human consumption)	Total – 4 OR B_1 – 2	10 (Dried vine fruit, currants, raisins, sultanas)
	Nuts, dried fruits (to be processed)	Total – 10 OR B_1 – 5	
Algeria	Nuts and peanuts	Total – 20 OR B_1 – 10	
Egypt	Peanuts	Total – 10 OR B_1 – 5	
Israel	Nuts, peanuts, figs and their products	Total – 15 OR B_1 – 5	
Jordan	Almonds, pistachio and pine nuts	Total – 30 OR B_1 – 15	
Morocco	Pistachio, peanuts and almonds	B_1 – 1	
Turkey	Nuts, dried fruits and their products	Total – 10 OR B_1 – 5	10 (Dried raisins)
Tunisia	All food products	B_1 – 2	

partitioning the total error into sampling, sample preparation and analytical error components; and determining the functional relationships between the variance components and aflatoxin concentration. The mycotoxin concentration of a bulk lot can not be determined at 100% certainty because of the variation associated with each step of the mycotoxin test procedure. In this study, the sources of error in the aflatoxin sampling procedure and different homogenization methods (wet and dry) for hazelnut were determined as well as the distribution of aflatoxin in hazelnut lots. Then sampling, sample preparation and analytical steps were responsible for 99.4, 0.4 and 0.2%, respectively, of the observed variability (Özay *et al.*, 2006).

Regulations for mycotoxins

Mycotoxin problems differ by geographic region and by crop. The Food and Agriculture Organization (FAO) updated information on worldwide mycotoxin regulations in 2003 (FAO, 2004). This survey found that at least 98 countries had regulations for food and/or animal feed in 2002, an increase of ~ 30% compared to 1995. The regulatory requirements differ substantially by country, although several regulations have been (or are being) harmonized in countries belonging to various economic communities. In the EU Mediterranean countries, some countries apply different limits for aflatoxin contamination (Table 2) and more specific and harmonized regulations are needed.

Research needs on mycotoxins

- Controlling and managing mycotoxins in an effective way requires science-based data. Therefore, national and regional surveillance studies of specific commodities are needed. The contamination process, when and under what conditions mycotoxin contamination may occur, needs to be defined by systematic studies during pre- and post-harvest stages. Prevention measures also should be defined for each critical step. In the field and farm, Good Agricultural Practices (GAP) are effective in preventing or reducing preharvest contamination, *e.g.*, pest and disease control, and proper irrigation practices. Correct harvesting time and good harvesting techniques also should be defined for the crops. The stage at which mycotoxins are most likely to be produced is during the long drying period before storage, so appropriate (simple, inexpensive and effective) drying techniques need to be developed. In addition, optimal storage conditions and parameters should be defined to prevent mycotoxin formation during storage, *i.e.*, GMP (Good Manufacturing Practices), GHP (Good Hygienic Practices), GSP (Good Storage Practices). More systematically identifying Hazard Analysis and Critical Control Points (HACCP) is the most effective tool to use to manage mycotoxins as part of a food safety management system (FAO, 2001).

- Existing ecological conditions may favor the occurrence of mixtures of mycotoxins in agricultural products and foods. There are limited data on the co-occurrence of mycotoxins on commodities grown in the region.
- Establishment of a database on fungi and mycotoxins present in the region would help both scientists and regulatory bodies.
- Studies of exposure assessment, a part of risk assessment, are needed in the region. Sound data on the occurrence of mycotoxins grown and consumed in the region and the food intake by local populations are needed. There are some exposure studies for ochratoxin A in some countries.

Discussion

Mycotoxin contamination of foods and feeds is a world-wide problem. Populations in the less-developed parts of the world have a higher risk of mycotoxin exposure than do those living in developed countries, *e.g.*, the European Union or the United States, with high food safety standards. Some recommendations for effective action to prevent and control mycotoxin contamination include:

- Improved knowledge of the occurrence of mycotoxins and preventive measures (preharvest, drying, storage, processing, packaging, and transportation) that help control them.
- Mycotoxin prevention requires national/regional/international approaches and multidisciplinary efforts. Multinational efforts including all parties (producer, processor, trader, and retailer) are required.
- In the same way, adoption of the European Union standards in the region is an urgent necessity. Appropriate sampling procedures, reliable analytical techniques and reasonable regulations could eliminate many trade problems. More trade will contribute to the development of the regional economy by utilizing high quality agricultural resources for export purposes and for protecting public health.
- Education and training of technical staff involved in the food chain is always a necessity. Producers, processors, and traders should participate in training programs to implement mycotoxin-free production systems.
- Scientific studies of the development/improvement of alternative drying techniques (solar or mechanical) should be conducted and the results applied in the food system.
- GAP, GHP, GMP, GSP and more systematically HACCP procedures should be implemented systematically to manage mycotoxins.
- Cooperative efforts by all stakeholders (national, regional, international) including scientific, economic and regulatory bodies are needed to prevent and control mycotox-

in contamination. Sharing knowledge, assistance and advice on mycotoxin prevention helps all countries produce safe agricultural products, increase health standards and reduce the level of contaminated crops. The MycoGlobe project is a good example of this kind of collaboration and cooperation.

References

Abdelhafez, A.I.I. and Saber, S.M. (1993) Mycoflora and mycotoxin of hazel nut (*Corylus-Avellana L.*) and Walnut (*Juglans-Regia* L.) seeds in Egypt. *Zentralblatt fürMikrobioogie* 148, 137-147.

Battilani, P. and Pietri, A. (2002) Ochratoxin A in grapes and wine. *European Journal of Plant Pathology* 108, 639-643.

Bellí, N., Ramos, A.J., Coronas, I., Sanchis, V. and Marin S. (2005) *Aspergillus carbonarius* growth and ochratoxin A production on a synthetic grape medium in relation to environmental factors. *Journal of Applied Microbiology* 98, 839-844.

Boutrif, E., Jemmali, M., Pohland, A.E. and Campbell, A.D. (1977*a*) Aflatoxin in Tunisian Aleppo pine nuts. *Journal of the Association of Official Analytical Chemists* 60, 747-748.

Boutrif, E., Jemmali, M., Campbell, A.D. and Pohland, A.E. (1977*b*) Aflatoxin in Tunisian foods and foodstuffs. *Annalesde la Nutition et de l'Alimentation* 31, 431-434.

Boyacıoğlu, D. and Gönül, M. (1990) Survey of aflatoxin contamination of dried figs grown in Turkey in 1986. *Food Additives and Contaminants* 7, 235-237.

Cano, G. and Sanchez, M.T. (2004) The Euro-Mediterranean Food-Processing Sector: Industrial and Technological Outlook. Red tematica frutas vegetales, Universidad de Cordoba, Spain

Düzbastılar, M. and Arasiler, Z. (1992) R&D Project No 23 *TARİŞ Araştıra Geliştirme Müdürlüğü Bornova*, Izmir, Turkey.

Düzbastılar, M. and Arkan, N. (1992) R&D Project No. 44 *TARİŞ Araştıra Geliştirme Müdürlüğü Bornova*, Izmir, Turkey.

FAO (United Nations Food and Agriculture Organization). (2001) *Manual on the Application of the HACCP System in Mycotoxin Prevention and Control. Food and Nutrition Paper* 73. Rome, Italy.

FAO. (2004) *Worldwide Regulations for Mycotoxins in Food and Feed in 2003. Food and Nutrition Paper* 81. Rome, Italy.

Jimenez, M., Mateo, R., Querol, A., Huerta, T. and Hernandez, E. (1991) Mycotoxin and mycotoxigenic molds in nuts and sunflower seeds for human consumption. *Mycopathologia* 115, 121-127.

Özay, G. and Alperden, I. (1991) Aflatoxin and ochratoxin A contamination of dried figs (*Ficus carina L.*) from the 1988 crop. *Mycotoxin Research* 7, 85-91.

Özay, G., Aran, N. and Pala, M. (1995) Influence of harvesting and drying techniques on microflora and mycotoxin contamination of figs. *Nahrung* 39, 156-165.

Özay, G., Seyhan, F., Yilmaz, A., Whitaker, T.B., Slate, A.B. and Giesbrecht, F. (2006) Sampling hazelnuts for aflatoxin: Uncertainty associated with sampling, sample preparation, and analysis. *Journal of the Association of Official Analytical Chemists International* 89, 1004-1011.

Shenasi, M., Aidoo, K.E. and Candlish, A.A.G. (2002) Microflora of date fruits and production of aflatoxin at various stages of maturation. *International Journal of Food Microbiology* 79, 113-119.

Stefanaki, I., Foufa, E., Tsatsou-Dritsa, A. and Dais, P. (2003) Ochratoxin A concentration in Greek domestic wines and dried vine fruits. *Food Additives and Contaminants* 20, 74-83.

Steiner, W.E., Rieker, R.H. and Battaglia, R. (1988) Aflatoxin contamination in dried figs: Distribution and association with fluorescence. *Journal of Agricultural and Food Chemistry* 36, 88-91.

Whitaker, T.B. (2003) Standardization of mycotoxin sampling procedures: an urgent necessity. *Food Control* 14, 233-237.

Zohri, A.A. and Abdelgawad, K.M. (1993) Survey of mycoflora and mycotoxins of some dried fruits in Egypt. *Journal of Basic Microbiology* 33, 279-288.

13

Between Emerging and Historical Problems: An Overview of the Main Toxigenic Fungi and Mycotoxin Concerns in Europe

Antonio F. Logrieco* and Antonio Moretti

Abstract

Contamination by toxigenic fungi and their mycotoxins of agricultural products grown in Europe is a problem, and dramatic differences in environmental conditions in various European countries significantly influences the distribution of toxigenic fungi and related mycotoxicological risks. Emerging problems due to climate change and new mycotoxin/ commodity combinations increase these concerns. Trans-global transposition and trade exchanges of plant products also contribute significantly to the spread of toxigenic fungi worldwide and represent an important source of inoculum for new plant diseases in Europe and for broadening the genetic diversity of already existing local populations. We illustrate the major problems associated with mycotoxin contamination of agricultural products in Europe by focusing on three crops and their relevant diseases: (i) wheat with Fusarium head blight and *Penicillium* infections, (ii) maize with Fusarium ear rot and *Aspergillus* infections, and (iii) grapes with black Aspergillus rot.

Introduction

Twenty-five percent of the world's food crops are estimated to be affected by mycotoxins (Charmley *et al.*, 1995). The main factor influencing the geographic distribution of toxigenic fungi and related mycotoxin accumulation is climate. Environmental variation at different geographic locations, particularly temperature, humidity and rainfall, influence the growth, survival, dissemination and incidence of toxin-producing organisms and toxin accumulation in crops (Doohan *et al.*, 2003). However, understanding the influence of climatic factors on fungal diseases is difficult, as fungal diseases may result from members of a single species or from multiple species in a disease complex. Moreover, the optimal environmental conditions for each fungal species may vary depending on their mode of reproduction and the conditions required for vegetative growth and toxin production, which need not be the same. The main causal agents of toxin accumulation in cereals and grapes belong to species of *Aspergillus*, *Fusarium* and *Penicillium.* Some of these species are strictly asexual while others reproduce both sexually and asexually with varying relative frequency.

The migration of agriculturally important toxigenic fungi through trade may be an important source of inoculum for new plant diseases in Europe and for increasing the genetic diversity of already existing local populations. All of these factors may influence the spread of toxigenic fungi, their growth and the accumulation of their mycotoxin products on crops. A better understanding of the role of the environment and the increased sensitivity of toxin assays has recently added to the endemic problems some emerging ones. In this chapter we consider three important European crops of economic importance that also have high risk of contamination by mycotoxins (Logrieco and Visconti, 2004).

Wheat

Fusarium head blight and related toxins

Fusarium *species*

Several *Fusarium* species are widespread pathogens of soft and durum wheats worldwide, including all European cereal-growing areas where they may cause stem and ear rots and severely reduce crop yields. In addition, several *Fusarium* strains can produce mycotoxins either in preharvest infected plants in the fields or in stored grain (Bottalico, 1998). The occurrence of mycotoxins in wheat is of great concern worldwide, because their presence in processed feeds and foods seems almost unavoidable. Virtually all of the wheat grown in Northern Europe is at least to some extent contaminated by deoxynivalenol, whereas wheat grown in southern parts of the continent is less contaminated by *Fusarium* toxins (Bottalico and Perrone, 2002). To the most common and best known mycotoxin contaminants of wheat, *i.e.*, trichothecenes, zearalenone and moniliformin, are now being added new problems, *e.g.*, the cyclic hexadepsipeptides enniatins and beauvericin (Logrieco *et al.*, 2002a; Uhlig *et al.*, 2006).

Fusarium species that cause Fusarium head blight of wheat can affect individual kernels, single spikelets or entire heads. Infected spikelets first appear water-soaked and then lose their chlorophyll and become straw-colored. In warm, humid weather, pinkish-red mycelia and conidia develop abundantly in the infected spikelets, and the infection spreads to adjacent spikelets. Infected kernels become shriveled and discolored with a white, pink, or light brown scaly appearance as a result of mycelia growing from the pericarp.

The species of *Fusarium* most commonly associated with Fusarium head blight of wheat in Europe are *Fusarium graminearum* (and its widespread teleomorph *Gibberella zeae*), *Fusarium culmorum*, *Fusarium avenaceum* (teleomorph *Gibberella avenacea*), and *Fusarium poae* (Xu *et al.*, 2005). Other, less-frequently isolated species are *Fusarium cerealis* (syn. *Fusarium crookwellense*), *Fusarium equiseti* (teleomorph *Gibberella intricans*), *Fusarium sporotrichioides* and *Fusarium tricinctum*, with other species of the genus found on an irregular basis (Bottalico, 1998). An important etiological characteristic of Fusarium head blight in Europe is the co-occurrence or the quick succession of several *Fusarium* species often referred to as a "species complex." Many different *Fusarium* species can be recovered from grain harvest in a limited geographic area. However, only a small number of species are regarded as pathogenic and these species usually co-exist with other less pathogenic or opportunistic *Fusarium* species that also can produce considerable amounts of mycotoxins. Therefore, the mycotoxin content of a contaminated crop is due not only to the common pathogenic *Fusarium* species, but also to other opportunistic species in the species complex. The distribution and predominance of Fusarium head blight pathogens and related toxins also are influenced by agronomic and genetic factors. Climate, particularly temperature and moisture, are important, so the species composition of the Fusarium head blight species complex may vary significantly across Europe (Xu *et al.*, 2005), with the fungal species present and the climate not necessarily independent variables.

Geographic distribution of Fusarium *species*

Of the four *Fusarium* species most commonly associated with Fusarium Head Blight of wheat in Europe, *F. graminearum* is the most common in continental and moist-warm climates, *e.g.*, Central, Eastern and Southern Europe, although it also occurs frequently in maritime and cooler northwestern areas, where *F. culmorum* and *F. avenaceum* are more commonly found. Strains of *F. graminearum* and *F. culmorum* have been classified into two toxigenic types based on the trichothecene toxin they produce – either deoxynivalenol or nivalenol. The nivalenol chemotype may be more common amongst *F. graminearum* strains associated with maize (Miller *et al.*, 1991). The deoxynivalenol and nivalenol chemotypes of *F. graminearum* are not evenly distributed worldwide, and the ecological differences in chemotype distribution may be important in determining specific regional grain contamination risks. The chemotypes of *F. culmorum* also differ in aggressiveness in plant pathogenicity assays (Bottalico and Perrone, 2002).

In the coolest areas of Europe, *i.e.*, the Scandinavian regions, *F. avenaceum*, *F. poae* and *F. langsethiae* are the most common species, and all have interesting and important taxonomic problems. Within section *Sporotrichiella*, *Fusarium langsethiae*, formerly termed "powdery *F. poae*," recently was elevated to species status to distinguish it from *F. poae sensu stricto*. *Fusarium langsethiae* synthesizes T-2 toxin and its deacetylated derivative, HT-2 toxin, in Norwegian cereals (Torp and Nirenberg, 2004). Although this species is morphologically similar to *F. poae*, its mycotoxin profile is similar to that of *F. sporotrichioides*. In Europe, this species usually occurs in Scandinavia, where the climatic conditions commonly are cooler and wetter. The acceptability of *F. avenaceum* as a homogeneous species is a matter of debate. Leslie and Summerell (2006) include all strains of both *F. avenaceum* and *F. arthrosporioides* in *F. avenaceum*. Yli-Mattila *et al.* (2004) distinguished *F. avenaceum* from *F. arthrosporioides*, but the morphological and molecular similarities of the two species make it very difficult to correctly distinguish these species. Another argument for the taxonomical distinctness of *F. arthrosporioides* is that the strains assigned to this species can produce much higher levels of enniatins under both natural and *in vitro* conditions than do strains assigned to *F. avenaceum* (Jestoi *et al.*, 2004; Uhlig *et al.*, 2006).

Southern Europe. The species profile of strains causing Fusarium Head Blight in southern regions of Europe varies annually depending on environmental conditions. In general, Fusarium Head Blight incidence is low and in the most southern regions of Italy and Spain, the disease is absent. In northern portions of Italy, Spain, Portugal, and in the south of France, *F. graminearum* and *F. poae* are the dominant species. There are reports of *F. avenaceum* and *F. culmorum* being present and associated with diseased plants in Southern Europe, but such reports are not common and often are not well-substantiated suggesting that these two species are less common and are problematic only if the weather conditions are right (Bottalico and Perrone, 2002; Xu *et al.*, 2005).

Central-Eastern Europe. *Fusarium graminearum* is the most common species, but *F. avenaceum*, occurs at significant frequency in more central regions, *e.g.*, Austria and Switzerland (Bottalico and Perrone, 2002). In eastern countries, *e.g.*, Hungary, *F. graminearum* usually is the dominant species although *F. poae* may occur in warmer, dryer seasons (Bottalico and Perrone, 2002; Xu *et al.*, 2005).

Maritime areas of Northern and Western Europe. In the cooler maritime areas of Northwestern Europe, including the Netherlands, Belgium, England, Scotland and northwestern areas of France, the four *Fusarium* species commonly associated with Fusarium head blight, *F. culmorum*, *F. graminearum*, *F. avenaceum* and *F. poae,* all occur at significant levels. However, more recent surveys describe an increasing frequency of *F. culmorum* and a greater importance of *F. poae* and *F. avenaceum*, especially in years less conducive to *F. graminearum* infection (Waalwijk *et al.*, 2003; Xu *et al.*, 2005).

Northern and Eastern Europe. In Poland, surveys of Fusarium head blight conducted in the late 1990s and covering various climatic areas, found that *F. poae* dominated, with *F. sporotrichioides* reemerging in importance. In Russia, Fusarium head blight of wheat is very widespread and losses can reach 25-50%. In the northwestern and in central regions (Moscow area) of Russia, the most frequently isolated species were *F. culmorum*, *F. avenaceum*, *F. tricinctum*, *F. poae*, and *F. sporotrichioides*, with *F. graminearum* almost completely absent. In the warm, humid areas of the Southeastern European part of Russia, Belorussia and the Ukraine, *F. graminearum* dominates (Bottalico and Perrone, 2002). Strains of *F. poae* isolated from the coolest areas of Russia, where the environment is similar to that in Scandinavia, need to be evaluated to see if they are *F. langsethiae*.

Scandinavia. This geographic area has the coolest conditions in Europe. The *Fusarium* species most frequently encountered in surveys of wheat in these countries is *F. avenaceum*/*F. arthrosporioides*, together with species from section *Sporotrichiella*, *e.g.*, *F. poae*, *F. langsethiae*, *F. tricinctum* and *F. sporotrichioides*. *Fusarium culmorum* occurs when drier and colder weather that is less conducive to *F. avenaceum* occurs (Logrieco *et al,*. 2002*a*; Jestoi *et al.*, 2004; Uhlig *et al.*, 2006).

Mycotoxins

Deoxynivalenol, other trichothecenes and zearalenone are endemic contaminants of wheat in Europe, while moniliformin and hexadepsipeptides are emerging mycotoxins that can occur at high levels in cereals, particularly in cooler European regions.

Deoxynivalenol. The mycotoxins most frequently encountered in Fusarium head blight of wheat throughout Europe are the trichothecenes, and in particular deoxynivalenol. This toxin occurs across the continent. Its occurrence is correlated with environmental conditions, but this toxin has been found in wheat continent-wide. Schothorst and van Egmond (2004) found numerous samples positive for deoxynivalenol in all monitored European countries except Italy. However, some reports from Italy in 1998–2000 indicated that deoxynivalenol may be a common wheat contaminant there as well. In particular, deoxynivalenol contamination decreased from northern to central Italian regions, with very little, if any, deoxynivalenol detected in southern parts of the country where the disease is almost completely absent. Agronomic (crop rotation) and genetic (resistant genotypes) strategies that prevent the accumulation of this toxin in wheat kernels should be pursued simultaneously and combined with forecasting models to predict Fusarium head blight incidence and the potential for deoxynivalenol production (Xu, 2003).

Other trichothecenes. Nivalenol and fusarenon X, often co-occur with deoxynivalenol and are commonly reported from throughout Europe in ears of cereals affected by Fusarium head blight. Their presence has been attributed to the infection of wheat by strains of *F. graminearum* and *F. culmorum* with nivalenol chemotypes. Nivalenol and fusarenon X also may be produced by *F. poae* in Sweden and other northern countries, and by *F. cerealis* in Central and Northeastern Europe. Nivalenol in wheat samples from Poland usually is associated with *F. poae*, while in the United Kingdom, the highest levels of nivalenol are associated with *F. poae* and *F. culmorum* (Bottalico and Perrone, 2002; Tomczak *et al.*, 2002). Schothorst and van Egmond (2004) reported that ~15% of 2200 samples of wheat from across Europe were positive for nivalenol and that most of these samples were from Scandinavian countries.

Epidemics of *F. sporotrichioides* and *F. langsethiae* in cold European locations may lead to contamination with T-2 toxin derivatives, such as T-2, HT-2 and T-2 triol. In Poland, scabby wheat grains contain T-2 and HT-2 toxins if the grain is infected with *F. sporotrichioides* (Bottalico and Perrone, 2002). Curiously, the highest levels of T-2 toxin have been reported from wheat collected in warmer countries such as Italy, France and Portugal (Schothorst and van Egmond, 2004).

Zearalenone. Zearalenone is produced primary by *F. graminearum* and *F. culmorum* and is a common co-contaminant with deoxynivalenol and its derivatives. Zearalenone is one of the most common mycotoxins associated with Fusarium head blight of wheat in Europe. Austrian pre-harvest hard wheat ears are infected primary by *F. graminearum*, but also with lower numbers of *F. culmorum*. Wheat from Slovakia with Fusarium head blight symptoms was infected primarily with *F. graminearum*, and wheat samples from Germany and the Netherlands all were contaminated with zearalenone (Bottalico, 1998).

Moniliformin. Severe infection by *F. avenaceum*, *F. tricinctum*, and to a lesser extent *Fusarium subglutinans*, in Central and Northeastern European countries, could result in the accumulation of moniliformin in scabby grain (Jestoi *et al.*, 2004; Uhlig *et al.*, 2004). In particular, in Poland, significant levels of moniliformin were found in scabby kernels obtained in 1998/1999 from ears of wheat heavily infected with *F. avenaceum* (Tomczak *et al.*, 2002). Moniliformin also was reported in freshly harvested durum wheat in Austria. In all of these surveys, the moniliformin content of the kernels was correlated with the presence of *F. avenaceum* (Bottalico and Perrone, 2002).

An emerging problem – hexadepsipeptides. Enniatins and the chemically closely related beauvericins are cyclic hexadepsipeptides and selective cation ionophores. Beauvericin is cytotoxic to mammalian cell tissues and causes apoptosis in both murine and human cell lines (Logrieco *et al.*, 2002*b*). Beauvericin co-occurred with enniatins in Finnish samples of wheat kernels heavily contaminated with *F. avenaceum*, *F. poae* and *F. tricinctum* (Logrieco *et al.*, 2002*a*; Jestoi *et al.*, 2004). In addition, most wheat samples heavily colonized by *F. avenaceum*/*F. arthrosporioides*, collected in two years (2001/2) in Norway also were contaminated with high levels of beauvericin and enniatins (Uhlig *et al.*, 2006). Thus, the contamination of cereals with cyclic hexadepsipeptides is an important emerging toxicological problem.

In the same report (Uhlig *et al.*, 2006) there was a significant correlation between the concentration of enniatins and moniliformin. This correlation is important from a toxicological point of view as it means that several different and potentially hazardous fungal metabolites may be present simultaneously in the same sample. Beauvericin and moniliformin are metabolites with toxic activities, but there are no data on the toxicological effects of enniatins on humans and animals. Thus, it is not possible to determine if the levels of enniatins found in the European surveys, are a safety concern for human and animal health. Further studies are needed to identify the possible toxic effects of enniatins, and, whether their co-occurrence with moniliformin, beauvericin and trichothecenes has a synergistic effect on human and animal toxicity.

Penicillium and ochratoxin A

Fungal agents

Ochratoxin A is nephrotoxic, immunotoxic, teratogenic and has been classified by the International Agency for Research on Cancer (IARC) as a possible human carcinogen (group 2B) (IARC, 1993). The major fungal species producing ochratoxin A in wheat kernels are *Aspergillus ochraceus* and *Penicillium verrucosum*. These toxigenic species are considered to be opportunistic or saprophytic and may colonize cereals before harvest under favorable climatic conditions. *Aspergillus ochraceus* and *P. verrucosum* differ with respect to their ecological niches and their geographic distributions. *Aspergillus ochraceus* grows at warmer temperatures and at water activities (a_w) as low as 0.80. It can affect wheat in warmer, subtropical and tropical parts of the world. *Penicillium verrucosum* grows well below 30°C and at $a_w \geq 0.80$ and usually is found in wheat in cool temperate regions of Northern Europe and North America, although it may occasionally be found in Mediterranean regions.

Ochratoxin A: Natural occurrence on wheat

Ochratoxin A-producing molds grow and produce toxin when the moisture content of the grain is high at harvest and during subsequent drying and storage. Grain should be dried to < 18% moisture content as quickly as possible, with drying continued until the moisture content is ≤ 14.5%. The only producer of ochratoxin A reliably recorded from cereals in Europe is *P. verrucosum* (Lund and Frisvad, 2003). The maximum tolerable limits adopted by European Union countries for raw cereal grain contamination with ochratoxin A is 5 ng/g. Ochratoxin A often occurs above this limit in wheat grown in Europe, with the biggest problems usually occurring in Northern European countries (Rizzo *et al.*, 2002). Data from Croatia, Denmark, Germany, Italy, Norway, Poland, Spain, Sweden, Switzerland, the United Kingdom and Yugoslavia, confirm the differential geographic associated risks with ochratoxin A contamination (Puntaric *et al.*, 2001; Rizzo *et al.*, 2002; Lund and Frisvad, 2003).

Table 1. *Fusarium* mycotoxins occurring in cereals of European Countries (after SCOOP, 2003).

Toxin	Number of Participating Countries[1]	Number of samples examined	% Positive samples
Trichothecenes – Type A			
T-2 toxin	8	3,490	20
HT-2 toxin	6	3,032	14
T-2 triol	2	1,389	6
Neosolaniol	2	1,323	1
Trichothecenes – Type B			
Deoxynivalenol	11	11,022	57
Nivalenol	7	4,166	16
3-Acetyl-deoxynivalenol	6	3,721	8
15-Acetyl-deoxynivalenol	3	1,954	20
Fusarenon X	3	1,872	10
Zearalenone	9	5,018	32
Fumonisins			
B_1	9	3,863	46
B_2	6	1,010	42
B_3	1	239	36

[1]The 13 participating European countries were: Austria, Belgium, Denmark, England, Finland, France, Germany, Ireland, Italy, Netherlands, Norway, Portugal and Sweden.

Maize

Fusarium ear rot and related toxins

Fusarium *species*

Several *Fusarium* species are widespread pathogens of maize in temperate and semitropical areas, including all European maize-growing areas, and cause severe reductions in crop yield. In wheat, as in maize, these species may co-occur or follow one another in quick succession, and it is not uncommon to isolate several different species from a single infected kernel. In addition to the small number of toxigenic species that also are regarded as pathogenic, there are several less-pathogenic or opportunistic *Fusarium* species that also can produce considerable amounts of toxins both in preharvest infected plants and in stored grain. Thus, the toxigenic profile of a contaminated crop is determined not only by the pathogenic species present, but also by the co-occurring opportunistic species (Bottalico, 1998).

The incidence of Fusarium mycotoxins, *e.g.*, deoxynivalenol and fumonisin B_1, in maize kernels probably exceeds 25% (Tables 1 and 2) (Charmley *et al.*, 1995; Logrieco *et al.*, 2002*c*). The relationship between the ecological distribution of *Fusarium* species and their relative mycotoxin profiles can be used to predict which mycotoxins are most likely to be formed when conducive conditions such as tillage practices, host genotypes and environmental conditions occur.

Table 2. Cereal commodities contaminated by *Fusarium* mycotoxins in Europe (after SCOOP, 2003; participating countries: Austria, Belgium, Denmark, England, Finland, France, Germany, Ireland, Italy, Netherlands, Norway, Portugal and Sweden).

Toxin	Grain and feed/food (positive samples)
Trichothecenes – Type A	
T-2 toxin	Maize (28%), wheat (21%) & oats (21%)
HT-2 toxin	Oats (41%), maize (24%) & rye[1] (17%)
Trichothecenes – Type B	
Deoxynivalenol	Maize (89%) & wheat[1] (61%)
Nivalenol	Maize (35%), oat (21%) & wheat[1] (14%)
3-Acetyl-deoxynivalenol	Maize (27%), wheat[1] (8%)
Zearalenone	Maize (79%), maize bran (51%), maize by-products (53%), wheat (30%), wheat flour (24%), wheat by-products (11%) & cereal for baby food (23%)
Fumonisins	
B_1	Maize (66%), maize bran (79%), maize by-products (31%), maize flakes (46%) & wheat (79%)
B_2	Maize (51%)

[1]Kernels and flour.

There are two kinds of maize ear rot, roughly differentiated as "red ear rot" or "red fusariosis", caused mainly by species of the *Discolor* section, and "pink ear rot" or "pink fusariosis", caused mainly by representatives of the *Liseola* section (Bottalico, 1998). The distribution and prevalence of the different *Fusarium* species causing these two kinds of ear rot disease are largely governed by environmental conditions, primarily temperature and humidity (Bottalico, 1998). In general, "red fusariosis" is particularly severe in years and locations characterized by frequent rainfalls and low temperatures during summer and early fall, while "pink fusariosis" prevails in the drier and warmer climates of Southern Europe (Bottalico, 1998). Among the species causing maize "red ear rot" the most common ones are *F. graminearum*, *F. culmorum*, and to a lesser extent *F. cerealis* and *F. avenaceum*. The species most frequently isolated from maize "pink ear rot" are *Fusarium verticillioides* (teleomorph *Gibberella moniliformis*), *Fusarium proliferatum* (teleomorph *Gibberella intermedia*), and *F. subglutinans* (teleomorph *Gibberella subglutinans*). All of these species can produce several mycotoxins in naturally infected maize kernels, including trichothecenes, zearalenone, and fumonisins. Other potentially toxic metabolites produced by *Fusarium* spp., *e.g.*, beauvericin and moniliformin, also have been found on maize although the toxicity of these compounds needs to be better evaluated. For information on the main chemical and toxic features of the mycotoxins produced by *Fusarium* species commonly occurring on maize, see Logrieco *et al.* (2002*c*) or Desjardins (2006).

Geographic distribution of Fusarium *species*

Central and Northern Europe – Maize "red ear rot". The main pathogen of maize red ear rot, is *F. graminearum*. Reports from Austria, Slovenia, Yugoslavia, Poland, Czech Republic, and Romania (Logrieco *et al.*, 2002*c*) clearly indicate that *F. graminearum* is increasingly common in Central and Northern Europe. *Fusarium graminearum* occasionally is associated with many other *Fusarium* species, whose occurrence and prevalence vary by region and year, depending primarily on climatic conditions (temperature and rain) and less importantly on tillage practices (crop rotation, fertilization and planting area). *Fusarium graminearum*, *F. culmorum* and *F. cerealis* co-occur more frequently in Central Europe. *Fusarium sporotrichioides* also can cause maize red ear rot, but only rarely cause epidemics. This species occurs in the northern and colder regions of Europe where maize cultivation is common, *e.g.*, Poland, and is better adapted to extremely cold environmental conditions than are other *Fusarium* species. In northern Italy, only in unusually cold seasons does *F. graminearum* occur at high incidence on maize kernels. A very rare report of *F. graminearum* from Southern Italy was due to maize grown as a second crop that was harvested at the end of November under cooler and wetter conditions than usual (Logrieco *et al.*, 2003).

Central and Southern Europe – Maize "pink ear rot". Maize "pink ear rot" is common in Southern and Central European areas. The species most commonly associated with pink ear rot were *F. verticillioides*, *F. subglutinans* and to a lesser extent *F. proliferatum*. *Fusarium proliferatum* was more common in Southern Europe, and was displaced by *F. subglutinans* in central areas where the latter predominates as the primary causal agent of pink ear rot and usually is isolated much more frequently than is *F. verticillioides*. *Fusarium proliferatum* commonly co-occurs with *F. verticillioides* in Italy (Logrieco *et al.*, 2002*c*), but in Austria, Croatia, Hungary, and the Slovak Republic, *F. proliferatum* is only rarely isolated (Logrieco *et al.*, 2002*c*). The unusually dry and warm summers that occurred in the late 1990s increased the incidence of *F. proliferatum* in Central Europe and the species was reported from the Slovak Republic in 1996 (Srobarova *et al.*, 2002) and from Austria (Logrieco *et al.*, 2002*c*). In Yugoslavia (Logrieco *et al.*, 2002*c*) during a three-year survey (1994–96) of maize ears *Fusarium* species present at harvest included *F. verticillioides*, *F. subglutinans*, and *F. proliferatum*. In Italy, where environmental conditions often are conducive to a high incidence of maize pink ear rot, *F. verticillioides* dominates, with *F. proliferatum* co-occurring in southern areas in over 60% of the infected kernels and then declining in frequency from the central to the northern parts of the country (Logrieco *et al.*, 2002*c*). However, *F. proliferatum* is assumed to be more widespread than reported since it can be misidentified relatively easily as *F. verticillioides*.

Mycotoxins

Mycotoxin production depends on temperature, substrate and a_w, however, the optimum climatic conditions for mycotoxin production in grain varies by genus, species and isolate. Most studies indicate that high moisture and warm temperatures favor the production of all classes of mycotoxins. However, production of type A trichothecenes by *F. sporotrichioides* requires moderate temperatures and low moisture rather than warmer temperatures and higher a_ws (Mateo *et al.*, 2002). The production of type B trichothecenes is favored when infected grain is stored under warm humid conditions (Martins and Martins, 2002). The conditions for optimal production of zearalenone also are species, isolate and substrate specific, and may differ from those that are optional for trichothecene production. Reports of an optimum temperature for zearalenone production are confusing and vary depending on the isolate and the substrate (Doohan *et al.*, 2003). Fumonisins commonly are produced in maize infected by *F. verticillioides* and *F. proliferatum*, species that usually grow better at higher temperatures. The optimal a_w for the production of fumonisins seems to be high ($a_w \sim 0.98$), but varies depending on temperature, which may be optimal across a relatively wide range, *e.g.*, 15-30°C (Marín *et al.*, 1999).

Trichothecenes and zearalenone. Epidemics of maize red ear rot, induced by *F. graminearum*, *F. culmorum* and *F. cerealis* usually lead to the widespread occurrence of zearalenone, deoxynivalenol, and to a lesser incidence of nivalenol and fusarenon X. The severity of the disease is not always proportional to the amount of toxin produced, and unexpectedly high levels of toxins may occur in the absence of severe disease symptoms. A survey in 1977 found that zearalenone occurred in maize ears infected by *F. graminearum* when the maize was grown as a second crop in southern Italy and harvested when weather conditions were cooler and wetter than usual. The occurrence of nivalenol and fusarenon X in Europe is most closely related to the presence of *F. cerealis* rather than the presence of *F. graminearum* as reported in Austria, Finland, Germany, Poland, and Yugoslavia (Logrieco *et al.*, 2002*c*), since *F. graminearum* nivalenol-chemotype strains are rare. Finally, accumulation of T-2 toxin in maize ears usually is related to epidemics of *F. sporotrichioides* as seen in Poland when record low temperatures occurred (Logrieco *et al.*, 2002*c*).

Fumonisins. Fumonisins usually are formed in plants infected with *F. verticillioides* or *F. proliferatum* and are commonly found in maize prior to harvest. The incidence of fumonisin B_1 in maize and maize-based foods and feed in Europe is increasing and problematic (SCOOP, 2003). In general, fumonisin B_1 in maize is most important in Southern Europe, *e.g.*, Portugal, France, Spain, Croatia and Italy (Logrieco *et al.*, 2002*c*). In Italy, high levels of fumonisins (up to 250 μg/g) were more often associated with pink ear rot than they were in other European countries (Bottalico, 1998). Moreover, an investigation carried out over a five-year period (1995-1999) by Pietri *et al.* (2004) found that fumonisin B_1 is the most common mycotoxin in northern Italian maize and that contamination with other mycotoxins usually is low, with the exception of 1996, a year in which high levels of deoxynivalenol contamination also occurred.

Fumonisin B_1 levels are significantly lower in Central to Northeastern Europe (Bottalico, 1998). In some northern areas, under very favorable conditions, fumonisins can accumulate to significant levels (up to 30 μg/g), as reported for the Slovak Republic in 1998, and may be accompanied by severe epidemics of *F. verticillioides* and *F. proliferatum* (Srobarova *et al.*, 2002). Thus, the risk of fumonisin B_1 contamination is highest in Southern Europe, where both the major fungal producers of fumonisin B_1, *F. verticillioides* and *F. proliferatum*, co-occur.

Moniliformin. In infected maize ears, the main moniliformin-producing *Fusarium* species are *F. subglutinans*, and *F. proliferatum*. *Fusarium subglutinans* is widespread in Central to Northern Europe, whereas *F. proliferatum* is increasingly common in South and Central European maize-growing areas (Logrieco *et al.*, 1995). In Italy, high levels of moniliformin were found in maize ears co-infected with *F. proliferatum* and *F. verticillioides* (Logrieco *et al.*, 2002*c*). Moniliformin also was found commonly in maize ears infected with *F. subglutinans* from Central and Northern European countries, including Austria and Poland, with very high amounts of moniliformin found in years with severe epidemics of *F. subglutinans*. The high incidence of moniliformin in maize ear rot from Austria and Poland also may be related to the occurrence of *F. avenaceum*, although this species rarely occurs at high frequencies on maize kernels (Logrieco *et al.*, 2002*c*). The co-occurrence of *F. avenaceum* and *F. subglutinans* in maize in Central and Northern Europe increases the risk of moniliformin accumulation in infected ears. Thus, moniliformin may be the most commonly expected mycotoxin associated with maize ear rot in Central Europe.

Beauvericin. In maize ears infected with *F. verticillioides*, *F. subglutinans* and *F. proliferatum*, beauvericin co-occurred with fumonisin B_1 produced by *F. verticillioides* or *F. proliferatum*, and/or moniliformin produced by *F. subglutinans* and *F. proliferatum*. In particular, beauvericin was reported in Italy, Poland and Austria (Logrieco *et al.*, 2002*c*), and in the Slovak Republic (Srobarova *et al.*, 2002). However, beauvericin is presumed to be more widespread than recorded, since it is produced not only by *F. proliferatum* and *F. subglutinans*, but also by several other *Fusarium* species that occur less frequently on maize including *F. tricinctum* (Logrieco *et al.*, 1998) and *F. poae* (Chelkowski *et al.*, 2007).

Aspergillus **maize infection and aflatoxins: A future problem for Southern Europe?**

Aflatoxins

Aflatoxins are produced by some strains of *A. flavus*, which synthesize aflatoxin B_1 and aflatoxin B_2, and by most strains of *A. parasiticus*, which also synthesize aflatoxin G_1 and aflatoxin G_2. Of these four main aflatoxins, aflatoxin B_1 and aflatoxin G_1, occur most frequently and at the highest levels in plant products. Aflatoxin-producing strains of *A. flavus* and *A. parasiticus* are distributed worldwide in soil and air. Some strains are plant pathogens and can infect maize in the field and then colonize harvested or stored maize kernels where aflatoxins accumulate. The contamination of maize by aflatoxins is a worldwide problem of particular concern in tropical and sub-tropical areas, although *A. flavus* and *A. parasiticus* also may colonize maize before harvest under favorable climatic conditions in the United States and Southern Europe. Hence, crops may be contaminated with aflatoxins, which leads to the contamination of processed maize-based foods and feeds. Such contamination is dangerous because aflatoxins are mutagenic, teratogenic and carcinogenic compounds that are classified by the IARC as Group 1 compounds, *i.e.*, compounds known to be carcinogenic to humans (IARC, 1993).

Occurrence of aflatoxins and aflatoxin producers on maize in Southern Europe

European Union legislation limits maize intended for feed to < 20 ng/g aflatoxin B_1 and maize intended for human consumption to < 2 ng/g aflatoxin B_1. Although some contamination of maize in the field always occurs (Pietri *et al.*, 2004), the occurrence of aflatoxins in excess of the European Union limits in maize kernels and maize-based products is rare in Italy. During the 2003 cropping season, environmental conditions during maize cultivation enabled high levels of maize contamination by aflatoxins in northern Italy. The weather was very hot and dry for several months, which water-stressed the maize plants and resulted in a very early harvest of very dry kernels. Thus, the levels of the aflatoxin B_1 contamination in 2003 were 5-70 times higher then the levels observed from 1995-2000. Over 25% of the flour samples intended for polenta preparation analyzed between November 2003 and

June 2004, were contaminated by aflatoxin B_1 at levels > 2 ng/g legal limit. Moreover, since this maize also was used to feed dairy cattle, the milk produced was highly contaminated by aflatoxin M_1 leading to the dumping of thousands of tons of contaminated milk (Pinelli *et al.*, 2005). Such incidents confirm the widespread occurrence of aflatoxigenic strains of *Aspergillus* in Italy and stress the importance of good agricultural practices (GAPs) both in the field and post-harvest. As the global climate changes to become hotter and drier, the possibility that all of Southern Europe may face a constant threat of aflatoxin contamination increases.

Grape

Black *Aspergillus* grape rot and ochratoxin A

Occurrence of ochratoxin A in grapes and grape derivatives

In 1966, Zimmerli and Dick (1996) reported a high incidence of ochratoxin A contamination in Swiss table wine that sparked widespread interest in ochratoxin A in wine and grape juice. The more southerly in Europe the wine originated the higher the incidence and level of ochratoxin A contamination, particularly in red wines. These findings resulted in a major panic amongst consumers, wine producers and national and European regulatory authorities. A number of research programs resulted to increase knowledge of the frequency and levels of contamination, to identify the etiological fungal agents, and to develop possible tools to control ochratoxin A contamination in grapes and grape derivatives. Confirmation of high levels of ochratoxin A in wine produced in the most Southern European regions was provided by market surveys of retail wines in Germany, Italy, Spain, and UK (Battilani and Pietri, 2002). In addition to wine, ochratoxin A also was found in dessert wines, grape juice (especially from red grapes), vinegar (especially balsamic vinegar), and dried vine fruits (Battilani and Pietri, 2002). There have been numerous reports since the mid-1990s of ochratoxin A in wines at levels above the 2 μg/l legal limit. Contamination is related to geographic origin and wine color with the highest ochratoxin A contamination in southern products, and with red wines more contaminated than rosé wines and rosé wines more contaminated than white wines (Battilani and Pietri, 2002).

Fungal producers of ochratoxin A in grapes

Ochratoxin A is produced mainly by species in the genera *Aspergillus* and *Penicillium*. Initially, *A. ochraceus* and *P. verrucosum* were considered to be responsible for the production of the toxin in grapes, similar to the problems in wheat. However, several recent investigations have shown that ochratoxin A in wine is due to contamination by black *Aspergilli*, primarily strains of *A. carbonarius* and others belonging to the *Aspergillus niger* species aggregate that cause a "black rot" disease (Battilani and Pietri, 2002). The disease is particularly severe in the warmer grape-producing southern portions of Spain, France, Italy, and Greece. The black, heavily melanized spores are highly resistant to sunlight and survive sun-drying. The fungi survive on plant debris in the soil and the conidia are disseminated in vineyards by air currents. These fungi infect via injuries caused by careless handling, insects and previous fungal pathogens, usually in mature berries, and then spread throughout the bunch.

The taxonomy of *Aspergillus* section *Nigri* is not resolved and the correct identification of the species responsible for ochratoxin A accumulation in grapes remains problematic (Perrone *et al.*, 2006). This point is critical since each species usually has a specific toxin profile and accurate identifications could be used to predict potential toxicological risks. Recently, DNA sequence polymorphisms have been used to improve identifications and help distinguish species in this section (Perrone *et al.*, 2004; Patino *et al.*, 2005). Other studies have combined ochratoxin A production with DNA analysis, to better identify the fungal producers of ochratoxin A (Cabañes *et al.*, 2002; Perrone *et al.*, 2006). In particular, Perrone *et al.* (2006) used Amplified Fragment Length Polymorphisms with genomic DNA sequences from and toxin production by 77 black aspergilli strains from grape berries, to show that *Aspergillus tubingensis* is a distinct ochratoxin A-producing species. Thus, *A. tubingensis, A. carbonarius* and *A. niger* all can lead to ochratoxin A contamination of wines. *Aspergillus carbonarius* is currently viewed as the most important fungal source of ochratoxin A contamination since it occurs the most frequently and has the highest proportion of toxigenic strains in field populations (Battilani *et al.*, 2003; Perrone *et al.*, 2006).

Future perspectives

The many reports from Europe on food and feed contamination by fungi show that the food and feed can be colonized by many fungi with a broad range of mycotoxin production capabilities that are adapted to different environmental niches. The biodiversity of these fungal species requires a complex response from plant pathologists and that particular attention be paid to different toxicological problems related to the specific fungal contamination. The correct identification of the toxigenic fungi contaminating feeds and foods in different parts of Europe is important for studying the levels of the interaction between toxigenic fungi and host plants, and for obtaining a clear picture of the toxicological risks to humans and domesticated animals. Therefore, developing easy to use tools for the rapid identification of toxigenic fungi and the detection of mycotoxins is a major challenge for scientists working in this field. Moreover, as the profile of fungi colonizing crop plants is heavily dependent on environmental conditions, developing risk assessment models for mycotoxin contamination from data collected at numerous sites in Europe for various crops is an important goal. Finally, potential additive and/or synergistic effects and risks resulting from exposure to multiple mycotoxins need to be evaluated to determine if risks beyond those already identified need to be considered in developing management practices or setting regulatory limits for these compounds in feeds and foods.

References

Battilani, P., Giorni, P. and Pietri, A. (2003) Epidemiology of toxin producing fungi and ochratoxin A ocurrence in grape. *European Journal of Plant Pathology* 109, 715-722.

Battilani, P. and Pietri, A. (2002) Ochratoxin A in grapes and wine. *European Journal of Plant Pathology* 108, 639-643.

Bottalico, A. (1998) *Fusarium* diseases of cereals: Species complex and related mycotoxin profiles. *European Journal of Plant Pathology* 80, 85-103.

Bottalico, A. and Perrone, G. (2002) Toxigenic *Fusarium* species and mycotoxins associated with head blight in small-grain cereals in Europe. *European Journal of Plant Pathology* 108, 611-624.

Cabañes, F.J., Accensi, F., Bragulat, M.R., Abarca, M.L., Minguez, S. and Pons, A. (2002) What is a source of ochratoxin A-contamination in wine? *International Journal of Food Microbiology* 79, 213-215.

Charmley, L.L, Trenholm, H.L., Prelusky, D.A. and Rosenberg, A. (1995) Economic losses and decontamination. *Natural Toxins* 3, 99-203.

Chelkowski, J., Ritieni, A., Wisniewska, H., Mulè, G. and Logrieco, A. (2007) Occurrence of toxic hexadepsipeptides in preharvest maize ear rot infected by *Fusarium poae* in Poland. *Journal of Phytopathology* 155, 8-12.

Desjardins, A.E. (2006) Fusarium *Mycotoxins: Chemistry, Genetics and Biology*. APS Press, St. Paul, Minnesota.

Doohan, F.M., Brennan, J. and Cooke, B.M. (2003) Influence of climatic factors on *Fusarium* species pathogenic to cereals. *European Journal of Plant Pathology* 109, 755-768.

International Agency for Research on Cancer (IARC). (1993) *IARC Monographs on the Evaluation of Carcinogenic Risks to Humans. Vol. 56. Some Naturally Occurring Substances: Food Items and Constituents, Heterocyclic Aromatic Amines and Mycotoxins*. International Agency for Research on Cancer, Lyon, France, pp. 397-444; 445-466; 467-488.

Jestoi, M., Rokka, M., Yli-Mattila, T., Parikka, P., Rizzo, A. and Peltonen, K. (2004) Presence and concentrations of the *Fusarium*-related mycotoxins beauvericin, enniatins and moniliformin in Finnish grain samples. *Food Additives and Contaminants* 21, 794-802.

Leslie, J.F. and Summerell, B.A. (2006) *The* Fusarium *Laboratory Manual*. Blackwell, Ames, Iowa.

Logrieco, A. and Visconti, A., eds. (2004) *Overview of Toxigenic Fungi and Mycotoxins in Europe*. Kluwer Academic Publishers, Dordrecht, The Netherlands.

Logrieco, A., Moretti, A., Ritieni, A., Bottalico, A. and Corda, P. (1995) Occurrence and toxigenicity of *Fusarium proliferatum* from preharvest maize ear rot, and associated mycotoxins, in Italy. *Plant Disease* 79, 727-731.

Logrieco, A., Moretti, A., Castella, G., Kostecki, M., Golinski, P. Ritieni, A. and Chelkowski, J. (1998) Beauvericin production by *Fusarium* species. *Applied and Environmental Microbiology* 64, 3084-3088.

Logrieco, A., Rizzo, A., Ferracane, L. and Ritieni, A. (2002*a*) Occurrence of beauvericin and enniatins in wheat affected by *Fusarium avenaceum* head blight. *Applied and Environmental Microbiology* 68, 82-85.

Logrieco, A., Moretti, A., Ritieni, A., Caiaffa, M.F. and Macchia, L. (2002*b*) Beauvericin: Chemistry, biology and significance. In: Upadhyay, R.K. (ed.) *Advances in Microbial Toxin Research and its Biotechnological Exploitation*. Kluwer Academic, New York, pp. 23-30.

Logrieco, A., Mulè, G., Moretti, A. and Bottalico, A. (2002*c*) Toxigenic *Fusarium* species and mycotoxins associated with maize ear rot in Europe. *European Journal of Plant Pathology* 108, 597-609.

Logrieco, A., Bottalico, A., Mulè, G., Moretti, A. and Perrone, G. (2003) Epidemiology of toxigenic fungi and their associated mycotoxins for some Mediterranean crops *European Journal of Plant Pathology* 109, 645-667.

Lund, F. and Frisvad, J.C. (2003) *Penicillium verrucosum* in wheat and barley indicates presence of ochratoxin A. *Journal of Applied Microbiology* 95, 1117-1123.

Marín, S., Homedes, V., Sanchis, V., Ramos, A.J. and Magan, N. (1999) Impact of *Fusarium moniliforme* and *Fusarium proliferatum* colonization of maize on calorific losses and fumonisin production under different environmental conditions. *Journal of Stored Food Products* 35, 15-26.

Martins, M.L. and Martins, H.M. (2002) Influence of water activity, temperature and incubation time on the simultaneous production of deoxynivalenol and zearalenone in corn (*Zea mays*) by *Fusarium graminearum*. *Food Chemistry* 79, 315-318.

Mateo, J.J., Mateo, R. and Jimenez, M. (2002) Accumulation of type A trichothecenes in maize, wheat and rice by *Fusarium sporotrichioides* isolates under diverse culture conditions. *International Journal of Food Microbiology* 72, 115-123.

Miller, J.D., Greenhalgh, R., Wang, Y.Z. and Lu, M. (1991) Trichothecene chemotype of three *Fusarium* species. *Mycologia* 83, 121-130.

Patino, B., Gonzàlez-Salgado, A., Gonzàlez-Jaèn, M.T. and Vàzquez, C. (2005) PCR detection assays for ochratoxin-producing *Aspergillus carbonarious* and *Aspergillus ochraceus* species. *International Journal of Food Microbiology* 104, 207-214.

Perrone, G., Susca, A., Stea, G., and Mulè, G. (2004) PCR assay for identification of *Aspergillus carbonarius* and *Aspergillus japonicus*. *European Journal of Plant Pathology* 110, 641-649.

Perrone, G., Mulè, G., Susca, A., Battilani, P., Pietri, A. and Logrieco, A. (2006) Ochratoxin A production and amplified fragment length polymorphism analysis of *Aspergillus carbonarius* and *Aspergillus tubingensis* and *Aspergillus niger* strains isolated from grapes in Italy. *Applied and Environmental Microbiology* 72, 680-685.

Pietri, A., Bertuzzi, T., Pallaroni, L. and Piva, G. (2004) Occurrence of mycotoxins and ergosterol in maize harvested over five years in Northern Italy. *Food Additives and Contaminants* 21, 479-487.

Pinelli, C., Scianchi, L. and Venè, F. (2005) Aflatossine nells filiera del latte: Programmi e metodologie di prevenzione (Aflatoxins in milk chain: Prevention programs and methodologies). *Scienza e Tecnica Lattiero-Casearia* 56, 37-46.

Puntaric, D., Bosnir, J., Smit, Z., Skes, I. and Baldaic, Z. (2001) Ochratoxin A in corn and wheat: Geographical association with endemic nephropathy. *Croatian Medical Journal* 42, 175-180.

Rizzo, A., Eskola, M. and Atroshi, F. (2002) Ochratoxin A in cereals, foodstuffs and human plasma. *European Journal of Plant Pathology* 108, 631-637.

Schothorst, R.C. and van Egmond, H.P. (2004) Report from SCOOP task 3.2.10 Collection of occurrence data of *Fusarium* toxins in food and assessment of dietary intake by the population of EU member states – Subtask: Trichothecenes. *Toxicology Letters* 153, 133-143.

SCOOP Report on Task for Scientific Cooperation. (2003) *Collection of Occurrence Data of* Fusarium *Toxins in Food and Assessment of Dietary Intake by the Population of the EU Member States* – Final Report. Task 3.2.10 European Community, Health and Consumer Protection Directorate General, Brussels, Belgium.

Srobarova, A., Moretti, A., Ferracane, R., Ritieni, A. and Logrieco, A. (2002) Toxigenic *Fusarium* species of *Liseola* section in pre-harvest ear rot, and associated mycotoxins in Slovakia. *European Journal of Plant Pathology* 108, 299-306.

Tomczak, M., Wisniewska, H. Stêpieñ, L., Kostecki, M., Chelkowski, J. and Golinski, P. (2002) Deoxynivalenol, nivalenol and moniliformin occurrence in wheat samples with scab symptoms in Poland (1998-2000). *European Journal of Plant Pathology* 108, 625-630.

Torp, M. and Nirenberg, H.I. (2004) *Fusarium langsethiae sp. nov.* on cereals in Europe. *International Journal of Food Microbiology* 95, 247-256.

Uhlig, S., Torp, M., Jarp, J., Parich, A., Gutleb, A.C. and Krska, R. (2004) Moniliformin in Norwegian grain. *Food Additives and Contaminants* 21, 598-606.

Uhlig, S., Torp, M. and Heier, B.T. (2006) Beauvericin and enniatins A, A_1, B and B_1 in Norwegian grain: A survey. *Food Chemistry* 94, 193-201.

Waalwijk, C., Kastelein, P., de Vries, I., Kerényi, Z., van der Lee, T., Hesselink, T., Kohl, J. Kema, G (2003) Major changes in *Fusarium* spp. in wheat in Netherlands. *European Journal of Plant Pathology* 109, 743-754.

Xu, X.M. (2003) Effects of environmental conditions on the development of *Fusarium* head blight of cereals. *European Journal of Plant Pathology* 109, 683-689.

Xu, X.M., Parry, D.W., Nicholson, P., Thomsett, M.A., Simpson, D., Edwards, S.G., Cooke, B.M., Doohan, F.M., Brennan, J.M., Monaghan, S., Moretti, A., Tocco, G., Mulè, G., Hornok, L., Giczey, G. and Tatnell, J. (2005) Predominance and association of pathogenic fungi causing Fusarium ear blight in wheat in four European countries. *European Journal of Plant Pathology* 112, 143-154.

Yli-Mattila, T, Paavanen-Huhtala, T., Parikka, P., Konstantinova, P. and Gagkaeva, T.Y. (2004) Molecular and morphological diversity of *Fusarium* species in Finland and northwestern Russia. *Mycological Research* 110, 573-585.

Zimmerli, B. and Dick, R. (1996) Ochratoxin A in table wine and grape-juice: Occurrence and risk assessment. *Food Additives and Contaminants* 13, 655-668.

14

The Impact of Mycotoxins in Animal Feeds

Johanna Fink-Gremmels*

Abstract

Exposure of animals to one or more mycotoxins present in their diet may induce acute signs of intoxication. More importantly in daily practice are the consequences of exposure to lower (sub-clinical) levels that affect performance and feed efficiency, resulting in significant economic losses. Undesirable residues of mycotoxins and/or their metabolites have been found in a wide range of food products, but the contribution of residues in animal-derived products to human exposure appears to be very low. The secondary effects of feed contamination with mycotoxins, *e.g.*, reduced immune competence and increased susceptibility to infectious diseases, can account for some of the increased use of antimicrobials in animal flocks. The increase of antimicrobial resistance in large animal operations and the related transfer of resistant organisms to humans are of increasing public health concern.

Introduction

Numerous feed materials are susceptible to mold invasion resulting in the contamination of raw materials and mixed feeds with mycotoxins. Mycotoxin formation can occur at the pre-harvest stage, as well as at postharvest stages and during storage. Adverse health effects associated with the contamination of feed materials by mycotoxins comprise acute intoxications, which can be life-threatening. More frequent are sub-acute (and often sub-clinical) alterations of organ function that impair weight gain and reduce productivity. Such reductions account for significant economic losses worldwide in agricultural production (CAST, 2003; Wu, 2007).

The prevalence of molds is associated with the local climate, and differs by continent and region. Many *Aspergillus* species require high temperatures for toxin production, whereas *Fusarium* and *Penicillium* species commonly are found in more temperate climates. The increasing global trade in feed materials has changed feed production practices. Large farms use less locally produced cereals and grains and have moved towards the use of mixed feeds that are composed of a variety of materials. Hence feed composition largely depends on the global market price of individual feed components (Binder *et al.*, 2007). Thus, animals frequently are exposed to complex mixtures of toxins, with the clinical signs of exposure and intoxications both variable and difficult to diagnose.

Exposure of farm animals to mycotoxins may result in a deposition of mycotoxins and/or their metabolites in animal tissues, and hence human exposure to undesirable residues. The concentrations of mycotoxins found in edible tissues, milk and eggs generally are low, so research has focused on mycotoxins that are potential human carcinogens. As a precautionary measure, maximum permissible levels in feeds have been set for these toxins, including aflatoxins, ochratoxin A and fumonisins in many countries (FAO, 2004). Guidance levels have been prescribed for individual animal species, taking into account differences in feed composition and consumption patterns. These levels vary significantly between countries, and often reflect regional differences in mycotoxin occurrence and farming practices.

Despite these regulatory attempts, the level of animal protection remains controversial. Farmers demand both optimal feed quality for their animals and a fair price for their products. In years in which the mycotoxin burden is high, a large percentage of the harvest does not meet the quality standards set for marketing and contaminated feed materials are used at the farm level, with negative consequences for the livestock on the farm due to long-term chronic exposures. Moreover, countries exporting feed materials experience significant trade barriers for their products, as exemplified by the rejection by the EU of shipments of peanuts and cereal grains contaminated with aflatoxins. Of increasing concern is the use of the plant residues that remain following biofuel production, which are less controlled prior to harvest for fungal invasion and mycotoxin levels and which can impose a considerable mycotoxin burden when they enter the feed chain.

The adverse health effects of mycotoxins on farm animals depend on the chemical nature of the mycotoxin and its mode of action (Fink-Gremmels, 1999; Hussein and Brasel, 2001). A major problem is the level of uncertainty in the assessment of chronic (long-term) exposure and exposure to mixtures of toxins, that represent real life situations. The available experimental data are derived from controlled experiments conducted under experimental conditions with healthy animals that are exposed to a single toxin. This setting clearly is artificial and often fails to identify subclinical effects and adverse health effects associated with a biological interaction of infectious agents and toxic compounds. The objective of this brief review is to summarize some of the major recognized mycotoxicoses in animals and to identify emerging risks associated with mycotoxin exposure of farm animals.

Common mycotoxicoses in farm animals

The most commonly described mycotoxins associated with acute and chronic intoxication of animals are aflatoxins, ochratoxins, fumonisins, trichothecenes, zearalenone and the ergot alkaloids. The impact of the presence of mycotoxins in animals feeds has been reviewed by the European Food Safety Authority (EFSA), who summarized the available data on mycotoxin exposure and adverse effects in farm animals (EFSA, 2004*a*,*b*,*c*, 2005*a*,*b*,*c*).

Aflatoxins

Aflatoxins occur worldwide on peanuts, maize and numerous other important feed commodities (Scudamore *et al*., 1997; Galvano *et al*., 2001; Thirumala-Devi *et al*., 2002; van Egmond *et al*., 2007). As in laboratory animals and humans, aflatoxins affect liver function in all farm animal species. Following acute exposure, cellular necrosis of parenchymal cells in the liver, proliferation of the bile ducts and jaundice are the most prominent clinical signs of aflatoxicosis. The subsequent impairment of liver function results in loss of feed intake, reduced weight gain, and reduced productivity including reduced egg or milk production. The liver lesions often can be diagnosed by visual inspection of the internal organs, as the liver is pale and enlarged and may show signs of neoplastic alterations. The animal species most sensitive to aflatoxins are poultry and fish. In poultry, the co-occurrence of cyclopiazonic acid, produced by the same fungal species, adds muscle weakness and signs of cardiotoxicity to the clinical presentations of exposed animals.

Aflatoxin exposure also results in effects on the immune system. Long-term exposure to low mycotoxin concentrations, which may produce no prominent signs of toxicity, can impair immune system functions. Subsequently, a lower immune response to vaccination programs and increased susceptibility to infectious diseases are noticed, as both, humoral and cellular immune responses are affected. Secondary effects include increased prevalence of *Salmonella* spp. and *Campylobacter* spp. infections in animals, unpredictable outbreaks of viral diseases and coccidiosis. Consequently, affected flocks require more medication, which in turn increases the risk for undesirable residues and the emergence of resistance to antimicrobials in poultry units and the products derived from the animals. The relationship between exposure to aflatoxins and the prevalence of hepatitis virus infections in human populations has been the subject of various epidemiological investigations, *e.g.*, Groopman and Kensler (2005).

Ochratoxins

The primary target organ of aflatoxicosis is the liver, while ochratoxins, primarily ochratoxin A, which is produced by various *Penicillium* and *Aspergillus* spp. that occur on grains and cereals, usually affect kidney functions. The signs of ochratoxin intoxication often are mild, and the initial increase in water intake and excretion (polyuria – polydipsia syndrome), hypothermia and reduced weight gain is often not noted in large animals units. The most susceptible animal species are pigs, followed by poultry. In contrast to aflatoxins, mortality usually is low, but like aflatoxins, ochratoxins affect the immune system and reduce the health status of a herd.

Ochratoxin A has a short half-life in poultry, but a much longer half-life in pigs. Following exposure, ochratoxin A can be detected in pig blood serum where it is bound to proteins. The high serum protein binding is one of the explanations for the longer biological half-life. In addition to blood serum, high concentrations of ochratoxin A are found in the liver and the kidneys. The persistence of residues in these organs has resulted in many investigations of the contribution of residues in animal-derived tissues to overall human exposure to ochratoxins. A recent overall exposure assessment of European consumers (EFSA, 2004*a*), however, concluded that the contribution from animal-derived products does not exceed 10%, as the major sources of human exposure are plant-derived products, including cereals, nuts, wines, grape juice, dried fruits and coffee.

In humans, ochratoxin A is excreted with mother's milk, resulting in exposure of the newborn. The same route of excretion also is expected in pigs. Various clinical case reports, describe general weakness, splay-leg and gangrene-like alterations in peripheral tissues, *e.g.*, ears and tails, as associated with the presence of ochratoxin A in feeds, but experimental data confirming a causal relationship are lacking. Exposure of newborn piglets to ochratoxin A also is expected to affect the immune system and to result in increased susceptibility to infectious diseases (Stoev *et al.*, 2000).

Fumonisins

Fumonisin B_1 was first described as the causal agent of equine leukoencephalomalacia, a fatal disease in horses (Kellerman *et al.*, 1990). In pigs, the clinical signs of fumonisin intoxication are entirely different, and comprise pulmonary hypertension and severe pulmonary edema that may result in acute mortality (Harrison *et al.*, 1990). In poultry, exposure to fumonisins has been described to cause a toxic feed syndrome (spiking mortality) characterized by poor performance, multifocal hepatic necrosis and paralysis. In ruminants, as well as in laboratory animal species, the earliest signs of fumonisin exposure are renal lesions. The diversity of target organs of toxicity is incompletely understood, although inhibition of the enzyme ceramide synthetase (a N-acetyl transferase) is commonly assumed to underlie the diverse adverse effects of fumonisin (Merrill *et al.*, 2001). The rate of absorption of fumonisin B_1, following ingestion with feed, is very limited. The concentrations in feed materials, particularly in maize and products thereof, however, are much higher than that of other toxins, and a commonly accepted tolerable threshold level is 1000 ng/g in products for human consumption. Fumonisins also are known as immunomodulating agents that can decrease the vaccinal antibody titers in pigs (Taranu *et al.*, 2005).

Fumonisins are produced primarily by *Fusarium verticillioides* [formerly termed *Fusarium moniliforme* (Seifert *et al.*, 2003)] and *Fusarium proliferatum*. These *Fusarium* species are common in tropical and subtropical areas and temperate regions with warm growing seasons. In temperate climatic zones with cooler growing seasons, however, other *Fusarium* species such as *Fusarium graminearum, Fusarium culmorum, F. proliferatum* and *Fusarium subglutinans* are more prevalent. In these cooler regions, fumonisin B_1 contamination is reduced and toxins produced by other *Fusarium* spp. become more important and more prominent. For example, *F. graminearum* is an important pathogen of maize, wheat, barley and oats, and is responsible for the production of both deoxynivalenol and zearalenone in these grains (D'Mello and Macdonald, 1997).

Trichothecenes

Deoxynivalenol is the most prominent toxin of a large group of epoxy-trichothecenes that share an epoxide group at position C12-C13. More than 180 trichothecenes have been described (Desjardins, 2006), and are produced not only by *Fusarium* species, but also by members of other fungal genera such as *Trichoderma*, *Stachybotrys*, *Verticimonosporium* and *Myrothecium*. Trichothecenes are divided into three broad classes based on their chemical characteristics. Group A trichothecenes include T-2 toxin, HT-2 toxin, neosolaniol, monoacetoxy scirpenol and diacetoxyscirpenol, whereas type B trichothecenes include deoxynivalenol, nivalenol, 3- and 15-acetyl-deoxynivalenol and fusarenon X. A third group, the macrocyclic trichothecenes, contains toxins such as satratoxins, verrucarins and roridins, which usually are produced by *Stachybotrys* species.

Type A trichothecenes, *e.g.*, T-2 toxin, were first detected as dermatotoxins that cause mucosal and skin lesions due to the reactivity of their epoxide moiety. They also inhibit protein and DNA synthesis and hence are associated with reduced cell proliferation in the bone marrow (resulting in pancytopenia), the lymphatic system (resulting in reduced immune competence through inhibition of the cellular and humoral immune response), as well as in the gastrointestinal tract, where primary necrosis and reduced cell proliferation result in hemorrhages and a non-specific inflammatory response that affect nutrient transport and utilization (Glenn, 2007).

These effects are exerted to a varying degree by all trichothecenes. A specific effect of deoxynivalenol is its obvious interaction with the dopaminergic system of the central nervous system, which can induce nausea and vomiting. This interaction resulted in the now rarely used term vomitoxin being given to deoxynivalenol to describe this clinically prominent effect. Deoxynivalenol is one of the most frequently occurring mycotoxins in cereal grains in Northern Europe, and appears to occur most frequently in common wild grasses, and products thereof, *e.g.*, hay and silage. Thus, next to pigs, which are the most sensitive animal species, cattle and wild ruminants, *e.g.*, red deer, are exposed to large quantities of this mycotoxin, due to their high consumption of grass. Under these circumstances, the ability of the rumen to degrade deoxynivalenol into less toxic metabolites is exhausted, and intact deoxynivalenol reaches the small intestines, where it is rapidly absorbed. Following absorption, deoxynivalenol has biphasic effects on the immune system, where it acts as an immune stimulant and pro-inflammatory agent at low concentrations, and as an immunosuppressant at higher levels. The balance between activation and suppression of the immune system depends on the health status of the animal. Thus, various reports have focused on the interaction between pathogens or products therefrom, *e.g.*, the bacterial lipopolysaccharide (LPS), with deoxynivalenol (Pestka *et al.*, 2004).

Recent evidence (Miller, 2002; Promputtha *et al.*, 2007) suggests an interaction between *Fusarium* species, and the presence of endophytes in grasses. This interaction might explain the increasing frequency of deoxynivalenol-contaminated small grains and seeds, and also provide a target for intervention strategies aimed at reducing the incidence of deoxynivalenol in foods and feed materials.

Zearalenone

Zearalenone, which also is produced by *F. graminearum* and related species, differs entirely in its chemical structure from the trichothecenes. The open lactone ring has many structural characteristics of the natural steroid estradiol. Zearalenone also is recognized by mammalian estrogen receptors. By acting as a mixed agonist/antagonist on these receptors, zearalenone induces a hyperestrogenic syndrome, with vulvovaginitis, enlarged uteri, enlarged nipples, and ovarian cysts. Young pigs are particularly sensitive and may be born with vulvovaginitis following intra-uterine exposure, and develop reproductive disorders at puberty. Co-exposure of animals to both zearalenone and deoxynivalenol seems to accelerate reproductive failure.

Zearalenone also interacts with the enzymes that are involved in steroid synthesis, *e.g.*, hydroxysteroid dehydrogenases. The dual mode of action, receptor binding and interaction with the synthesis of steroids, qualifies zearalenone as an endocrine disruptor. More research is needed to understand the long-term and transgenerational effects of this toxin in animals and probably humans (Fink-Gremmels, 2007).

Ergot alkaloids

Ergot alkaloids were amongst the first mycotoxins associated with a human mycotoxicosis, *cf.*, medieval reports of the Holy Fire or St. Anthony's Fire, a gangrenous disease in humans. Ergot alkaloids cause vasoconstriction, which reduces blood flow and oxygen transport in the terminal circulation of legs, tail and ears, resulting in necrosis. These compounds also cause abortions by reducing the blood supply to a gravid uterus. Ergot alkaloids are produced by *Claviceps* spp. that transform developing kernels of cereal grains (and grasses) into enlarged dark sclerotia. Ergot alkaloids also are produced as secondary metabolites by many species of the genera *Epichloë* and *Neotyphodium.* These endophytes often invade cold season grasses, which they colonize symbiotically. These fungi provide the plant with peramines, a class of compounds that protects the seeds and the living plant from insect damage while increasing heat tolerance (Faeth *et al.*, 2006). The most prominent animal diseases associated with endophytes are tall fescue toxicoses, which are associated with exposure to ergovaline and other alkaloids produced by *Neotyphodium coenophialum* in tall fescue (*Festuca arundinacea*), a grass species that covers millions of acres of farm land. In a similar way, the production of lolitrem B by *Neotyphodium lolii*, which colonizes rye grass (*Lolium perenne*) is responsible for a common neurotoxic syndrome in sheep, cattle and horses, generally referred to as rye grass staggers disease (Tor-Agbidye *et al.*, 2001; Fink-Gremmels, 2005).

The aim of this very brief summary was to provide an indication of the type of mycotoxicoses that occur in farm animals, with special reference to Europe. However, even more important than these acute intoxications, which occur only incidentally and affect only a small percentage of the animal population, are chronic mycotoxicoses. In the absence of obvious clinical symptoms of intoxication, chronic exposure to mycotoxins via contaminated feed materials often goes undiagnosed, as the entire herd is affected with some degree of decreased performance and reproductive capacity. Exposure to these lower levels originates in many cases from on-farm products as well as from mixed feeds. Critical products are maize and maize products, *e.g.*, maize-cob mix and silage, cereal grains, gluten, legumes, sorghum, peanuts, soybean products, cotton and sunflower seeds, straw, grass, hay and other forages that are given to animals in large quantities, very often without any control for the presence of mycotoxins. Chronic exposure is associated with two major effects: economic losses due to reduced performance and fertility, and increased risk to public health, due to higher disease prevalence in affected herds.

Nutrient utilization and effects of mycotoxins on the gut barrier

Based on the observation that mycotoxins reduce weight gain and feed conversion, mycotoxins were hypothesized to have a direct effect on the intestines (Li *et al.*, 2005). A malabsorption syndrome is a common result of exposure to aflatoxins, ochratoxins or trichothecenes. Measurable signs of the malabsorption syndrome are decreased transport of soluble nutrients and (fat-soluble) vitamins. Aflatoxin B_1 and ochratoxin A were the first toxins known to induce hypocarotinoedemia and reduced tissue levels of α-tocopherol (vitamin E). Comparable effects were seen in chickens, following exposure to T-2 toxin and aurofusarin. The reduced levels of vitamins with antioxidant properties are associated with an increase of reactive oxygen species and cellular oxidative stress, which impairs both normal cellular function and differentiated traits, such as intracellular killing of pathogens by cells of the immune system (see below).

Mycotoxins affect not only the transport of nutrients and vitamins, but also impair the barrier function of the gut (Bouhet and Oswald, 2005). State of the art experiments with

Caco cell monolayers and isolated segments of the intestines demonstrated that various mycotoxins, including ochratoxin A, deoxynivalenol and fumonisin B_1, alter the transmembrane electric resistance (Sergent *et al.*, 2006). These changes in membrane integrity not only contribute to the malabsorption syndrome (and hence the impaired nutrient transport), but also may increase the passage rate for bacterial toxins, including the LPS of Gram-negative bacteria and facilitate the migration of pathogens from the gut lumen into the systemic circulation. Such migration might contribute to the local inflammatory reaction in the gastrointestinal tract and the increased susceptibility of animals to bacterial and fungal infections following exposure to these mycotoxins (Islam and Pestka, 2006).

Chronic mycotoxicoses and immunosuppression

While growth retardation and reduced productivity are of economic importance, the intrinsic activity of many mycotoxins on the immune system of the animals is of increasing concern. The presence of moderate to low amounts of mycotoxins in daily feed rations increases the susceptibility of animals to viral, bacterial and parasitic diseases (Bondy and Pestka, 2000). This increased susceptibility requires increased therapeutic intervention with antibiotics and antiparasitic drugs. These interventions increase the costs for animal health care and the use of anti-infective agents, particularly antibiotics, at the farm level with a concomitant increase in the risk of induction and spread of antimicrobial resistance. The immunosuppressive effect of mycotoxins also may result in incomplete protection of farm animals following vaccination against viral diseases, as antibody formation is impaired. The impaired immune competence of animals following long-term exposure to mycotoxins has been hypothesized to facilitate the emergence and re-emergence of pandemic viral diseases in animals, including swine fever and avian influenza.

Measurable signs of impairment of innate immunity include the inhibition of phagocytosis by macrophages, the induction of an inflammatory response by different classes of cytokines, and activation of the complement cascade, which commonly follows exposure to various trichothecenes. The acquired immunity comprises B-lymphocyte-dependent immunoglobulin production, as well as the activation of T lymphocytes. The response of the immune system to individual toxins varies. For example, aflatoxins affect primarily the cellular immune response to pathogens and phagocytic cell functions. Various experiments with chickens, but also with laboratory animal species, and *ex vivo* experiments with human and rodent cells show that aflatoxin B_1 decreases the number of circulating T lymphocytes and splenic cell counts. It also reduces the phagocytic activity of the macrophages, reduces total complement activity, and decreases the production of pro-inflammatory cytokines. Inhibition of phagocytosis and reduced intracellular killing of pathogens make animals more susceptible to acute bacterial and fungal infections.

Ochratoxin A primarily affects the antibody-producing cells and decreases the synthesis of immunoglobulins. Impairment of cell-mediated immunity is associated with a decrease in circulating lymphocytes, monocytes and macrophages, and challenge experiments in immunized animals with specific pathogens, *e.g.*, *Pasteurella multocida*, demonstrated the impaired acquired immunity following vaccinations. These effects also are observed in new-born animals that were exposed prenatally to ochratoxin A. Higher sensitivity of animals to a bacterial challenge with *Listeria monocytogenes* or *Salmonella typhimurium* also occurs following exposure to T-2 toxin. T-2 toxin is a Type A trichothecene that affects bone marrow cells, resulting in a pancytopenia with reduced leukocyte counts, reduced blastogenic transformation of lymphocytes, and reduced response to mitogens of peripheral lymphocytes, thymic and spleen cells.

The immunomodulatory effects of deoxynivalenol and fumonisins remain the subject of intense research. Short-term exposure to deoxynivalenol seems to activate many immune functions, *e.g.*, the expression of pro-inflammatory cytokines, while long-term exposure to moderate or low levels suppresses the response to pathogens and induces autoimmune-like effects, such as the production and renal disposition of IgA. Total IgM and IgG levels also decrease. Mitogen-induced lymphocyte proliferation is induced at low concentrations, but reduced at higher concentrations. A comparable time and dose-dependent deregulation of immune functions also is observed following fumonisin B_1 exposure. Fumonisins induce the production of nitric oxide in various animal models with and without an LPS challenge. Nitric oxide is a strong vasodilator, which could contribute to the cardiovascular changes known to occur in pigs following exposure to fumonisin B_1, and to the leukoencephalomalacia seen in horses.

In conclusion, these few examples show that exposure to mycotoxins can both activate and suppress the immune system. General responses to mycotoxins, *e.g.*, the modulation of cell proliferation, impairment of protein synthesis, induction of oxidative stress and induction of apoptosis also may contribute to the immunomodulatory effects of mycotoxins (for a review see Bouhet and Oswald, 2005).

Induction of oxidative stress and cancer

Induction of cellular oxidative stress, resulting from an increase in the production of oxygen and hydrogen radicals, and a depletion of cellular defense mechanisms such as glutathione, is a common following exposure to many mycotoxins (Surai and Dvorska, 2005). The pro-oxidant effect of mycotoxins is dose- and time-dependent and may go beyond glutathione depletion to include changes in the activity of glutathione peroxidase, catalase and superoxide dismutase. For example, glutathione depletion in the liver is one of the earliest signs of exposure to mycotoxins such as aflatoxins and ochratoxins, but the response is transient. After repetitive exposure to low amounts of mycotoxins, measurable glutathione levels increase again, as a sign of adaptation. Cellular oxidative stress and enhanced radical production cause lipid peroxidation and cellular necrosis. Many mycotoxins including the trichothecenes, aflatoxins and ochratoxins also affect the transport of tocopherol, carotinoids and ascorbic acid (as mentioned above), which encourages lipid peroxidation, impairs vital cellular functions and induces cell necrosis and apoptosis. Particularly vulnerable cells are those with a high energy demand as well as rapidly proliferating cells, such as enterocytes and cells of the immune system. The degree of oxidative stress is assumed to be one of the key factors contributing to the immunosuppressive effect of mycotoxins.

Reactive oxygen species also can interact with nuclear DNA to form 8-deoxyguanosine or O^6-methylguanine, which are indicative of oxidative damage to DNA. Formation of 8-deoxyguanosine is associated with mutations resulting from G/T transitions, so reactive oxygen species may ultimately induce tumor formation. Therefore, oxidative DNA damage is considered a prominent non-genotoxic mechanism in determining the tissue specific risk of cancer development. This type of non-genotoxic mechanism has been proposed for ochratoxin A and fumonisin B_1, whereas aflatoxins can induce mutational events directly since its exo-epoxide can interact directly with DNA to form a stable adduct. Moreover, mutations in the tumor suppressor gene *p53* have been observed following exposure to aflatoxins.

In farm animals, the induction of cancer by mycotoxins is rarely seen, with the exception of liver cancer induced in fish species by aflatoxins, as during the relatively short life span of these animals, neoplastic lesions are not of clinical importance.

Effects of mycotoxins on reproduction and fertility

One of the most prominent toxins associated with impairment of the reproductive cycle, is zearalenone. Pigs are the most sensitive animal species, and this sensitivity is related to the metabolic conversion of zearalenone into its α-hydroxy metabolite, α-zearalenol, which has a higher estrogenic potency than does the parent molecule. Ruminants are less sensitive to zearalenone as α-zearalenol is produced in the rumen by the rumen microorganisms (Malekinejad *et al.*, 2006). Intestinal absorption of this hydroxy-metabolite is assumed to be rather low and the α-zearalenol is thought to be reconverted in the liver to zearalenone and/or to α-zearalenol or other secondary metabolites with lower estrogenic potential. The fate of zearalenone in horses remains to be evaluated, but zearalenone affects oocyte maturation *ex vivo*, suggesting a role for zearalenone in fertility disorders in mares (Minervini *et al.*, 2006).

Research on the effect of mycotoxins on reproduction has focused on females, particularly on the impairment of oocyte maturation and early embryogenesis. This focus is consistent with reports from veterinary practice that describe reduced fertility or reduced litter sizes as a consequence of mycotoxin exposure. Male reproduction and sperm quality also may be affected following exposure to mycotoxins. However, very few *in vivo* investigations have addressed this topic, *e.g.*, evaluating the effects of ochratoxin A exposure on sperm quality in boars (Biro *et al.*, 2003). The various *in vitro* experiments showing the susceptibility of spermatocytes to mycotoxins provide only limited information, as the concentrations used cannot be correlated to actual exposure levels or to mycotoxin levels in the testicles or the seminal fluid.

Experiments in rodents have shown that some mycotoxins are teratogens at high concentrations. Comparable effects in farm animals cannot be entirely excluded since the time window between exposure, in the vulnerable phase of the first trimester of pregnancy, and visible signs during delivery, is quite long. In daily practice, malformed animal neonates generally are euthanized, and records of their occurrence are very incomplete.

A mycotoxin with a distinct mechanism of action is ergovaline, the best investigated ergot alkaloid in cases of fescue (grass) intoxications. Ergovaline activates dopaminergic receptors and acts as an anti-prolactin agent. Thus, milk production following the *partus* is absent or delayed and can result in potentially fatal starvation of foals and calves if undiagnosed.

Risk assessment of mycotoxins in animal feed materials

The adverse biological effects exerted by mycotoxins have been subjected to various risk assessment strategies, most of them specifically addressing potential risk for humans. EFSA has generated a series of evaluations of mycotoxins, including aflatoxins, ochratoxins, fumonisins, ergot alkaloids and zearalenone, that specifically address adverse animal health effects (EFSA 2004*a*,*b*,*c*, 2005*a*,*b*,*c*). These evaluations were intended to provide the scientific rationale for setting guidance levels for these mycotoxins in animal feed materials that are produced in, or enter the European market. These evaluations confirm that the exposure of human consumers to mycotoxin residues in edible products of farm animals is very low. At the same time, they also clearly indicate that current knowledge of exposure and of dose-effect relationships in farm animals is limited. Data on the occurrence of mycotoxins in feed materials reported officially to the EU are scarce, and in many cases it was not clear whether the commodities evaluated were intended for human consumption or for animal feed(s). Data from the feed industry (voluntary quality control programs) generally remain unpublished. A recent survey demonstrated the worldwide occurrence of mycotoxins in animal feeds (Binder *et al.*, 2007), but the number of samples analyzed is too small for detailed exposure assessments for individual animal species.

Exposure assessment for farm animals also is complicated by the large variation in the composition of animal diets. Mixed feeds for poultry and pigs are produced on a large scale by commercial feed businesses. These feeds are composed to meet the nutritional requirements of the target animals (or age category) in terms of major feed constituents, such as proteins and fat (convertible energy), to which minerals and vitamins are added. A standardized diet does not exist, as the production of mixed feeds is a day-to-day business, largely influenced by the price for individual feed materials on the world market. The situation for ruminants is even more complex. For these species, the nutritional requirements are strongly influenced by the production stage, *e.g.*, high-milking cows versus dry cows or steers for fattening, and no standardized model is available that would cover these differences at the level of individual feed materials. Moreover, ruminants generally receive a diet that is composed of forages and concentrates, with the level of the concentrates varying from a few percent up to more than 70% in high yielding dairy cows. The heterogeneous distribution of mycotoxins within any commodity complicates this situation even further.

The large uncertainties in exposure assessment also make it difficult to establish realistic maximum tolerance levels in individual feed components for all animal species. The first feed legislation worldwide addressed the levels of aflatoxins B_1 and total aflatoxins in the feed for dairy cows. These levels were established because aflatoxin M_1, an aflatoxin metabolite produced in the liver, can be carried over into milk, and not because of adverse clinical signs in cattle. The current limits on aflatoxins in feed have recently been challenged, due to the high variability in diet composition, particularly the amount of concentrates, which may contain aflatoxins. Differences in the rate of carry-over of aflatoxin M_1 into milk, which seems to involve active transport mechanisms, may also be important and dependent upon the physiological and overall health status of individual animals (EFSA, 2004*c*).

In pig diets, both subclinical and transgenerational effects are of major concern, and these are difficult to analyze in common exposure models. Exposure of a sow to zearalenone may have consequences for the next generation, as intra-uterine exposure of piglets might have an impact in later stages of life, particularly the onset of puberty, the first pregnancy and lactation (Fink-Gremmels, 2007). The intensity of adverse effects of deoxynivalenol and other trichothecenes seems to be modulated by infectious agents, and hence varies from farm to farm. This variation explains why the European Union at present has set only guidance levels for mycotoxins in feeds. Evaluation of these levels over a period of several years might result in a more accurate estimate of the adverse health effects.

Conclusions and future perspectives

The objective of this brief overview was to provide a short introduction to the adverse effects and the risk assessment of mycotoxins in animal feeds. This overview is limited to the most prominent toxins, *i.e.*, toxins that have been addressed in legal actions and/or have a role in international trade disputes. From the veterinary point of view, mycotoxicoses in animals remain a prominent differential diagnosis at the farm level, and are a subject of increasing concern in terms of animal health and well being. The perception of the risk for consumers associated with the presence of mycotoxins in animal feeds is moving from the direct risk of toxicologically active residual amounts in milk, meat and eggs to the indirect risks, *e.g.*, the dissemination of diseases by animals that have inadequate immune competence due to mycotoxin exposure, and to the risks associated with the ever increasing risk of overuse of antimicrobial agents. The resulting induction of antibiotic resistant mutations in microorganisms is a matter of increasing public health and environmental concern. In addition to these concerns, other fungal secondary metabolites in addition to the well-established mycotoxins, are emerging and warrant further characterization.

References

Binder, E.M., Tan, L.M., Chin, L.J., Handl, J. and Richard, J. (2007) Worldwide occurrence of mycotoxins in commodities, feed and feed ingredients. *Animal Feed Science and Technology* 137, 265-282.

Biro, K., Barna-Vetro, I., Pecsi, T., Szabo, E., Winkler, G., Fink-Gremmels, J. and Solti, L. (2003) Evaluation of spermatological parameters in ochratoxin A-challenged boars. *Theriogenology* 60, 199-207.

Bondy, G.S. and Pestka J.J. (2000) Immunomodulation by fungal toxins. *Journal Toxicology Environmental Health B, Critical Reviews* 3, 109-43.

Bouhet, S. and Oswald, I. (2005) The effects of mycotoxins, fungal food contaminants on the intestinal epithelial cell-derived innate immune response. *Veterinary Immunology and Immunopathology* 108, 199-209.

CAST. (2003) *Mycotoxins: Risk in plant, animal and human systems.* Council for Agricultural Science and Technology, Ames, Iowa, pp. 58-85.

D'Mello, J.P.F. and Macdonald A.M.C. (1997) Mycotoxins. *Animal Feed Science and Technology* 69, 155-166.

Desjardins, A.E. (2006) Fusarium *Mycotoxins: Chemistry, Genetics and Biology*. APS Press, St. Paul, Minnesota.

EFSA. (2004*a*) The European Food Safety Agency. Opinion of the CONTAM Panel related to ochratoxin A as undesirable substance in animal feed. www.efsa.europa.eu/science/contam/contam_ opinions 645_en.html.

EFSA. (2004*b*) The European Food Safety Agency. Opinion of the CONTAM Panel related to deoxynivalenol as undesirable substance in animal feed. www.efsa.europa.eu/science/contam/con tam_ opinions 478_en.html.

EFSA. (2004*c*) The European Food Safety Agency. Opinion of the CONTAM Panel related to aflatoxins as undesirable substances in animal feed. www.efsa.europa.eu/science/contam/contam_ opinions 294_en.html

EFSA. (2005*a*) The European Food Safety Agency. Opinion of the CONTAM Panel related to fumonisins as undesirable substance in animal feed. www.efsa.europa.eu/science/contam/contam_ opinions 1037_en.html

EFSA. (2005*b*) The European Food Safety Agency. Opinion of the CONTAM Panel related to zearalenone as undesirable substance in animal feed. www.efsa.europa.eu/science/contam/contam_ opinions 527_en.html

EFSA. (2005*c*) The European Food Safety Agency. Opinion of the CONTAM Panel related to ergot as undesirable substance in animal feed. www.efsa.europa.eu/science/contam/contam_opinions 941_en.html

Faeth, S.H., Gardner, D.R., Hayes, C.J., Jani, A., Wittlinger, S.K. and Jones, T.A. (2006) Temporal and spatial variation in alkaloid levels in *Achnatherum robustum*, a native grass infected with the endophyte *Neotyphodium*. *Journal of Chemical Ecology* 32, 3007-3024.

FAO. (2004) *Worldwide regulations for mycotoxins in food and feed in 2003. A compendium. FAO Food and Nutrition Paper No 81*. FAO, Rome, Italy.

Fink-Gremmels, J. (1999) Mycotoxins: Their implications for human and animal health. *Veterinary Quarterly* 21, 115-120.

Fink-Gremmels, J. (2005) Mycotoxins in forages. In: Diaz, D.E. (ed.) *The Mycotoxin Blue Book*. Nottingham University Press, Nottingham, UK, pp. 249-268.

Fink-Gremmels, J. (2007) Clinical effects and biochemical mechanisms associated with exposure to the mycoestrogen zearalenone. *Animal Feed Science and Technology* 137, 326-342.

Galvano, F., Galofaro, V., Ritieni, A., Bognanno, M., de Angelis, A. and Galvano, G. (2001) Survey of the occurrence of aflatoxin M_1 in dairy products marketed in Italy: Second year of observation. *Food Additives and Contaminants* 18, 644-646.

Glenn, A.E. (2007) Mycotoxinogenic *Fusarium* species in animal feed. *Animal Feed Science and Technology* 137, 213-240.

Groopman, J.D. and Kensler, T.W. (2005) Role of metabolism and viruses in aflatoxin-induced liver cancer. *Toxicology and Applied Pharmacology* 206, 131-137.

Harrison, L.R., Colvin, B.M., Green, J.T., Newman, L.E. and Cole, J.R. (1990) Pulmonary edema and hydrothorax in swine produced by fumonisin B_1, a toxic metabolite of *Fusarium moniliforme*. *Journal of Veterinary Diagnostic Investigations* 2, 217-221.

Hussein, H.S. and Brasel, J.M. (2001) Toxicity, metabolism and impact of mycotoxins on humans and animals. *Toxicology* 167, 101-134.

Islam, Z. and Pestka, J.J. (2006) LPS priming potentiates and prolongs proinflammatory cytokine response to the trichothecene deoxynivalenol in the mouse. *Toxicology Applied Pharmacology* 211, 53-63.

Kellerman, T.S., Marasas, W.F.O., Thiel, P.G., Gelderblom, W.C.A., Cawood, M. and Coetzer, J.A.W. (1990) Leukoencephalomalacia in two horses induced by oral dosing of fumonisin B_1. *Onderstepoort Journal of Veterinary Research* 57, 269-275.

Li, M., Cuff, C.F. and Pestka, J.J. (2005) Modulation of murine host response to enteric reovirus infection by the trichothecene deoxynivalenol. *Toxicological Science* 87, 134-145.

Malekinejad, H., Maas-Bakker, R. and Fink-Gremmels, J. (2006) Species differences in the hepatic biotransformation of zearalenone. *Veterinary Journal* 172, 96-102.

Merrill, A.H., Jr., Sullards, M.C., Wang, E., Voss, K.A. and Riley, R.T. (2001) Sphingolipid metabolism: Roles in signal transduction and disruption by fumonisins. *Environmental Health Perspectives* 109, 283-289.

Miller, J.D. (2002) Aspects of the ecology of *Fusarium* toxins in cereals. *Advances in Experimental Medicine and Biology* 504, 19-27.

Minervini, F., Giannoccaro, A., Fornelli, F., Dell'Aquila, M.E., Minoia, P. and Visconti, A. (2006) Influence of the mycotoxin zearalenone and its derivatives (α- and β-zearalenol) on apoptosis and proliferation of cultured granulosa cells from equine ovaries. *Reproductive Biology and Endocrinology* 4, 62- final page.

Pestka, J.J., Zhou, H.-R., Moon, Y. and Chung, Y.J. (2004) Cellular and molecular mechanisms for immune modulation by deoxynivalenol and other trichothecenes: Unraveling the paradox. *Toxicological Letters* 153, 61-73.

Promputtha, I., Lumyong, S., Dhanasekaran, V., McKenzie, E.H., Hyde, K.D. and Jeewon, R. (2007) A phylogenetic evaluation of whether endophytes become saprotrophs at host senescence. *Microbial Ecology* 53, 579-90.

Scudamore, K.A., Hetmanski, M.T., Chan, H.K., and Collins, S. (1997). Occurrence of mycotoxins in raw ingredients used for animal feeding stuffs in the United Kingdom in 1992. *Food Additives and Contaminants,* 14, 157-173.

Seifert, K.A., Aoki, T., Baayen, R.P., Brayford, D., Burgess, L.W., Chulze, S., Gams, W., Geiser, D., de Gruyter, J., Leslie, J.F., Logrieco, A., Marasas, W.F.O., Nirenberg, H.I., O'Donnell, K., Rheeder, J.P., Samuels, G.J., Summerell, B.A., Thrane, U. and Waalwijk, C. (2003) The name *Fusarium moniliforme* should no longer be used. *Mycological Research* 107, 643-644.

Sergent, T., Parys, M., Garsou, S., Pussemier, L., Schneider, Y.L. and Larondelle, Y. (2006) Deoxynivalenol transport across human intestinal Caco-2 cells and its effects on cellular metabolism at realistic intestinal concentrations. *Toxicological Letters* 164, 167-76.

Stoev, S.D., Goundasheva, D., Mirtcheva, T., and Mantle, P.G. (2000). Susceptibility to secondary bacterial infections in growing pigs as an early response in ochratoxicosis. *Experimental Toxicology and Pathology* 52, 287-296.

Surai, P.F. and Dvorska, J.E. (2005) Effects of mycotoxins on antioxidant status and immunity. In: Diaz, D.E. (ed.) *The Mycotoxin Blue Book.* Nottingham University Press, Nottingham, UK, pp. 93-138.

Taranu, I., Marin, D.E., Bouhet, S., Pascale, F., Bailly, J.-D., Miller, J.D., Pinton, P. and Oswald, I. (2005) Mycotoxin fumonisin B_1 alters the cytokine profile and decreases the vaccinal antibody titer in pigs. *Toxicological Sciences* 84, 301-307.

Thirumala-Devi, K., Mayo, M.A., Reddy, G. and Reddy, D.V. (2002) Occurrence of aflatoxins and ochratoxin A in Indian poultry feeds. *Journalof Food Protection* 65, 1338-1340.

Tor-Agbidye, J., Blythe, L.L. and Craig, A.M. (2001) Correlation of endophyte toxins (ergovaline and lolitrem B) with clinical disease: Fescue foot and perennial ryegrass staggers. *Veterinary and Human Toxicology* 43, 140-146.

van Egmond, H.P., Schothorst, R.C., and Jonker. M.A. (2007) Regulations relating to mycotoxins in food: Perspectives in a global and European context. *Analytical and Bioanalytical Chemistry* 389, 147-157.

Wu, F. (2007) Measuring the economic impacts of *Fusarium* toxins in animal feeds. *Animal Feed Science and Technology* 137, 363-374.

Mycotoxin Detection Methods

15

Overview of Detection Methods for Mycotoxins

Michelangelo Pascale* and Angelo Visconti

Abstract

Several sensitive, accurate analytical methods based on chromatographic or immunochemical techniques are available for the major mycotoxins occurring in agricultural and food commodities. In addition, novel technologies including fluorescence polarization immunoassays, molecularly imprinted polymers, infrared spectroscopy, capillary electrophoresis and surface plasmon resonance biosensors have recently been suggested for use in mycotoxin detection. This review evaluates classical and emerging methods for the analysis of mycotoxins in foods and feedstuffs with an emphasis on their advantages and limitations.

Introduction

Mycotoxin contamination of agricultural and food commodities and beverages pose a risk to human and animal health due to the toxic effects of these naturally occurring contaminants. The major mycotoxins of worldwide concern are: aflatoxins – aflatoxin B_1, B_2, G_1, G_2, and M_1, ochratoxin A, fumonisins – fumonisins B_1, B_2 and B_3, deoxynivalenol, zearalenone, T-2 and HT-2 toxins. Exposure to these substances can lead to acute or chronic diseases and in some cases death. Consequently, sensitive and accurate methods of analysis are needed to gather adequate information on the levels of exposure to mycotoxins and to assess the relevant toxicological risk for humans and animals. In addition, analytical methods should allow the measurement of such contaminants at levels lower than the legal limits fixed by the European Union or other national or international regulations with good accuracy and precision to enable monitoring programs and international trade security.

Analytical methods for mycotoxins (Table 1) in feeds and foodstuffs generally require toxin extraction from the matrix with an adequate extraction solvent, a clean-up step intended to eliminate interference from the extract, and, finally, detection/determination of the toxin by suitable analytical instruments/technologies. Clean-up is essential for the analysis of mycotoxins at trace levels, and involves the use of solid phase extraction, and multifunctional or immunoaffinity columns. Different chromatographic methods are commonly used for quantitative determination of mycotoxins, including high performance liquid chromatography (HPLC) coupled with ultraviolet (UV), fluorescence or mass spectrometry (MS) detection, gas chromatography coupled with electron capture detection, flame ionization detection or MS detection, and thin-layer chromatography. In addition, commercial immunometric assays, such as enzyme-linked immunosorbent assays (ELISA) or membrane-based immunoassays, frequently are used for screening purposes as well.

Table 1. Advantages and disadvantages of classical methods for mycotoxin analysis.

Method	Advantages	Disadvantages
TLC	Simple, inexpensive and rapid Can be used for screening Simultaneous analysis of multiple mycotoxins Sensitive for aflatoxins & ochratoxin A	Poor sensitivity (for some mycotoxins) Poor precision Adequate separation may require two-dimensional analysis Quantitative only when used with a densitometer
GC	Simultaneous analysis of multiple mycotoxins Good sensitivity May be automated (autosampler) Provides confirmation (MS detector)	Expensive equipment Specialist expertise required Derivatization required Matrix interference problems Non-linear calibration curve Drifting response Carry-over effects from previous sample Variation in reproducibility & repeatability
HPLC	Good sensitivity Good selectivity Good repeatability May be automated (autosampler) Short analysis times Official methods available	Expensive equipment Specialist expertise required May require derivatization
LC/MS	Simultaneous analysis of multiple mycotoxins Good sensitivity (LC/MS/MS) Provides confirmation No derivatization required	Very expensive Specialist expertise requested Sensitivity relies on ionization technique Matrix assisted calibration curve (for quantitative analysis)
ELISA	Simple sample preparation Inexpensive equipment High sensitivity Simultaneous analysis of multiple samples Suitable for screening Limited use of organic solvents Visual assessment	Cross-reactivity with related mycotoxins Matrix interference problems Possible false positive/negative results Confirmatory LC analysis required Critical quantitation near regulatory limits Semi-quantitative (visual assessment)
Rapid tests	Simple and fast (5-10 min) No expensive equipment required Limited use of organic solvents Suitable for screening purposes Can be used *in situ*	Qualitative or semi-quantitative (cut off level) Possible false positive/negative results Cross-reactivity with related mycotoxins Matrix interference problems Lack of sensitivity near regulatory limits

TLC = Thin Layer Chromatography; GC = Gas Chromatography; HPLC = High Performance Liquid Chromatography; LC/MS = Liquid Chromatography/Mass Spectrometry; ELISA = Enzyme-Linked Immunosorbent Assay; Rapid tests = membrane-based card test; antibody-coated tube; immunodot cup test.

Recently, a variety of emerging methods (Table 2) have been proposed for mycotoxin analyses. They are based on novel technologies including a lateral flow device, fluorescence polarization immunoassay, infrared spectroscopy, capillary electrophoresis, fiber optic immunosensors, molecularly imprinted polymers, and biosensors based on surface plasmon resonance, quartz crystal microbalance and screen-printed carbon electrodes. This chapter describes classical and emerging methods for the analysis of mycotoxins in foods, feeds and beverages, with emphasis on their advantages and limitations.

Sample preparation

Sampling

Due to the heterogeneous contamination of mycotoxins in grains, sampling is the largest source of variation associated with the analysis of these naturally occurring contaminants. Some mycotoxin-sampling plans have been established, primarily for aflatoxins, to reduce the variation of mycotoxin test results (Whitaker, 2004). The sampling aspects are discussed in more detail elsewhere in this book (Miraglia *et al.*, Chapter 16).

Extraction and clean-up

Mycotoxins commonly are extracted from ground solid matrices by shaking or blending with mixtures of water or other polar solvents, most commonly methanol or acetonitrile. Liquid samples, *e.g.*, fruit juices or milk, are extracted by liquid-liquid partitioning, although this procedure is now being replaced by solid phase or immunoaffinity column extraction/clean-up.

Purification of the extract is an essential step in the analysis of mycotoxins, especially when chromatographic techniques are used for their determination at trace levels. Solid phase extraction, multifunctional clean-up columns, *e.g.*, MycoSep® or Trichothecenes P&EP, and immunoaffinity columns, frequently are used to clean up extracts of food products and animal feeds. Silica gel, Florisil®, SAX, C_8 and C_{18} reversed-phase media have been used as the stationary phase of solid phase extraction columns that perform well in the removal of interfering substances that carry over from the original matrices. MycoSep® columns are one of the most commonly used commercially available columns. These columns often are used for the simultaneous and rapid clean up of type A- and type B-trichothecenes. Immunoaffinity columns are based on monoclonal or polyclonal antibodies and increasingly are being used for mycotoxin analyses. The advantages of antibodies, with respect to other methods of purification, include: (i) provision of clean extracts due to the specificity of the antibody, (ii) utility in complex matrices, *e.g.*, feeds, coffee and cheeses, (iii) good precision, accuracy and sensitivity for analytical methods, (iv) rapid clean up, and (v) limited use of organic solvents. Extracts after clean up with immunoaffinity columns are analyzed by either an HPLC equipped with a UV or fluorescence detector or by a dedicated fluorometer. Immunoaffinity columns are commercially available for aflatoxins (B_1, B_2, G_1, G_2 and M_1), ochratoxin A, fumonisins, zearalenone, deoxynivalenol, and T-2 and HT-2 toxins and have been used to simultaneously detect the presence of these toxins by HPLC with fluorescence detection with good accuracy and precision (Gobel and Lusky, 2004; Sizoo and van Egmond, 2005; Visconti *et al.*, 2005).

Table 2. Advantages and disadvantages of emerging methods for mycotoxin analysis

Method	Advantages	Disadvantages
LFD	Rapid No clean up No expensive equipment Easy to use No specific training required Quantitative for aflatoxins	Semi-quantitative (visual assessment) Cross-reactivity with related mycotoxins Validation required for additional matrices
FPIA	Rapid No clean up required Validated for DON in wheat	Inconsistent with ELISA or HPLC analyses (except for DON) Poor sensitivity in some cases Cross-reactivity with related mycotoxins
IR spectroscopy (NIR, MIR)	Rapid Non-destructive measurement No extraction or clean up Easy operation	Expensive equipment Calibration model must be validated Knowledge of statistical methods Poor sensitivity
Capillary Electrophoresis	Rapid Limited organic solvent use Good resolution of analyte from interfering substances Good sensitivity (fluorescence capillary electrophoresis)	Expensive equipment Alternative to HPLC Clean-up may be required
Immunosensors/ Biosensors (SPR, FOI, QCM, SPCE)	Rapid No clean up procedure	Extract clean-up needed to improve sensitivity Cross-reactivity with related mycotoxins No labeling for detection
MIP	Low cost Stable Reusable	Non-imprinted polymers may perform similarly Poor selectivity

LFD = Lateral Flow Device; FPIA = Fluorescence Polarization Immunoassay; IR = Infrared Spectroscopy (NIR, near-infrared; MIR, mid-infrared); Immunosensors/Biosensors = Surface Plasmon Resonance (SPR); Fiber Optic Immunosensors (FOI); Quartz Crystal Microbalance (QCM); Screen-Printed Carbon Electrodes (SPCE); MIP = Molecularly Imprinted Polymer.

Classical technologies for detecting/quantifying mycotoxins

Thin-layer chromatography (TLC)

TLC is a simple, cost-effective technique often used as a mycotoxin screening assay when low detection limits are not required. A paper highlighting the status of TLC methods for mycotoxins in various sample matrices was published in a special issue of the *Journal of Chromatography A* (Lin *et al.*, 1998). Although TLC is a powerful tool for the simultaneous analysis of multiple samples for multiple mycotoxins, it cannot be used for sensitive or precise measurements unless densitometric analyses are performed. Highly reproducible, reliable results can be obtained if an autospotter is used to apply the samples. TLC can be used without cleaning up the extract, but extract purification prior to spotting increases sensitivity (Stroka *et al.*, 2000). A reversed-phase TLC method has been validated for the measurement of fumonisin B_1 in maize at μg/g levels (Shephard and Sewram, 2004), and a two-dimensional high performance thin layer chromatography (HPTLC) method has been developed for the determination of ochratoxin A at 5 ng/g in green coffee beans (Ventura *et al.*, 2005).

Gas chromatography (GC)

Gas chromatographic methods based on flame ionization detection (FID), electron-capture detection (ECD) and MS detection are the most widely used methods for quantitative determination of trichothecenes (mainly type A) in foods and feedstuffs (Krska *et al.*, 2001). These methods require preliminary clean-up of extracts by charcoal-alumina, Florisil®, silica gel or MycoSep® columns and pre-column derivatization of the purified extract with specific reagents to increase the volatility and sensitivity of the toxins. A comparative interlaboratory study, funded by the European Union, evaluated the performance of methods for analyzing trichothecenes with GC and found that improvements in methods are needed in toxin recovery and in the accuracy and precision of the measurements. The main problems came from compounds in the matrix that increased the trichothecene response (up to 120%) and resulted in non-linear calibration curves, drifting responses, carry-over or memory effects from previous samples, and high variation in terms of reproducibility and repeatability (Petterson and Langseth, 2002).

High performance liquid chromatography (HPLC)

HPLC coupled with UV, a diode array detector (DAD) or a fluorescence detector (FD) currently is the most widely used technique for the identification of the major mycotoxins in food commodities. Aflatoxin M_1, ochratoxin A, zearalenone, patulin and deoxynivalenol are routinely analyzed by HPLC/FD or HPLC/UV(DAD) with good accuracy and precision. HPLC/FD is highly sensitive, selective and repeatable, so specific labeling reagents have been developed, and are commercially available, for the derivatization of non-fluorescent mycotoxins to form fluorescent derivatives. Either pre-column derivatization, with trifluoroacetic acid (TFA), or post-column derivatization, with Br or I, can be used to identify aflatoxins B_1, B_2, G_1 and G_2, whereas pre-column derivatization with OPA reagent is required for fumonisins B_1, B_2 and B_3, and T-2 and HT-2 toxins require pre-column derivatization with 1-anthroylnitrile after purification of the extracts with immunoaffinity columns, solid phase extraction or MycoSep® columns.

Several HPLC methods for identifying various mycotoxins in a number of foods have been validated by collaborative studies, for which performance characteristics such us accuracy, repeatability, reproducibility, detection and quantification limits were established. These methods have been adopted as official or standard methods by the AOAC International or the European Standardization Committee. In particular, methods for measuring aflatoxins in maize, raw peanuts and peanut butter (AOAC Official Method 991.31), aflatoxin B_1 and total aflatoxins in peanut butter, pistachios, figs and paprika (999.07), ochratoxin A in barley (2000.03), aflatoxin M_1 in milk (2000.08), aflatoxin B_1 in baby food (2000.16), ochratoxin A in roasted coffee (2000.09), ochratoxin A in wine and beer

(2001.01), fumonisins B_1 and B_2 in maize flour and maize flakes (2001.04), aflatoxins in animal feed (2003.02) and ochratoxin A in green coffee (2004.10) that use immunoaffinity column clean-up and HPLC/FD have been approved as official methods by AOAC International (http://www.aoac.org). In addition, HPLC/immunoaffinity column methods have been validated for the measurement of deoxynivalenol in cereals and cereal products, zearalenone in barley, maize, wheat flour, polenta and maize-based baby food, aflatoxins in hazelnut paste and ochratoxin A in cocoa powder (Brera *et al.*, 2005; MacDonald *et al.*, 2005*a*,*b*; Senyuva and Gilbert, 2005).

Liquid chromatography/mass spectrometry (LC/MS)

Liquid chromatography coupled with mass spectrometry has been used for many years mainly as technique for mycotoxin confirmation. At the present time, LC/MS is the most promising technique for simultaneously screening, identifying and measuring a large number of mycotoxins.

Advances in mycotoxin detection by hyphenated chromatographic techniques/mass spectrometry have been reviewed recently (Sforza *et al.*, 2006). The following mycotoxins were examined: patulin, aflatoxins, ochratoxin A, zearalenone and its metabolites, trichothecenes and fumonisins. HPLC with tandem mass spectrometry (HPLC-MS/MS) and an Atmospheric Pressure Chemical Ionization or Electro-Spray Ionization interface was used for the simultaneous determination of the major type A- and type B-trichothecenes and zearalenone in cereals and cereal-based products at trace levels (Berthiller *et al.*, 2005*a*). HPLC-MS/MS also has been used for the simultaneous identification of nine mycotoxins – aflatoxins B_1, B_2, G_1, G_2 and M_1, ochratoxin A, mycophenolic acid, penicillic acid and roquefortine C – in cheese, for the identification of 18 mycotoxins and metabolites – ochratoxin A, zearalenone, α-zearalenol, β-zearalenol, α-zearalanol, β-zearalanol, fumonisins B_1 and B_2, T-2 toxin, HT-2 toxin, T-2 triol, diacetoxyscirpenol, 15-monoacetoxyscirpenol, deoxynivalenol, 3-acetyl deoxynivalenol, 15-acetyl deoxynivalenol, deepoxy-deoxynivalenol and aflatoxin M_1 – in milk, and for the determination of zearalenone and its metabolites in various biological matrices (Kleinova *et al.*, 2002; Kokkonen *et al.*, 2005; Sorensen and Elbaek, 2005). HPLC-MS/MS also is a powerful technique for the determination of masked deoxynivalenol, *i.e.*, deoxynivalenol-glucosides, in wheat. Masked mycotoxins are mycotoxins conjugated to more polar substances, *e.g.*, glucose, not detected by routine analytical methods even though they can release their toxic precursors after hydrolysis (Berthiller *et al.*, 2005*b*).

Accuracy, precision, and sensitivity of LC/MS methods may vary depending on the mycotoxin, matrix and instrument with the sensitivity of the method depending on the ionization technique used. Quantitative measurement of mycotoxins by LC/MS often is unsatisfactory due to matrix effects and ion suppression. Purification of extracts by MycoSep® or immunoaffinity columns generally is needed prior to MS detection.

Enzyme-linked immunosorbent assays (ELISAs)

Immunological assays have been used to successfully detect mycotoxins since the late 1970s (Pestka *et al.*, 1995). Several microtiter plate- or membrane-based ELISAs that use monoclonal or polyclonal antibodies against mycotoxins currently are available commercially for qualitative, semi-quantitative and quantitative analysis of the major known mycotoxins from a number of food matrices. In general, ELISAs do not require clean-up procedures and the extract containing the mycotoxin is analyzed directly. Even though they often lack accuracy at very low concentrations (competitive assays) and are limited in the range of matrices examined, immunoassays provide fast, inexpensive screening assays. However, matrix interference or the presence of structurally related mycotoxins can interfere with binding of the conjugate and the antibody leading to mistakes in quantitative measurements of mycotoxins. ELISA kits should be used routinely only for the analysis of matrices that have been extensively tested. Confirmatory analyses by more robust methods, *e.g.*, HPLC with an immunoaffinity column clean up or LC-MS, are required for contamination levels that approach the legal limit.

Some immunoassays have been validated by collaborative studies and adopted by the AOAC International as official methods for determination of aflatoxin B_1/total aflatoxins and zearalenone in food and feed matrices (AOAC Official Methods No. 989.06, No. 990.32, No. 990.34, No. 991.45, No. 993.16 and No. 994.01). Nevertheless, the use of ELISAs to detect mycotoxins at contamination levels approaching the legal limits is inappropriate since these assays were validated at levels much higher than the legal limits. In addition, ELISAs have less precision than either TLC or HPLC methods. A direct ELISA has been validated for measuring total fumonisins, *i.e.*, fumonisins $B_1 + B_2 + B_3$, in maize at levels > 1.0 μg/g (AOAC Official Methods No. 2001.06) with good precision.

Flow-through enzyme immunoassays for field use have been developed for the rapid detection of aflatoxin B_1, aflatoxin M_1, fumonisins, ochratoxin A, zearalenone and T-2 toxin. A quick qualitative (visual) discrimination between positive and negative samples can be made at levels close to the regulatory limits. Rapid, flow-through tests have been validated for several matrices and provide results consistent with those obtained by HPLC. This assay was both accurate and reliable giving no false compliant and only a few false non-compliant results (Papens *et al.*, 2004).

The U.S. Department of Agriculture – Grain Inspection, Packers and Stockyards Administration (USDA-GIPSA) has implemented a program (GIPSA Directive 9181.2) to verify the performance of antibody-based rapid kits for the quantification of mycotoxins in grain and grain products (http://archive.gipsa.usda.gov/reference-library/directives/9181-2.pdf). A number of kits for aflatoxins, deoxynivalenol and fumonisins have been verified and selected (http://archive.gipsa.usda.gov/tech-servsup/metheqp/testkits.pdf).

Emerging technologies for mycotoxin analyses

Lateral flow devices (LFDs)

A lateral-flow device, also called an immunochromatographic test, is a rapid immunoassay based on the interaction between specific antibodies, immobilized on a membrane strip, and antibody-coated dyed receptors, *e.g.*, latex or colloidal gold, that react with the analyte to form an analyte-receptor complex. Competitive LFDs rely on the competition of the analyte, *e.g.*, a mycotoxin, in solution for the binding sites of the labeled receptor. The test line contains an analyte-conjugate attached to the membrane that binds unbound receptor to form a colored analyte-conjugate receptor complex. The control line includes a specific antibody attached to the membrane that binds with the labeled receptor. Upon binding, the control line changes to a colored signal. If the signal in the test line is missing or weakly visible, the test indicates that the analyte is present in a sufficient amount (positive test). If the test line signal is clearly visible, the test indicates that the analyte is not present in the extract (negative test). The benefits of LFDs include user-friendly format, rapid response and price. These features make strip tests ideal for applications such as "on-site" detection of environmental and agricultural analytes.

A one-step LFD has been developed in a competitive immunoassay format, for the determination of ochratoxin A in fungal cultures. The use of an immunoaffinity column clean-up of grain extracts allowed LFDs to be used to determine the presence of ochratoxin A at low levels (Danks *et al.*, 2003).

Lateral flow immunochromatographic assays are commercially available for the determination of aflatoxin B_1 in maize and peanuts (Rosa®aflatoxin – Charm, USA; AgraStrip – Romer Labs, USA; Reveal®aflatoxin – Neogen, USA), aflatoxin M_1 in milk (Rosa®aflatoxin M1 – Charm, USA) and DON in wheat (RIDA®Quick DON – Biopharm, Germany; Reveal®DON – Neogen, USA). The Rosa®aflatoxin test was the first quantitative lateral flow test developed that was approved by USDA-GIPSA for determination of aflatoxins in maize samples at levels > 1.0 ng/g.

Fluorescence polarization immunoassay (FPIA)

FPIA is a simple technique that measures interactions between a fluorescently labeled antigen and a specific antibody. The technology, first developed in the 1970s, has long been used in human and veterinary diagnostics. Fluorescence polarization is based on differences in the rate of rotation between a free fluorescently labeled antigen and the same antigen bound to a specific antibody. Adaptation of these assays to mycotoxin analyses is still in progress. FPIAs have been used for the rapid determination of aflatoxins, zearalenone, fumonisins and deoxynivalenol in solution, although low accuracy and sensitivity were problems when these assays were used with cereal samples (Maragos and Kim, 2004). Recently, an optimized FPIA has been developed for rapid screening of deoxynivalenol in wheat and derivative products (pasta and semolina). In a comparison with a widely used HPLC/immunoaffinity method, FPIA was a rapid, inexpensive alternative to the more robust chromatographic methods for the determination of deoxynivalenol in wheat and derivative products (Lippolis *et al.*, 2006).

Infrared spectroscopy

The use of near-infrared (NIR) transmittance spectroscopy and mid-infrared (MIR) spectroscopy with attenuated total reflection for the rapid determination of deoxynivalenol in kernels of wheat and maize has been evaluated by observing changes in protein, lipid and carbohydrate contents. Principal component analyses, and cluster analyses or partial least square regression models were used to identify samples contaminated with > 310 ng/g deoxynivalenol by MIR and to predict deoxynivalenol content at levels close to the legal limit in the European Union, *i.e.*, 1,750 ng/g, by NIR. The applicability of NIR reflectance spectroscopy, combined with multivariate statistical methods, was recently evaluated for its ability to predict the incidence of fungal infection in maize and fumonisin B_1 content by analyzing 280 naturally and artificially contaminated maize samples. The NIR methodology was used to monitor mold contamination in post-harvest maize and to distinguish lots contaminated with fumonisin B_1 from those that were not contaminated, although the limit of detection of the method was not reported (Berardo *et al.*, 2005).

Capillary electrophoresis

Capillary electrophoresis is an analytical technique that allows good separation of mycotoxins from potential interfering species present in the extract on the basis of electrical charge. Capillary electrophoresis methods have been developed for various mycotoxins, including aflatoxins, citrinin, deoxynivalenol, fumonisins, moniliformin, ochratoxins, penicillic acid, roridin A, sterigmatocystin, and zearalenone by using UV/visible detection. Recently, fluorescence-based capillary electrophoresis methods have been developed for several mycotoxins allowing their detection at levels commonly found in naturally contaminated food samples. Capillary zone electrophoresis with laser-induced fluorescence was used to measure ochratoxin A in roasted coffee, maize and sorghum after tandem clean-up (silica and immunoaffinity columns) of extracts for the analysis of aflatoxin B_1 in maize, and for the measurement of fumonisin B_1 in maize after immunoaffinity column clean-up and derivatization with fluorescein isothiocyanate (Corneli and Maragos, 1998). A capillary zone electrophoresis-diode array detection method was developed for the measurement of moniliformin in maize. Capillary zone electrophoresis methods are comparable in sensitivity, precision and accuracy to HPLC methods (Maragos, 2004). The use of less expensive capillaries, the absence of organic solvents during the detection step, and shorter analysis times, all make capillary-zone electrophoresis methods viable alternatives to those requiring HPLC.

Fiber-optic immunosensors

Evanescent wave-based fiber-optic immunosensors have been developed for the detection of fumonisin B_1 and aflatoxin B_1 in maize. A competitive format was used to measure fumonisin B_1 in solution by immobilizing fumonisin monoclonal antibodies on an optical fiber and measuring the competition between fumonisin B_1 and a fumonisin B_1 fluorescently labeled probe for binding to the fiber. A detection limit of 3.2 μg/g was observed. The use of an immunoaffinity column for clean-up of the extracts increased the sensitivity of the method (detection limit 0.4 μg/g). A non-competitive assay was used for aflatoxin B_1 that takes advantage of the native fluorescence of this mycotoxin. The sensor could detect 2 ng/ml of aflatoxin B_1 in phosphate buffered saline solution. Problems due to refractive index-related effects were observed in the presence of organic solvents, which reduced the specificity of the assay (Maragos and Thompson, 1999).

Biosensors

A biosensor is an analytical device that incorporates a specific biological element, *e.g.*, an antibody, that creates a recognition event and a physical element that transduces the recognition event into an acoustic, electrical or optical signal. Immunochemical biosensors that use surface plasmon resonance, quartz crystal microbalance and screen-printed carbon electrodes have been described for the detection of mycotoxins.

Competitive surface plasmon resonance-based immunoassays have been used for rapid screening of aflatoxin B_1, zearalenone, ochratoxin A, fumonisin B_1 and deoxynivalenol in naturally contaminated matrices (Daly *et al.* 2000; Dunne *et al.*, 2005; Tudos *et al.*, 2003; van der Gaag *et al.*, 2003). Surface plasmon resonance biosensors measure refractive index changes that occur on a metal film and provide a signal that is positively correlated with the mass density changes on the metal surface. Mycotoxins were chemically modified and then immobilized on the surface of the sensor chip. A fixed concentration of specific antibody is mixed with sample extract containing the mycotoxin to be detected and the mixture is passed over the sensor surface where non-conjugated antibodies are bound to the mycotoxin on the surface. Regeneration of the surface of the sensor chip allows reuse of the same chip up to 100 times. All toxins were detected at very low levels with good accuracy and precision. The use of a surface plasmon resonance equipment with four flow cells enables the detection of four mycotoxins in a single measurement (van der Gaag *et al.*, 2003).

A quartz crystal microbalance biosensor has been developed for the rapid measurement of ochratoxin A in liquid matrices, *e.g.*, water or juices, at concentrations ranging from 4 to 50 ng/ml without a clean-up step (Visconti and de Girolamo, 2005).

Competitive electrochemical ELISAs based on disposable screen-printed carbon electrodes have been developed for quantitative determination of ochratoxin A in wheat and in wine. Detection limits were 0.4 ng/g and 0.9 ng/ml, respectively. Results from screen-printed carbon electrodes and HPLC/immunoaffinity column clean-up methods for naturally contaminated samples were correlated (Alarcòn *et al.*, 2006).

All of these biosensors have the potential to be very convenient in terms of time of analysis, but may require coupling with a clean-up technique for optimal effectiveness.

Molecularly imprinted polymers

Molecularly imprinted polymers are cross-linked polymers that are thermally, photochemically or electrochemically synthesized by the reaction of a monomer, *e.g.*, methacrylic acid, 4-vinyl pyridine or pyrrole, and a cross-linker, *e.g.*, ethylene glycol dimethacrylate, divinyl benzene, in the presence of an analyte, *e.g.*, a mycotoxin, or mimic compounds, *i.e.*, a "dummy", used as a template. After polymerization, the analyte is removed leaving specific recognition sites inside the polymer. Molecularly imprinted polymers provide biomimetic recognition elements capable of selective binding/rebinding to the analyte with efficiencies comparable to those of antibody-antigen interactions. The development of molecularly imprinted polymers for mycotoxins is very attractive due to their low costs, easy preparation, high chemical stability and long shelf-life. The synthesis of molecularly imprinted polymers with high affinity for deoxynivalenol, zearalenone and ochratoxin A has been reported (Weiss *et al.*, 2003; Zhou *et al.*, 2004; Visconti and de Girolamo, 2005; Yu *et al.*, 2005). These polymers have been used as the stationary phase in chromatographic applications or for the preparation of solid phase extraction columns for use in sample clean-up although, in a few cases, non-imprinted polymers, *i.e.*, polymers synthesized without a mycotoxin template, performed similarly to molecularly imprinted polymers. Recently, a molecularly imprinted polypyrrole film was synthesized on the sensor of a miniaturized surface plasmon resonance device for detection of ochratoxin A in wheat (Yu and Lai, 2005).

Conclusions

Several analytical methods for the measurement of mycotoxins occurring in foods, feeds and beverages have been developed and continuously improved in order to satisfy performance criteria (accuracy and precision) and to reliably reach the low detection limits needed for risk assessment studies. Currently, immunoaffinity column clean-up coupled with HPLC is the most powerful technique for the measurement of the major known mycotoxins occurring in agricultural and food commodities. HPLC/immunoaffinity column methods have good sensitivity and comply with the standards established by international organizations, *e.g.*, AOAC International and the European Standardization Committee. A number of official/validated methods based on HPLC/FD or UV and immunoaffinity column clean-up are currently available for the determination of the major mycotoxins in several food matrices that fulfill European Union and other international regulations. Various immunological assays, both ELISAs and other rapid antibody-based tests, also are available for measuring mycotoxins in different foods and generally used for screening purposes. These methods often require confirmatory analyses by HPLC/immunoaffinity columns or LC-MS. Methods for the simultaneous detection of multiple mycotoxins are highly desirable for screening purposes. HPLC coupled with mass spectrometry detection seems to be the most promising technique for the simultaneous determination and identification of a large number of mycotoxins. Finally, several novel technologies, which often are combined with immunochemical assays, have been proposed for the rapid analysis of mycotoxins in foods and beverages, but further investigations are required to validate them and to determine their applicability to real samples, especially at levels close to legal limits.

References

Alarcòn, S.H., Palleschi, G., Compagnone, D., Pascale, M., Visconti, A. and Barna-Vetró, I. (2006) Monoclonal antibody based electrochemical immunosensor for the determination of ochratoxin A in wheat. *Talanta* 69, 1031-1037.

Berardo, N., Pisacane, V., Battilani, P., Scandolara, A., Pietri, A. and Marocco, A. (2005) Rapid detection of kernel rots and mycotoxins in maize by Near-Infrared Reflectance spectroscopy. *Journal of Agricultural and Food Chemistry* 53, 8128-8134.

Berthiller, F., Schuhmacher, R., Buttinger, G. and Krska, R. (2005*a*) Rapid simultaneous determination of major type A- and B-trichothecenes as well as zearalenone in maize by high performance liquid chromatography-tandem mass spectrometry. *Journal of Chromatography A* 1062, 209-216.

Berthiller, F., Dall'Asta, C., Schuhmacher, R., Lemmens, M., Adam, G. and Krska, R. (2005*b*) Masked mycotoxins: determination of a deoxynivalenol glucoside in artificially and naturally contaminated wheat by liquid chromatography-tandem mass spectrometry. *Journal of Agricultural and Food Chemistry* 53, 3421-3425.

Brera, C., Grossi, S. and Miraglia, M. (2005) Interlaboratory study for ochratoxin A determination in cocoa powder samples. *Journal of Liquid Chromatography and Related Technologies* 28, 35-61.

Corneli, S. and Maragos, C.M. (1998) Capillary electrophoresis with laser-induced fluorescence: method for the mycotoxin ochratoxin A. *Journal of Agricultural and Food Chemistry* 46, 3162-3165.

Daly, S.J., Keating, G.J., Dillon, P.P., Manning, B.M, O'Kennedy, R., Lee, H.A. and Morgan, M.R.A. (2000) Development of surface plasmon resonance-based immunoassay for aflatoxin B_1. *Journal of Agricultural and Food Chemistry* 48, 5097-5104.

Danks, C., Ostoja-Starzewska, S., Flint, J. and Banks, J.N. (2003) The development of a lateral flow device for the discrimination of OTA producing and non-producing fungi. *Aspects of Applied Biology* 68, 21-28.
Dunne, L., Daly, S. Baxter, A., Haughey, S. and O'Kennedy, R. (2005) Surface plasmon resonance-based immunoassay for the detection of aflatoxin B_1 using single-chain antibody fragments. *Spectroscopy Letters* 38, 229-245.
Gobel, R. and Lusky, K. (2004) Simultaneous determination of aflatoxins, ochratoxin A and zearalenone in grains by new immunoaffinity column/liquid chromatography. *Journal of AOAC International* 87, 411-416.
Kleinova, M., Zollner, P., Kahlbacher, H., Hochsteiner, W. and Lindner, W. (2002) Metabolic profiles of the mycotoxin zearalenone and of the growth promoter zeranol in urine, liver and muscle of heifers. *Journal of Agricultural and Food Chemistry* 50, 4769-4776.
Kokkonen, M., Jestoi, M. and Rizzo, A. (2005) Determination of selected mycotoxins in mouldy cheeses with liquid chromatography coupled to tandem with mass spectrometry. *Food Additives and Contaminants* 22, 449-456.
Krska, R., Baumgartner, S. and Joseph, R. (2001) The state-of-the-art in the analysis of type-A and -B trichothecene mycotoxins in cereals. *Fresenius Journal of Analytical Chemistry* 371, 285-299.
Lin, L., Zhang, J., Wang, Y. and Chen, J. (1998) Thin-layer chromatography of mycotoxins and comparison with other chromatographic methods. *Journal of Chromatography A* 815, 3-20.
Lippolis, V., Pascale, M. and Visconti, A. (2006) Optimization of a fluorescence polarization immunoassay for rapid quantification of deoxynivalenol in durum wheat based products. *Journal of Food Protection* 69, 2712-2719.
MacDonald, S.J., Chan, D., Brereton, P., Damant, A. and Wood, R. (2005*a*) Determination of deoxynivalenol in cereals and cereal products by immunoaffinity column cleanup with liquid chromatography: Interlaboratory study. *Journal of AOAC International* 88, 1197-1204.
MacDonald, S.J., Anderson, S., Brereton, P., Wood, R. and Damant, A. (2005*b*) Determination of zearalenone in barley, maize and wheat flour, polenta, and maize-based baby food by immunoaffinity column cleanup with liquid chromatography: Interlaboratory study. *Journal of AOAC International* 88, 1733-1740.
Maragos, C.M. (2004) Detection of moniliformin in maize using capillary zone electrophoresis. *Food Additives and Contaminants* 21, 803-810.
Maragos, C.M., and Kim, E.-K. (2004) Detection of zearalenone and related metabolites by fluorescence polarization immunoassay. *Journal of Food Protection* 67, 1039-1043.
Maragos, C.M. and Thompson, V.S. (1999) Fiber-optic immunosensor for mycotoxins. *Natural Toxins* 7, 371-376.
Papens, C., de Saeger, S., Sibanda, L., Barna-Vetro, I., Leglise, I., van Hove, F. and van Peteghem, C. (2004) A flow-through enzyme immunoassay for the screening of fumonisins in maize. *Analytica Chimica Acta* 523, 229-235.
Pestka, J.J., Abouzied M.N. and Sutikno. (1995) Immunological assays for mycotoxin detection. *Food Technology* 49, 120-128.
Petterson, H. and Langseth, W. (2002) Intercomparison of trichothecenes analysis and feasibility to produce certified calibrants and reference material. European Commission BCR Information Project reports EUR 20285/1 EN (part I) and EUR 20285/2 EN. http://www.plant.wageningen-ur.nl/projects/fusarium.
Senyuva, H.Z. and Gilbert, J. (2005) Immunoaffinity column cleanup with liquid chromatography using post-column bromination for determination of aflatoxins in hazelnut paste: Interlaboratory study. *Journal of AOAC International* 88, 526-535.
Sforza, S., Dall'Asta, C. and Marchelli, R. (2006) Recent advances in mycotoxin determination in food and feed by hyphenated chromatographic techniques/mass spectrometry. *Mass Spectrometry Reviews* 25, 54-76.

Shephard, G.S. and Sewram, V. (2004) Determination of the mycotoxin fumonisin B_1 in maize by reversed-phase thin-layer chromatography: A collaborative study. *Food Additives and Contaminants* 21, 498-505.

Sizoo, E.A. and van Egmond, H.P. (2005) Analysis of duplicate 24-hour diet samples for aflatoxin B_1, aflatoxin M_1 and ochratoxin A. *Food Additives and Contaminants* 22, 163-172.

Sorensen, L.K. and Elbaek, T.H. (2005) Determination of mycotoxins in bovine milk by liquid chromatography tandem mass spectrometry. *Journal of Chromatgraphy B* 820, 183-196.

Stroka, J., van Otterdijk, R. and Anklam, E. (2000) Immunoaffinity column clean-up prior to thin-layer chromatography for the determination of aflatoxins in various food matrices. *Journal of Chromatography A* 904, 251-256.

Tudos, A.J., van den Bos, E.R.L. and Stigter, E.C.A. (2003) Rapid surface plasmon resonance-based inhibition assay of deoxynivalenol. *Journal of Agricultural and Food Chemistry* 51, 5843-5848.

van der Gaag, B., Spath, S., Dietrich, H., Stigter, E., Boonzaaijer, G., van Osenbruggen, T. and Koopal, K. (2003) Biosensors and multiple mycotoxin analysis. *Food Control* 14, 251-254.

Ventura, M., Anaya, I., Broto-Puig, F., Agut, M. and Comellas, L. (2005) Two-dimensional thin-layer chromatographic method for the analysis of ochratoxin A in green coffee. *Journal of Food Protection* 68, 1920-1922.

Visconti, A. and de Girolamo, A. (2005) Fitness for purpose – Ochratoxin A analytical developments. *Food Additives and Contaminants* 22, 37-44.

Visconti, A., Lattanzio, V.M., Pascale, M. and Haidukowski, M. (2005) Analysis of T-2 and HT-2 toxins in cereal grains by immunoaffinity clean-up and liquid chromatography with fluorescence detection. *Journal of Chromatography A* 1075, 151-158.

Weiss, R., Freudenschuss, M., Krska, R. and Mizaikoff, B. (2003) Improving methods of analysis for mycotoxins: Molecularly imprinted polymers for deoxynivalenol and zearalenone. *Food Additives and Contaminants* 20, 386-395.

Whitaker T.B. (2004) Sampling for mycotoxins. In: Magan, N. and Olsen, M. (ed.), *Mycotoxins in Food Detection and Control*. Woodhead Publishing Ltd, Cambridge, United Kingdom, pp 69-87.

Yu, J.C.C. and Lai, E.P.C. (2005) Interaction of ochratoxin A with molecularly imprinted polypyrrole film on surface plasmon resonance sensor. *Reactive and Functional Polymers* 63, 171-176.

Yu, J.C.C., Krushkova, S., Lai, E.P.C. and Dabek-Zlotorzynska, E. (2005) Molecularly-imprinted polypyrrole-modified stainless steel frits for selective solid phase preconcentration of ochratoxin A. *Analytical and Bioanalytical Chemistry* 382, 1534-1540.

Zhou, S.N., Lai, E.P.C. and Miller, J.D. (2004) Analysis of wheat extracts for ochratoxin A by molecularly imprinted solid-phase extraction and pulsed elution. *Analytical and Bioanalytical Chemistry* 378, 1903-1906.

16

Mycotoxin Concentration Data Quality: The Role of Sampling

Marina Miraglia*, Barbara de Santis, Elena Pannunzi, Francesca Debegnach and Carlo Brera

Abstract

Sound strategies for the meaningful evaluation of mycotoxin contamination should include: (i) identification of the sites (*where*) and time (*when*) that lots are to be sampled, and (ii) a protocol to ensure that the test sample is representative of the lot from which it was taken. In this chapter we detail the European norms for sampling for aflatoxins in cereals, milk, dried fruits, dried figs, peanuts, nuts, spices and baby foods, for ochratoxin A in cereals, coffee, dried fruits, spices, wine, dried vine fruits and baby foods, and for *Fusarium* toxins in cereals and baby foods.

Introduction

The availability of high quality concentration residue data is an indispensable tool for making food safety assessments. However, the determination of the concentration of a given analyte is the final step of a chain of activities (Fig. 1); the first step of which is the selection of the time and location of sampling. The appropriateness and the reliability of each step are crucial to the reliability of the conclusions to be drawn from the data.

These steps are important for most of the food safety issues, but they have particular importance for mycotoxins because of the peculiarities of this type of contamination, the most important of which is the heterogeneous distribution of mycotoxins in the lots. Therefore, the concepts of "fit for purpose sampling" and of "representativeness of the laboratory sample" must be considered by all those who collect or use such data from food and feed, *e.g.*, scientists, risk managers, risk assessors and risk communicators. Cooperation and collaboration amongst and between those who collect and use these data will increase the confidence in the data and its suitability for regulatory and scientific purposes.

"Fit for purpose" sampling for mycotoxins

A "fit for purpose" sampling for mycotoxins requires decisions on *"where and when"* to collect samples, *e.g.*, *where*: field, ship, silo, processing step or retail; *when*: identification

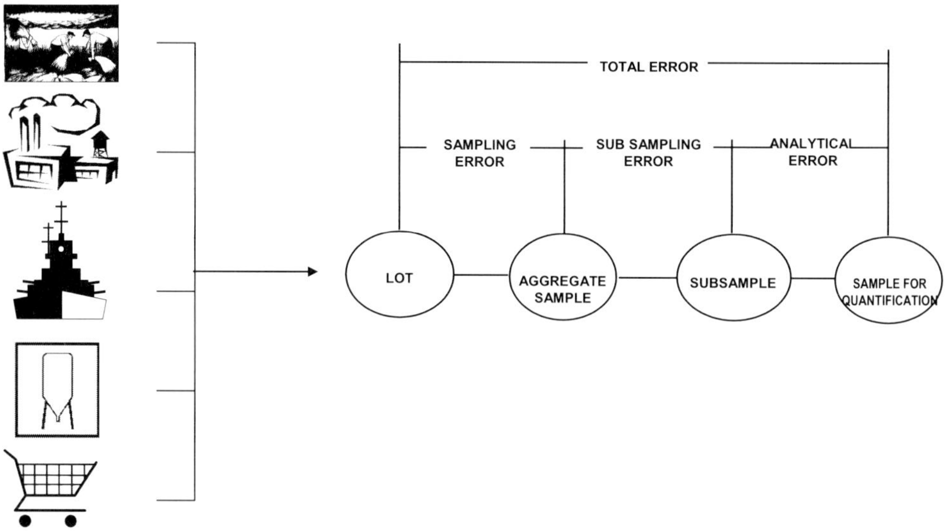

Figure 1. Sources of errors in each step of the analytical chain.

of a time relative to a benchmark such as harvest; and *"how many"* and "*how large*" the sampling for any particular samples should be. Mycotoxins have their own individual characteristics that depend on the potentially contaminated food matrix, the producing fungus, when contamination is most likely to occur, *e.g.*, in field, storage or shipping; geographic origin, the stability of the toxin in the matrix; the need to test the sample for more than one toxin, potential decontamination processes, the projected end use of the product, and the possibility that other compounds might be present that would complicate the analytical processes.

Concentration data for the presence of mycotoxins in foods and feeds are needed by different categories of people depending on the objective of the assessment. Thus there is no single sampling protocol for mycotoxins and each set of data should be derived from *ad hoc* collected samples, with the sampling process taking into account the decisions that will be based on the mycotoxin contamination data once the data have been obtained.

Risk managers require a variety of data/information on the status of contamination of both raw materials and final products to take adequate preventive/corrective actions, and must assume each set of samples has been properly drawn. For example, risk managers need reliable data on the contamination of samples collected from imported goods. Results based on these samples will influence economic, trade and political decisions and will drive the frequency and level of appropriate official controls. The location and time of taking samples for regulatory purposes should be selected based on knowledge of existing and emerging problems in the country or area of origin. Data on the effects of processing on mycotoxins should be given to risk managers to determine if different maximum limits should be set for raw and processed products. Such differentiated regulations could reduce the detrimental economic impact that could result from unnecessarily stringent limits on raw materials. Health perspective risk managers also need data on finished food and feed products to provide policy makers with information on compliance with maximum limits, and to confirm both compliance with and appropriateness of food safety schemes.

Table 1. Subdivision of lots into sublots, number of incremental samples and weight of aggregate sample to be taken and number of laboratory samples for aflatoxins in dried figs, peanuts, nuts and spices.

Commodity	Lot weight (t)	Weight (t) or number of sublots	Number of incremental samples	Aggregate sample weight (kg)	Number of laboratory samples from aggregate sample
Dried figs	≥15	15 – 30	100	30	3
Peanuts, pistachios, brazil nuts and other nuts	≥500	100	100	30	3
	> 125 – < 500	5 sub-lots	100	30	3
	≥ 15 – ≤ 125	25	100	30	3
Dried figs, peanuts, pistachios, brazil nuts and other nuts	> 10 – < 15	–	100	30	3
	> 5.0 – ≤ 10	–	80	24	3
	> 2.0 – ≤ 5.0	–	60	18 – < 24	2
	> 1.0 – ≤ 2.0	–	40	12	2
	> 0.5 – ≤ 1.0	–	30	9 – < 12	1 (no subdivision)
	> 0.2 – ≤ 0.5	–	20	6	1 (no subdivision)
	> 0.1 – ≤ 0.2	–	15	4.5	1 (no subdivision)
	≤0.1	–	10	3	1 (no subdivision)
Spices	≥ 15	25	100	10	1 (no subdivision)
	> 10 – < 15	–	100	10	1 (no subdivision)
	> 5.0 – ≤ 10	–	80	8	1 (no subdivision)
	> 2.0 – ≤ 5.0	–	60	6	1 (no subdivision)
	> 1.0 – ≤ 2.0	–	40	4	1 (no subdivision)
	> 0.5 – ≤ 1.0	–	30	3	1 (no subdivision)
	> 0.2 – ≤ 0.5	–	20	2	1 (no subdivision)
	> 0.1 – ≤ 0.2	–	15	1.5	1 (no subdivision)
	> 0.01 – ≤ 0.1	–	10	1	1 (no subdivision)
	≤0.01	–	5	0.5	1 (no subdivision)

Risk assessors usually are involved in the generation and scientific analysis of most data sets on mycotoxins. Risk assessors also regularly perform and promote research activities whose goal is to evaluate the impact of mycotoxins on human and animal health. Such data are crucial to risk managers who must develop the necessary responses to various levels of contamination. These objectives are amenable to many risk assessment strategies, *e.g.*, point estimates, probabilistic approaches, total diet, duplicate diet, *etc.*, all of which require reliable concentration data on mycotoxins. High quality concentration data is a prerequisite for all further decisions and analyses and should use a statistical sampling plan that focuses on *which* samples, *where* and *when* they should be taken and the number of samples required to ensure a representative analysis.

More recently [Pan-European proactive identification of emerging risks in the field of food production (PERIAPT): http://www.periapt.net/], risk assessors were assigned roles in identifying and evaluating emerging risks of mycotoxin contamination, including re-emerging mycotoxins, less investigated compounds, new sources of contamination attributable to changes in climate or diet and contamination due to reduced hygienic conditions or to calamities, *e.g.*, war, in the country of origin. In all cases, samples taken to evaluate toxin levels should be selected only after the additional information on possible contamination problems has been incorporated into the sampling plan.

Risk communication remains a very broad and misunderstood concept, and involves all stakeholders. Stakeholders need information on mycotoxin levels that are consistent with their needs. For example, producers of raw products need data from samples collected before harvesting and at various stages of processing to identify an appropriate destination and to ensure that any necessary treatment/remediation has been successfully accomplished. Consumer organizations and food processors also need to know toxin levels for materials that enter a particular process as well as levels in the final product on the shelf.

Representativeness of laboratory samples

The soundness of residue data for mycotoxins relies both on "fit for purpose" sampling and on the strategy employed for taking a sample that is quantitatively representative of the entire lot. The primary concern in this process is that mycotoxins often have a peculiar, non-homogeneous distribution both in the bulk lot and in the product as it is packaged for retail sale. The sampling step of the process has by far the largest source of error. Many exhaustive reviews have been published since the 1960s, most of them based on studies by Whitaker and colleagues (Whitaker and Wiser, 1969; Whitaker *et al.,* 1974, 1976, 1979, 1994, 1998, 2000; Whitaker, 2004; Vargas *et al.*, 2006; Miraglia *et al.*, 2006).

Many studies are available that evaluate the distribution of contamination in raw agricultural commodities, that quantify the variance of each sampling step, and develop the Operative Characteristics curves required for an operative sampling plan. An Operative Characteristics curve provides an estimate of the performance of a given sampling plan. This curve is based on a well-defined sampling plan, has an accept/reject limit, and estimates the risk to the seller/exporter and to the buyer/importer that the lot will be misclassified.

The characteristics of a sampling plan generally are chosen based on the final objective of the test. Thus a sampling plan adopted by industry may differ from one used for official regulatory control. Both sampling theory and the pragmatic limitations of sampling have been taken into consideration in developing sampling plans that depend on the crop, the mycotoxin and the global perspective of the organization that will use the results acquired.

We describe the sampling procedures in force in the European Union in some detail and provide cross references to sampling plans in *Codex Alimentarius*. The U.S. Department of Agriculture and the U.S. Pistachio Industry have adopted specific sampling plans for raw shelled peanuts destined for further processing and for pistachio nuts, respectively.

European Union

Since 1998, the European Union has enacted a package of legislative provisions (European Commission 1998, 2001, 2002*a,b*, 2003, 2004, 2005*a,b*, 2006) that define the sampling procedures for the most prominent mycotoxins, *i.e.*, aflatoxins, ochratoxin A, *Fusarium* toxins and patulin, in many raw and processed products. This package represents one of the more comprehensive tools for sampling mycotoxins worldwide. These procedures were developed for the regulatory control process and are coupled with the maximum tolerable levels for mycotoxins set at European levels. Most of these sampling plans account for theoretical sampling issues, with some also considering the practicalities of the sampling process.

Regulation 401/2006/EC (European Commission, 2006) is a unified provision that contains sampling procedures for all regulated mycotoxins in food, both in bulk and in packages, and that harmonizes sampling protocols for regulatory purposes amongst all EU members (Tables 1-3). Regulated mycotoxins are: aflatoxins in cereals, milk, dried fruits and figs, spices, baby food, peanuts, pistachios and other nuts; ochratoxin A in cereals, roasted coffee, dried fruits, spices, wine, dried vine fruits and baby food; and *Fusarium* toxins in cereals and baby foods.

Sampling in bulk

A procedure common to all of the EU Directives/Regulations is the subdivision of the lots into a number of sublots. The number of sublots depends on the size of the lot, the matrix being examined, the manner in which incremental samples are taken from a sublot and the way in which the samples from the sublots are pooled to form an aggregate sample, from which the test sample(s) is derived. The procedures can be used in both static and dynamic sampling protocols.

For aflatoxins, ochratoxin A and *Fusarium* toxins in cereals and cereal products, and for aflatoxins and ochratoxin A in dried fruit and derived products, a lot is accepted if the test sample derived from the aggregate sample is less than the maximum limit once corrections for recovery and measurement uncertainty have been applied. A lot is rejected if the test sample exceeds the maximum limit beyond reasonable doubt once corrections for recovery and measurement uncertainty have been applied.

In samples of dried figs, peanuts and nuts for aflatoxin B_1 and for total aflatoxins (Table 1), the size of an aggregate sample and of incremental samples (up to 30 kg and 300 g respectively) is specified that it is higher than that for the food products listed in Table 2. The number of incremental samples to be taken depends on the weight of the lot, with a minimum of 10 and a maximum of 100.

For aflatoxins in food products (Table 1), and for lots $>$ 15 tons, before grinding, the aggregate sample (= 30 kg) from each sublot must be mixed together and then divided into three 10 kg laboratory samples from which test samples are taken. Subdivision of the aggregate sample is not required for dried figs, peanuts or other nuts subjected to further sorting or to other physical treatment, or if equipment capable of homogenizing the samples is not available. If a lot weighs $\leq$ 15 tons, then the number of incremental samples, the weight of aggregate samples and the number of laboratory samples to be taken from aggregate samples, as well as the acceptance criteria on the basis of the analysis of the aggregate sample/laboratory samples, depends on the size of the lot and on the final destination of the product. The rationale for sampling procedures and acceptance criteria for aflatoxins in spices (Table 1) is similar to that applied to cereals and dried fruits. The aggregate sample for milk and milk products, infant formulae and follow-on formulae (Table 2), including infant milk and follow-on milk, must be at least 1 kg or 1 l, unless such samples are not available, *e.g.*, if the sample consists of a single bottle.

Table 2. Minimum number of incremental samples to be taken from a lot for aflatoxin M_1 in milk and milk products, patulin in apple products, and ochratoxin A in wine, grape juice and must.

Form of commercialization	Volume (l) or weight (kg) of lot	Minimum number of incremental samples	Minimum volume (l) or weight (kg) of aggregate sample
Aflatoxin M_1 in milk and milk products, infant formulae and follow-on formulae including infant milk and follow-on milk			
Bulk	–	3 – 5	1
Bottles/packages	≤ 50	3	1
Bottles/packages	50 – 500	5	1
Bottles/packages	> 500	10	1
Patulin in products from apple and ochratoxin A in wine, grape juice & grape must			
Bulk (fruit juice, spirit drinks, cider, wine)	_	3	1
Bottles/packages (fruit juice, spirit drinks, cider)	≤ 50	3	1
Bottles/packages (fruit juice, spirit drinks, cider)	50 – 500	5	1
Bottles/packages (fruit juice, spirit drinks, cider)	> 500	10	1
Bottles/packages wine	≤ 50	1	1
Bottles/packages wine	50 – 500	2	1
Bottles/packages wine	> 500	3	1
Patulin in solid apple products, apple juice and solid apple products for infants and young children			
Bottles/packages	1 – 25	1 unit	1
Bottles/packages	26 – 100	≥ 2 units, ~5%	1
Bottles/packages	> 100	≤ 10 units, ~5%	1
Bulk	< 50	3	1
Bulk	50 – 500	5	1
Bulk	> 500	10	1

For ochratoxin A in roasted coffee and coffee products (Table 3) the weight of the incremental sample must be ≤ 100 g, depending on the size of the lot. For retail packages, the weight of the incremental sample depends on the weight of the package. If the package weighs more than 100 g, the aggregate sample will weigh more than 10 kg. If the weight of a single retail pack is > 100 g, then 100 g must be taken from each individual retail pack as an incremental sample. If the retail pack weighs less than 100 g, then one retail pack is considered an incremental sample, resulting in an aggregate sample that weighs < 10 kg.

When sampling for patulin in products derived from apples or for ochratoxin A in wine, grape juice, and grape must (Table 2), the aggregate sample must be at least 1 l, except when it is not possible, *e.g.*, when the sample consists of one bottle. The incremental samples, which often are a single bottle or a single package, must be of similar weight. The weight of an incremental sample must be at least 100 g, resulting in an aggregate sample of at least 1 l. Sampling for patulin in solid apple products, apple juice and solid apple products for

Table 3. Subdivision of lots into sublots, number of incremental samples to be taken from each sublot and weight of aggregate samples for aflatoxins, ochratoxin A and *Fusarium* toxins in cereals and cereal products, in dried fruit and derived products, and in coffee.

Lot weight (t)	Weight (t) or number of sublots	Number of incremental samples	Aggregate sample weight (kg)
Aflatoxins, ochratoxin A and* Fusarium *toxins in cereals and cereal products			
≥ 1500	500	100	10
> 300 – < 1500	3 sub-lots	100	10
≥ 50 – ≤ 300	100	100	10
> 20 – < 50	–	100	10
> 10 – ≤ 20	–	60	6
> 3 – ≤ 10	–	40	4
> 1 – ≤ 3	–	20	2
> 0.5 – ≤ 1	–	10	1
> 0.05 – ≤ 0.5	–	5	1
≤ 0.05	–	3	1
Ochratoxin A in roasted coffee beans, ground roasted coffee and soluble coffee			
OR			
Aflatoxins in dried fruit (except dried figs) and ochratoxin A in dried vine fruit and derived products			
≥ 15	15 – 30	100	10
> 10 – ≤ 15	–	100	10
> 5.0 – ≤ 10	–	80	8
> 2.0 – ≤ 5.0	–	60	6
> 1.0 – ≤ 2.0	–	40	4
> 0.5 – ≤ 1.0	–	30	3
> 0.2 – ≤ 0.5	–	20	2
> 0.1 – ≤ 0.2	–	15	1.5
≤ 0.1	–	10	1

infants and young children has to be performed in a particular way (Table 2). The weight of an incremental sample must be ≥ 100 g, and the aggregate sample ≥ 1 kg. Sampling for baby foods and processed cereal-based foods for infants and young children for aflatoxins, ochratoxin A, *Fusarium* toxins, and patulin should follow the format in Table 3 for lots < 50 t.

An ambiguous aspect of the described sampling procedure is the point in the food chain at which samples are taken. Regulation 401/2006/EC (European Commission, 2006) provides little guidance on this point. This problem can be particularly significant for raw products, since samples taken before or after procedures, such as cleaning could lead to different mycotoxin levels. For cereals, the maximum limit for *Fusarium* toxins is set before the first processing step; however for aflatoxins and ochratoxin A, no guidance is given. Therefore differences in the relative severity of contamination by these toxins may depend on the point in the food chain from which the sample was taken.

Sampling procedures for retail products also are available for derived products and compound foods with very small particle weights, *e.g.*, flour and peanut butter, in which the mycotoxin contamination is more homogeneously distributed, and for peanuts, dried figs and derived products traded in vacuum packages.

Codex Alimentarius sampling procedures

Codex Alimentarius has begun to develop sampling procedures for mycotoxins. For example, for raw shelled peanuts, *Codex* recommends a single 20 kg sample. The sample test results must be $\leq$ the accept/reject limit of 15 ng/g total aflatoxins for the lot be accepted (*Codex Alimentarius*, 2006). *Codex* also is considering sampling plans for aflatoxin contamination in almonds, Brazil nuts, hazelnuts and pistachios. In general *Codex* sampling plans are based on the principle that the importer and the exporter share the risk associated with sampling and the setting of the accept/reject limit.

Conclusions

Protocols aimed at the meaningful evaluation of mycotoxin contamination should identify the sites and the time (*where and when*) lots are to be sampled, and describe how a representative sample can be taken from the individual lots. Decisions on *"where and when"* lots should be sampled depend on the final target of the process of generating concentration data. These targets vary by country, policy and scientific expertise and infrastructure. European norms for taking representative test samples are amongst the most thorough and contribute greatly to the generation of good quality concentration data. A lack of representativeness of either the sampling lot or the test sample could result in errant conclusions regarding mycotoxin contamination levels and have serious consequences for both health and trade.

Acknowledgements

We thank Ms. Viviana Renzi for technical assistance.

References

Codex Alimentarius Commission. (2006) Joint FAO/WHO Food Standards Programme *Codex* Committee on Food Additives and Contaminants. 38th Session (The Hague, The Netherlands, 24-28 April 2006). Proposed draft sampling plan for aflatoxin contamination in almonds, brazil nuts, hazelnuts and pistachios. (N07-2004) CX/FAC 06/38/21 February 2006.

European Commission. (1998) Commission Directive 98/53/EC of 16 July 1998 laying down the sampling methods and the methods of analysis for the official control of the levels for certain contaminants in foodstuffs. *Official Journal of the European Union* L 201, 17.07.1998, pp. 93–101.

European Commission. (2001) Commission Regulation (EC) No 466/2001 of 8 March 2001 setting maximum levels for certain contaminants in foodstuffs. *Official Journal of the European Union* L 077, 16.03.2001, pp. 1–13.

European Commission. (2002*a*) Commission Directive 2002/26/EC of 13 March 2002 laying down the sampling methods and the methods of analysis for the official control of the levels of ochratoxin A in foodstuffs. *Official Journal of the European Union* L 075, 16.03.2002, pp. 38–43.

European Commission. (2002*b*) Commission Directive 2002/27/EC of 13 March 2002 amending Directive 98/53/EC laying down the sampling methods and the methods of analysis for the official control of the levels for certain contaminants in foodstuffs. *Official Journal of the European Union* L 075, 16.03.2002, pp. 44–45.

European Commission. (2003) Commission Directive 2003/78/EC of 11 August 2003 laying down the sampling methods and the methods of analysis for the official control of the levels of patulin in foodstuffs. *Official Journal of the European Union* L 203, 12.08.2003, pp. 40–44.

European Commission. (2004) Commission Directive 2004/43/EC of 13 April 2004 amending Directive 98/53/EC and Directive 2002/26/EC as regards sampling methods and methods of analysis for the official control of the levels of aflatoxin and ochratoxin A in food for infants and young children. *Official Journal of the European Union* L 113, 20.04.2004, pp. 14–16.

European Commission. (2005*a*) Commission Regulation (EC) No 856/2005 of 6 June 2005 amending Regulation (EC) No 466/2001 as regards *Fusarium* toxins. *Official Journal of the European Union* L 143, 07.06.2005, pp. 3–8.

European Commission. (2005*b*) Commission Directive 2005/38/EC of 6 June 2005 laying down the sampling methods and the methods of analysis for the official control of the levels of *Fusarium* toxins in foodstuff. *Official Journal of the European Union* L 143, 07.06.2005, pp. 18–26.

European Commission. (2006) Commission Regulation (EC) No 401/2006 of 23 February 2006 laying down the methods of sampling and analysis for the official control of the levels of mycotoxins in foodstuffs. *Official Journal of the European Union* L 70, 09.03.2006, pp. 12-34.

Miraglia, M., de Santis, B., Minardi, V., Debegnach, F. and Brera, C. (2006) The role of sampling in mycotoxin contamination: An holistic view. *Food Additives and Contaminants* 22 (Supplement 1), 31-36.

Vargas, E.A., Whitaker, T.B., dos Santos, E.A., Slate, A.B., Lima, F.B. and Franca, R.C.A. (2006) Design of a sampling plan to detect ochratoxin A in green coffee. *Food Additives and Contaminants* 23, 62-72.

Whitaker, T.B. and Wiser, E.H. 1969. Theoretical investigations into the accuracy of sampling shelled peanuts for aflatoxin. *Journal of the American Oil Chemists Society* 46, 377–379.

Whitaker, T.B., Dickens, J.W. and Monroe, R.J. (1974) Variability of aflatoxin test results. *Journal of the American Oil Chemists Society* 51, 214–218.

Whitaker, T.B., Whitten, M.E. and Monroe, R.J. (1976) Variability associated with testing cottonseed for aflatoxin. *Journal of the American Oil Chemists Society* 53, 502–507.

Whitaker, T.B., Dickens, J.W. and Monroe, R.J. (1979) Variability associated with testing corn for aflatoxin. *Journal of the American Oil Chemists Society* 56, 789–794.

Whitaker, T.B., Dowell, F.E., Hagler, W.M. Jr., Giesbrecht, F.G. and Wu, J. (1994) Variability associated with sampling, sample preparation, and chemically testing farmers' stock peanuts for aflatoxin. Journal *of AOAC International* 77, 107–116.

Whitaker, T.B., Truckess, M.W., Johansson, A.S., Giesbrecht, F.G., Hagler, W.M. Jr. and Bowman, D.T. (1998) Variability associated with testing shelled corn for fumonisin. *Journal of AOAC International* 81, 1162–1168.

Whitaker, T.B., Hagler, W.M. Jr, Giesbrecht, F.G. and Johansson A.S. (2000) Sampling, sample preparation, and analytical variability associated with testing wheat for deoxynivalenol. *Journal of AOAC International* 83, 1285–1292.

Whitaker, T.B. (2004) Sampling for mycotoxins. In: Magan, N., and Olsen M. (eds.) *Mycotoxins in Food: Detection and Control*. Woodhead Publishing Ltd., Cambridge, UK, pp. 69-87

17

Development of Quantitative Detection Methods for *Fusarium* in Cereals and Their Application

Cees Waalwijk*, Ineke M. de Vries, Jürgen Köhl, Xiude Xu, Theo A. J. van der Lee and Gert H. J. Kema

Abstract

We developed a multiplex PCR for the identification of the most frequently occurring species of *Fusarium* on wheat in Western Europe. Surveys suggest that *Fusarium graminearum* has replaced *Fusarium culmorum* as the dominant species on winter wheat in the Netherlands during the 1990s. Quantitative PCR was used to monitor *Fusarium* populations during the growing season. Populations on lower leaves consisted primarily of *Microdochium nivale*, but *F. graminearum* dominated in heads and on harvested grain, suggesting that the inoculum for the infection of the heads did not come from the leaves. A quantitative PCR recently developed for *Fusarium verticillioides* from maize may help to reduce the threat posed by fumonisin on large communities in Africa that rely on maize as a staple food.

Introduction

Fusarium diseases are a major cause of yield loss in many crops, including major staple foods, such as wheat (Bai and Shaner, 1994; Windels, 2000), rice (Abbas *et al.*, 1998), maize (Fandohan *et al.*, 2003) and banana (Ploetz, 1990; 2000). In some of these crops the losses due to reduced yield are increased since the causal agents also produce mycotoxins that reduce further the quality of the harvested crop. Mycotoxins have attracted considerable attention over the last years and regulatory limits have been established and enforced in the EU (European Commission, 2005) and several other countries, but not in many African countries (van Egmond, 2004).

Two major staple crops, wheat and maize are recurrently infected by various *Fusarium* species that can produce mycotoxins. In the United States, Fusarium head blight (FHB) of wheat is caused primarily by *F. graminearum*, but in other parts of the world a number of species may be involved (CWSCG, 1984; Parry *et al.*, 1995), both toxigenic and non-toxigenic. In Northern and Western Europe, non-toxigenic species, *e.g. Microdochium nivale* var. *majus*, are commonly found on wheat particularly early in the season (Turner *et al.,* 2002). The toxigenic species usually include *F. graminearum*, *F. culmorum* and *Fusarium avenaceum* and, depending on the region and the weather conditions, perhaps *Fusa-*

rium poae, *Fusarium sporotrichioides*, *Fusarium tricinctum* amongst others. Each of these toxigenic species has a unique secondary metabolite profile (Table 1). Multi-locus sequencing of six single-copy nuclear genes has identified a series of phylogenetic lineages within *F. graminearum* (O'Donnell *et al.*, 2000) that have been proposed to have species status (O'Donnell *et al.*, 2004). Trichothecene production is not correlated with phylogenetic lineage in *F. graminearum* (Ward *et al.*, 2002). Toxins produced by *F. graminearum* include zearalenone (Desjardins, 2006; Lysøe *et al.*, 2006) in addition to the type B trichothecenes nivalenol and deoxynivalenol. Other species found on wheat produce different toxins, *e.g.*, *F. poae* produces the type-A trichothecenes T-2 and HT-2 (Desjardins, 2006) and *F. avenaceum* produces moniliformin (Desjardins, 2006).

Fusarium infection of maize occurs as two distinct diseases that are caused by partially overlapping groups of species. Red ear rot, or Gibberella ear rot, occurs primarily in areas with high precipitation or lower temperature and usually is caused by *F. graminearum*, although other species also have been associated with the disease (Table 1). Pink ear rot, or Fusarium ear rot, occurs primarily on damaged kernels and is caused by a group of species including *F. subglutinans* and the fumonisin producers *F. verticillioides* and *Fusarium proliferatum* (Table 1). Toxigenic and non-toxigenic species can be found in the same field, in the same plant, on the same cob or even within the same kernel.

Qualitative detection

Traditionally the species occurring on cereals were characterized morphologically, but this is time-consuming and the results may be ambiguous, *e.g.*, *F. verticillioides* and *F. thapsinum*

Table 1. Toxigenic species of *Fusarium* associated with Fusarium Head Blight (FHB) of wheat and, Pink Ear Rot (PER) and Red Ear Rot (RER) of maize (after Bottalico and Perrone, 2002, and Logrieco *et al.*, 2002).

Species	Disease	Mycotoxin[a]
F. acuminatum	FHB	NEO, **T-2**
F. avenaceum	FHB, RER	BEA, ENN, **MON**
F. cerealis[b]	FHB	**FUS**, **NIV**, **ZEA**, ZOH
F. culmorum	FHB, RER	**DON**, NIV, **ZEA**, ZOH
F. equiseti	FHB, RER	**DAS**, ZEA, ZOH
F. graminearum	FHB, RER	**DON**, FUS, **NIV**, **ZEA**
F. poae	FHB	**BEA**, DAS, ENN, FUS, **NIV**
F. proliferatum	PER	**BEA**, **FUM**, **FUP**, **MON**
F. sporotrichioides	FHB	**HT-2**, NEO, **T-2**
F. subglutinans	PER	**BEA**, FUP, **MON**
F. tricinctum	FHB	BEA, ENN, **MON**
F. verticillioides	PER	**FUM**

[a]Toxins in bold are synthesized by most members of the corresponding species. Abbreviations: BEA – beauvericin, DAS – diacetoxyscirpenol, DON – deoxynivalenol, ENN – enniatins, FUM – fumonisins, FUP – fusaproliferin, FUS – fusarenon X, HT-2 – HT-2 toxin, MON – moniliformin, NEO – neosolaniol, NIV – nivalenol, T-2 – T-2 toxin, ZEA – zearalenone and ZOH – zearalenol.
[b]Also known as *F. crookwellense* (Leslie and Summerell, 2006).

Table 2. Number of isolates of *Fusarium* spp. collected in winter wheat in the Netherlands in 2000 and 2001.

Region	*F. avenaceum*		*F. culmorum*		*F. graminearum*		*M. nivale* var. *majus*	
	2000	2001	2000	2001	2000	2001	2000	2001
Southwest	0	0	18	9	25	10	5	7
East	1	0	4	1	14	6	3	0
Central	1	0	14	15	35	27	20	3
Northeast	1	0	4	12	36	29	6	0
Southeast	0	0	2	6	12	13	0	0

are hard to distinguish by morphology even though the ability of these species to produce fumonisins differs by several orders of magnitude. We developed a diagnostic PCR that rapidly identifies the most common species occurring on wheat (Waalwijk *et al.*, 2003). Primer sets that specifically amplify target sequences from one of *F. graminearum*, *F. culmorum*, *F. avenaceum*, *F. poae*, *F. proliferatum*, and the non-toxigenic species *M. nivale* var. *majus* and *Microdochium nivale* var. *nivale* were combined into a multiplex PCR containing 14 primers. We used this primer mixture to survey five geographic regions in the Netherlands in 2000 and 2001 (Waalwijk *et al.*, 2003; Table 2).

Fusarium graminearum was clearly the dominant species. In surveys made in the 1980s and 1990s, *F. culmorum* was the most common species, with occasionally high levels of *F. avenaceum* detected as well (de Nijs *et al.*, 1997). This difference suggests that a dramatic shift has occurred in the composition of the FHB complex in The Netherlands during the last decade. There is no clear cause for this shift, but increased production of maize may play an important role. As *F. graminearum* also is a pathogen on maize, rotations of maize with wheat allow the fungus to cycle from one host to the other. *F. culmorum* also can be found on maize in temperate climates (Munkvold *et al.*, 1998) but inoculum of *F. culmorum* is less likely to increase in the absence of wheat. Alternatively, *Gibberella zeae*, the teleomorph of *F. graminearum*, could affect inoculum pressure since it produces ascospores that are capable of traveling long distances (Fernando *et al.*, 1997; Maldonado-Ramirez and Bergstrom, 2001). Indeed, mutants of *F. graminearum* that cannot produce ascospores also cause less disease (Brown *et al.*, 2001*b*). Whether this results from a direct effect of the disruption on the mating capacity remains to be elucidated, since disruption of the mating type gene *mat1-2* in *F. verticillioides* also affects the expression of many genes (Waalwijk et al., 2006; Keszthelyi *et al.*, 2007). A third possibility is that *F. graminearum* has a higher temperature optimum for growth and sporulation than does *F. culmorum* (Doohan *et al.*, 2003). Thus, the increased incidence of *F. graminearum* might be an indicator of climate change occurring in Europe. However, in a comparative study of Dutch and Italian strains of both species both *F. culmorum* and *F. graminearum* had the same temperature requirements (Köhl *et al.*, unpublished).

The diagnostic multiplex PCR described above also was used for an inventory of isolates collected from wheat in China and France in 2002. All the French isolates had a PCR fragment unique for *F. graminearum*, but none of the Chinese isolates had this fragment. Instead 156/172 isolates had a fragment that was ~ 30 bp longer. This fragment was pre-

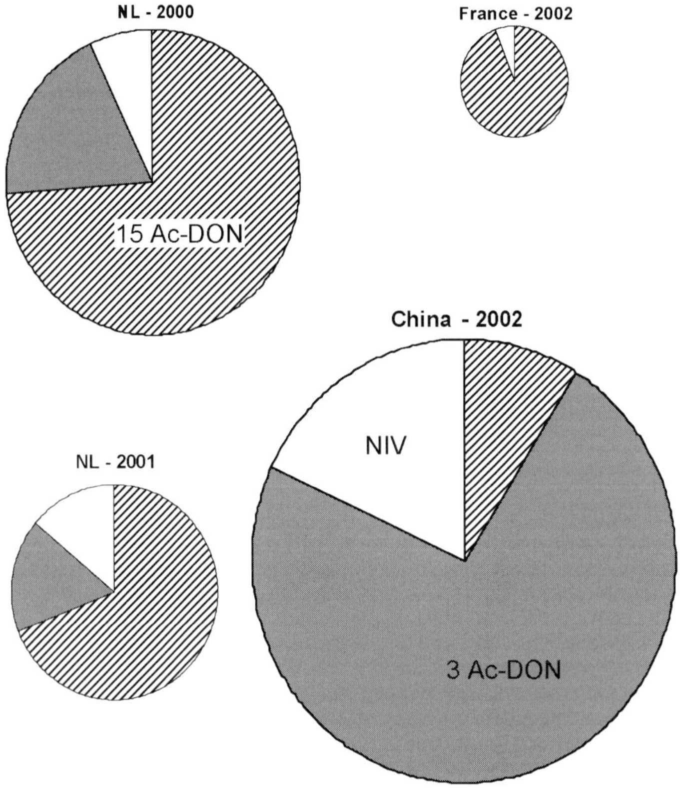

Figure 1. Distribution of chemotypes of *F. graminearum* lineage 6 in a single wheat field in Hubei province in China and *F. graminearum* lineage 7 in France and in two consecutive years in The Netherlands. Sizes of the circles are proportional to the size of the samples.

viously identified from Chinese isolates (Waalwijk *et al.*, 2003) that belong to *F. graminearum* lineage 6. The remaining 16 Chinese isolates produced the diagnostic *M. nivale* var. *majus* fragment (Waalwijk *et al.*, 2003).

Individual *F. graminearum* isolates produce either deoxynivalenol (DON) or nivalenol (NIV). The molecular basis for the production of these related trichothecenes has been studied extensively and the ability to produce either DON or NIV depends upon a mutation in one of the genes involved in the toxin biosynthetic pathway (Brown *et al.*, 2001*a*, 2002; Lee *et al.*, 2001, 2002). The *tri7* and *tri13* genes have been used by several groups to differentiate between DON- and NIV-producing strains of *F. graminearum* (Jennings *et al.*, 2004*a,b*; Waalwijk *et al.*, 2003). In the Netherlands we observed 76% and 68% DON producers in *F. graminearum* in 2000 and 2001, respectively. DON producers could be subdivided into those that produce 15-AcDON or 3-AcDON in an ~ 4:1 ratio (Fig. 1). A similar survey in the United Kingdom, that used specific primers to identify NIV, 15-AcDON and 3-AcDON found that 25% of the *F. graminearum* isolates and 43% of the F. *culmorum* isolates had the NIV chemotype (Jennings *et al.*, 2004*a,b*). The regular increase in the proportion of NIV producers observed in the United Kingdom, is consistent with a hypothesis that NIV producers might be becoming more dominant (Waalwijk *et al.*, 2003).

Quantitative detection of the Fusarium Head Blight complex

The multiplex detection tool is very useful for screening isolates recovered from the field, but it is still too time consuming for large scale epidemiological studies. We developed a quantitative real-time PCR assay (Waalwijk *et al.*, 2004*a*) based on TaqMan technology. For each of the common members of the FHB species complex, the fragments from the multiplex analysis were sequenced and these sequences used for the design of additional

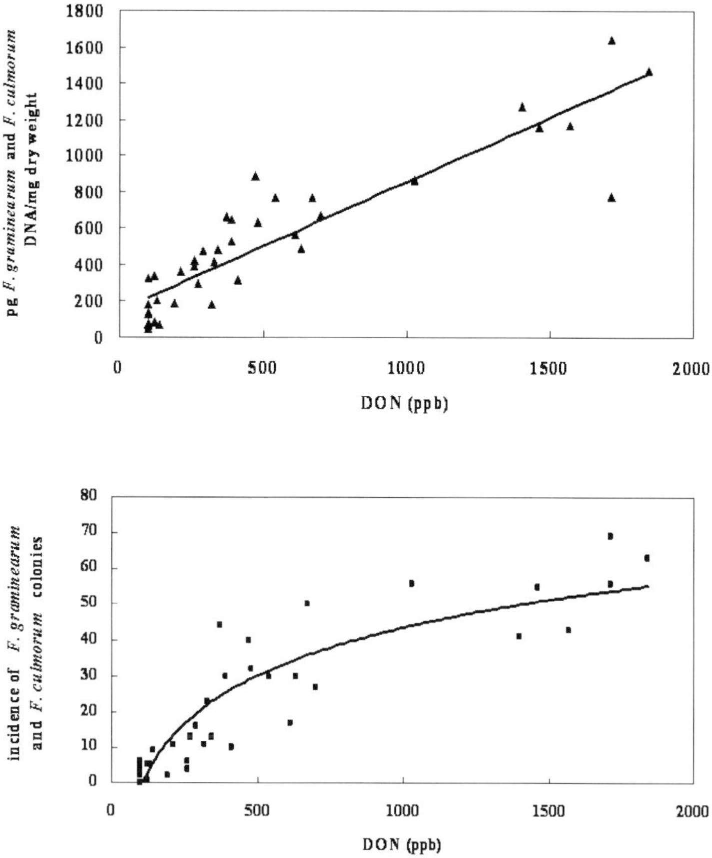

Figure 2. Top. Linear correlation ($y = 0.71x + 147$; $r^2 = 0.82$) between the concentration of DON and the total amount of DNA of DON producing species (*F. graminearum* and *F. culmorum*). **Bottom.** The correlation ($y = 19.3ln(x) - 89.7$; $r^2 = 0.82$) between the DON concentration in field samples and the number of DON-producing colonies among 100 kernels from the corresponding samples.

primers and internal probes. The assay also includes an internal standard that provides a control for the presence of inhibitors of the PCR reaction. Different probes are labeled with fluorescent dyes that fluoresce at different wave lengths, and the amount of fluorescence observed is a measure of the amount of DNA replication that has occurred. A serial dilution of DNA from a pure culture of *F. graminearum* had a high correlation ($r^2 = 0.99$) with the threshold cycle, and a dynamic range of four orders of magnitude (Waalwijk *et al.*, 2004*a*). Similar values were obtained for the other species and we could quantify accurately the amount of DNA from *F. graminearum*, *F. culmorum*, *F. avenaceum*, *F. poae*, *F. proliferatum* and *M. nivale* var. *majus* between 0.9 pg and 9000 pg per reaction. In some cases we could detect the subsequent dilution of 0.09 pg, but these results were not consistent. The genome size of the PH-1 sequenced isolate of *F. graminearum* is estimated as 36 Mb (Broad Institute, 2005), which equals 0.04 pg, so the detection limit of the *F. graminearum* TaqMan assay we developed is ~ 5 genome equivalents (Waalwijk *et al.,* 2004*a*).

These quantitative PCRs were used to study fungal population dynamics during the growing season in different geographic areas in the Netherlands. Although there were regional differences, the lower leaves contained primarily *M. nivale* var. *majus* at levels ranging from zero to 40 pg/mg sample (dry weight). However, in wheat heads from the same fields collected at growth stages 78 through 89 *M. nivale* var. *majus* was hardly detectable (< 3 pg/mg sample). Instead, *F. graminearum* dominated with *F. avenaceum* and/or *F. culmorum* present in some of the samples. The dominance of *F. graminearum* was even more striking at harvest with some samples containing up to 300 pg/mg sample. We validated the methodology by making similar analyses in the Netherlands in 2002, when most of the fields had a much broader spectrum of *Fusarium* species. In some fields nearly equal amounts of all five species were found at levels between 50 and 150 pg/mg of sample. In samples from France, the situation was dramatically different as *F. graminearum* was present almost exclusively in most samples at levels up to 22,000 pg/mg of sample (Waalwijk *et al.*, 2004*a*). The DON concentrations of a series of field samples also were compared with the combined quantities in these samples of *F. graminearum* and *F. culmorum* (as these are the DON producers in the FHB complex). The combined DNA quantities for *F. graminearum* and *F. culmorum* were correlated ($r^2 = 0.82$) with DON concentration (Fig. 2). A similar correlation ($r^2 = 0.82$) was observed between DON concentration and the total number of *F. graminearum* and *F. culmorum* colonies, as determined by plating 100 kernels per sample and morphological characterization of the colonies present (Fig. 2).

Recently, we also used quantitative PCR to evaluate the survival of species from the FHB complex in the field after harvest. The population dynamics of *Fusarium* spp. in crop residues of wheat left in the field after harvest was monitored by sampling at regular intervals, from harvest until mid-anthesis of the next season. We found that the fungal biomass of each of the species decreased significantly in nodes and internodes, but not in residues of stem bases, suggesting that stem bases are a major source of inoculum for the succeeding crop. At harvest, only 14% and 1% of the biomass of *F. avenaceum* or *F. graminearum*, respectively, found in haulm tissues was present in stem bases (Köhl *et al.*, unpublished). After field exposure for ten months, more than 85% of the biomass of the two species present in the haulm residues was found in the stem bases. Such knowledge of population dynamics of the head blight pathogens in different plant tissues is being used to develop preventive agronomic measures such as stubble treatments or soil tillage aimed at the reduction of inoculum sources of head blight pathogens.

Quantitative detection of *Fusarium* species in maize

Some of the *Fusarium* species that occur in maize consist in part of species also found on wheat (Table 1) and the quantitative tools we developed for these species can be directly employed in maize and maize products. Both *F. graminearum* and, to a lesser extent, *F. culmorum* can occur on maize (Munkvold *et al.*, 1998), but the greatest concern in maize is the presence of *F. verticillioides* and *F. proliferatum* because of their capacity to produce fumonisins primarily fumonisin B_1. These mycotoxins are associated with high incidences of esophageal cancer in Southern Africa, certain provinces in China (Chu and Li, 1994) and some areas in Italy and Iran (Shephard *et al.*, 2000). In most maize growing areas *F. verticillioides* is by far the most common species and the fungus may be living as an endophyte in maize (Miller, 2001). The subtle balance between plant and fungus can be disturbed by biotic and/or abiotic stresses, including insect damage and drought stress.

We developed a TaqMan PCR assay quantification for fumonisin-producing fungi that is based on one of the genes in the toxin biosynthetic pathway. Fumonisin biosynthesis requires at least 15 genes that are organized in a gene cluster found in the genome of both *F. verticillioides* and *F. proliferatum* (Proctor *et al.*, 2003; Waalwijk *et al.*, 2004*b*). Unlike the trichothecene toxins produced by *F. graminearum* on wheat, disruption of the ability to produce fumonisins does not reduce the virulence of the nonproducing strains (Desjardins and Plattner, 2000). A few naturally occurring nonproducing strains have been identified, and we included them in our analyses. The first gene in the fumonisin biosynthetic pathway is *fum1*, which encodes a polyketide synthase that is member of a family of PKS proteins. A conserved domain in this protein was used to design primers that can amplify DNA from both *F. verticillioides* and *F. proliferatum*. Fumonisin nonproducing isolates of *F. verticillioides* however, are negative in this test (Fig. 3). This PCR can discriminate between

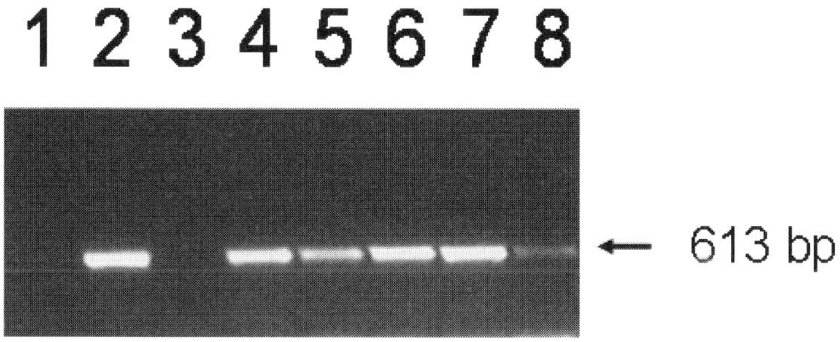

Figure 3. PCR amplification of *F. verticillioides* and *F. proliferatum* isolates by using primers designed to bind to one of the conserved regions of *fum1*. Lanes 1 and 3, fumonisin non-producing isolates of *F. verticillioides* Gf11 and Gf8 respectively (20); lanes 2, 6 and 7 fumonisin-producing *F. verticillioides* isolates Gf20, ITEM 2142 (FGSC 7600, A-0149) and ITEM 2143 (FGSC 7603, A-0999) and lanes 4, 5 and 8, fumonisin-producing isolates of *F. proliferatum* ITEM 2287, ITEM 2400 and ITEM 2148 (FGSC 7614, D-4853).

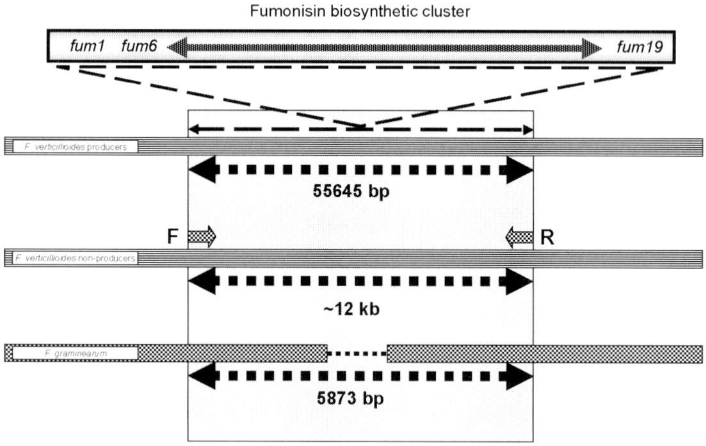

Figure 4. Schematic representation of the regions flanking the fumonisin gene cluster in fumonisin-producing isolates of *F. verticillioides* and *F. proliferatum* and the corresponding region from some non-producers of *F. verticillioides*. The genomic sequences of *F. graminearum* and *F. verticillioides* (Broad Institute, 2005) were used to align the gene clusters. Synteny in the flanking regions of both species extends for > 300 kb (data not shown).

strains that can and cannot produce fumonisins, unlike those based on sequences not directly related to the biosynthetic gene cluster (Mirete *et al.*, 2003; Bluhm *et al.*, 2004).

The structure and organization of the gene cluster in both species is very similar except for the flanking regions (Waalwijk *et al.*, 2004*b*). In *F. verticillioides*, the gene cluster is flanked by stretches of DNA that are contiguous in *F. graminearum*. In *F. proliferatum*, the gene cluster is adjacent to a region that resides in a very different part of the *F. graminearum* genome (Fig. 4). These differences suggest that the fumonisin gene cluster might be (part of) a mobile genetic element that can insert in and excise from various locations in the genome. Fumonisin nonproducing isolates of *F. verticillioides* could have lost the entire fumonisin gene cluster. We tested this hypothesis with primers designed to amplify the flanking regions with complete homology in *F. graminearum*. In *F. graminearum* these primers produced an amplicon of 5.8 kb and in some non-fumonisin producing isolates of *F. verticillioides* the fragment is 12 kb in length. The distance between the primers in fumonisin-producing strains of *F. verticillioides,* however, is > 55 kb which prevents amplification by PCR, and in these strains no amplification product results.

Homologous sequences of the *fum1* gene were identified by comparing the published sequences of the *F. verticillioides* and *F. proliferatum* genes with the corresponding regions in *F. globosum* and *F. nygamai*. These sequences were used to design a set of primers and a fluorescent probe for a Taqman assay that was then tested on a series of *Fusarium* species known to occur on maize. All isolates of *F. verticillioides*, *F. proliferatum*, *F. globosum* and *F. nygamai* were detected in this assay, but no signal was obtained for any of the tested strains of *F. equiseti*, *F. graminearum*, *F. oxysporum*, *F. semitectum* or *F. subglutinans*, suggesting that this PCR is diagnostic for fumonisin-producing *Fusarium* species.

Conclusions

Multiplex PCR is a powerful method to characterize the composition of disease complexes such as the FHB complex of wheat, and pink or red ear rot of maize. The emergence of new species in a complex can be identified easily as isolates that do not have fragments from the previously known species. The only requirement for including a species in the multiplex system seems to be the design of a species-specific amplicon of unique size, since combining multiple primers has not yet been limiting. Quantitative PCR allows the study of population dynamics of plant pathogens during the growing season, their survival on crop residues, and the competition between species within a disease complex, as well as monitoring the effects of agronomic measures such as fungicide treatments.

Acknowledgments

We thank Dr. M.-T. Gonzalez-Jaen (University of Madrid, Spain) for providing the non-producing isolates of *F. verticillioides*, Prof. W. F. O. Marasas (PROMEC, Capetown, South Africa) for providing strains of *F. globosum* and *F. nygamai*, and Drs. K. O'Donnell and T. Ward (USDA, Peoria, IL, USA) for sequence analyses of the Chinese, Dutch and French isolates of *F. graminearum sensu lato*. Xiude Xu was supported by a grant from the China Scholarship Council.

References

Abbas, H.K., Cartwright, R.D., Shier, W.T., Abouzied, M.M., Bird, C.B., Rice, L.G., Ross, P.F., Sciumbato, G.L., and Meredith, F.I. (1998) Natural occurrence of fumonisins in rice with Fusarium sheath rot disease. *Plant Disease* 82, 22-25.

Bai, G. and Shaner, G. (1994) Scab of wheat: Prospects for control. *Plant Disease* 78, 760-766.

Bluhm, B.H., Cousin, M.A. and Woloshuk, C.P. (2004) Multiplex real-time PCR detection of fumonisin-producing and trichothecene-producing groups of *Fusarium* species. *Journal of Food Protection* 67, 536-543.

Bottalico, A. and Perrone, G. (2002) Toxigenic *Fusarium* species and mycotoxins associated with head blight in small-grain cereals in Europe. *European Journal of Plant Pathology* 108, 611-624.

Broad Institute. (2005) Updated release of the sequence of the *F. graminearum* genome http://www.broad.mit.edu/annotation/fungi/fusarium/ and the *F. verticillioides* genome http://www.broad.mit.edu/annotation/fungi/fusarium_verticillioides/.

Brown, D.W., McCormick, S.P., Alexander, N.J., Proctor, R.H. and Desjardins, A.E. (2001*a*) A genetic and biochemical approach to study the trichothecene diversity in *Fusarium sporotrichioides* and *Fusarium graminearum*. *Fungal Genetics and Biology* 32, 121-133.

Brown, D.W., Desjardins, A.E., Yun, S.-H., Plattner, R., Lee, T., Dyer, R. and Turgeon, B.G. (2001*b*) Are *Gibberella zeae* sexual spores the critical inoculum to FHB epidemics? *Proceedings of the 2001 National Fusarium Head Blight Forum (Cincinnati, Ohio, USA)*, p. 104.

Brown, D.W., McCormick, S.P., Alexander, N.J., Proctor R.H., and Desjardins, A.E. (2002) Inactivation of a cytochrome P-450 is a determinant of trichothecene diversity in *Fusarium* species. *Fungal Genetics and Biology* 36, 224-233.

Chu, F.S. and Li, G.Y. (1994) Simultaneous occurrence of fumonisin B_1 and other mycotoxins in moldy corn collected from the People's Republic of China in regions with high incidences of esophageal cancer. *Applied and Environmental Microbiology* 60, 847-852.

CWSCG. (1984) *Fusarium* species, distribution and pathogenicity from scabby heads in China. *Journal of Shanghai Normal College* 3, 69-82.

de Nijs, M., Larsen, J., Gams, W., Rombouts, F.M. Wernars, K., Thrane, U. and Notermans, S.H.W. (1997) Variations in random polymorphic DNA patterns and secondary metabolite profiles within *Fusarium* species from cereals from various parts of The Netherlands. *Food Microbiology* 14, 449-459.

Desjardins, A. E. (2006) Fusarium *mycotoxins: Chemistry, Genetics and Biology*. APS Press, St. Paul, MN, USA.

Desjardins, A.E. and Plattner, R.D. (2000) Fumonisin B_1-nonproducing strains of *Fusarium verticillioides* cause maize (*Zea mays*) ear infection and ear rot. *Journal of Agricultural and Food Chemistry* 48, 5773–5780.

Doohan, F.M., Brennan, J. and Cooke, B.M. (2003) Influence of climatic factors on *Fusarium* species pathogenic to cereals. *European Journal of Plant Pathology* 109, 755-768.

European Commission. (2005) Commission regulation (EC) No 856/2005 of 6 June 2005 amending Regulation (EC) No 466/2001 as regards Fusarium toxins. http://europa.eu.int/eur-lex.

Fandohan, P., Hell, K., Marasas, W.F.O. and Wingfield, M.J. (2003) Infection of maize by *Fusarium* species and contamination with fumonisin in Africa. *African Journal of Biotechnology* 2, 570-579.

Fernando, W.G.D., Paulitz, T.C., Seaman, W.L., Dutilleul, P. and Miller, J.D. (1997) Head blight gradients caused by *Gibberella zeae* from area sources of inoculum in wheat field plots. *Phytopathology* 87, 414-421.

Jennings, P., Coates, M.E., Walsh, K., Turner, J.A. and Nicholson, P. (2004*a*) Determination of deoxynivalenol- and nivalenol-producing chemotypes of *Fusarium graminearum* isolated from wheat crops in England and Wales. *Plant Pathology* 53, 643-652.

Jennings, P., Coates, M.E., Turner, J.A., Chandler, E.A. and Nicholson, P. (2004*b*) Determination of deoxynivalenol and nivalenol chemotypes of *Fusarium culmorum* isolates from England and Wales by PCR assay. *Plant Pathology* 53, 182-190.

Keszthelyi, A., Jeney, A., Kerenyi, Z., Mendes, O., Waalwijk, C. and Hornok, L. (2007) Tagging target genes of the MAT1-2-1 transcription factor in *Fusarium verticillioides (Gibberella fujikuroi MP-A) Antonie van Leeuwenhoek* (published online doi: 10.1007/s10482-006-9123-5)

Lee, T., Han, Y.-K. Kim, K. Yun, S.-H. and Lee, Y.-W. (2002) *Tri13* and *tri7* determine deoxynivalenol and nivalenol producing chemotypes of *Gibberella zeae*. *Applied and Environmental Microbiology* 68, 2148-2154.

Lee, T., Oh, D.-W., Kim, H.-S., Lee, J., Kim, Y.-H., Yun, S.-H. and Lee, Y.-W. (2001) Identification of deoxynivalenol and nivalenol producing chemotypes of *Gibberella zeae* by using PCR. *Applied and Environmental Microbiology* 67, 2966-2972.

Leslie, J.F. and Summerell, B.A. (2006) *The* Fusarium *Laboratory Manual*. Blackwell Professional Publishing, Ames, Iowa.

Logrieco, A., Mulè, G., Moretti, G.A. and Bottalico, A. (2002) Toxigenic *Fusarium* species and mycotoxins associated with maize ear rot in Europe. *European Journal of Plant Pathology* 108, 597-609.

Lysøe, E., Klemsdal, S.S., Bone, K.R., Frandsen, R.J.N., Johansen, T., Thrane, U. and Giese, H. (2006) The *PKS4* gene of *Fusarium graminearum* is essential for zearalenone production. *Applied and Environmental Microbiology* 72, 3924-3932.

Maldonado-Ramirez, S.L. and Bergstrom, G. C. (2001) Temporal patterns of ascospore discharge by *Gibberella zeae* from colonized corn stalks under natural conditions. *Proceedings of the 2001 National Fusarium Head Blight Forum (Cincinnati, Ohio, USA)*, pp. 159-162.

Miller, J. D. (2001) Factors that affect the occurrence of fumonisin. *Environmental Health Perspectives* 109S, 321-324.

Mirete, S., Vazquez, M.G., Jurado, M. and Gonzalez-Jaen, M.T. (2003) Differentiation of *Fusarium verticillioides* from banana fruits by IGS and EF-1α sequence analyses. *European Journal of Plant Pathology* 110, 515-523.

Munkvold, G., Stahr, H.M., Logrieco, A., Moretti, A. and Ritieni, A. (1998) Occurrence of fusaproliferin and beauvericin in *Fusarium*-contaminated livestock feed in Iowa. *Applied and Environmental Microbiology* 64, 3923-3926.

O'Donnell, K., Kistler, H.C., Tacke, B.K. and Casper, H.H. (2000) Gene genealogies reveal global phylogeographic structure and reproductive isolation among lineages of *Fusarium graminearum*, the fungus causing wheat scab. *Proceeding of the National Academy of Sciences USA* 97, 7905-7910.

O'Donnell, K., Ward, T.J., Geiser, D.M., Kistler, H.C. and Aoki, T. (2004) Genealogical concordance between the mating type locus and seven other nuclear genes supports formal recognition of nine phylogenetically distinct species within the *Fusarium graminearum* clade. *Fungal Genetics and Biology* 41, 600–623.

Parry, D.W., Jenkinson, P. and MacLeod, L. (1995) Fusarium ear blight (scab) in small grain cereals – A review. *Plant Pathology* 44, 207-238.

Ploetz, R.C., ed. (1990) *Fusarium wilt of banana*. APS Press, St. Paul, Minnesota, USA.

Ploetz, R.C. (2000) Panama disease: A classic and destructive disease of banana. *Plant Health Progress*. Published online doi: 10.1094/PHP-2000-1204-01-HM.

Proctor, R.H., Brown, D.W., Plattner, R.D. and Desjardins, A.E. (2003) Co-expression of 15 contiguous genes delineates a fumonisin biosynthetic gene cluster in *Gibberella moniliformis*. *Fungal Genetics and Biology* 38, 237-249.

Shephard, G.S., Marasas, W.F.O., Leggott, N.L., Yazdanpanah, H., Rahimian, H., Safavi, N. (2000) Natural occurrence of fumonisins in corn from Iran. *Journal of Agricultural and Food Chemistry* 48, 1860–1864.

Turner, A.S., Nicholson, P., Edwards, S.G., Bateman, G.L., Morgan, L.W., Todd, A.D., Parry, D.W., Marshall, J. and Nuttall, M. (2002) Relationship between brown foot rot and DNA of *Microdochium nivale*, determined by quantitative PCR, in stem bases of winter wheat. *Plant Pathology* 51, 464-471.

van Egmond, H. (2004) Worldwide regulations for mycotoxins in food and feed in 2003. *FAO Food and Nutrition Paper* 81, 1-170.

Waalwijk, C., Kastelein, P., de Vries, P.M. Kerényi, Z., van der Lee, T.A.J., Hesselink, T., Köhl, J., and Kema, G.H.J. (2003) Major changes in *Fusarium* spp. in wheat in The Netherlands. *European Journal of Plant Pathology* 109, 743-754.

Waalwijk, C., Keszthelyi, A., van der Lee, T., Jeney, A., de Vries, I., Kerenyi, Z., Mendes, O., Hornok, L. (2006) Mating type loci in *Fusarium*: structure and function. *Mycotoxin Research* 22, 54-60.

Waalwijk, C., van der Heide, R., de Vries, P.M., van der Lee, T.A.J., Schoen, C., Costrel-Decorainville, G., Haeuser-Hahn, I., Kastelein, P., Köhl, J., Lonnet, P., Demarquet, T. and Kema, G.H.J. (2004*a*) Quantitative detection of *Fusarium* species in wheat using TaqMan. *European Journal of Plant Pathology* 110, 481-494.

Waalwijk, C., van der Lee, T.A.J., de Vries, P.M., Hesselink, T., Arts, J. and Kema, G.H.J. (2004*b*) Synteny in toxigenic *Fusarium* species: The fumonisin gene cluster and the mating type region as examples. *European Journal of Plant Pathology* 110, 533-544.

Ward, T.J., Bielawski, J.P., Kistler, H.C., Sullivan, E., and O'Donnell, K. (2002) Ancestral polymorphism and adaptive evolution in the trichothecene mycotoxin gene cluster of phytopathogenic *Fusarium*. *Proceedings of the National Academy of Sciences USA* 99, 9278-9283.

Windels, C.E. (2000) Economic and social impacts of *Fusarium* head blight: Changing farms and rural communities in the Northern Great Plains. *Phytopathology* 90, 17-21.

Mycotoxin Management

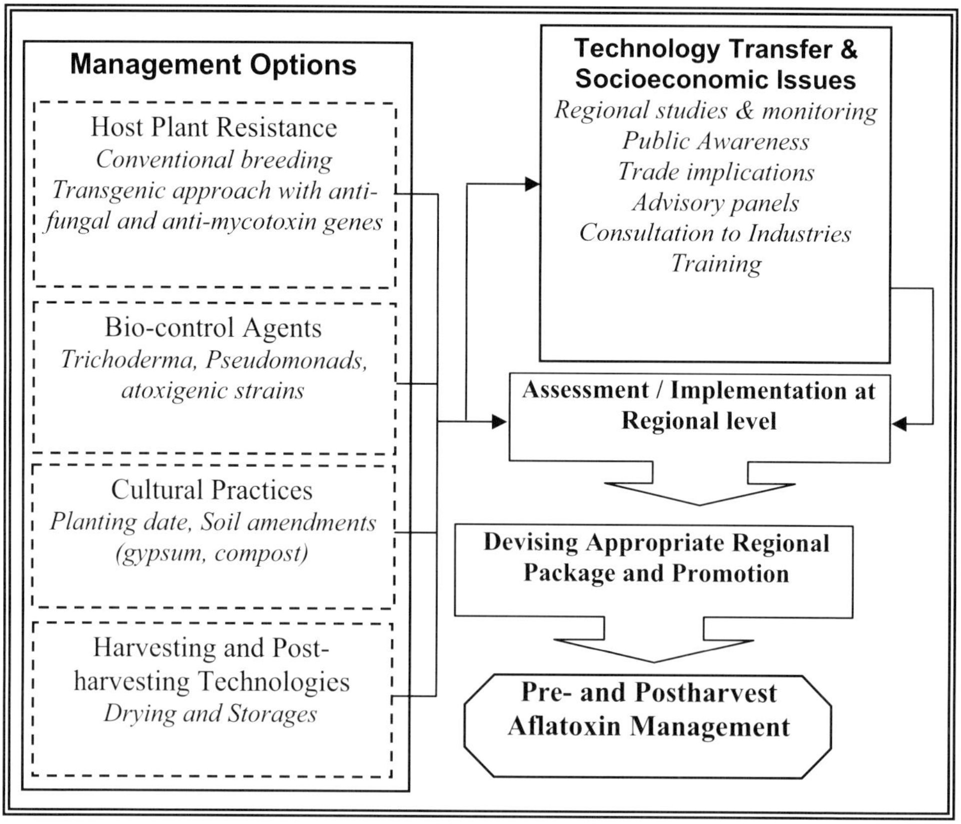

Figure 1. Schematic representation of ICRISAT's integrated management strategy for pre- and postharvest management of aflatoxins in peanut.

Aflatoxins, especially aflatoxin B_1 produced by *A. flavus* in peanut seeds, are potent carcinogens and immunosuppressive agents that are directly responsible for various human and animal diseases, and result in economic losses due to the lower market potential of contaminated peanut products (Reed and Kasali, 1989; Waliyar *et al.*, 2003*b*; Williams *et al.*, 2004). In West Africa, many individuals are not only malnourished but also chronically exposed to high levels of aflatoxins in their daily diet (Gong *et al.*, 2002; Williams *et al.*, 2004). Due to concerns over aflatoxin contamination, African exports of peanut have declined by 19% during the last three decades, even though peanut utilization for food in West Africa has increased by 209% during the same period of time (Freeeman *et al.*, 1999; Reveredo and Fletcher, 2002). Due to insufficient infrastructure for aflatoxin monitoring and a lack of awareness of the problem amongst farmers and the general populace, human and domesticated animal populations are at risk of health problems due to the consumption of aflatoxin-contaminated peanuts.

Although several *Aspergillus* spp. can produce aflatoxins, *A. flavus* is the most aggressive and most commonly occurring species on peanut. This fungus is saprophytic and survives in the soil as mycelium, conidia and sclerotia. It may infect peanut at the preharvest stages through the pegs that have penetrated into the soil. End of the season drought stress favors further fungal invasion and aflatoxin production in the seeds (Mehan *et al.*, 1991*b*; Cole *et al.*, 1995). In Sub-Saharan Africa, peanuts are grown as a rainfed crop, under subsistence farming conditions in poor soils, and end-of-season drought is common. These conditions are highly conducive to *A. flavus* infestation and toxin production. Prevention of fungal contamination under such conditions is a complex task and requires a series of strategies to reduce infection incidence and the proliferation of the fungi both preharvest, and postharvest. ICRISAT and its partners have promulgated an integrated research program to mitigate aflatoxin contamination through the development of genetic resistance, and cultural and biological controls. In this chapter we describe briefly the ICRISAT research on integrated approaches for pre- and postharvest management of aflatoxin contamination in peanut (Fig. 1).

Prevention and control strategies

Aflatoxin prevention strategies can be divided into primary and secondary categories (Table 1; Semple *et al.,* 1989). The primary prevention strategies include measures to stop fungal infestation and aflatoxin production, whereas secondary strategies focus on eliminating or limiting the spread of the fungal contaminant (Table 1). Field contamination is considered a significant source of inoculum in peanuts, thus efforts have focused on preharvest management through integrated approaches for controlling critical factors known to increase fungal infection and aflatoxin production (Cole *et al.*, 1989; Mehan *et al.*, 1991*a*). These practices include growing resistant cultivars, avoiding end-of-season drought stress, reducing inoculum sources, and avoiding pod damage through timely harvest and postharvest handling (Fig. 1). This research has been facilitated by sick plots containing highly aggressive, toxigenic strains of *A. flavus* that can be used to screen large amounts of germplasm (Waliyar *et al.*, 1994), and an inexpensive ELISA for quantitative estimation of aflatoxin concentration in peanut kernels (Ramakrishna and Mehan, 1993; Devi *et al.*, 1999).

Cultivar resistance through conventional and non-conventional approaches

There have been several efforts to breed resistant peanut cultivars for reduced aflatoxin contamination (Cole *et al.*, 1995). Earlier research on genetic improvement at ICRISAT focused on *in vitro* seed colonization by *A. flavus,* as it was then thought to be a priority postharvest problem (Mehan *et al.*, 1987). Subsequent field studies found that preharvest contamination by the fungi also occurs frequently and efforts were broadened to include breeding high-yielding varieties with resistance to seed infection and aflatoxin production (Mehan *et al.*, 1988, 1991*a,b*). These efforts identified some *A. flavus*-resistant germplasm sources and breeding lines, but the level of resistance in them was not sufficient to protect the crop from aflatoxin contamination under all conditions (Pettit, 1986; Waliyar *et al.*, 1994). These efforts have resulted in the identification of over 30 resistant/tolerant germplasm accessions (Pettit, 1986; Waliyar *et al.*, 1994; Upadhyaya *et al.*, 2002, 2004). Some of these genotypes, including 55-437, Tamnut 74, PI 365553 (resistant to seed infection); PI 337394 F, PI 337409, UF 71513, Ah 7223, J 11, U 4-47-7, Var 27, Faizpur and Monir 240-30 (resistant to *in vitro* seed colonization), and U 4-7-5 and VRR 245 (resistant to aflatoxin production) were used to develop elite aflatoxin resistant varieties, *e.g.*, ICGV # 88145, 89104, 91278, 91283 and 91284, which have been released as improved germplasm. Three lines, ICGV 87084, 87094 and

Table 1. Methods for minimizing pre- and postharvest aflatoxin contamination in peanuts.

METHOD	PURPOSE
Primary prevention: Minimize fungal infestation and aflatoxin contamination	
Cultivation of *A. flavus* resistant varieties	Limit fungal invasion and toxin production during crop growth
Control field infection by following appropriate phytosanitary measures	Limit fungal inoculum in the field
Seed treatment and fungicide application	Limit fungal invasion during crop growth
Appropriate scheduling of planting, harvest and post-harvest	Avoid drought and other abiotic stresses
Application of soil amendments, *e.g.*, gypsum, farmyard manure, *etc.*	Increase soil nutrients (especially calcium) and water-holding capacity; promotes growth of antagonistic native soil microflora
Lower moisture content of seed after harvest and during storage	Limit fungal invasion and growth during storage
Add preservatives to prevent insect infestation and fungal contamination during storage	Limit fungal invasion during storage
Secondary prevention: Eliminate or limit fungal contamination	
Sort contaminated pods and kernels	Reduce aflatoxin contamination in final product
Redry groundnut pods and kernels	Limit fungal growth and toxin synthesis
Appropriate storage conditions	Limit fungal growth and toxin synthesis
Detoxification of contaminated products	Chemical inactivation or binding of aflatoxins through the use of clay dietary supplements or ammoniation

87110 were resistant to seed infection in Niger, Senegal and Burkina Faso (Upadhyaya *et al.,* 2002). Despite considerable efforts at genetic enhancement, the level of resistance in them was not sufficient to protect the crop from aflatoxin contamination under all conditions and they exhibited high G×E interaction effects (Upadhyaya *et al.,* 2004).

Breeding activities continue with the goal of developing high-yielding varieties adapted to different agroecosystems with enhanced levels of resistance to *A. flavus* infection (resistant to fungal invasion and proliferation, and resistant to drought, soil pests and diseases). Biotechnological approaches to increase host plant resistance through the use of anti-fungal and anti-mycotoxin genes also have begun. This approach received a major boost with the successful establishment of peanut regeneration and transformation protocols (Sharma and Anjaiah, 2000), and led to the transformation of popular peanut cv. JL24 with a rice chitinase gene to help prevent invasion by fungal pathogens. These transgenic events have now advanced to the T_3 generation, with three events showing good resistance to *A. flavus* infection (< 10% infection) in *in vitro* seed inoculation tests (Sharma *et al.*, 2006).

Use of biocontrol agents

Fluorescent pseudomonads and several strains of *Trichoderma* species inhabit the rhizosphere of many crop plants and have been identified as potentially promising biocontrol agents against *A. flavus*. Since the beginning of the 21st century, a large number of *Trichoderma* (> 250) and *Pseudomonas* (> 100) isolates have been obtained from peanut rhizosphere and evaluated for their antagonism towards *A. flavus* and their ability to reduce preharvest kernel infection of peanuts (Thakur and Waliyar, 2005; Anjaiah *et al.*, 2006). Significant reduction of *A. flavus* populations and kernel infection occurred in both greenhouse and field experiments (Desai *et al.*, 2000; Kumar *et al.*, 2002; Thakur *et al.*, 2003). Two *Trichoderma* isolates, Tv 47 and Tv 23, and two bacterial isolates *P. cepacia* (B 33) and *P. fluorescens* (Pf 2), were effective in reducing aflatoxin content in the kernels. Efforts also are being made to identify atoxigenic strains of *A. flavus* that can be used to alter the population dynamics of toxigenic strains of *A. flavus* in risk-prone zones (Cotty *et al.*, Chapter 24). The effectiveness of the biocontrol agents still needs to be established under African field conditions and simple, cheap and effective formulations developed for use in farmers' fields. Integration of these biocontrol agents with host plant resistance and agronomic management would provide an environmentally-friendly option for the management of aflatoxin contamination in peanuts.

Control through cultural practices

During the growing period several factors influence fungal colonization and aflatoxin production, including the soil type and condition, rate of evapotranspiration, availability of viable spores, end-season drought stress, damage to peanut pods by soil pests, and mechanical damage during harvesting (Mehan *et al.*, 1991*a*, 1995; Nahdi, 1996; Waliyar *et al.*, 2003*a*). It is impossible to control all of these factors, but some cultural practices can greatly reduce the amount of fungal infection. Some of these practices include: summer plowing, selecting planting dates to take advantage of periods of higher rainfall and avoiding end of the season drought effects, seed dressing with systemic fungicides or biocontrol agents, maintaining good plant density in the fields, soil amendment with gypsum and farmyard manure, removing prematurely dead plants, managing pests and diseases, timely harvesting, excluding damaged and immature pods, drying pods quickly, controlling storage pests, and only storing pods/seeds with < 10% moisture content. The use of mechanical threshers and seed storage bins also can reduce aflatoxins in peanuts. Although most of these practices are cost-effective and practical under subsistence farming conditions, they remain largely unadopted by subsistence farmers due to various socio-economic constraints including farmers' attention to other revenue generating activities and a lack of appropriate structures for drying and storage.

Control through soil amendments

In studies conducted at ICRISAT research stations in Sadoré (Niger) and Samanko (Mali) from 1999-2001, application of lime (0.5 t/ha), farm yard manure (10 t/ha) and cereal crop residue (5 t/ha) at the time of sowing, either singly or in combinations of lime and farmyard manure, helped reduce *A. flavus* seed infection and aflatoxin contamination in peanut by 50-90% (data not shown). Lime, a source of calcium, enhances cell wall thickness and pod filling and decreases fungal infection (Rosolem *et al.*, 1997). Organic supplements, such as farmyard manure and crop residues, favor growth of native microbial antagonists and suppress soil- and seedborne infections (Karthikeyan, 1996). These three components also improve the water-holding capacity of the soil, minimizing the effect of end-of-the-season moisture stress, and thereby reduce the fungal colonization and aflatoxin accumulation in the peanut seeds. Lime and farmyard manure are cheap and easily available in most developing countries, including those in Sub-Saharan Africa.

Control through postharvest practices

Although the prevention of preharvest infection in peanuts is the best control strategy, aflatoxin also accumulates during harvesting and postharvest processing. We found in West Africa that aflatoxin accumulates, especially in susceptible cultivars, when pod removal from lifted plants is delayed. The toxin content in peanut kernels increased from 4 ng/g when tested immediately after lifting the plants to 6.8 ng/g two weeks after plant lifting in resistant cultivars 55-437 and J11. In the susceptible cultivars JL24 and Fleur 11, aflatoxin levels were 105 ng/g immediately after lifting and 270 ng/g two weeks after lifting. Farmers traditionally dry lifted plants in heaps, which increases humidity due to poor ventilation and favors fungal proliferation and toxin accumulation. To alleviate these problems, postharvest recommendations include removal of pods soon after lifting of plants. Replacing farmers' traditional practice of drying in heaps with a "batch" drying process (pods facing the sun for rapid drying and facilitating good aeration; *cf.*, Fig. 2) dramatically reduces aflatoxin accumulation. For example, in experiments conducted in farmers' fields at Kayes village (Mali), the mean aflatoxin level was 14.3 ng/g (range 2-21.5 ng/g) for peanuts dried in batches, and 46 ng/g (range 12-60 ng/g) for peanuts dried by traditional heaping method.

On-farm and household storage conditions available to subsistence farmers are inadequate for the safe storage of produce, *e.g.*, lack of clean storage bins and frequent pest infestations. These problems create conditions conducive to the accumulation of aflatoxin even in products that were uncontaminated when harvested. For example, toxin content increased in stored peanut kernels, cv. Fleur 11, from 84 ± 33 ng/g at harvest to 184 ± 54 ng/g and, 255 ± 78 ng/g, when tested after one and two months of on-farm storage, respectively. The development of adequate postharvest storage requires further study, including profiles of storage and practices in various regions, to identify methods that maintain proper temperature and humidity while providing protection from insects and other pests.

Creating awareness and technology transfer

The farming community and general public in developing nations are unaware of the hazards associated with aflatoxin contamination of food and feed and their implications for trade and for human and livestock health. Increased awareness will help increase the adoption of the technologies available to minimize mycotoxin contamination. Various information pathways are being used to increase awareness about the dangers of mycotoxins in food and feed. These include the distribution of flyers and brochures in national languages, *e.g.*, Bambara in Mali and Hausa in Nigeria, a farmer-participatory approach to technology evaluation and dissemination, and conducting training courses and workshops. To strengthen the local capacity to monitor aflatoxin contamination, infrastructure for aflatoxin diagnostics has been established in several countries (Malawi, Mali, Niger and Senegal) and personnel have been trained to manage these facilities locally (Waliyar *et al.*, Chapter 31).

Conclusions and perspectives

Aflatoxin contamination of peanut is widespread in developing countries of the world. Sub-Saharan Africa is particularly vulnerable to this problem due to persistent environmental con-

Figure 2. Postharvest methods for drying peanuts. (Left to right) Traditional heap drying and improved batch drying with pods exposed to the sun.

ditions that favor fungal colonization and proliferation in numerous crops and commodities. Efforts continue to find solutions for reducing the risk of aflatoxin contamination in staple foods, particularly among the resource poor of the world (Ortiz *et al.*, Chapter 35). Our work in Sub-Saharan Africa suggests that a holistic approach combining host plant resistance, crop management practices, postharvest technologies, capacity development and public awareness, can effectively reduce the risk of aflatoxin contamination in peanut. Presowing soil amendment with lime, timely harvesting and improved drying procedures alone can result in dramatic reductions in the levels of aflatoxin contamination. These approaches can be scaled-up in locations where aflatoxin contamination is a chronic problem.

There also is a need to increase genetic resistance of peanuts to *A. flavus* and to lower aflatoxin accumulation by using biotechnological tools when they are the most practical way to manage the aflatoxin problem. Biocontrol of aflatoxin contamination is a promising technology. However, before it can be implemented, a more thorough characterization of the diversity, abundance, and activities of the microbes in the peanut rhizosphere is needed to understand the mechanisms involved in host-pathogen-biocontrol agent interactions and to determine the suitability of these microbes for commercial production and for application under diverse agroecological conditions. Thorough studies also are needed to profile farmers' storage practices in various regions in developing countries, to determine the efficacy of different storage structures and practices that reduce aflatoxin contamination, and to identify storage technologies that minimize the risk of increasing aflatoxin contamination. Strengthening the local capacity to monitor aflatoxins is necessary for developing countries to regain the high-value export peanut trade. Such monitoring requires the availability of cheap, accurate and rapid-testing procedures, such as ELISA-based methods or other technologies that match location-specific needs with the socio-economic profiles of farmers in developing countries.

Establishing a prediction system to forecast the likely risk of aflatoxin contamination before or during the cropping season would facilitate implementation of appropriate management practices by the farmers. Development of these forecasting models requires quantitative data on the relationships amongst environmental and crop management factors, *A. flavus* infection and preharvest aflatoxin contamination. For broad-scale applications these models should be based on remote soil, water and temperature monitoring. The availability of such prediction models will help target improved aflatoxin management technologies on a regional scale.

References

Anjaiah, V., Thakur, R.P. and Koedam, N. (2006) Evaluation of bacteria and *Trichoderma* for biocontrol of pre-harvest seed infection by *Aspergillus flavus* in groundnut. *Biocontrol Science and Technology* 16, 431-436.

Cole, R.J., Sanders, T.H., Dorner, J.W. and Blankenship, P.D. (1989) Environmental conditions required to induce preharvest aflatoxin contamination of groundnuts: Summary of six years research. In: McDonald, D. and Mehan, V.K. (eds.). *Aflatoxin Contamination of Groundnuts: Proceedings of the International Workshop (6-9 October 1987, ICRISAT, Patancheru, India)*. ICRISAT, Patancheru, India, p. 279.

Cole, R.J., Dorner, J.W. and Holbroook, C.C. (1995) Advances in mycotoxin elimination and resistance. In: Pattee, H. E. and Stalker, H. T. (eds.). *Advances in Peanut Science.* American Peanut Research and Education Society, Stillwater, Oklahoma, pp. 456-474.

Desai, S., Thakur, R.P., Rao, V.P. and Anjaiah, V. (2000) Characterization of isolates of *Trichoderma* for biocontrol potential against *Aspergillus flavus* infection in groundnut. *International Arachis Newsletter* 20, 57-59.

Devi, K.T., Mayo, M.A., Reddy, K.L.N., Delfosse, P., Reddy, G., Reddy, S.V. and Reddy, D.V. R. (1999) Production and characterization of monoclonal antibodies for aflatoxin B_1. *Letters in Applied Microbiology* 29, 284-288.

Dorner, J.W., Cole, R.J. and Blankenship, P.D. (1992) Use of biocompetitive agent to control preharvest aflatoxin in drought stressed peanuts. *Journal of Food Protection* 55, 888-892.

FAO 2006. FAOSTAT-2007 Database, Food and Agriculture Organization of the United Nations. http://faostat.fao.org/.

Freeman, H.A., Nigam, S.N., Kelley, T.G., Ntare, B.R., Subrahmanyam, P. and Bougthon, B. (1999) *The World Groundnut Economy: Facts, Trends, and Outlook.* ICRISAT, Patancheru, India.

Gong, Y.Y., Cardwell, K., Hounsa, A., Egal, S., Turner, P.C., Hall, A.J. and Wild, C.P. (2002) Dietary aflatoxin exposure and impaired growth in young children from Benin and Togo: Cross sectional study. *British Medical Journal* 325, 20-21.

Holbrook, C.C., Wilson, D.M. and Matheroan M.E. (1995) An update on breeding peanut for resistance to preharvest aflatoxin contamination. In: *Proceedings of the USDA-ARS Aflatoxin Elimination Workshop (23-24 October 1995, Atlanta, Georgia)*, p. 3.

Karthikeyan, A. (1996) Effect of organic amendments, antagonist *Trichoderma viride* and fungicides on seed and collar rot of groundnut. *Plant Disease Research* 11, 72-74.

Kumar, K.V.K., Desai, S., Rao, V.P., Nur, H.A., Srilakshmi, P. and Thakur, R.P. (2002) Evaluation of an integrated management package to reduce preharvest seed infection by *Aspergillus flavus* in groundnut. *International Arachis Newsletter* 22, 42-44.

Mehan V.K., McDonald, D. and Rajagopalan, K. (1987) Resistance of peanut genotypes to seed infection by *Aspergillus flavus* in field trials in India. *Peanut Science* 14, 17-21.

Mehan, V.K., Rao, R.C.N. McDonald, D. and Williams, J.H. (1988) Management of drought stress to improve field screening of peanuts for resistance to *Aspergillus flavus*. *Phytopathology* 78, 659-663.

Mehan, V.K., McDonald, D., Haravu, L.J. and Jayanthi, S. (1991*a*) *The Groundnut Aflatoxin Problem, Review and Literature Database.* ICRISAT, Patancheru, India.

Mehan, V.K., Mayee, C.D., Jayanthi, S. and McDonald, D. (1991*b*) Preharvest seed infection by *Aspergillus flavus* group of fungi and subsequent aflatoxin contamination in groundnuts in relation to soil types. *Plant Soil* 136, 239-248.

Mehan, V.K., Ramakrishna, N., Rao, R.C.N. and McDonald, D. (1995) Preharvest aflatoxin contamination of groundnuts subjected to terminal drought stress in postrainy season. *Mycotoxin Research* 11, 103-109.

Nahdi, S. (1996) Drought stress and preharvest seed invasion of selected groundnut genotypes by *Aspergillus flavus* invasion and aflatoxin contamination. *Indian Phytopathology* 49, 52-56.

Ntare, B.R., Waliyar, F., Ramouch, M., Masers, E. and Ndjeungda, J. (2005) *Market Prospects for Groundnut in West Africa.* CFC Technical Paper no. 39. Common Fund for Commodities, Amsterdam, The Netherlands.

Payne, G.A. (1998) Process of contamination by aflatoxin producing fungi and their impacts on crops. In: Sinha, K.K. and Bhatnagar, D. (eds.), *Mycotoxins in Agriculture and Food Safety,* Marcel Dekker, Inc., New York, pp. 279-306.

Pettit, R.E. (1986) Incidence of aflatoxin in groundnuts as influenced by seasonal changes in environmental conditions – A review. In: Sivakumar, M. V. K., Virmani, S. M. and Beckerman, S. R. (eds.) *Agrometeorology of Groundnut, Proceedings of an International Symposium (21-26 August 1985, ICRISAT Sahelian Center, Niamey, Niger).* ICRISAT, Patancheru, India., pp. 163-174.

Ramakrishna, N. and Mehan, V.K. (1993) Direct and indirect competitive monoclonal antibody-based ELISA of aflatoxin B_1 in groundnut. *Mycotoxin Research* 9, 53-63.

Reveredo, C.L. and Fletcher, S. (2002) World peanut market: An overview of the past 20 years. *Research Bulletin Number 437*, Georgia Agricultural Experiment Station, Athens, Georgia.

Reed, J.D. and. Kasali, O.B. (1989) Risk to human health associated with consumption of groundnuts contaminated with aflatoxins. In: McDonald, D. and Mehan, V.K. (eds.) *Aflatoxin Contamination of Groundnut: Proceedings of the International Workshop (6-9 October 1987, ICRISAT, Patancheru, India).* ICRISAT, Patancheru, India, pp. 31-38.

Rosolem, C.A., Fernandez, E.M., Maringoni, A.C. and Oliveira, D.M.T. (1997) Fungus incidence on peanut grains as affected by drying method and Ca nutrition. *Field Crops Research* 52, 9-15.

Semple, R.L., Firo, A.S., Hicks, P.A. and Lozare, J.V. (1989*) Mycotoxin Prevention and Control in Food Grains.* UNDP/FAO/REGNET and the ASEAN Grain Postharvest Program, Bangkok, Thailand.

Sharma, K.K. and Anjaiah, V. (2000) An efficient method for the production of transgenic plants of peanut (*Arachis hypogaea* L.) through *Agrobacterium tumefaciens*-mediated genetic transformation. *Plant Science* 159, 7-19.

Sharma, K.K., Bhatnagar-Mathur, P., Sai-Vishnu-Priya, K., Anjaiah, V., Vadez, V., Waliyar, F., Kumar, P.L., Nigam, S.N. and Hoisington, D.H. (2006) Genetic engineering of groundnut for crop improvement. *Abstracts of the International Conference on Groundnut Aflatoxin Management and Genomics (5-9 November 2006, Guangzhou, Guangdong, China).* Guangdong Academy of Sciences, Guangzhou, China, p. 44.

Thakur, R.P., Rao, V.P. and Subramanyam, K. (2003) Influence of biocontrol agents on population density of *Aspergillus flavus* and kernel infection in groundnut. *Indian Phytopathology* 56, 454-458.

Thakur, R.P. and Waliyar, F. (2005) Biomanagement of preharvest aflatoxin contamination in groundnut, In: Sharma, H.C. and Gowda, C.L.L. (eds.). *Proceedings of IPM Research at ICRISAT: Present Status and Future Priorities (5 April 2005, ICRISAT, Patancheru, India).* ICRISAT, Patancheru, India, pp. 33-40.

Upadhyaya, H.D., Nigam, S.N., and Thakur, R.P. (2002) Genetic enhancement for resistance to aflatoxin contamination in groundnut. In: Waliyar, F., and Adomou, M. (eds.) *Summary Proceedings of the 7th ICRISAT Regional Groundnut Meeting for Western and Central Africa, (6-8 December 2000, Cotonou, Benin).* ICRISAT, Patancheru, India, pp. 29-36.

Upadhyaya, H.D., Nigam, S.N., and Waliyar, F. (2004) Aflatoxin contamination of groundnut: Conventional breeding for resistance. In: Robens, J. (ed.) *Proceedings of the 3rd Fungal Genomics, 4th Fumonisin, and 16th Aflatoxin Elimination Workshops (13-15 October 2003, Savannah, Georgia, USA).* USDA, ARS, Beltsville, Maryland, p. 55.

Waliyar, F., Ba, A., Hassan, H., Bonkoungou, S. and Bosc, J.P. (1994) Sources of resistance to *Aspergillus flavus* and aflatoxin contamination in groundnut genotypes in West Africa. *Plant Disease* 78, 704-708.

Waliyar, F., Traore, A. Fatondji, D. and Ntare, B.R. (2003*a*) Effect of irrigation interval, planting date and cultivar on *Aspergillus flavus* and aflatoxin contamination of peanut in a sandy soil of Niger. *Peanut Science* 30, 79-84.

Waliyar, F., Reddy, S.V., Subramanyam, K., Reddy, T.Y., Rama Devi, K., Crauford, P.Q. and Wheeler, T.R. (2003*b*) Importance of mycotoxins in food and feed in India. *Aspects of Applied Biology* 68, 147-154.

Williams, J.H., Phillips, T.D., Jolly, E.P., Stiles, J.K., Jolly, C.M. and Aggarwal, D. (2004) Human aflatoxicosis in developing countries: A review of toxicology, exposure, potential health consequences and interventions. *American Journal of Clinical Nutrition* 80, 1106-1122.

19

Pre- and Postharvest Management of Aflatoxin in Maize: An African Perspective

Kerstin Hell, Pascal Fandohan, Ranajit Bandyopadhyay*, Sebastian Kiewnick, Richard Sikora and Peter J. Cotty

Abstract

Pre- and postharvest contamination of aflatoxin in maize is a major health deterrent for people in Africa where maize production has increased dramatically. This chapter highlights management options for pre- and postharvest toxin contamination in maize. Sound crop management practices are an effective way of avoiding, or at least diminishing, infection by *Aspergillus flavus* and subsequent aflatoxin production. Pre- and postharvest practices that reduced aflatoxin contamination include: the use of resistant cultivars, harvesting at maturity, rapid drying on platforms to avoid contact with soil, appropriate shelling methods to reduce grain damage, sorting, use of clean and aerated storage structures, controlling insect damage, and avoiding long storage periods. These contamination reducing management practices are being tested in collaboration with farmers. Work continues on food basket surveys, the bio-ecology of aflatoxin production, developing biological control through a competitive exclusion strategy, reducing the impact of postharvest management practices on human blood toxin levels, and breeding to reduce the impact of mycotoxins on trade.

Introduction

In developing countries, many individuals are not only food insecure, but also are chronically exposed to high levels of mycotoxins in their diet. Food security exists when all people, at all times, have physical and economic access to sufficient, safe and nutritious food to meet their dietary needs and food preferences for an active and healthy life (FAO, 1996). Food safety results when microbial contaminants and chemical toxicants are present below tolerance levels in foods. Aflatoxin, a mycotoxin, compromises food security in the most vulnerable groups of people in Africa.

Aspergillus flavus, *Aspergillus parasiticus* and, rarely, *Aspergillus nomius* produce aflatoxins as secondary metabolites in agricultural products prone to fungal infection. Aflatoxins may cause liver cancer, suppressed immune systems, and retarded growth and development by contributing to malnutrition. Children are the most sensitive to the effects of aflatoxin-contaminated food. The effects of chronic exposure to aflatoxin are common in Africa, but acute toxicity, leading to death of humans, also has been reported (Azziz-

Baumgartner *et al.*, 2005). Some of the highest and most persistent human exposures to aflatoxin occur in West Africa, where nearly 99% of the children were positive for an aflatoxin biomarker (Gong *et al.*, 2002, 2004). Maize consumption is an important source of aflatoxin exposure for these children (Egal *et al.*, 2005).

Aflatoxin-producing fungi also cause direct economic losses by spoiling grain. Animals fed aflatoxin-contaminated grain have lower productivity and slower growth. Commodities contaminated with aflatoxins have a lower market value and often are consumed locally, since they cannot be exported. Levels of mycotoxins acceptable in foods in developed countries have been lowered, which can result in lowered export earnings by African countries that cannot comply with the stricter regulations. Overall costs for mycotoxin management and monitoring in the United States are estimated at between \$0.5 million to > \$1.5 billion for aflatoxin in maize and peanuts, fumonisin in maize, and deoxynivalenol in wheat (CAST, 2003).

In many parts of Africa maize has become the preferred cereal for food, feed and industrial use, displacing traditional cereals such as sorghum and millets. Maize production in Sub-Saharan Africa tripled from the early 1960s to late 1990s because of nearly 2-fold increase in area under cultivation and a > 40% increase in productivity. The greatest gains occurred in West Africa (350% for production, 64% for productivity and 170% for area), particularly in Nigeria where the increases were 385% for production, 46% for productivity and 231% for area (FAOSTAT, 2003). Consequently, maize consumption is high in Africa, ranging from 85 kg/year per person in Eastern and Southern Africa to 105 kg/year per person in West Africa (FAO, 2005). Maize is one of the cereals most susceptible to aflatoxin contamination (Wilson *et al.*, 2006). High consumption of maize coupled with frequent and elevated aflatoxin levels, leads to a high aflatoxin risk. The development and dissemination of aflatoxin management practices are essential to reduce exposure to aflatoxins by consumers and producers dependent on maize for food and income generation. In this chapter, we briefly describe the prevalence and distribution of aflatoxin contamination in West Africa and different management approaches that can be used to reduce aflatoxin contamination in maize, with emphasis on smallholder farmers in Africa.

Prevalence and distribution of aflatoxins in West Africa

Aflatoxin production depends on factors such as: water stress, high-temperature (> 32°C) stress, insect damage to the host plant, susceptible crop growth stages, poor soil fertility, high crop density, and weed competition (Bruns, 2003). Thus, the extent of aflatoxin contamination varies with geographic location, agricultural and agronomic practices, and the susceptibility of cultivars to fungal invasion during preharvest, storage, and/or processing.

In West Africa, agroecological zones are distinguished on the length of the growing period, *i.e.*, the period that water is available for crop production in well-drained soils. This period is a function of precipitation, evaporation, and available water in the soil. In Benin, aflatoxin contamination at the beginning of storage was highest in the Southern Guinea Savanna agroecological zone [moist grassland or derived forest with a 9-month rainy season; see Sétamou *et al.* (1997) for more details], where > 50% of the stores were contaminated with a mean aflatoxin level of 77 ng/g (Hell *et al.*, 2003). Six months after storage, both the incidence of contamination and the level of aflatoxin present in the maize samples had increased in all zones (from the southern coast to the north of Benin with decreasing rainfall

Table 1. Farming practices associated with high and low aflatoxin levels in stored maize in Benin.

Lower Aflatoxin Levels	Higher Aflatoxin Levels
Production Practices	
Crop rotation	Maize mono cropping
Local variety in South	Improved variety in South
Improved variety in North	Local variety in North
Maize in mixed cropping	Cowpea, peanut or cassava intercrop
Diammonium phosphate fertilizer	No fertilizer
Farmers aware of incomplete husk cover	Maize is damaged in the field
Harvest Practices	
Harvest at crop maturity	Delayed harvest
Harvest of maize with the husk	Harvest maize in heaps; cobs shelled later
Sun drying on platform	"Field" drying on the plant
Drying of maize without the husk	Delayed drying
Immediate removal of damaged cobs	No sorting at harvest
Storage Practices	
Cleaning of the storage structure	No preparation of the storage structure
Maize stored for 3-5 months	Maize stored for 8-10 months
Smoke or insecticide use	No insect control
Maize stored in aerated stores	Maize stored in poorly aerated stores

from south to north: Forest Mosaic Savanna, Southern Guinea Savanna, Northern Guinea Savanna and Sudan Savanna), but the increase varied with year, season and zone (Hell *et al.*, 2003). After six months of storage, > 57% of the maize samples from the Sudan Savanna had levels of aflatoxin ranging from 52 to 220 ng/g. In the other agroecological zones toxin contamination ranged between 8 and 80 ng/g.

In Nigeria, the percentage of stores contaminated with aflatoxin was similar to that in Benin, but mean levels of contamination were much higher (Udoh *et al.*, 2000). As in Benin, the Southern Guinea Savanna and Sudan Savanna zones in Nigeria had significantly higher aflatoxin contamination than did the other agroecological zones. In West Africa, aflatoxin contamination levels measured in maize sold to the public were high and ranged from 0.4 to 490 ng/g in Ghana, 0.7 to 110 ng/g in Togo, and 0.2 to 120 ng/g in Benin (James *et al.*, 2007). In the same study, 40% of the samples from the Southern Guinea Savanna exceeded the 20 ng/g internationally recommended safety limit.

The Southern Guinea Savanna appears to be the agroecological zone, in which aflatoxin contamination is the highest (Hell *et al.*, 2003). This zone has a bi-modal rainfall pattern with the first crop being harvested at the beginning of the second rainy season which makes drying the crop difficult. The second crop often does not get enough rain and high insect pressure increases the likelihood of aflatoxin contamination.

IITA's approach to mycotoxin management in Africa is based on questionnaires and surveys about farmers' management practices (Table 1) that were related to aflatoxin contamination in Benin (Hell *et al.*, 2000b; 2003) and Nigeria (Udoh *et al.*, 2000). The questionnaire and survey information were used to design and conduct on-farm trials to identify technologies that could significantly reduce toxin content (Hell *et al.*, 2005). Strategies tested include the use of resistant and/or tolerant varieties, insect management practices, appropriate post-harvest handling (sorting, cleaning, drying, good packaging, application of hygiene, use of appropriate storage systems, appropriate transportation means), awareness and sensitization.

Preharvest crop management practices

Developing strategies for the prevention or reduction of aflatoxins requires a good understanding of the factors that influence the infection process and the conditions that influence toxin formation. Soil type and condition and the availability of viable spores, are important factors (Horn, 2003). Environmental factors that favor *A. flavus* infection in the field include high soil and/or air temperature, drought stress, nitrogen stress, crowding of plants and conditions that aid the dispersal of conidia during silking (Diener *et al.*, 1987). Factors that influence the incidence of fungal infection include the presence of invertebrate vectors, grain damage, oxygen and carbon dioxide levels in stores, inoculum load, substrate composition, fungal infection levels, prevalence of toxigenic strains and microbiological interactions (Horn, 2003). Crop rotation and management of crop residues also are important in controlling *A. flavus* infection in the field.

Tillage practices, crop rotation, fertilizer application, weed control, late season rainfall, irrigation, wind and pest vectors all can affect the source and level of fungal inoculum maintaining the disease cycle in maize (Diener *et al.*, 1987). When maize was intercropped with cowpea the likelihood of aflatoxin contamination increased (Hell, 1997). In Africa, crops are cultivated under rainfed conditions, with low levels of fertilizer and little or no pesticide application. These conditions promote *A. flavus* infection of fertility stressed plants, and any action taken to reduce the probability of silk and kernel infection will reduce aflatoxin contamination.

Insects vector fungi and cause damage that allow fungal access to grain and other crop tissues thereby increasing the chances of aflatoxin contamination (Sétamou *et al.*, 1998). Incidence of the insect borer *Mussidia nigrivenella*, was positively correlated with aflatoxin contamination of maize in Benin. When loose-husked maize hybrids are used, the chance of insect damage and aflatoxin contamination increases.

Research in progress will develop host-plant and biocontrol options for preharvest management of aflatoxin. Maize genotypes with aflatoxin resistance have been identified in West and Central Africa (Brown *et al.*, 2001) and these sources of resistance are being used in a breeding program to develop aflatoxin-resistant, high-yielding cultivars adapted to tropical Africa (Menkir *et al.*, Chapter 23). The biocontrol principle of competitive exclusion of toxigenic strains of *A. flavus* by atoxigenic strains (Cotty *et al.*, Chapter 24) has been used in the United States to reduce aflatoxin contamination of cotton (Cotty, 1994), peanut (Dorner *et al.*, 1998) and maize (Abbas *et al.*, 2006). A similar approach was attempted in Benin (Cardwell and Cotty, 2000) that was further expanded in Nigeria (Bandyopadhyay *et al.*, 2005). Presently, four atoxigenic strains are being field-tested in Nigeria for their potential to control aflatoxin in maize. Adding resistant cultivars and biocontrol to the currently available technologies for the reduction of aflatoxin contamination would significantly reduce aflatoxin levels.

Table 2. Occurrence (%) of some toxigenic fungal species in maize grains following seven days of drying with the indicated drying method.

Drying method	*Aspergillus*	*Fusarium*	*Penicillium*	Others
Cobs on stalk in the field	4.7 ab[1]	99 a	41.7 a	5.3 a
Sun drying; cobs on the ground	21 a	95 a	44 a	10 a
Sun drying; cobs on a platform	2.0 b	86 b	4.7 b	2.7 a
Sun drying; cobs on a plastic sheet	18 a	33 c	9.7 b	4.7 a

[1]Means within a column followed by the same letter are not significantly different based on the Student-Neuman, Keuls test ($P < 0.05$). There were 12 replications per treatment.

Harvest and drying management practices

Timing of harvest greatly affects the extent of aflatoxin contamination. Extended field drying of maize increased insect infestation and fungal contamination. Delayed harvest increased mold incidence, insect damage and aflatoxin levels (Kaaya *et al.*, 2005). Aflatoxin levels increased 4-fold and more than 7-fold when maize harvest was delayed by 3 and 4 weeks, respectively, after maturity (Kaaya *et al.*, 2005). Moisture content was reduced when harvest was delayed, but the grain did not dry to the required safe storage moisture content of 15%. Fungal growth and mycotoxin production can occur within a few days if the grain is not properly dried and cooled before it is stored.

After harvest, maize grain should be dried to a safe level to stop fungal growth. Aflatoxin contamination can increase ten-fold in three days if maize grain is not dried properly (Tanboon-ek, 1989). A common recommendation is that harvested field crops should be dried as quickly as possible to safe moisture levels of 10-13% for cereals and 7-8% for oil seeds. Farmers also are advised to dry grain outside the field and off the ground to reduce fungal contamination during drying. Dry grains keep longer, are rarely attacked by insects, and usually do not support mold growth, since the free water required for their development is not available. Drying in Africa usually is solar-based, and often takes longer to reach a "safe" moisture level. When high rainfall occurs at harvest, farmers may stack cobs with the stalk to shield the products from rain, pile grains in a home yard under cover, dry grains over a kitchen fire, or mix moist and dry grains. Drying the grain on a raised drying platform often reduces contamination by toxigenic fungi (Table 2). Sometimes drying is not completed before storage. In Benin, drying for 3-6 days during the driest part of the year, *e.g.* humidity as low as 20%, resulted in whole yam tuber chips with a moisture content of 20%. Thus, drying was not complete, but most farmers were unaware of this problem (Mestres *et al.*, 2004). Simple devices should be developed so that African farmers can determine if their products have reached a safe moisture level.

Postharvest crop management practices

Aflatoxin is preferably controlled in the standing crop, since contamination of harvested cobs increases with storage time. Aflatoxin contamination in Africa is compounded by excessive heat, high humidity, lack of aeration in the storage area, and insect and rodent damage. The first step to reduce aflatoxin levels is to sort cobs that are damaged, insect infested, have an incomplete husk cover, or contain moldy grains from the rest of the grain. This grain should be consumed last, if it is consumed at all, and kept apart from the grain to be stored for the long-term. Sorting is an efficient way to reduce aflatoxin levels in stored maize, although the percentage of cobs sorted out varies widely by farmer, and may depend on both personal judgment and economic status.

To reduce aflatoxin contamination after sorting, maize cobs should be stored in a well-ventilated drying bin. From time to time the grain quality must be checked and insect infestation controlled. If high insect infestation levels are found, then the maize cobs should be shelled, the bad grains removed, and the good grains put in bags, preferably bags made of jute. Farmers in Africa increasingly store grains in polypropylene bags, but the poor aeration in these bags may encourage fungal growth and aflatoxin production, in grains not dried to a safe level (Udoh *et al.*, 2000; Hell *et al.*, 2000*b*).

The storage form (cobs or shelled grain) of maize influences contamination by toxigenic fungi. Mora and Lacey (1997) found higher levels of aflatoxigenic fungi in maize that was shelled immediately after harvest than in maize kernels that were left on the cob through drying. Shelling maize by beating cobs in a bag with a stick injures the kernels and facilitates fungal infection of the grain. Damaged maize kernels are prone to high levels of aflatoxin contamination, as are maize cobs that are threshed with mechanical shellers (Fandohan *et al.*, 2006).

The type of storage also influences aflatoxin levels, and the types of storage structures and their placement vary across the agroecozones in West Africa. Traditional storage methods are of two types: (i) temporary storage, used primarily for drying, and (ii) long-term storage structures made from plant materials (wood, bamboo or thatch), clay or bags (Fiagan, 1995). Maize stored as grain had the highest levels of *A. flavus*, reaching a maximum of 32% infected kernels in bags and 30% infected kernels in clay stores after four months of storage. The incidence of *A. flavus* in maize kernels stored on the cob with the husk was low and $< 1.3\%$ irrespective of the storage structure (Hell, 1997).

Disinfestation management methods

Insect infestation is related to aflatoxin contamination both preharvest (Sétamou *et al.*, 1997) and postharvest (Hell *et al.*, 2000*a*). Insect species correlated with high levels of aflatoxin in West Africa include Coleopteran and Lepidopteran insect species, and the role of specific species, *e.g.*, *Mussidia nigrivinella* (Lepidoptera: Pyralidae), in the transmission of fungal spores has been determined (Sétamou *et al.*, 1998). Measures to reduce insect infestation postharvest either through the application of commercial insecticides in storage or through the installation of barriers that protect the cob against infestation either in the field or in store are being tested in West Africa. Use of a prophylactic pesticide, especially at the beginning of storage when pest incidence is low, often is not cost efficient (Meikle *et al.*, 2002). Instead a decision tree approach, such as the one outlined by Meikle *et al.* (2002), to control pest infestation while incorporating decision-making on reducing mycotoxin contamination should be followed to monitor commodity product quality during storage.

There are several methods to control insect and fungal development once they have infested the stored commodities. The use of insecticides and fungicides in Africa is limited by their availability in remote rural areas. African farmers often use methods such as smoking to reducing moisture content and insect damage. The efficacy of smoking in controlling insect infestation is comparable to that of Actellic, *i.e.*, Pirimiphos-methyl (Daramola, 1986).

Between four and 12% of the farmers in Nigeria use smoke to preserve their grain and reduce the aflatoxin levels (Udoh *et al.*, 2000).

Many farmers use local plant products, either in their pure form or as oil or water extracts to control insects. *Ocimum gratissimum*, *Aframonium* spp., *Zingiber officinalis*, *Xylopia aethiopica*, *Monodera myristica*, *Ocimum basilicum*, *Tetrepleura tetraptera* and *Piper guineense* all have been tested for their ability to inhibit the mycelial growth of *A. flavus* (Cardwell and Dongo, 1994). Aqueous extracts of a mixture of dried fruits of *X. aethiopica* and *P. guineense* inhibit the growth of all tested maize pathogens. Essential oils from *Azadirachta indica* and *Morinda lucida* inhibit the growth of toxigenic *A. flavus* and significantly reduced aflatoxin synthesis in inoculated maize grains (Bankole, 1997). Essential oils from *O. gratissimum*, *Thymus vulgaris* and *Cymbopogon citratus* prevented conidia germination and the growth of *F. verticillioides*, *A. flavus* and *A. fumigatus* (Nguefack *et al.*, 2004). Ground *Aframomum danielli* (Zingiberaceae) can control molds and insect infestation in stored maize and soybeans for up to 15 month under ambient conditions in southwestern Nigeria (Adegoke *et al.*, 2000). Further tests are needed to determine the inhibition mechanism(s) and to identify the active ingredient of the natural products that inhibit fungal growth, before definitive statements can be made on the role of natural botanical products in controlling postharvest aflatoxin contamination.

Removing aflatoxin through physical separation and hygiene

The distribution of aflatoxin on a maize cob or in a grain lot is very heterogeneous with large quantities of the toxin concentrated in just a few or a small percentage of the kernels (Whitaker, 2003). The highest concentrations of aflatoxin usually are found on heavily molded and/or damaged kernels. Sorting out physically damaged and infected grains (based on their coloration, odd shapes, shriveled and reduced size) from the intact commodity can reduce aflatoxin levels by 40-80% (Park, 2002). Sorting can be done manually or with electronic sorters, which are used to reduce aflatoxin contamination in peanuts, Brazil nuts, almonds and pistachio. However, the extent and method of sorting required to attain satisfactory reduction in aflatoxin levels of agricultural products acceptable to the subsistence African farmers and consumers remains unknown.

Clearing the remains of the previous harvest and destroying infested crop residues are basic sanitary measures that also reduce grain deterioration in the field and in storage. Cleaning storage areas prior to filling them with the new harvest reduced aflatoxin levels (Hell *et al.*, 2000*a*). Keeping the area surrounding the storage facility clean reduces infestation with insects that take refuge in host plants near the storage facility. Storage of healthy cobs after separating heavily damaged maize cobs, *i.e.*, those that have more than 10% ear damage due to insects also reduces aflatoxin levels (Sétamou *et al.*, 1998). Finally, levels of mycotoxins in contaminated commodities prior to consumption may be reduced by food processing methods such as wet and dry milling, grain cleaning, canning (autoclaving), roasting, baking, frying, alkali cooking (nixtamalization), extrusion cooking, *etc.* There are diverse traditional food processing methods that significantly reduce the amount of aflatoxin in food prepared from maize and peanuts in different parts of Africa. Some of these techniques have been identified and described, *cf.*, Fandohan *et al.* (Chapter 26). Further evaluations of these processing techniques on aflatoxin levels are needed to identify methods that expose consumers to the least amount of aflatoxin.

Dietary change, dietary interventions and detoxification

High incidences of mycotoxin-associated diseases have been recorded in areas where maize and peanuts are dietary staples. Thus, one approach is to reduce the frequent consumption of these "high risk" foods by consuming a more varied diet. In parts of China, individuals that change their diet from maize to rice reduce their risk of aflatoxin exposure (Yu, 1995). People in developed countries experience a low risk of mycotoxin contamination primarily due to a diverse diet that contains foods from a range of climatic zones in which crops are produced with varying risks of mycotoxin exposure. Many of these foods are produced under excellent sanitary conditions, with only a small proportion of at-risk foods used for human consumption, unlike developing countries in which most people eat the same staple at most meals.

The toxic effects of mycotoxins may be limited by natural or synthetic agents such as antioxidants, *e.g.*, selenium, vitamins and provitamins, food components, *e.g.*, phenolic compounds, coumarin, chlorophyll and its derivatives, fructose and aspartame, medicinal herbs and plant extracts, and mineral and biological binding agents, *e.g.*, hydrated sodium calcium aluminosilicate, bentonites, zeolites, activated carbons, bacteria, and yeast (Farombi, 2006) Chemoprevention can block, retard or even reverse the carcinogenic effect resulting from mycotoxin exposure (Farombi, 2006). Oltipraz, a drug used against schistosomiasis, is a potent inducer of enzymes that detoxify carcinogens including aflatoxins. Another potential group of chemopreventive agents are natural components in fruits and vegetables, such as chlorophyll, which are found in low concentrations in balanced diets. The tight binding of chlorophyll or chlorophyllin, a semi-synthetic mixture of sodium copper salts derived from chlorophyll, to potential carcinogens may interfere with their absorption from the gastrointestinal tract and reduce the amount of the toxin that reaches susceptible tissues (Egner *et al.*, 2003).

Another approach widely used in the feed industry is to mix clay minerals with the animal feed. The clay selectively binds aflatoxins tightly to prevent their absorption in the gastrointestinal tracts and the clay-aflatoxin complex is eliminated from the body (Afriyie-Gyawu *et al.*, Chapter 25). Such adsorbents act more as prophylactics than as curative remedies.

Some mycotoxins can be destroyed chemically with calcium hydroxide, monoethylamine, ozone or ammonia. For example, ammoniation degrades 95-98% of the aflatoxin B_1 present. This process is not effective against other toxins, however, and the treated grain can be used only as animal feed. For a detoxification method to be acceptable, it must be efficient, safe and cost effective while safeguarding nutritional quality.

Outlook for aflatoxin management strategies for maize from Africa

Aflatoxin contamination of agriculture commodities is gaining public prominence in Africa. This toxin is now perceived to have many more health effects than previously thought (Williams *et al.*, 2004). Aflatoxins appear to be much more pervasive than previously thought, with a large percentage of foods and a high percentage of the population in Africa affected. The negative impact of chronic exposure of aflatoxins on human health and nutrition has been overlooked even though it has serious effects on children's growth and development. Prevention through preharvest and postharvest control is the first step in ensuring a safe final product.

IITA has developed a management package to control aflatoxin contamination from the field to the consumer. Component technologies in this package effectively lower toxin levels and are accessible to farmers. Key components of this package are insect control from the field through the end of storage, timely harvest, suitable sanitary conditions during postharvest operations, speedy grain drying prior to storage, selection of wholesome cobs for storage, use of appropriate storage structures to avoid insect infestation and grain rewetting, and sorting of the grain prior to its consumption. Inclusion of biocontrol agents and/or resistant cultivars, as available, in the package should reduce aflatoxin contamination even further. The impact of this package of technologies on child health is being evaluated in collaboration with many national programs in Africa.

The export potential of primary raw and processed crops from Africa remains effectively unrealized, and the institutions that monitor food safety in Africa are very weak. New approaches, tools and coalitions to manage mycotoxin are needed. Aflatoxins have received the most attention thus far, but studies are needed on other mycotoxins, *e.g.*, fumonisins as well.

Acknowledgments

We thank BMZ and GTZ for supporting this research through Project No. 2002.7860.6 – 001.00.

References

Abbas, H.K., Zablotowicz, R.M., Bruns, H.A. and Abel, C.A. (2006) Biocontrol of aflatoxin in corn by inoculation with non-aflatoxigenic *Aspergillus flavus* isolates. *Biocontrol Science and Technology* 16, 437-449.

Azziz-Baumgartner, E., Lindblade, K., Gieseker, K., Rogers, H.S., Kieszak, S., Njapu, H., Schleicher, R., McCoy, L.F., Misore, A., DeCock, K., Rubin, C., Slutsker, L. and the Aflatoxin Investigative Group. (2005) Case-control study of an acute aflatoxicosis outbreak, Kenya 2004. *Environmental Health Perspectives* 113, 1779-1783.

Adegoke, G.O., Iwahasi, H., Komatsu, Y., Obuchi, K. and Iwahasi, Y. (2000) Inhibition of food spoilage yeasts and aflatoxigenic moulds by monoterpenes of the spice *Aframonium danielli. Flavour Fragrance Journal* 15, 147-150.

Bandyopadhyay, R.,Kiewnick, S., Atehnkeng, J., Donner, M., Cotty, P.J. and Hell, K. (2005) Biological control of aflatoxin contamination in maize in Africa. *Proceedings of the Tropentag Conference on International Research for Development, (11-13 October 2005, Hohenheim, Germany).* http://www.tropentag.de/2005/abstracts/full/398.pdf.

Bankole, S.A. (1997) Effect of essential oil from two Nigerian medicinal plants (*Azadirachta indica* and *Morinda lucida*) on growth and aflatoxin B_1 production in maize grain by a toxigenic *Aspergillus flavus. Letters in Applied Microbiology* 24, 190-192.

Brown, R.L., Chen, Z.-Y., Menkir, A., Cleveland, T.E., Cardwell, K.F., Kling, J. and White, D.G. (2001) Resistance to aflatoxin accumulation in kernels of maize inbreds selected for ear rot resistance in West and Central Africa. *Journal of Food Protection* 64,396-400.

Bruns, H.A. (2003) Controlling aflatoxin and fumonisin in maize by crop management. *Journal of Toxicology: Toxin Reviews* 22, 153-173.

Cardwell, K.F., and Cotty, P.J. (2000) Interactions among U.S. and African *Aspergillus* spp. strains: Influence on aflatoxin production. *Phytopathology* 90, S11.

Cardwell, K. F., and Dongo, L. (1994) Effects of extracts from nine plant species found in Africa on the mycelia growth of *A. flavus* Link *ex* Fries. In: Highley, E.; Wright, E.J.; Banks, H.J.; Champ, B.R. (eds.), *Stored Product Protection, Proceedings of the 6th International Working Confe-*

rence on Stored-Product Protection (17-23 April 1994, Canberra, Australia). CAB International, Wallingford, United Kingdom.

CAST. (2003) *Mycotoxins: Risks in Plant, Animal, and Human Systems. Task Force Report No. 139.* Council for Agricultural Science and Technology, Ames, IA.

Cotty, P.J. (1994) Influence of field application of an atoxigenic strain of *Aspergillus flavus* on the population of *A. flavus* infecting cotton bolls and on the aflatoxin content of cottonseed. *Phytopathology* 84, 1270-1277.

Daramola, A.M. (1986) Corn ear worm infestation of seven maize cultivars and control in Southwestern Nigeria. *Insect Science and its Application* 7, 49-52.

Diener, U.L., Cole, R.J., Sanders, T.H., Payne, G.A., Lee, L.S. and Klich, M.A. (1987) Epidemiology of aflatoxin formation by *Aspergillus flavus*. *Annual Review of Phytopathology* 25, 249-270.

Dorner, J.W., Cole, R.J. and Blankenship, P.D. (1998) Effect of inoculum rate of biological control agents on preharvest aflatoxin contamination of peanuts. *Biological Control* 12, 171-176.

Egal, S., Hounsa, A., Gong, Y.Y., Turner, P.C., Wild, C.P., Hall, A.J., Hell, K. and Cardwell, K.F. (2005) Dietary exposure to aflatoxin from maize and groundnut in young children from Benin and Togo, West Africa. *International Journal of Food Microbiology* 104, 215-224.

Egner, P.A., Munoz, A. and Kensler, T.W. (2003) Chemoprevention with chlorophyllin in individuals exposed to dietary aflatoxin. *Mutation Research* 523-524, 209-216.

Fandohan, P., Ahouansou, R., Houssou, P., Hell, K., Marasas, W.F.O. and Wingfield, M.J. (2006) Impact of mechanical shelling and dehulling on *Fusarium* infection and fumonisin contamination in maize. *Food Additives and Contaminants* 23, 451-421.

FAO. (1996) *Rome Declaration on World Food Security and World Food Summit Plan of Action. (World Food Summit 13-17 November 1996)*. FAO, Rome, Italy.

FAO. (2005) Food outlook: Global information and early warning system on food and agriculture (GIEWS) No. 4 December 2005.

FAOSTAT (2003) Food Supply and Commodity Balance Data. http://faostat.fao.org/.

Farombi, E.O. (2006) Aflatoxin contamination of foods in developing countries: Implications for hepatocellular carcinoma and chemopreventive strategies. *African Journal of Biotechnology* 5, 1-14.

Fiagan, Y.S. (1995) Le système de stockage du maïs en milieu paysan Beninois: Bilan et perspectives. [Maize storage in rural Benin: present and future situation]. In *Production et Valorisation du Mais a L'echelon Villageois en Afrique de l'Ouest; Actes du Seminaire "Mais prospere" (25-28 January 1994, Cotonou, Benin)*. FAO, Rome, Italy, p. 201-211; http://www.fao.org/docrep/X5158F/x5158f00.htm.

Gong, Y.Y., Cardwell, K.F., Hounsa, A., Egal, S., Turner, P.C., Hall, A.J. and Wild, C.P. (2002) Dietary aflatoxin exposure and impaired growth in young children from Benin and Togo: Cross sectional study. *British Medical Journal* 325, 20-21.

Gong, Y.Y., Hounsa, A., Egal, S., Turner, P.C., Sutcliffe, A.E., Hall, A.J., Cardwell, K.F. and Wild, C.P. (2004) Postweaning exposure to aflatoxin results in impaired child growth: A longitudinal study in Benin, West Africa. *Environmental Health Perspectives* 112, 1334-1338.

Hell, K. (1997) *Factors Contributing to the Distribution and Incidence of Aflatoxin Producing Fungi in Stored Maize in Benin*. Ph.D. Dissertation, University of Hannover, Germany.

Hell, K., Cardwell, K.F., Sétamou, M. and Schulthess, F. (2000*a*) Influence of insect infestation on aflatoxin contamination of stored maize in four agroecological regions in Benin. *African Entomology* 8, 169-177.

Hell, K., Sétamou, M., Cardwell, K.F. and Poehling, H.-M. (2000*b*) The influence of storage practices on aflatoxin contamination in maize in four agroecological zones in Benin, West Africa. *Journal of Stored Products Research* 36, 365-382.

Hell, K., Cardwell, K.F. and Poehling, H.-M. (2003) Distribution of fungal species and aflatoxin contamination in stored maize in four agroecological zones in Benin, West-Africa. *Journal of Phytopathology* 151, 690-698.

Hell, K., Bandyopadhyay, R., Kiewnick, S., Coulibaly, O., Menkir, A. and Cotty, P. (2005) Optimal management of mycotoxins for improving food safety and trade of maize in West Africa. In

Tielkes, E., Hülsebusch, C., Häuser, I., Deininger, A., Becker, K. (eds) *The Global Food & Product Chain — Dynamics, Innovations, Conflicts, Strategies: International Research on Food Security, Natural Resource Management and Rural Development; Book of Abstracts/Tropentag 2005 Stuttgart-Hohenheim.* Universität Hohenheim, Tropenzentrum, Centre for Agriculture in the Tropics and Subtropics, Stuttgart, Germany; http://www.proceedings2005.tropentag.de, p. 369.

Horn, B.W. (2003) Ecology and population biology of aflatoxigenic fungi in soil. *Journal of Toxicology – Toxin Reviews* 22, 351-379.

James, B., Adda, C., Cardwell, K., Annang, D., Hell, K., Korie, S., Edorh, M., Gbeassor, F., Nagatey, K. and Houenou, G. (2007) Public information campaign on aflatoxin contamination of maize grains in market stores in Benin, Ghana and Togo. *Food Additives and Contaminants* 24, 1283-1291.

Kaaya, A.N., Warren, H.L., Kyamanywa, S. and Kyamuhan, W. (2005) The effect of delayed harvest on moisture content, insect damage, moulds and aflatoxin contamination of maize in Mayuge district of Uganda. *Journal of the Science of Food and Agriculture* 85, 2595-2599.

Meikle, W.G., Markham, R.H., Nansen, C., Holst, N., Degbey, C.P., Azoma, K. and Korie, S. (2002) Pest management in traditional maize stores in West Africa: A farmer's perspective. *Journal of Economic Entomology* 95, 1079-1088.

Mestres, C., Bassa, S., Fagbohoun, E., Nago, M., Hell, K., Vernier, P., Champiat, D., Hounhouigan, J. and Cardwell, K.F. (2004) Yam chip food chain: hazardous practices and presence of aflatoxins in Benin. *Journal of Stored Products Research* 40, 575-585.

Mora, M., and Lacey, J. (1997) Handling and aflatoxin contamination of white maize in Costa Rica, *Mycopathologia* 138, 77-89.

Nguefack, J., Leth, V., Amvam Zollo, P.H. and Mathur, S.B. (2004). Evaluation of five essential oils from aromatic plants of Cameroon for controlling food spoilage and mycotoxin producing fungi. *International Journal of Food Microbiology* 94, 329-334.

Park, D.L. (2002) Effect of processing on aflatoxin. *Advances in Experimental Medicine and Biology* 504, 173-179.

Sétamou, M., Cardwell, K.F., Schulthess, F. and Hell, K. (1997) *Aspergillus flavus* infection and aflatoxin contamination of pre-harvest maize in Benin. *Plant Disease* 81, 1323-1327.

Sétamou, M., Cardwell, K.F., Schulthess, F. and Hell, K. (1998) Effect of insect damage to maize ears, with special reference to *Mussidia nigrivenella* (*Lepidoptera: Pyralidae*), on *Aspergillus flavus* (*Deuteromycetes: Monoliales*) infection and aflatoxin production in maize before harvest in the Republic of Benin. *Journal of Economic Entomology* 91, 433-438.

Tanboon-ek, P. (1989) Control of aflatoxin in maize. In: Semple, R.L., Frio, A.S., Hicks P.A. and Lozare, J.V. (eds.) *Mycotoxin Prevention and Control in Food Grains*. FAO, Rome, Italy, pp. p. 166-172.

Udoh, J.M., Cardwell, K.F., and Ikotun, T. (2000) Storage structures and aflatoxin content of maize in five agro-ecological zones of Nigeria. *Journal of Stored Products Research* 36, 187-201.

Whitaker, T.B. (2003) Standardization of mycotoxin sampling procedures: An urgent necessity. *Food Control* 14, 233-237.

Williams, H.J., Phillips, T.D., Jolly, P.E., Stiles, J.K., Jolly, C.M. and Aggarwal, D. (2004) Human aflatoxicosis in developing countries: a review of toxicology, exposure, potential health consequences, and interventions. *American Journal of Clinical Nutrition* 80, 1106-1122.

Wilson, J.P., Jurjevic, Z., Hanna, W.W., Wilson, D.M., Potter, T.L. and Coy, A.E. (2006) Host-specific variation in infection by toxigenic fungi and contamination by mycotoxins in pearl millet and corn. *Mycopathologia* 161, 101-107.

Yu, S.Z. (1995) Primary prevention of hepatocellular carcinoma. *Journal of Gastroenterology and Hepatology* 10, 674-682.

20

Management of Ochratoxin A in the Cocoa Supply Chain: A Summary of Work by the CAOBISCO/ECA/FCC Working Group on Ochratoxin A

Martin Gilmour and Marianne Lindblom*
on behalf of the Working Group on Ochratoxin A in cocoa

Abstract

The presence of ochratoxin A in cocoa products has been reported. This chapter describes work by the European chocolate and cocoa industry and trade to determine where ochratoxin A enters the cocoa supply chain and how to minimize it. Ochratoxin A can be found in beans from most producing countries. Ochratoxin A is produced in the beans in the producing countries and toxin levels do not seem to increase during shipping or during storage in consuming countries. Ochratoxin A levels increase at the end of the main West-African cocoa harvest when damaged pods are implicated in the contamination. When ochratoxin A contamination occurs, the toxin is located mainly on the shell of the bean, so a large portion of the toxin present is removed during bean processing. Analysis of cocoa products on the European market confirms that only low levels of ochratoxin A are present in cocoa-containing products as consumed.

Introduction

Ochratoxin A, sometimes abbreviated as OTA, is a secondary metabolite produced by some *Aspergillus* and *Penicillium* spp. Ochratoxin A is found in a range of foods such as cereals, dried fruits, grape juice, coffee, cocoa, wine and beer with cereals providing the largest contribution to the intake of ochratoxin A in Europe (DG Health and Consumer Protection, 2002). Ochratoxin A's toxicological effects have been evaluated on several occasions by the Joint FAO/WHO Expert Committee on Food Additives (JECFA) and the EU Scientific Committee for Foods (SCF) (DG Health and Consumer Protection, 1998; JECFA, 2001). SCF concluded in 1998 that ochratoxin A possesses carcinogenic, nephrotoxic, teratogenic, immunotoxic and possibly neurotoxic properties. Based on this evaluation SCF recommended that ochratoxin A exposure should be reduced as much as possible ensuring that exposures are < 5 ng/kg bw per day (DG Health and Consumer Protection, 1998). Since 1998 further research, especially on

 Mycotoxins: Detection Methods, Management, Public Health and Agricultural Trade (eds. J. F. Leslie *et al.*).

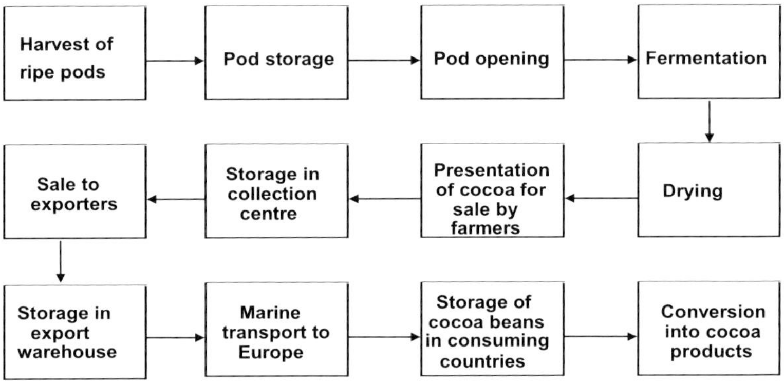

Figure 1. Major steps in the cocoa supply pathway.

the mechanism(s) of ochratoxin A-induced carcinogenicity has been conducted. Based on the new research the European Food Safety Agency (EFSA) re-evaluated risks from ochratoxin A in 2006 and established a Tolerable Weekly Intake of 120 ng/kg bw and recommended that all efforts should be made to continue to reduce ochratoxin A levels in foods (EFSA, 2006).

Cocoa production and supply chain

Cocoa beans are a key raw material for the manufacture of chocolate. Approximately 65% of the world supply of cocoa beans comes from West Africa, especially Côte d'Ivoire, Ghana and Nigeria. Cocoa also is produced in Asia and Latin America. As cocoa is a crop produced by smallholders, it is a valuable non-perishable cash crop for hundreds of thousands of farmers in cocoa-producing countries and is of great importance to the economies of these countries.

Cocoa pods are harvested and broken manually. The wet cocoa beans from the pods are fermented, usually under banana leaves, and then dried in the sun. The fermented and dried cocoa beans are collected from farms and buying stations, checked for quality, graded and put into export bags. Most of the cocoa beans are exported to Europe and North America to be made into cocoa liquor, cocoa butter and cocoa powder (Fig. 1).

The CAOBISCO, ECA and FCC work program on ochratoxin A

The presence of low levels of ochratoxin A in cocoa or cocoa products was reported in the late 1990s (Engel, 2000). In 1999, CAOBISCO (The Association of the Chocolate, Biscuit and Confectionery Industries of the EU), ECA (European Cocoa Association) and FCC (Federation for Cocoa Commerce) agreed to initiate work on ochratoxin A in cocoa and a joint working group was established. The objective of the work was to "specify as precisely as possible conditions favoring ochratoxin A production in cocoa and identify solutions to minimize its production".

The stage at which ochratoxin A contamination of cocoa occurs and conditions that stimulate (or discourage) ochratoxin A production were not known at this time. Studies were therefore initiated at different steps in the supply chain (Fig. 1), *e.g.*, fermentation, drying, storage in export warehouses, marine shipment, storage in warehouses in Europe, and the conversion to cocoa products and chocolate, were evaluated. Samples were collected from each step and analyzed for ochratoxin A level, to identify the steps at which ochratoxin A contamination occurred. The program activities were built on knowledge and experience from other crops contaminated with ochratoxin A or other mycotoxins. The bulk of the research in producing countries was contracted to CIRAD, France. Work also was contracted to CABI Bioscience and University of Aberystwyth, U.K. Studies of the processing steps in Europe were carried out by the cocoa and chocolate industry.

Sampling and Analysis

Mycotoxins are not homogenously distributed in commodities and therefore special sampling plans are defined for regulatory purposes. The purpose of the work described here was to identify the critical steps for production of ochratoxin A and therefore a more pragmatic approach to sampling was implemented. In the studies of cocoa bean processing at cocoa farms and of beans in collection centers where the batches were small (at maximum a few hundred kg), the samples were ~1 kg. If the samples were wet, they were immediately dried in an oven. All samples were sent to Europe for analysis.

Studies of storage in export warehouses in Africa and in warehouses in Europe were made on batches of 16 big bags (1 t each) or 100 standard bags (60 kg each). A sample of 1 kg was taken from each big bag and mixed to form a composite sample. From the standard bags, samples of ~100 g were taken from each of the 100 bags and mixed to make a composite sample. The composite samples were run through a splitter and reduced to a sample of 1 kg which was used as the analytical sample. Arbitrage samples, *i.e.*, samples taken routinely for quality control, from lots of cocoa beans arriving in Europe were used to survey the amount of ochratoxin A in cocoa beans. The arbitrage samples are composite samples of subsamples taken randomly from at least 30% of the bags in the lot. In the finished products (chocolate, cocoa drink powder and cocoa powder) the batches are expected to be homogenous with each finished sample weighing 100-200 g.

Ochratoxin A analyses were made by laboratories that had participated in a proficiency test organized by the working group and had a *z*-score between ±2. The proficiency test was made with one sample each of cocoa beans and cocoa powder with ochratoxin A levels 1.1 ng/g and 2.5 ng/g, respectively. The principle analytical steps used by the laboratories involved were: extraction of the toxin, sample clean up on an immunoaffinity column, and HPLC on reversed phase column with fluorescence detection. The laboratories worked with different minimal detection limits, 0.1, 0.2 or 0.5 ng/g. When calculating averages, half of the limit of detection value was used for the samples that had ochratoxin A present below the detection limit.

Cocoa bean processing in producing countries

Some initial studies were conducted both in Ghana and Côte d'Ivoire, but after the initial phase the work concentrated in Côte d'Ivoire.

Fresh beans

As with other fruits, the beans inside an undamaged pod were thought to be sterile until the pod wall is broken. Analysis of beans from nine healthy pods are consistent with this hypothesis, and ochratoxin A was not detected in any of the samples.

Mycoflora of cocoa samples collected in Ghana and Côte d'Ivoire

During field visits to Ghana and Côte d'Ivoire in November 2000, samples for mycological studies were collected from every stage of the cocoa processing system: fermentation heaps, drying beans, and bagged beans at both village and regional depot stores. In addition, swabs were taken from bamboo drying mats, tarpaulin drying sheets, plantain leaves covering the fermentation heaps, and air samples taken around fermentation heaps. Thirty-seven samples were analyzed for their fungal flora. All *Penicillium* and *Aspergillus* isolates were identified to species, and all other genera were ignored as not being ochratoxin A producers.

Yeast and fungi other than *Aspergillus* and *Penicillium* dominated the samples, but no ochratoxin A-producing *Penicillium* species were found. In samples from wooden boxes (sometimes used at larger farms for fermentation), bamboo mats and tarpaulin sheets, mostly field fungi were found, but in one sample *Aspergillus niger*, which can produce ochratoxin A, also was found. *Aspergillus niger* also was found in air samples around fermentation heaps after 3 days of fermentation and in some samples of visibly moldy beans.

Some of the samples of beans taken during drying and storage of dried beans had high ochratoxin A levels, but from only one of these samples was a species capable of producing ochratoxin A (*Aspergillus carbonarius*) recovered. Other non-ochratoxin A producing fungi were found on most of the samples. This pattern is not unusual since fungal vegetative growth often stops after toxin is produced since the environmental conditions may no longer be suitable for fungal growth, *e.g.,* when a product is dried. However, the mycotoxin produced will remain in the product even after the death of the organism that produced it. This limited mycological survey of cocoa beans and farm environment shows that while yeasts and fungi that do not produce ochratoxin A dominate, fungi that can produce ochratoxin A also are present in cocoa bean samples, in the on-farm environment and on the equipment.

Ochratoxin A development during fermentation and drying at a large commercial farm in Côte d'Ivoire

The first study of ochratoxin A development during fermentation and drying was conducted on a commercial farm (about 70 hectares). The samples were fermented in bags (micro-fermentation) placed in the middle and on the top of wooden fermentation boxes. The boxes had no visible mold contamination. Samples were taken after 2, 4 and 6 days of fermentation and sun dried. When the fermentation was complete, the content of the boxes was dried at two different depths (3 and 8 cm) in drying beds. The experiments were repeated four times between November 2000 and February 2001. No ochratoxin A was detected in any of the samples. Ochratoxin A also was not detected in samples from any of five micro-fermentation tests carried out with beans from black pod (*Phytophtora*) infected pods.

Fermentation boxes normally are covered, *e.g.*, with jute bags. Two trials were made in boxes that were not covered. Samples were taken after 2, 4 and 6 days of fermentation. In trials in February, 1.6 ng/g of ochratoxin A was detected after 2 days with a slight increase to 2.5 ng/g after 6 days. Repeat trials in June produced no detectable ochratoxin A.

These experiments indicate that very little ochratoxin A is produced during well-controlled fermentations in big boxes followed by sun drying.

Ochratoxin A development under smallholder conditions in Côte d'Ivoire

From February to July 2001, 62 samples of beans produced under smallholder conditions were analyzed for ochratoxin A. Unlike the results from the experiments from industrial scale fermentation in large boxes, the beans produced under small holder conditions (heap fermentation, small batches) contained ochratoxin A in many of the samples. The ochratoxin A level was > 0.5 ng/g in 24 of the samples (39%) and > 2 ng/g in 11 samples (18%). Higher levels of ochratoxin A were found in February and March than in the later April to July period.

Based on these results, we focused on ochratoxin A formation at the smallholder level. During the 2001/02 season a general survey of cocoa post-harvest practices was conducted and 168 samples were taken at small farms in three cocoa producing regions of Côte d'Ivoire (two sites in the West region, two sites in the Centre-West region and one site in the East region). Sixty-three of the 168 samples (38%) had > 0.5 ng/g ochratoxin A and 28 samples (17%) had > 2 ng/g of ochratoxin A. These results are very similar to those from the samples taken in spring 2001. Ochratoxin A was not found in 48 of the samples (26%). Very few samples were severely contaminated and only seven samples contained > 10 ng/g ochratoxin A. The average ochratoxin A level increased steadily from 1.4 ng/g in November to a maximum of 4.1 ng/g in February. In March the ochratoxin A level fell to < 1.0 ng/g ochratoxin A and stayed at these low levels through June. Clear conclusions regarding the regional distribution of ochratoxin A contamination cannot be drawn from the results.

For each field sample of dried cocoa beans produced by the farmer, control samples also were collected just after fermentation. The control samples were frozen immediately, transported to the lab, and dried. The only difference between the control samples and the field samples was the drying method. Twenty-four of the control samples were analyzed for ochratoxin A, all of which had less, but still significant, ochratoxin A contamination than did the corresponding samples dried on the farm. Thus ochratoxin A may occur prior to the drying stage. From these studies we concluded that:

- Ochratoxin A often is present in beans that have just been fermented and usually increases during the drying process. The initial colonization of cocoa beans by ochratoxin A producing fungi probably occurs between the time the pod is opened and the end of the fermentation.
- In Côte d'Ivoire, the end of the main harvest season (January and February) corresponds to the end of the dry season and appears to be the most critical period for ochratoxin A contamination. Thus, the more difficult fermentation conditions that occur at that time, due to climatic conditions and the nature of the mucilage, could facilitate mold growth and ochratoxin A production.
- Drying conditions alone are not responsible for the ochratoxin A level, which depends on interactions between harvesting, fermentation and drying conditions.

Ochratoxin A levels at the end of the main crop

A working hypothesis for higher ochratoxin A levels at the end of the main crop was developed, based on the following premise: smaller amounts of beans are available resulting in smaller fermentation heaps; the pulp is drier resulting in drier heaps due to the dry season; there is a large variation in ambient day/night temperature influencing the temperature in the heaps; there could be a long time between harvest and pod opening in order to collect sufficient amount of beans for a proper heap. Trials were conducted to test the factors associated with this hypothesis in February, March and April 2003.

Three treatments were set up to test the influence of heap size and of moisture: a) fermentation in a small heap (less than 50 kg of fresh beans), b) fermentation in a large heap (more than 200 kg of fresh beans), and c) fermentation in a small heap with wetted beans. The trial heaps were inoculated with moldy beans to strengthen eventual effects. The size of the inoculum was 1% (w/w) and it consisted of beans from damaged pods with internal mold. The first trial was conducted in March with a 3-day fermentation period, and was repeated in April with 5-day fermentations. All trial heaps fermented normally and no significant difference was found in the temperature profile of the heaps. None of the samples from these trials (four samples per heap in March and five samples per heap in April) had more than traces of ochratoxin A. It seems that neither the heap size nor the moisture in the heap have a significant influence.

The effect of pod storage time was studied in four experiments with healthy and damaged pods (damage by rodents, by cutlasses during harvesting and/or by *Phytophthora*). The pods were stored separately for 5 days or 4 weeks before fermentation. In each trial, beans from 1500 pods were used. Samples were taken 1 and 3 days into fermentation, from the middle and edges of the heap and immediately sun dried. The trials were started in February and beans after 5 days pod storage were fermented in February and March. Beans from the pods that had been stored for 4 weeks were fermented in March and April.

Ochratoxin A levels (Table 1) were higher in beans from damaged pods than in beans from healthy pods after 5 days of pod storage. Contamination of the beans from damaged pods began on the first day of fermentation, with a higher contamination levels in the middle of the heap. Three days into the fermentation, the trend was reversed and contamination was clearly greater at the edges. This reversal was accompanied by considerable mold growth on the surface of the heap. Five days into the fermentation, the ochratoxin A content increased further. Only traces of ochratoxin A were found after fermentation and drying of beans from healthy pods stored for 5 days.

After 4 weeks of pod storage the ochratoxin A levels were low and there was only a small difference between beans from healthy and damaged pods. During the longer storage, the inside of the pods dried out around the beans. It may be that the environment became unsuitable for the growth of ochratoxin A producing molds on the beans during this time. In these trials molds were recovered from placentas and beans affected by an undetermined black rot and from fermented beans severely contaminated with ochratoxin A. *Aspergillus* strains were recovered from five of the samples, and strains from two of the five samples produced ochratoxin A. Thus, ochratoxin A-producing *Aspergillus* strains are present on placenta and bean debris affected by black rot.

Effects of pod damage

Delaying pod opening does not affect ochratoxin A development if the pods are healthy. If the pods are damaged, then storage increases ochratoxin A content. Thus pod damage may contribute to the ochratoxin A peak found at the end of the main harvest in West Africa, where all damaged pods are routinely included in the end-of-harvest picking.

Table 1. Influence of pod types and storage time on ochratoxin A levels in the middle and at the edge of fermentation heaps during trials conducted in February, March and April.

Pod storage time	Pod type	Fermentation time (days)	Heap location	Ochratoxin A (ng/g)		
				February	March	April
5 days	Damaged	1	Middle	2.3	5.4	-[a]
			Edge	1.9	0.1	-
		3	Middle	5.4	14.4	-
			Edge	31.5	18.6	-
	Healthy	1	Middle	0.1	tr[b]	-
			Edge	tr	tr	-
		3	Middle	tr	0.1	-
			Edge	tr	0.3	-
4 weeks	Damaged	1	Middle	-	0.3	0.7
			Edge	-	0.5	0.3
		3	Middle	-	0.3	0.7
			Edge	-	1.6	tr
	Healthy	1	Middle	-	0.1	0.7
			Edge	-	0.1	0.1
		3	Middle	-	0.1	0.7
			Edge	-	0.2	0.5

a – no data
b – trace

Studies were set up during the 2003/04 harvest season to determine if the type of damage to the pods was correlated with the final level of ochratoxin A. The proportion of healthy and damaged pods and of different pod defects was determined by counting and characterizing pods from five farms at three times during the main crop harvest (December, January and February). Damaged pods were evaluated before and after pod opening to determine if they had internal damage. About 60% of the pods were healthy. Pods with only external damage represented ~25% of the total. These pods had been damaged by the machetes used when the pods were harvested or from the use of a machete to pick up harvested pods. The remaining 15% were affected by pests or were mummified or moldy; most of these pods had internal mold.

For fermentation trials, damaged pods were divided into four groups: moldy pods (black pod and other rot diseases), pods that had been left at previous collections and had dried on the tree ("mummified" pods), pods damaged by insects, and pods with external physical damage (rodent attack, machete or harvesting hook injury). For each category, an equal number of damaged and healthy pods were mixed and stored for 7 days before pod opening. The resulting pod heap was divided in two, with one half fermented by farmers' usual methods and the other by a standard practice which differed from the farmers' method as the placenta was removed and the heap was turned after 48 hours. Control samples consisting only of healthy pods also were prepared with both fermentation methods. After fermentation the beans were sun dried. The experiment was repeated four times during the harvesting period.

Higher ochratoxin A levels were found in beans from damaged pods that were stored for about one week. Beans from physically damaged pods, *e.g.*, from machetes, rodents, *etc.*, had the highest level of contamination, ~ 20 ng/g. Ochratoxin A levels in the moldy (~ 7 ng/g), insect damaged (~ 4 ng/g) and mummified (~ 3 ng/g) pods were substantially less than that found in the physically damaged pods, but generally greater than that found in the control un-

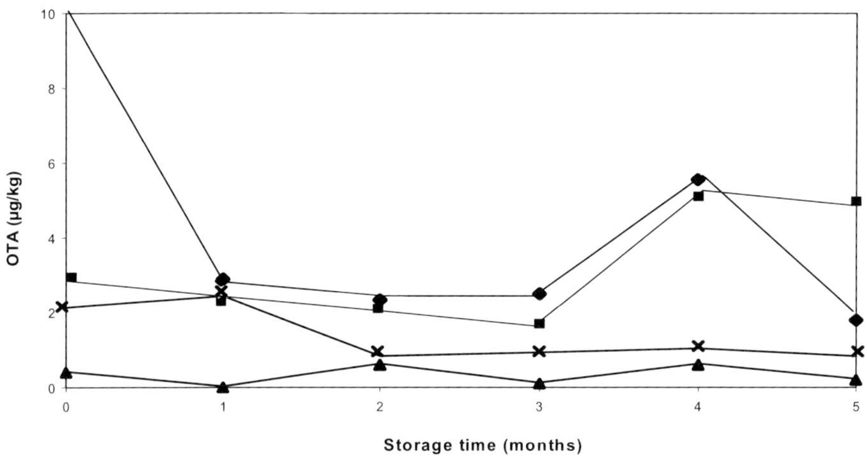

Figure 2. Ochratoxin A in cocoa beans stored in bags in Africa or Europe. Africa: Big bag (♦); Europe: Big bag (■), 60 kg bag, lot 1 (▲), 60 kg bag, lot 2 (×).

damaged pods (~ 2 ng/g). The two fermentation methods had no significant differential effect based on the type of damage. Due to the variation between trials and the limited number of trials, clear conclusions about the fermentation method cannot be made but the results indicate that if the pods are damaged the fermentation method has a limited effect.

Storage of beans in export warehouses and in Europe

Storage trials were conducted in export warehouses in Africa and in warehouses in Europe. In one trial one lot of 32 tons was split in two, with one half stored in an export warehouse in Africa and the other half shipped in bulk by container to Europe where it was stored in an open top container. Samples were taken monthly. There was considerable variation in the ochratoxin A level between samples, but there was no trend towards an increase with time (Fig. 2). The variation is probably due to nonhomogeneous distribution of the toxin in the lot since each composite sample consisted of only 16 subsamples. The results of storage of the two bean lots in 60 kg bags in Europe also showed no increase with time and less variation (Fig. 2). In these tests the composite samples were made from 100 subsamples and the effect of the nonhomogeneous distribution probably is smaller. Thus, ochratoxin A levels do not increase during storage in export warehouses in either Africa or Europe.

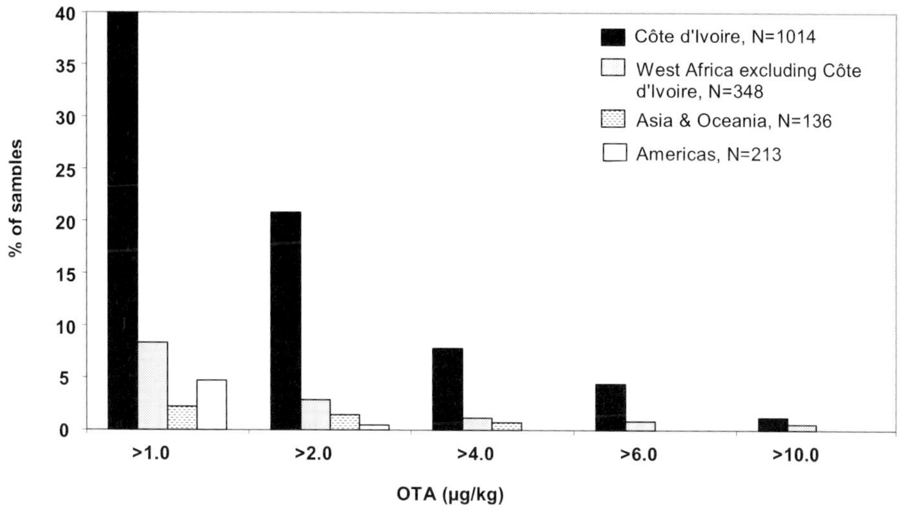

Figure 3. Ochratoxin A levels in cocoa beans imported into Europe from different regions of the world.

OTA levels in cocoa beans

For six seasons, starting in 1999, the European industry has analyzed samples of imported cocoa beans from different origins. Statistical sampling plans were not used for the selection of which lots to analyze and not all producing countries were included. Therefore the results presented (Fig. 3) are indicative, rather than definitive. Ochratoxin A contaminated cocoa beans were found in all cocoa-producing regions. Higher toxin levels were more frequent in beans from Côte d'Ivoire, but the number of analyzed samples from Côte d'Ivoire is about three times larger than from any other region.

The standard method for evaluating the quality of cocoa beans is a cut test in which the interior of a certain number of beans that have been longitudinally cut are visually inspected (Wood and Lass, 1985). One of the parameters in the cut test is visibly moldy beans. Cut tests were made on a large number of cocoa bean samples, which also were analyzed for ochratoxin A. There was no significant correlation between the number of moldy beans identified in the cut test, or any other quality parameter, and ochratoxin A contamination. The molds scored in a cut check apparently are not the species responsible for ochratoxin A contamination.

Industrial conversion of beans to cocoa products

Cocoa beans are not eaten as such, but instead undergo an industrial conversion (Fig. 4) before consumption. During this industrial processing the a_w is $\ll 0.8$, which is too low for ochratoxin A production. The first steps in processing are roasting and removal of the shell. During roasting the final bean temperature reaches 100-120°C and the duration is 15-70 min (Minifie, 1982). Boudra *et al.* (1995) showed that ochratoxin A is heat stable and that at most 20% of the toxin in wheat was decomposed by dry heat at 100°C for 160 min or 150°C for 32 min. Thus, the roasting of cocoa beans is not expected to significantly reduce the level of ochratoxin A present.

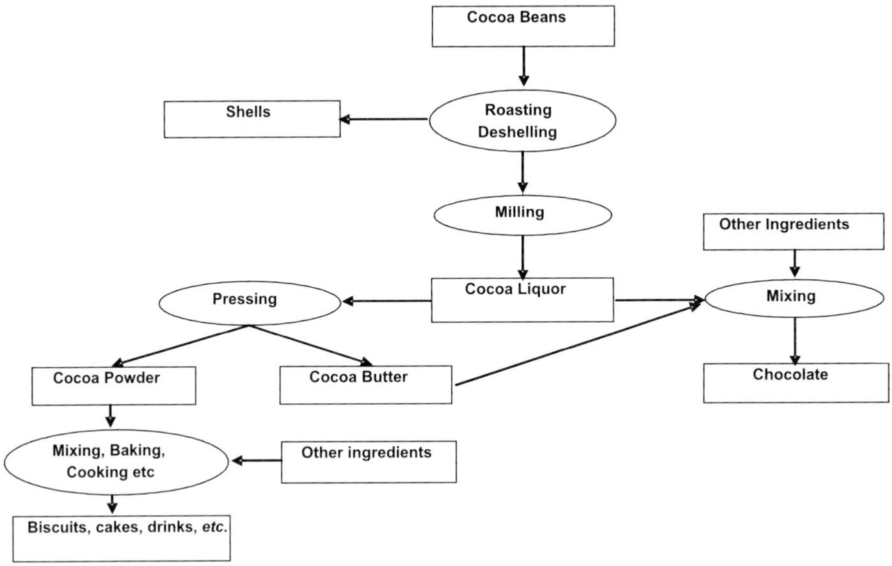

Figure 4. Steps in the industrial processing of cocoa beans. Processing studies were made at the roasting/deshelling and pressing stages.

The amount of ochratoxin A in the shell (discarded and not used for milling) and in the kernel (nib fraction; used for milling) of cocoa beans, was determined by analyzing samples of the two fractions (about 200 g shell and 1 kg nibs) taken simultaneously from industrial winnowers. Based on the measured concentrations and the assumption that the shell fraction represents 12% of the bean (Fincke, 1965), the ochratoxin A concentration in the whole beans (before removal of the shell fraction) was calculated. The calculated ochratoxin A concentration in the whole beans was between 0.3 and 3.0 ng/g. An average of 48% (range 25-72%) of the ochratoxin A in the beans is removed with the shell fraction.

The nibs are milled to form cocoa liquor, a viscous liquid containing ~ 50% fat. The cocoa liquor can be mixed with other ingredients to produce chocolate or it can be "pressed" to produce cocoa butter and cocoa powder. After pressing, all of the ochratoxin A originally in the nibs is recovered in the cocoa powder. Ochratoxin A has not been found in the cocoa butter fraction (and is not expected there), since ochratoxin A is not lipophilic. During other processing steps, *e.g.*, milling, conching, and storage of cocoa liquor and chocolate, the temperature is less than that needed to decompose ochratoxin A, so no reductions in toxin levels are expected to occur during these steps. The cocoa liquor or cocoa powder is mixed with other ingredients to make consumer products and this "dilution" further reduces the ochratoxin A levels in the products consumed.

OTA levels in cocoa and chocolate products

Levels of ochratoxin A in samples of cocoa and chocolate products produced in Europe have been measured by industry for several years. The chocolate samples measured were solid milk or dark chocolate products without inclusions or fillings, *i.e.*, there was no ochra-

Table 2. Ochratoxin A levels in chocolate and cocoa products.

	Milk chocolate	Dark chocolate	Cocoa drink powder	Cocoa powder
% of samples ≤ 0.1 ng/g	76	44	55	3
% of samples ≤ 0.5 ng/g	94	88	96	35
% of samples < 1.0 ng/g	98	96	100	60
% of samples ≥ 1.0 ng/g	2	4	0	40
% of samples ≥ 2.0 ng/g	0	0	0	12
Average (ng/g)	0.16	0.26	0.20	1.0
Total number of samples	228	536	247	1189

toxin A contamination from any source other than the cocoa (Table 2). The ochratoxin A levels are low in chocolate and in cocoa drink powder, with an average between 0.16 and 0.26 ng/g. More than 40% (44 to 76%) of the samples were below 0.1 ng/g. Similar results have been reported from Germany and Spain (Burdaspal and Legarda, 2003; Engel, 2000). The levels are higher in cocoa powder with an average of 1.0 ng/g. This result is expected since cocoa powder is a concentrated cocoa solids fraction. Cocoa powder is used as a flavoring ingredient, usually at levels < 5%, in products such as biscuits, cakes, desserts, ice cream, *etc.*

Towards preventive measures

Our results imply that interventions to reduce contamination with ochratoxin A need to occur at the farm level if the toxin levels are to be significantly reduced. These preliminary studies show that ochratoxin A contamination starts between the on-tree/harvesting to pre-fermentation stages, and that damaged pods are a major part of the problem. Further research is needed at the smallholder level to determine the contribution made by each class of damaged pod and to identify farm level interventions that can reduce ochratoxin A contamination. Developing an appropriate, simple method for the farmers to use will be difficult. In a smallholder crop, *e.g.* cocoa, farmers and farm laborers harvest the crop and process it for sale. Thus, these practitioners are the ones who will require training and must be convinced to change their practices. Communicating with and training hundreds of thousands of smallholder cocoa farmers, often in remote areas, is a daunting task. The national agricultural research centers in the cocoa producing countries must be involved, and could work through one of the several farmer organization strengthening projects currently underway in West Africa. Even with the help of these organizations, however, the education process will take time due to the unorganized nature of the cocoa smallholder producer base.

Conclusions

Fresh wet cocoa beans from undamaged cocoa pods do not contain measurable ochratoxin A, but this toxin is found in fermented, dried cocoa beans delivered to the end users. Fungi that produce ochratoxin A are present in the on-farm environment and on the equipment used in the post-harvest treatment of cocoa beans, which enables contamination very early in the supply chain. At the smallholder level, ochratoxin A can be detected in beans that have just been fermented, an indication that the initial inoculation occurred before or during the fermentation. The drying procedure for cocoa beans may play a role in ochratoxin A development, but does not seem to be the main source of contamination. Instead, poor drying appears to allow further increase in toxin levels in already-contaminated beans.

Ochratoxin A contamination levels in smallholder bean samples vary within the cropping season, probably due to post-harvest practices that enable better survival and proliferation of the toxin-producing fungi, *e.g.*, longer pod storage times and lower fermentation heap temperatures. Undamaged pods do not appear to be significantly contaminated with ochratoxin A even if they are stored. Any type of damage to the integrity of the pod wall can result in the growth of fungi that produce ochratoxin A, but mechanical or rodent damages appears to be the most significant. The type of fermentation procedure does not appear to have a major effect. Instead the presence of beans from damaged pods in the fermentation heap appears to be the primary determinant of ochratoxin A contamination.

Consistent with this conclusion are data showing that further increases in ochratoxin A levels are not found in samples of cocoa beans taken at stages later in the supply chain. The processing of cocoa beans into the raw materials for chocolate, either roasting or other heat treatments, do not destroy ochratoxin A, but ~50% of the contaminating toxin is physically removed when the shells are removed from the beans. The resulting cocoa raw materials are rarely consumed in a concentrated form and are almost always mixed with other ingredients, *e.g.*, milk and sugar, which dilutes any remaining toxin even further. This dilution helps explain the low levels of ochratoxin A found in cocoa and chocolate products (44-76% samples below 0.1 ng/g), and underlines the relatively minor contributions that cocoa and chocolate products make to human dietary ochratoxin A exposure.

Acknowledgements

We thank the CAOBISCO, ECA, and FCC Ochratoxin A Working Group for their active participation in this study, and CIRAD, CABI Bioscience and the University of Aberystwyth for carrying out the research.

References

Boudra, H., Le Bars, P. and Le Bars, J. (1995) Thermostability of ochratoxin A in wheat under two moisture conditions. *Applied and Environmental Microbiology* 61, 1156-1159.

Burdaspal, P.A., and Legarda, T.M. (2003) Ochratoxin A in samples of different types of chocolate and cacao powder, marketed in Spain and fifteen foreign countries. *Alimentaria* 40, 143-153.

DG Health and Consumer Protection. (2002) Report of experts participating in Task 3.2.7. Assessment of dietary intake of ochratoxin A by the population of EU Member States. http://ec.europa.eu/food/food/chemicalsafety/contaminants/task_3-2-7_en.pdf.

DG Health and Consumer Protection, Scientific Committee on Food. (1998) Opinion on Ochratoxin A. http://ec.europa.eu /food/fs/sc/scf/out14_en.html.

EFSA. (2006) Opinion of the scientific panel on contaminants in the food chain on a request from the Commission related to ochratoxin A in food. *The EFSA Journal* 365, 1-56.

Engel, G. (2000) Ochratoxin A in sweets, oil seeds and dairy products. *Archiv für Lebensmittelhygiene* 51, 98-101.

Fincke, H. (1965) Rohstoffe der Kakaoerzeugnisse In: Fincke, H. (ed.), *Handbuch der Kakaoerzeugnisse*, 2nd ed. Springer-Verlag, Berlin/Heidelberg/New York, pp. 66-67.

JECFA. (2001) Ochratoxin A. http://www.inchem.org/documents/jecfa/jecmono/v47je04.htm.

Minifie, B.W. (1982). Cocoa processes. In: Minifie, B.W. (ed.) *Chocolate, Cocoa and Confectionery: Science and Technology*, 2nd ed. AVI Publishing Company, Westport, Connecticut, p. 31.

Wood, G.A.R. and Lass, R.A. (1985) Appendix 2. In: Wood, G.A.R. and Lass, R.A. (eds.), *Cocoa* 4th ed. Longman Scientific and Technical, New York, p. 604

21

Prevention of Ochratoxin A in Grapes and Wine

Paola Battilani*

Abstract

Ochratoxin A has been frequently detected in grapes and its derived products since 1996, with contamination limits fixed by the European Commission for dried vine fruits, wine and grape juice to protect consumers. Ochratoxin A is produced in vineyards, primarily by *Aspergillus carbonarius*. Meteorological conditions are a major factor in determining risk areas and years. The cropping system also can significantly influence the final toxin content in bunches. Ochratoxin A can increase post harvest, during drying or standing before crushing, but this toxin is not synthesized during wine making.

Introduction

Mycotoxins, as toxic secondary metabolites produced by fungi, have been known since 1960 when aflatoxin was first described, but their effects have been studied in depth only since 1970. Ochratoxins were first isolated from *Aspergillus ochraceus* in 1965 and were named after the producing fungus. There are three main ochratoxins, A, B and C, that differ slightly in structure, but ochratoxin A is considered the only true toxin. Both *A. ochraceus* and *Penicillium verrucosum* are known to synthesize ochratoxin A in cereals (CAST, 2003). Ochratoxin A was first reported in wine and grape juice in 1996 by Zimmerli and Dick (1996). They suggested that processed grape products might be sources of the toxin in human diets and confirmed their hypothesis through a survey of products from Swiss markets. Many additional surveys followed, mainly in Europe, which confirmed that ochratoxin A was a problem in Southern Europe in rosé and red wines, and to a lesser extent in white wine (reviewed in Battilani *et al.*, 2004*b*).

Ochratoxin A was detected in all of the surveys, but the incidence of contaminated samples varied widely. The maximum ochratoxin A level reported for red wine was > 3 μg/l in countries such as Spain (Bellí *et al.*, 2004*a*; Blesa *et al.*, 2004), Italy and Morocco (reviewed in Battilani *et al.*, 2004*b*). The year of grape production and the associated meteorological conditions always were relevant as was latitude; higher latitudes generally were associated with lower ochratoxin A content (Battilani *et al.*, 2004*b*; Fig. 1).

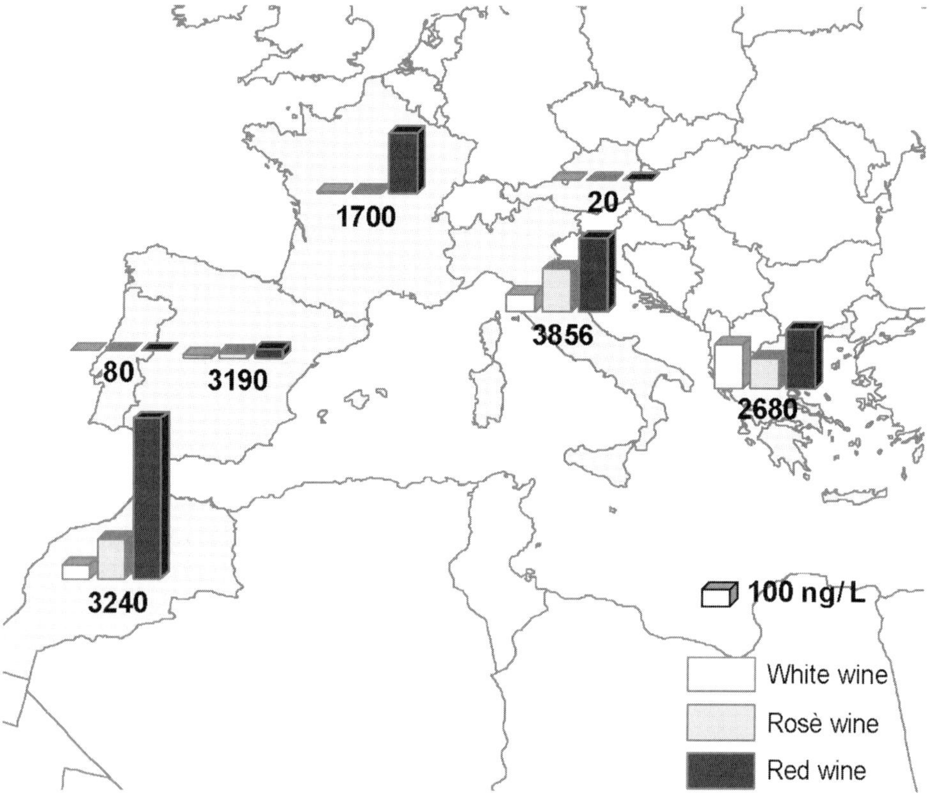

Figure 1. Mean ochratoxin A content (ng/l) in white, rosé and red wines sampled in different countries in the Mediterranean basin. Bars represent the mean value and numbers the maximum value reported (after Battilani *et al.*, 2004*b*).

Red wine and grape juice had higher contamination levels, > 5 μg/l, as did dried vine fruits, *e.g.*, currants contained up to 53 ng/g of ochratoxin A as did balsamic vinegar at levels up to 4.3 μg/l (Battilani *et al.*, 2004*b*; Table 1). The presence of ochratoxin A in grapes and products derived from them was confirmed in studies aimed at identifying the factors related to its production and ensuring compliance with European Commission regulation 466/2001 for dried vine fruits (10 ng/g) and regulation 123/2005 for wine and grape juice (2 μg/l).

Fungi responsible for the production of ochratoxin A in grapes

The first studies of ochratoxin A in grapes and wine evaluated fungi known to produce ochratoxin A in cereals that also were found in grapes. Subsequent surveys, however, shifted this focus. *Aspergillus* section *Nigri*, the "black Aspergilli", includes all of the fungi responsible for ochratoxin A production in grapes (Battilani and Pietri, 2002; da Rocha Rosa *et al.*, 2002; Sage *et al.*, 2002), but *A. ochraceus* was detected occasionally and at a relatively low incidence (Battilani *et al.*, 2003*b*; Bellí *et al.*, 2006).

Table 1. Maximum reported levels of ochratoxin A (ng/g) in grape juice, vinegar and dried vine (Battilani *et al.*, 2004*b*).

Product	Type	Ochratoxin A
Grape juice	White	1.3
	Red	5.3
Vinegar	White and Red	1.9
	Balsamic	4.4
Dried vine	Currant	53
	Sultana	25
	Raisin	30

The taxonomy of this group of fungi depends on the shape of the conidial heads with the distinction between uni- and biseriate conidial heads of particular importance (Raper and Fennell, 1965). Among the uniseriates, *Aspergillus* aculeatus and *Aspergillus japonicus* have been isolated from grapes. Identification of these species based on morphology is not easy, but is not particularly relevant since their ability to produce ochratoxin A has never been confirmed. Among the biseriates, *Aspergillus carbonarius* is easily distinguished by its large conidia. A high percentage of these strains (Teren *et al.*, 1996; Heenan *et al.*, 1998), or all biseriate strains according to other authors (Cabañes *et al.*, 2002; Sage *et al.*, 2002), are ochratoxin A producers. *Aspergillus carbonarius* strains that do not produce ochratoxin A are now regarded by some as a new species, *Aspergillus ibericus* (Serra *et al.*, 2006*b*). *Aspergillus carbonarius* is now confirmed as the primary producer of ochratoxin A in grapes in Europe and Israel (Battilani *et al.*, 2006*c*; Bejaoui *et al.*, 2006; Bellí *et al.*, 2006; Guzev *et al.*, 2006; Serra *et al.*, 2006*b*; Tjamos *et al.*, 2006).

Several other species also are in the biseriate group, but there is no consensus regarding their identification based on morphology. Molecular analyses distinguished two groups – *niger* and *tubingensis* – but did not clearly distinguish two species (Accensi *et al.*, 1999; Perrone *et al.*, 2006*b*). The *niger* group includes a low percentage of ochratoxin A producers (5-10%; Bau *et al.*, 2006), but all of the members of the *tubingensis* group produced the toxin (Perrone *et al.*, 2006*a*). Collectively these groups/species are termed the *Aspergillus niger* species aggregate or species complex.

The dynamics of black Aspergilli in vineyards

Black Aspergilli usually overwinter in soil (Kazi *et al.*, 2003), but they are present on berries from setting and their incidence increases as the fruit grows and matures, with their numbers peaking at ripening. By using a common protocol in several European countries and Israel (www.ochra-wine.com), the presence of black Aspergilli in grape berries was confirmed beginning at setting. The incidence of strains from *Aspergillus* Section *Nigri* was low, ~2% at setting increasing to ~5% as the berry increases to pea-size (Battilani *et al.*, 2006*a*). From early veraison to ripening the frequency of *Aspergillus* Section *Nigri* strains generally increased, except in Italy where the highest level of infected berries was reached at early veraison. The incidence of berries infected by black Aspergilli at early veraison

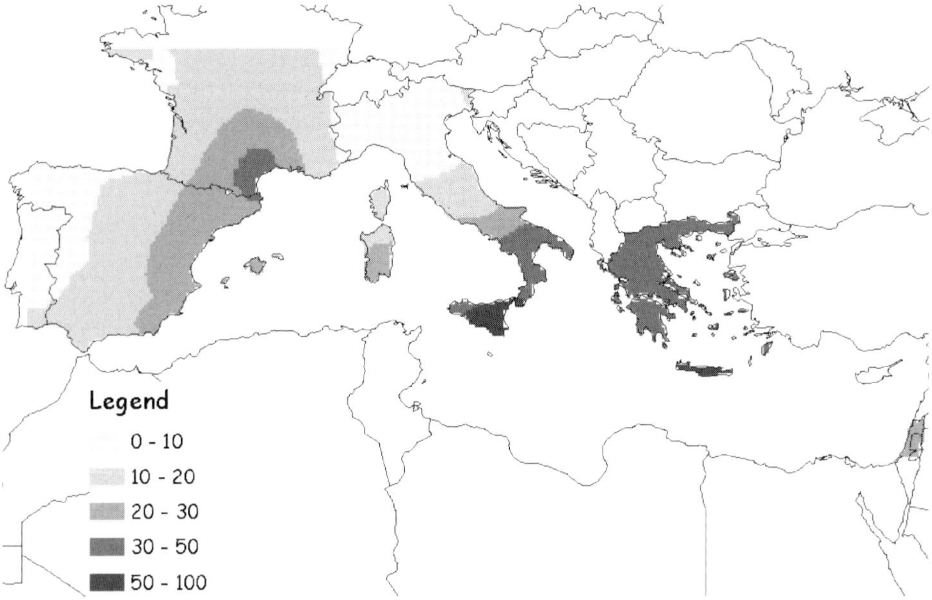

Figure 2. Predictive map for *Aspergillus carbonarius* at harvesting in 2003. Different shades of gray represent different incidences of *A. carbonarius* or black Aspergilli.

was < 25% in most countries, but in Italy and Israel it was > 30%. At harvest, the incidence of black Aspergilli was particularly high in Greece, France and Israel, where fungi were isolated from > 50% of the berries, while < 20% of the berries were infected in Spain.

Among black Aspergilli isolated from berries, uniseriate isolates were the least common. They were never detected in Greece, were found only sporadically in Portugal, were found on ~ 1% of the berries in Spain and on about 10% of the berries in the other participating countries at harvest. Members of *A. niger* species aggregate were the principal group in all growth stages, including ripening. In all countries and years isolates from the *A. niger* species aggregate were more common than isolates of either uniseriate *Aspergillus* sp. or *A. carbonarius*. The incidence of the *A. niger* species aggregate varied widely with the highest incidence detected in France and Israel with ~ 60% of the berries infected.

Aspergillus carbonarius was isolated in all the countries and years evaluated, but its incidence was < 10%, with high variation among vineyards from Italy, Spain and Portugal. The highest incidence of *A. carbonarius* occurred at ripening in France. In experimental studies of *A. carbonarius* in the field in Australia, the conduciveness of the ripening period to infection by this fungus was confirmed under natural conditions. Mean *A. carbonarius* counts decreased between pre-bunch closure and veraison, and then increased between veraison and pre-harvest, with fungal incidence dependent on the risk of berry splitting (Leong *et al.*, 2006).

The incidence of berries infected by black Aspergilli at harvesting is significantly related to latitude and longitude, with positive West → East and North → South gradients. Predictive maps of infected berry incidence had the same trend in different years, but the infection incidence was highest in the hottest and driest year. Thus, the meteorological conditions are a part of the explanation of the spatial distribution of black Aspergilli in the Mediterranean basin (Battilani *et al.*, 2006*a*).

The incidence of *A. carbonarius* was evaluated more critically in the WINE-OCHRA RISK project (Battilani *et al.*, 2006*a*; Table 2) because of its known ability as an ochratoxin A producer. The incidence of *A. carbonarius* varied by country and by year. Portugal and Spain had the lowest incidence, ~10% in 2001 and 2002 and slightly higher in Spain in 2003. In Italy and Israel the incidence was ~20%. In Greece in 2003 and in France the incidence of *A. carbonarius* was > 40%. These data could be entered into ArcView (2002), a geostatistic data analysis program, to generate a map predictive of the incidence of *A. carbonarius* (Fig. 2). The incidence of *A. carbonarius* increased from West → East and from North → South within Europe. All of the samples from 2003 positive for ochratoxin A at harvest (ochratoxin A > 0.3 ng/g) were collected from areas with an incidence of *A. carbonarius* that was > 30% (Battilani *et al.*, 2006*c*; Bejaoui *et al.*, 2006; Guzev *et al.*, 2006; Serra *et al.*, 2006*b*).

Ochratoxin A production

All of the studies identified *A. carbonarius* as the main ochratoxin A producer, because of the high percentage of *A. carbonarius* strains that could produce toxin and the relatively high levels of toxin produced relative to other strains of black Aspergilli. In the WINE-OCHRA RISK project, 70-100% of the *A. carbonarius* strains could produce ochratoxin A when grown *in vitro* and evaluated with high performance liquid chromatography (HPLC), while only 2-20% of the strains from the *A. niger* species aggregate produced ochratoxin A (Battilani *et al.*, 2006*c*, Bejaoui *et al.*, 2006; Bellí *et al.*, 2006; Serra *et al.*, 2006*b*). Most of the ochratoxin A producing strains were recovered at harvest or between veraison and ripening.

Ochratoxin A is produced while the grapes are in the vineyards and is not normally detected before early veraison (Battilani *et al.*, 2004*a*). Bunches without visible symptoms may contain ochratoxin A, but berries with visible black molds were more contaminated. The presence of ochratoxin A in berries only during ripening is not due to inadequacy of the substrate at earlier stages. *Aspergillus carbonarius* produced ochratoxin A when the fungus was inoculated on homogenized wine grapes at different growth stages and belonging to different varieties. Ochratoxin A production was positively correlated with total acidity and negatively correlated with sugar content, which suggests that ochratoxin A biosynthesis is potentially higher in earlier growth stages (Serra *et al.*, 2006*b*). I think that the fungus has problems penetrating the berry at the earlier stages and that the lack of fungal growth

Table 2. Percent incidence of *Aspergillus carbonarius* and black Aspergilli isolated from grapes at harvest in different European countries and Israel (after Battilani *et al.*, 2006*a*).

Country	*2001*	*2002*	*2003*
France	59	38	37
Israel	20	15	23
Italy	20	19	18
Portugal	11	7	11
Spain	10	7	18

within the berries at these early stages under field conditions can explain the low, or non-existent levels of ochratoxin A identified in berries from these early growth stages.

Ecology of black Aspergilli

Fungi in *Aspergillus* Section *Nigri* can grow between 10 and 37°C, with 30-37°C considered optimal. Fungal growth and spore production are very limited at 10°C, and occur only when the available water (a_w) is between 0.90 and 0.93. *Aspergillus niger* can produce $1\text{-}2\times10^6$ conidia/cm^2 of culture surface area when growing *in vitro* on a synthetic medium with an a_w of 0.93-0.97 (relative humidity = 93-97%) at 30-35°C (Parra and Magan, 2004). *Aspergillus carbonarius* produces no more than a third this number of conidia when cultured under similar conditions (Battilani *et al.*, 2006*d*). The ability to produce large numbers of spores is common to all of the strains tested. Spore germination is rapid, and occurs within 24 hours with an a_w of 0.90-0.99 and a temperature of 25-35°C, although germination can occur between 10 and 40°C. The optimal growth temperature for *A. carbonarius* is between 20 and 30°and between 30 and 35°C for members of the *A. niger* species aggregate.

Available water also is very important for black Aspergilli. The optimal a_w for growth is 0.98, which is similar to the the a_w in berries during ripening. Linear growth rate varies by species, with members of the *A. niger* species aggregate the fastest and *A. carbonarius* strains the slowest (Mitchell *et al.*, 2003, 2004; Bellí *et al.*, 2004*b*). *Aspergillus carbonarius* grew best at an a_w of 0.98-0.99, while growth by uniseriate strains or strains from the *A. niger* species aggregate were unchanged in growth across the a_w interval of 0.90-0.995 (Mitchell *et al.*, 2003; Bellí *et al.*, 2004*c*, 2005). Growth was never observed at an a_w of 0.85 and only a few isolates could grow at an a_w of 0.89.

The optimal conditions for growth and for ochratoxin A production were not the same. *Aspergillus carbonarius* could produce ochratoxin A at a_ws between 0.92 and 0.99 although the best toxin production occurred when the a_w was between 0.95 and 0.99 (Bellí *et al.*, 2004*b*, 2005; Mitchell *et al.*, 2004). Optimal ochratoxin A production occurred at 15-20°C and decreased significantly at 30-37°C (Mitchell *et al.*, 2004; Bellí *et al.*, 2005).

Vineyard features and management

Meteorological conditions are important in determining if ochratoxin A will form in grapes, although proximity to the sea and the cropping system also are important. In Italy in 1999-2000, differences were detected in ochratoxin A content from three neighboring vineyards that were managed similarly with either Malvasia nera or Negroamaro grape varieties. In 1999, ochratoxin A levels were 0, low or high in the three vineyards, but the toxin was not detected in 2000 (Battilani *et al.*, 2003*a*). Discriminant analysis based on a summation of degree-days and rain in late August-early September may be used to predict ochratoxin A presence in the vineyard (Battilani *et al.*, 2006*c*).

The cropping system clearly has an effect, as do other factors, *e.g.*, grape variety, trellising system and geographic location, whose effects may be confounded, and their individual effects difficult to resolve. Grape varieties are not universally susceptible to ochratoxin A production either *in vitro* (Battilani *et al.*, 2004*a*) or in field trials. In the field, the trellising system also may influence the incidence of black Aspergilli and the amount of ochratoxin A contamination. In particular, bunches that are closer to the soil appear more contaminated, even if a significant effect on the amount of ochratoxin A present has not yet been proven. The type of soil also can contribute significantly to the level of contamination with black Aspergilli, with clay soil being the most conducive.

Role of pests and diseases

Black Aspergilli are considered saprophytes, responsible for secondary rot, and wounds, of both mechanical and biological origin, are major entry sites. Data on the role of pests and diseases are limited, but good management of them in vineyards certainly results in a decrease in ochratoxin A content at harvest, as can be seen from a comparison of neighboring vineyards, managed with different approaches to crop protection (Kappes *et al.*, 2006).

Lobesia botrana (*Lepidoptera: Tortricidae*) is the major grape berry moth in vineyards of Southern Europe, where it usually completes 3-4 generations a year, depending on the weather conditions in late summer. First generation larvae damage flowers, while the succeeding larval generations damage berries at different stages of maturity. Ochratoxin A content in berries and pest damage are correlated, probably due to wounds caused by pests and spore dissemination. Larvae can act as vectors by trapping conidia in the cuticle ornamentation, moving to healthy host plants, and facilitating colonization by tunneling berries (Cozzi *et al.*, 2006).

Powdery mildew is the most conducive pathogen for black Aspergilli. Berries infected with powdery mildew often are misshapen, have rusty spots on the surface or split open during ripening when inoculum of black Aspergilli is readily available.

Control of *Aspergillus carbonarius*

Mycotoxin-producing fungi are not easy to control with fungicides because their reaction to the fungicide may be the opposite of that expected. Active ingredients reported as effective in reducing both fungal growth and ochratoxin A content in bunches include: mepanipyrim, pyrimethanil, fluazinam, iprodione and a mixture cyprodinil and fludioxonil. Only the cyprodinil/fludioxonil combination was effective in field trials in France, Spain, Greece and Italy. The treatment was most effective 21 days before harvest (stage D). A second treatment at the earlier veraison stage (stage C) was recommended under high risk conditions. This combination of active ingredients was originally developed for the control of grey mold, caused by *Botrytis cinerea*, with the same application schedule (Kappes *et al.*, 2006). Various biological control agents have been considered, including yeasts that occur naturally on grapes. So far, isolates of *Cryptococcus laurentii* and *Aureobasidium pullulans* are the most promising (Bleve *et al.*, 2006).

Post harvest and wine making

Ochratoxin A is produced in vineyards, but its level may increase in bunches after harvest but before processing. The optimal conditions for minimizing ochratoxin A production during this time are not known. The temperature in the storage bin is important as is the length of drying time and the time that the grapes remain in the bins. Maintaining the bins at temperatures unfavorable for fungal growth and reducing the time before crushing both help prevent increases in ochratoxin A levels.

Ochratoxin A production during wine making has not been demonstrated. There is, however, a dynamic of ochratoxin A content in must and wine that enables toxin levels to vary with processing (Fig. 3). When bunches are crushed, the first step in wine making, most of the ochratoxin A is released into the must. In the maceration step, additional ochratoxin A is released, followed by an increase in toxin content for at least five days. Subsequent operations reduce ochratoxin A content, with the size of the decrease dependent upon several factors. During red wine making, ochratoxin A content is reduced by fermentation both *in vitro* and in commercial scale experiments. The efficacy of both alcoholic and malolactic fermentations in reducing ochratoxin A depends upon the yeasts or bacterial strains involved and the level of contamination (Grazioli *et al.*, 2006). The effect of clarification on ochratoxin A level depends solely upon the method or product used, with carbon based adjuvants usually the most effective (Silva *et al.*, 2003). Finally, the ochratoxin A content is influenced by the individuals controlling the process.

In white wine-making, ochratoxin A content always decreases with the extent of the decrease dependent on choices made for the yeasts doing the alcoholic fermentation and the adjuvants added for clarification. Possible corrective actions during wine-making could be based on the use of carbon-based adjuvants for clarification because of their ability to absorb the toxin. Although these adjuvants are a good corrective tool in white wine making, more care must be exercised with red wines. A dosage of 10 g/l/hr is useful for reducing ochratoxin A while maintaining quality, with higher dosages having a negative effect on polyphenols and color. The use of yeasts or bacterial strains that can reduce ochratoxin A during alcoholic or malolactic fermentations require further evaluation.

Critical control points in grape production and processing

Critical control points can be defined both during grape production and processing. Early veraison is a crucial stage at which the berry status needs to be controlled, by checking for mechanical or insect damage and for visible black mold. Closer to ripening these visual checks should be repeated and accompanied by ochratoxin A analysis, particularly if the vineyard is located in a high risk area, if it is a high risk year, or if black mold is visible. Knowing the level of ochratoxin A present near ripening enables better post-harvest management. Ochratoxin A is not evenly distributed in vineyards, so the sampling protocol used is crucial and must result in a representative sample that neither oversamples nor misses potential hot spots (Battilani *et al.*, 2006*b*). During wine making, control points could occur at all operational units, but if the must has low ochratoxin A content, then the wine can be considered safe. An additional control after the alcoholic or malolactic fermentation can be added if the ochratoxin A level in the must is above the legal limit.

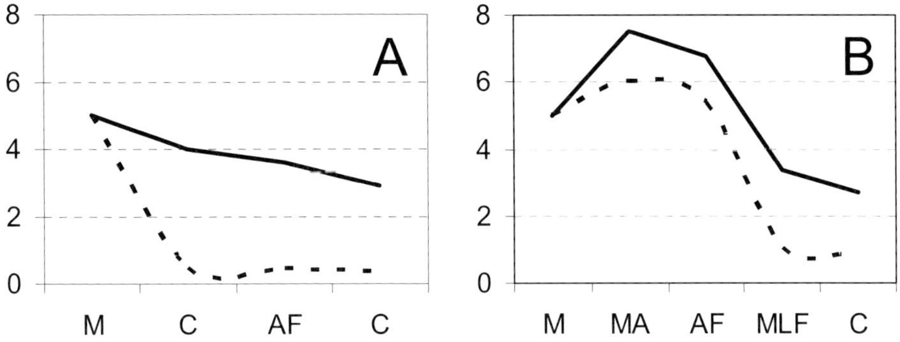

Figure 3. Simulated dynamics of ochratoxin A content (μg/l) during wine making: A – White wine, B – Red wine. The dotted lines represent optimal cases, when all unit operations are managed to minimize ochratoxin A content. Worst case scenarios are represented by the solid lines. M = must; MA = maceration; C = clarification; AF = alcoholic fermentation; MLF = malolactic fermentation [after Grazioli *et al.* (2006) and Silva *et al.* (2003, 2005)].

Good agriculture (GAP) and manufacturing (GMP) practices

Following GAP and GMP protocols can minimize the amount of ochratoxin A present in the final product. Good management of the vineyard that follows correct fertilizing, irrigation and trellising practices is essential for minimizing ochratoxin A contamination. Controlling pests and diseases that can damage the berries is of particular importance. Fungicide selection should favor those with ingredients active against black Aspergilli, especially if high risk conditions occur. Harvesting must be at ripening and overripening must be avoided, especially if damaged berries with visible black mold are present. The interval between harvesting and processing should be minimized to prevent further fungal growth and ochratoxin A biosynthesis by *A. carbonarius* in the detached bunches. The temperature during this time should be between 15 and 30°C.

Elimination of bunches with visible black mold is strongly advised, not only for high quality production, but also to reduce ochratoxin A contamination. The addition of carbon-based adjuvants may be necessary if the ochratoxin A level in the must is high. For red wine, the must should be treated and for white wines the adjuvant(s) should be added during clarification. Ochratoxin A contamination can be reduced further by adding yeast or lactic acid bacteria that can either adsorb or degrade ochratoxin A during fermentation. Finally, normal procedures for sanitizing materials and machines must be followed.

Conclusions

Ochratoxin A is produced in vineyards, primarily by *A. carbonarius*. Meteorological conditions play the most important role in determining risk areas and years, and the prediction of risk levels can be used to design meaningful practices that reduce the risk of ochratoxin A contamination. Farmers should follow GAP vineyard management practices and pay special attention to pest and disease control in high risk areas and years. Collaboration between farmers and wine makers to minimize the time that the grapes stand in the bins before crushing also is important. The last stage for prevention of ochratoxin A contamination occurs when the grape bunches are crushed, as ochratoxin A is not synthesized during the later stages of wine making. Adjuvants can be used, as necessary, to reduce the amount of ochratoxin A present in the final product.

Although ochratoxin A has been identified only recently in grape-derived products, there are many possible ways that the toxin content can be managed and reduced, if not eliminated. Further study to clarify the role of grape variety, trellising systems, the length of time between harvest and crushing, and the drying conditions, are needed to develop a Decision Support System and increase the safety of products for consumers.

Acknowledgments

Data reported were collected through the WINE-OCHRA RISK project (EC, Vth FP, QLK1-2001-01761). I thank J. Cabañes, Z. Kozakiewicz, A. Lebrihi, A. Lichter, N. Magan, G. Mulè, V. Sanchis, E. Tjamos, A. Venâncio and the members of their research teams for data collection, and C. Barbano, P. Giorni, S. Formenti, A. Pietri and A. Silva for their support of the activities managed through my laboratory.

References

Accensi, F., Cano, J., Figuera, L., Abarca, M.L. and Cabañes, F.J. (1999) New PCR method to differentiate species in the *Aspergillus niger* aggregate. *FEMS Microbiology Letters* 180, 191-196.

ArcView Ver. 8.2 2002. ESRI, Redlands, California.

Battilani, P. and Pietri, A. (2002) Ochratoxin A in grape and wine. *European Journal of Plant Pathology* 108, 639-643.

Battilani, P., Giorni, P. and Pietri, A. (2003*a*) Epidemiology of toxin producing fungi and ochratoxin A occurrence in grape. *European Journal of Plant Pathology* 109, 715-722.

Battilani, P., Pietri, A., Giorni, P., Bertuzzi, T., Languasco, L. and Kozakiewicz, Z. (2003*b*) Occurrence of ochratoxin A producing fungi in grape grown in Italy. *Journal of Food Protection* 66, 633-636.

Battilani, P., Logrieco, A., Giorni, P., Cozzi, G., Bertuzzi, T. and Pietri, A. (2004*a*) Ochratoxin A production by *Aspergillus carbonarius* on some grape varieties grown in Italy. *Journal of the Science of Food and Agriculture* 84, 1736-1740.

Battilani, P., Pietri, A. and Logrieco, A. (2004*b*) Risk assessment and management in practice: Ochratoxin in grapes and wine. In: Magan, N. and Olsen, M. (eds.) *Mycotoxins in Food: Detection and Control*, Woodhead Publishing, Cambridge, United Kingdom, pp. 244-261.

Battilani, P., Barbano, C., Marín, S., Sanchis, V., Kozakiewicz, Z. and Magan, N. (2006*a*) Mapping of *Aspergillus* Section *Nigri* in Southern Europe and Israel based on geostatistical analysis. *International Journal of Food Microbiology* 111F, S72-S82.

Battilani, P., Barbano, C., Rossi, V., Bertuzzi, T. and Pietri, A. (2006*b*) Spatial distribution of ochratoxin A (OTA) in vineyard and sampling design to assess must contamination. *Journal of Food Protection* 69, 884-890.

Battilani, P., Giorni, P., Bertuzzi, T., Formenti, S. and Pietri, A. (2006*c*) Black Aspergilli and ochratoxin A in grapes in Italy. *International Journal of Food Microbiology* 111F, S53-S61.

Battilani, P., Pollastro, S., Silva, A. and Fretra, F. (2006*d*) Cause e azioni di prevenzione della contaminazione dei vini da ocratossina A. Quaderni dei georgofili "Micotossine e alimentazione umana e zootecnica. Società editrice fiorentina, Firenze, Italy, pp. 87-126.
Bau, M., Castella, G., Bragulat, M.R. and Cabañes, F.J. (2006) RFLP characterization of *Aspergillus niger* aggregate species from grapes from Europe and Israel. *International Journal of Food Microbiology* 111F, S18-S21.
Bejaoui, H., Mathieu, F., Taillandier, P. and Lebrihi, A. (2006) Black Aspergilli and ochratoxin A production in French vineyards. *International Journal of Food Microbiology* 111F, S46-S52.
Bellí, N., Marín, S., Duaigues, A., Ramos, A.J. and Sanchis, V. (2004*a*) Ochratoxin A in wines, musts and grape juices from Spain. *Journal of the Science of Food and Agriculture* 84, 591-594.
Bellí, N., Marín, S., Sanchis, V. and Ramos, A.J. (2004*b*) Influence of water activity and temperature on growth of isolates of *Aspergillus* section *Nigri* obtained from grapes. *International Journal of Food Microbiology* 96, 19-27.
Bellí, N., Ramos, A.J., Sanchis, V. and S. Marín, S. (2004*c*) Incubation time and water activity effects on ochratoxin A production by *Aspergillus* section *Nigri* strains isolated from grapes. *Letters in Applied Microbiology* 38, 72-77.
Bellí, N., Ramos, A.J., Coronas, I., Sanchis, V. and Marín, S. (2005) *Aspergillus carbonarius* growth and ochratoxin A production on a synthetic grape medium in relation to environmental factors. *Journal of Applied Microbiology* 98, 839-844.
Bellí, N., Bau, M., Marín, S., Abarca, M.L., Ramos, A.J., and Bragulat, M.R. (2006) Mycobiota and ochratoxin A producing fungi from Spanish wine grapes. *International Journal of Food Microbiology* 111F, S40-S45.
Blesa, J., Soriano, J.M., Molto, J.C. and Manes, J. (2004) Concentration of ochratoxin A in wines from supermarkets and stores of Valencian Community (Spain). *Journal of Chromatography* A 1054, 397-401.
Bleve, G., Grieco, F., Cozzi, G., Logrieco, A. and Visconti, A. (2006). Isolation of epiphytic yeasts with potential for biocontrol of *Aspergillus carbonarius* and *A. niger* on grape. *International Journal of Food Microbiology* 108, 204-209.
Cabañes, F.J., Accensi, F., Bragulat, M.R., Abarca, M.L., Castella, G., Minguez, S. and Pons, A. (2002) What is the source of ochratoxin A in wine? *International Journal of Food Microbiology* 79, 213-215.
Council for Agricultural Science and Technology (CAST). (2003) *Mycotoxins: Risks in Plant, Animal and Human Systems*. Ames, Iowa.
Cozzi, G., Pascale, M., Perrone, G., Visconti, A. and A. Logrieco, A. (2006) Effect of *Lobesia botrana* damage on black Aspergilli rot and ochratoxin A content in grapes. *International Journal of Food Microbiology* 111F, S88-S92.
da Rocha Rosa, C.A., Palacios, V., Combina, M., Fraga, M.E., de Oliveira Rekson, A., Magnoli, C.E., and Dalcero, A.M. (2002) Potential ochratoxin A producers from wine grapes in Argentina and Brazil. *Food Additives and Contaminants* 19, 408-414.
Grazioli, B., Fumi, M.D., Galli, R. and Silva, A. (2006) The role of processing on ochratoxin A content in Italian must and wine: A study on naturally contaminated grapes. *International Journal of Food Microbiology* 111F, S93-S96.
Guzev, L., Danshin, A., Ziv, S. and Lichter, A. (2006) Occurrence of ochratoxin A producing fungi in wine and table grapes in Israel. *International Journal of Food Microbiology* 111F, S67-S71.
Heenan, C.N., Shaw, K.J. and Pitt, J.I. (1998) Ochratoxin A production by *Aspergillus carbonarius* and *A. niger* isolates and detection using coconut cream agar. *Journal of Food Mycology* 1, 67-72.
Kappes, M.E., Serrati, L., Drouillard J.B., Cantus, J.M. and Kazantzidou, M. (2006) A crop protection approach to *Aspergillus* and OTA management in Southern European vineyards. *Infowine, Proceedings of "International Workshop – Ochratoxin A in grapes and wine: Prevention and control" (Marsala, Italy)*, pp. 8-9.

Kazi, B.A., Emmett, R.W., Clarke, K. and Nancarrow, N. (2003) Black *Aspergillus* moulds in Australian vineyards. *Proceedings of the International Congress of Plant Protection (Auckland, New Zealand)*, p. 119.

Leong, S.L., Hocking, A.D. and Scott, E.S. (2006) Survival and growth of *Aspergillus carbonarius* on wine grapes before harvest. *International Journal of Food Microbiology* 111F, S10-S17.

Mitchell, D., Aldred, D. and Magan, N. (2003) Impact of ecological factors on growth and ochratoxin A production by *Aspergillus carbonarius* from different regions of Europe. *Aspects of Applied Biology* 68, 109-116.

Mitchell, D., Parra, R., Aldred, D. and Magan, N. (2004) Water and temperature relations of growth and ochratoxin A production by *Aspergillus carbonarius* strains from grapes in Europe and Israel. *Journal of Applied Microbiology* 97, 439-445.

Parra, R. and Magan, N. (2004) Modeling the effect of temperature and water activity on growth of *Aspergillus niger* strains and applications for food spoilage moulds. *Journal of Applied Microbiology* 97,429-438.

Perrone, G., Mulè, G., Battilani, P., Pietri, A. and Logrieco, A. (2006*a*) Ochratoxin A production by *Aspergillus carbonarius* and *A. tubingensis* strains isolated from grapes in Italy. *Applied and Environmental Microbiology* 72, 680-685.

Perrone, G., Susca, A., Epifani, F. and Mulè, G. (2006*b*) AFLP characterization of Southern Europe population of *Aspergillus* Section *Nigri* from grapes. *International Journal of Food Microbiology* 111F, S22-S27.

Raper, K.B. and Fennell, D.I. (1965) *The Genus* Aspergillus. Williams and Wilkins, Baltimore.

Sage, L., Krivoboc, S., Delbos, E., Seigle-Murandi, F. and Creppy, E.E. (2002) Fungal flora and ochratoxin A production in grapes and musts from France. *Journal of Agricultural and Food Chemistry* 50, 1306-1311.

Serra, R., Cabañes, F.J., Perrone, G., Castellá, G., Venâncio, A., Mulè, G. and Kozakiewicz, Z. (2006*a*) *Aspergillus ibericus*: A new species of the section *Nigri* isolated from grapes. *Mycologia* 98, 295-306.

Serra, R., Mendonça, C. and Venâncio, A. (2006*b*) Ochratoxin A occurrence and production in Portuguese wine grapes at various stages of maturation. *International Journal of Food Microbiology* 111F, S35-S39.

Silva, A., Galli, R., Grazioli, B. and Fumi, M.D. (2003) Metodi di riduzione di residui di ocratossina A nei vini. *Industria Bevande* 32, 467-472.

Silva, A., Grazioli, B., Galli, R., Fumi, M.D. (2005) Fate of ochratoxin A in red winemaking. (Destino dell'ocratossina A nella vinificazione in rosso). *Rapporti ISTISAN Istituto Superiore di Sanita* (05/42): 249-252.

Teren, J., Varga, J., Hamari, Z., Rinyu, E. and Kevei, F. (1996) Immunochemical detection of ochratoxin A in black *Aspergillus* strains. *Mycopathologia* 134, 171-176.

Tjamos, S.E., Antoniou, P.P. and Tjamos, E.C. (2006) *Aspergillus* spp., distribution, population composition and ochratoxin A production in wine-producing vineyards in Greece. *International Journal of Food Microbiology* 111F, S61-S66.

Zimmerli, B. and Dick, R. (1996) Ochratoxin A in table wine and grape-juice: Occurrence and risk assessment. *Food Additives and Contaminants* 13, 655-668.

22

Molecular Approaches to Development of Resistance to Preharvest Aflatoxin Contamination

Deepak Bhatnagar*, Kanniah Rajasekaran, Jeffrey W. Cary, Robert Brown, Jiujiang Yu and Thomas E. Cleveland

Abstract

Aflatoxins are extremely carcinogenic secondary metabolites produced by *Aspergillus flavus* when this fungus invades crops such as maize, cotton, tree nuts and peanuts. Adverse health effects from the ingestion of these toxins have caused regulatory agencies throughout the world to limit the amount of aflatoxins that are permitted in food or feed that is available for sale. This results in undue economic burden on the grower. New biotechnologies such as: (i) the use of disarmed, non-toxigenic biocompetitive strains of *A. flavus* in biocontrol of aflatoxin contamination, and (ii) identification of plant constituents that disrupt aflatoxin biosynthesis or fungal growth and their use in new biochemical marker-based breeding strategies to enhance resistance in crops to aflatoxin, could potentially save the agricultural industry in the United States alone hundreds of millions of dollars. In addition to the above mentioned studies, USDA, Agricultural Research Service scientists and other researchers have been studying the genetics of the biosynthetic pathway of this important toxin in order to understand how and why this fungus makes aflatoxins. With this knowledge, strategies are being developed to interrupt aflatoxin synthesis, thereby preventing aflatoxin contamination of crops. The fungal genome of *A. flavus* has been sequenced to understand the regulation of aflatoxin formation by environmental factors. This information is being used to assist in the development of host-resistance against aflatoxin contamination by studying the effects of various physiological parameters, *e.g.*, drought stress on gene expression in toxigenic fungi. The toxin biosynthetic machinery in the fungus also can be disrupted, which may enable the development of atoxigenic fungal strains that can be used as "designer" biocontrol agents.

Introduction

Aflatoxins are natural poisons produced predominantly by two common fungi, *Aspergillus flavus* and *Aspergillus parasiticus*. *A. flavus* is most commonly found in agricultural fields, and when it invades crops like maize, peanuts, cotton, and tree nuts, it produces aflatoxins if the environmental conditions permit. Chronic, as well as sporadic, aflatoxin contamination in a variety of field crops and agricultural commodities worldwide has had a serious impact on the economics and food safety of these products (Jelinek *et al.*, 1989; Henry *et al.*, 2002). The number of deaths in Indonesia due to aflatoxin-induced liver cancer is estimated at 20,000/year (reviewed in Pitt, 2000). The short-term toxicity of aflatoxins and the chronic exposure of humans to these compounds in foods leading to liver cancer or death have been well established (CDC, 2004; Azziz-Baumgartner *et al.*, 2005). Aflatoxins, furanocoumarin compounds, are the most widely studied of all mycotoxins because of their potent toxicity and carcinogenicity (Eaton and Groopman, 1994; Cary *et al.*, 2000; Bhatnagar *et al.*, 2002), and the most significant research progress towards controlling mycotoxins has been made with aflatoxins. A hundred countries are known to have regulations limiting mycotoxin levels with 61 having specific regulatory levels for total aflatoxins in foodstuffs, 60 for aflatoxin M_1 in milk and 39 having some regulations for aflatoxins in feedstuffs (FAO, 2004, van Egmond *et al.*, 2007). The Food and Drug Administration in the United States has set limits of 20 ng/g total aflatoxins for interstate commerce of food and feed and 0.5 ng/g of aflatoxin M_1 in milk.

The limits on aflatoxin contamination can result in severe economic losses, *i.e.*, > $250 million in direct losses to farmers (Richard and Payne, 2003). FDA economists (Vardon *et al.*, 2003) estimated the annual cost of aflatoxin contamination in the United States at ~ $500 million through two categories of loss, market rejection and animal health impacts. Wu (2004) suggested that the total economic impact of aflatoxins also should include many other factors, *e.g.*, export market losses, sampling and testing costs, costs to food processors, grocery markets and consumers, and human health effects. Therefore, it is imperative for researchers worldwide to develop strategies for effective control of aflatoxin contamination of crops. Up to $18 million per year in cottonseed losses by Arizona farmers alone could be prevented by the use of new information on aflatoxin contamination. Strategies to minimize or control aflatoxin contamination will economically benefit the agricultural industry in developed countries, where regulatory agencies follow strict guidelines against the sale of contaminated commodities, but also will increase food safety for populations in developing countries where such regulations, if they exist, may not be strictly enforced.

Three lines of defense against toxin contamination

Researchers trying to eliminate preharvest aflatoxin contamination are basically engaged in developing three lines of defense that will work together in concert. There is no "magic bullet" for solving the aflatoxin contamination problem, so several strategies must be utilized simultaneously to ensure a healthy crop, free of aflatoxins. The first line of defense attempts to keep the toxigenic fungus from reaching the crop. This goal is being achieved by biological control (preferably with native, atoxigenic strains), and effective, targeted cultural management practices. The second line of defense occurs if the toxigenic fungus does reach the crop. Attempts are being made to prevent the fungus specifically from growing in the seed. Strategies to achieve this goal include enhancing host resistance either by breeding for resistance or by genetically engineering host/non-host resistance traits into the crops affected by preharvest toxin contamination. The third line of defense is activated if the fungus grows into the seed. Once the fungus has colonized the seed, it is undesirable for the fungus to make toxins in the seed, *e.g.*, in the embryo in the case of maize. Once again, preventing the fungus from making the toxin can be achieved by enhancing host resistance and targeting factors that inhibit toxin synthesis.

Aspergillus flavus invades crops as diverse as maize, cottonseed, peanuts and tree nuts, and has been isolated from soils across the globe. The common feature in aflatoxin conta-

mination is the fungus. Therefore, to understand the preharvest aflatoxin contamination process, it is important to understand the genetic make-up and the gene expression profile of the fungus under various environmental conditions, during host-fungal interaction as well as during aflatoxin production.

Preharvest aflatoxin contamination process

Aspergillus flavus is capable of using a broad range of organic resources. In addition to being a saprophyte, this organism also is an opportunistic pathogen of plants, insects, and vertebrates including humans and domestic animals (reviewed in Brown *et al.*, 1998). In agricultural fields, during hot dry conditions, *A. flavus* populations increase on crop debris, on senescent or dormant tissues, and on damaged or weakened crops. Crops grown in these fields become associated with large populations of *A. flavus* that may remain associated with the crop throughout crop maturation, harvest and storage. Although the source of the initial infecting inoculum remains a subject of debate, the contamination process can be divided into two phases, one occurring during crop maturation and the other after maturation. Most fungal contamination appears to be associated with damaged crop components. Damage can occur either through mechanical damage, *e.g.*, insect activity, or through environmental or chemical stress on the host plant. Crop components damaged prior to maturity are, apparently, more susceptible to contamination. If damage and infection occur at the appropriate stage of maturity, very high levels of contamination (> 100,000 ng/g) can result. Highly contaminated seeds, nuts, or kernels are relatively rare, but drastically influence the overall contamination level of the crop.

Control of preharvest contamination

Cultural practices

Farming practices that include disease and insect management in field crops, crop rotation and adequate irrigation, use of quality seed, and altering the time of planting and harvest have had some impact on the control of aflatoxin contamination (reviewed in Bhatnagar *et al.*, 1994). Maintaining good cultural and management practices that promote the general health of crops can reduce but not eliminate preharvest aflatoxin contamination. For example, insect resistant germplasm, *e.g.*, maize transformed with the gene encoding *Bacillus thuringiensis* crystal protein (Bt maize), has reduced the level of insect damage commonly associated with increased mycotoxin contamination. Irrigation of peanuts essentially prevents aflatoxin contamination of this crop, probably by preventing the drought stress known to induce aflatoxin contamination in peanuts. However, optimization of management practices to control aflatoxin contamination is not always possible due to production cost, geographic location, and the nature of the production system for the crop. In addition, even the best management practices are sometimes negated by biotic and abiotic factors that are difficult or impossible to control. Therefore, there is an urgent need to develop and integrate additional easy-to-utilize strategies to control preharvest aflatoxin contamination.

Biological control

Microorganisms often have been suggested as agents of control for aflatoxin contamination. The best biocompetitive agent to control *A. flavus* in the field is atoxigenic strains of *A. flavus*, because these strains, as compared to other potential microbial biocompetitive agents, are adaptable to the same environmental conditions as are the toxigenic strains and would be biologically active at the same time as well (Chang *et al.*, 2005; Ehrlich and Cotty, 2004). *Aspergillus flavus* need not synthesize aflatoxins to infect a plant, and there is no relationship between the production of high levels of aflatoxins and strain virulence.

In greenhouse and field experiments, where developing cotton bolls or developing maize ears were wounded and then inoculated with different combinations of toxigenic and non-toxigenic strains, the presence of the non-toxigenic strains reduced preharvest aflatoxin contamination by 80-90% (Cotty and Bayman, 1993). Very significant reductions in aflatoxin levels also were obtained in peanuts when a non-aflatoxin-producing strain of *Aspergillus parasiticus* was added to the soil in peanut plots (Dorner, 2004). Significant reductions in aflatoxin contamination also were obtained when non-toxigenic strains of *A. flavus* inhabiting wheat kernels as a substrate were applied to cotton rows in Arizona cotton-growing areas with a high incidence of aflatoxin contamination (Cotty, 1994). The ability of non-toxigenic strains to interfere with aflatoxin contamination of various crops may, thus, have real practical value and is discussed in more detail elsewhere in this volume (Cotty *et al.*, Chapter 24).

Aspergillus flavus usually becomes associated with crops in the field during crop development and remains associated during harvest, storage and processing. Thus, applying non-toxigenic strains to agricultural fields prior to crop development also may provide postharvest protection from contamination. Non-toxigenic strains applied either prior to harvest or after harvest can reduce aflatoxin contamination of maize, even when toxigenic strains are associated with the crop prior to application.

Enhancement of host resistance

Host resistance can result by identifying biochemical markers that affect either fungal growth or toxin production. These markers can be found in the host plant or non-host plants by either traditional plant breeding methods or by molecular analyses of individual plants.

Enhancement of maize resistance through molecular breeding strategies

Efforts in the United States. Several resistant inbred lines among the 31 tested in Illinois field trials (Campbell and White, 1995), have been incorporated into a breeding program whose major objective is to improve elite Midwestern maize lines such as B73 and Mo17 by determining the mode of inheritance of resistance to aflatoxin contamination in crosses with B73 and/or Mo17 (White *et al.* 1995*a*, 1998; Hamblin and White, 2000; Walker and White, 2001; Maupin *et al.*, 2003). In several highly resistant inbreds, *i.e.*, Tex6, LB31, CI2, Oh513 and MI82, resistance was inherited as a genetically dominant character. Overall, the results indicate that selection for resistance to *Aspergillus* ear rot and aflatoxin production should be effective. The frequency distribution of ear rot ratings and aflatoxin production in F_3 families and in backcrosses to the susceptible self families of Mo17 × Tex6 and B73 × LB31 suggest that the development of resistant inbreds for use in breeding commercial hybrids also should be possible (White *et al.*, 1995*b*).

Chromosome regions associated with resistance to *A. flavus* and inhibition of aflatoxin production in maize have been identified through Restriction Fragment Length Polymorphism (RFLP) analysis in three "resistant" lines (R001, LB31, and Tex6) in the Illinois breeding program, after mapping populations were developed using B73 and/or Mo17 elite inbreds as the "susceptible" parents (White *et al.,* 1995*a*, 1998). In some cases, chromosomal regions were associated with resistance to *Aspergillus* ear rot and not aflatoxin inhibition, and *vice versa*, although some chromosomal regions were associated with both traits. This pattern suggests that the genetic control of these two traits may be at least partially distinct. Variation exists in the chromosomal regions associated with *Aspergillus* ear rot and aflatoxin inhibition in different mapping populations, suggesting that different genes/alleles for resistance may occur in the different resistance germplasms. RFLP technology may enable the strategy of pyramiding different types of resistances into commercially important genetic backgrounds, while avoiding the introduction of undesirable traits.

Quantitative Trait Loci (QTL) mapping of resistance also has been made by using a mapping population created from a resistant inbred Mp313E and a susceptible one, Va35 (Davis *et al.*, 1999). In this case, regions were identified on the chromosomes associated with resistance to aflatoxin contamination. Similar work in another program could ultimately facilitate the pyramiding of insect and fungal resistance genes into commercial germplasm (Guo *et al.*, 2000).

USDA-African collaborative efforts. A collaborative research effort between the International Institute of Tropical Agriculture (IITA) in Ibadan, Nigeria and the USDA-ARS in New Orleans is aimed at developing maize inbred lines resistant to aflatoxin contamination and to the identification of the resistance genes in these lines (Brown *et al.*, 2003; Menkir *et al.*, Chapter 23). Initially, a number of inbreds, adapted to the savanna and mid-altitude ecological zones of West and Central Africa were selected for moderate to high levels of resistance to ear rot under conditions of severe natural infection in their respective areas of adaptation (Brown *et al.*, 2001). The major ear rot-causing fungi in these environments include *Aspergillus, Botryodiplodia, Diplodia,* and *Fusarium.* To determine the potential of these lines to resist aflatoxin production by *A. flavus*, they were sent to SRRC-ARS for screening in a laboratory-based kernel screening assay (Brown *et al.*, 1995). A relatively large number of these lines accumulated low, or lower, levels of aflatoxin than did a resistant control line from the United States. These potentially aflatoxin-resistant lines from IITA and a number of resistant lines from the United States were used as parents for crosses. IITA developed a large number of inbred lines from backcross populations with 75:25 US: IITA germplasm and F_1 crosses with 50:50 US:IITA germplasm. A number of lines in both backgrounds have been developed and screened with the kernel screening assay. These lines, now in S6 and S7 generations, also are being evaluated for agronomic characteristics (Brown *et al.*, 2005). After confirming the promising phenotypes in both a kernel screening assay and field tests, the best lines with good agronomic features and resistance to aflatoxin accumulation will be released as potential sources of genes for resistance to aflatoxin production in breeding programs (Menkir *et al.*, 2006; Chapter 23).

Biochemical marker identification

Biochemical markers can be identified in many ways. Proteins inhibitory to fungal growth or toxin production can be identified by comparing protein profiles of resistant and susceptible germplasms or by following the inhibitory trait in protein fractionation. Alternatively, the levels of metabolites from various biochemical pathways with the desired effect on toxin production/growth or fungal development can be altered. Ultimately the task is to identify the proteins responsible for generating these metabolites and the genes that encode their synthesis. For example, several cotton boll volatiles have been identified that inhibit fungal growth with or without affecting toxin synthesis (Greene-McDowelle *et al.*, 1999; Wright *et al.*, 2000). A number of these metabolites, can alter fungal growth and fungal development, including inhibiting sporulation.

A number of these metabolites come from lipid metabolism, particularly products of the lipoxygenase (LOX) pathway that is involved in fungal development (Keller *et al.*, 1999, 2005; Yu and Keller, 2005). Earlier work from that lab suggested that linoleic acid contributes to the overall fitness of *Aspergillus*, and more recent work from the lab involves characterization of the lipoxygenase gene from peanut and maize that were responsive to *Aspergillus* infection. These genes could be used as biochemical markers. LOX involvement in resistance may not be simple. In soybeans resistant to aflatoxin accumulation, the resistance was more a function of seed viability and seed coat integrity than the presence of LOX. Whether the resistance mechanism in soybean is applicable to other crops, *e.g.*, maize and peanuts, is not yet known.

There also are metabolites from other plants, *e.g.*, walnuts, which have potential as endogenous constituents of tree nuts as inhibitors of growth of *Aspergillus flavus* or aflatoxigenesis (Mahoney and Molyneux, 2004). Bioactive constituents, situated in specific tissues of a tree nut, may have a protective function dependent on both location and concentration. The primary barrier to attack by external organisms is the hull of the walnut, which is known to contain a structurally related series of naphthoquinones, including 1, 4-naphthoquinone, juglone, 2-methyl-1,4-naphthoquinone, and plumbagin. All of the quinones can delay fungal spore germination at lower levels and can completely inhibit growth at higher levels. The most potent compounds were 2-methyl-1,4-naphthoquinone and plumbagin, which were similar to each other in activity, *i.e.*, they delayed germination for 40 hours at 20 μg/g and inhibited growth at 50 μg/g.

Identification of resistance-associated proteins (RAPs) through proteomics

Developing resistance to fungal infection in wounded and intact maize kernels would accelerate solving the aflatoxin problem (Payne, 1992). Studies of subpericarp (wounded kernel) resistance in maize has identified corresponding resistance mechanisms. In a number of studies resistance associated proteins (RAPs) were identified (Brown *et al.*, 2004). A proteomics (Wilkins *et al.*, 1996; Pennington *et al.*, 1997) approach has been used to identify additional RAPs by systematically evaluating proteins separated by 2-D electrophoresis. This approach takes advantage of the increased reproducibility, map resolution, reliability and accuracy that 2-D PAGE offers over 1-D gel electrophoresis (Gorg *et al.*, 1988) to detect post- and cotranslational modifications that cannot be predicted from a DNA sequence. By comparing aflatoxin-resistant and aflatoxin-susceptible lines proteins associated with resistance, as well as the genes encoding them, can be identified thereby facilitating marker-assisted breeding and/or genetic engineering efforts. Endosperm and embryo proteins from several resistant and susceptible genotypes were compared and over a dozen constitutively expressed protein spots, either unique or 5-fold up-regulated in resistant lines (Mp420 and Mp313E), were identified and analyzed with ESI-MS/MS after in-gel digestion with trypsin (Chen *et al.*, 2000, 2002).

Table 1. Examples of resistance-associated proteins (RAP) identified through proteomics.

Protein	Resistant genotype[a]	Putative function
Trypsin inhibitor (14 kDa)	MI82; CI2; T115	Antifungal; anti-amylase
Glyoxalase I	Mp420; Mp313E	Stress-related; antitoxigenic
Pathogenesis related protein (PR-10)	Mp420; Mp313E; GT-MAS:gk	Antifungal; ribonucleolytic
Peroxiredoxin antioxidant (PER1)	Mp420; Mp313E	Peroxidase; stress related
Zeamatin	GT-MAS:gk	Antifungal

[a]Maize genotypes in which an association with resistance has been identified; other resistant genotypes are involved as well.

Both constitutive and induced proteins are required for kernel resistance to aflatoxin production (Chen *et al.*, 2001). A major difference between resistant and susceptible maize genotypes is the relatively high level of antifungal proteins constitutively expressed by the resistant lines. These constitutive proteins may delay fungal invasion, and subsequent aflatoxin formation, until infection-induced antifungal proteins are synthesized. The identified proteins can be grouped into three categories based on their peptide sequence homology: (i) storage proteins, *e.g.*, globulins and late embryogenesis abundant proteins, (ii) stress-responsive proteins, *e.g.*, aldose reductase, a glyoxalase I protein and a 16.9 kDa heat shock protein, and (iii) antifungal proteins, including the pathogenesis-related protein PR-10 (Chen *et al.*, 2006).

The proteomes of near-isogenic maize lines from the same backcross in the IITA-USDA collaborative project differing significantly in aflatoxin accumulation were analyzed and proteins identified in all three of the predicted categories (Brown *et al.*, 2003). RAPs are more easily identified in these lines as the confounding effects of differences in the genetic backgrounds of the lines are absent.

Previously, most RAPs have had antifungal activities (Table 1). Increased temperature and drought, which often occur together, are major factors associated with aflatoxin contamination of maize kernels (Payne, 1992, 1998). If drought stress is imposed during grain filling, then dry matter accumulation is reduced in kernels (Payne, 1992, 1998) which often results in cracks in the seed and provides an easy entry site for fungi and insects. Possession of unique or increased levels of hydrophilic storage or stress-related proteins by resistant lines could increase protein synthesis and host defenses under stress conditions. Thus, commercially-useful, aflatoxin-resistant maize lines may include not only antifungal proteins but also high level expression of stress-related proteins. Further studies including physiological and biochemical characterization, genetic mapping, plant transformation using RAP genes, RNAi gene silencing and marker-assisted breeding should clarify the roles of stress-related RAPs in kernel resistance.

The direct involvement of glyoxalase I, a purported stress-related aflatoxin resistance protein, has been evaluated (Chen *et al.*, 2004). The substrate for glyoxalase I, methylglyoxal, is a potent cytotoxic compound produced from glycolysis and photosynthesis intermediates, glyceraldehydes-3-phosphate and dihydroxyacetone phosphate. Methylglyoxal induces aflatoxin production through up-regulation of aflatoxin biosynthetic pathway transcripts, including the *AflR* regulatory gene. Thus, glyoxalase I may directly affect resistance by removing the methylglyoxal aflatoxin-inducing substrate.

Genetic engineering for resistance to Aspergillus flavus

Breeding disease-resistant crops is very time consuming, especially in perennial crops such as tree nut crops, and is not readily applicable to countering the evolution of new virulent fungal races. The problem is exacerbated in developing resistance to saprophytic fungi such as *Aspergillus* spp. The specificity of the antifungal factors found in genotypes of maize or peanuts that are naturally resistant to *Aspergillus* is not known. In maize, the resistance often is polygenic (White *et al.*, 1999), and in other crops, *e.g.*, cotton, there are no known naturally resistant varieties to *Aspergillus* (Wilkins *et al.*, 2000; Zipf and Rajasekaran, 2003). Antifungal genes also can be expressed in plastids, which may increase the expression level due to the increased number of transgenes and prevent transgene escape through pollen (Daniell *et al.*, 2005).

Availability of transgenic varieties with antifungal traits will be extremely valuable in cotton breeding. Disease-resistant transgenic crops could control mycotoxin-producing organisms, *e.g.*, *A. flavus*, *A. parasiticus* and *Fusarium* spp., as well as other economically important plant diseases. Transgenic crops resistant to aflatoxin-producing fungi offer environmentally safe alternatives to current chemical pesticide practices in addition to lower levels of aflatoxin contamination. Given the information currently available on host proteomics and pathogen genomics and field ecology, novel aflatoxin contamination control strategies can be developed by countering the toxin biosynthetic and contamination processes.

Transgenic maize varieties resistant to *Aspergillus* spp. are still in experimental stages. Indeed, no transgenic varieties resistant to plant pathogenic fungi are available, indicating the complexity of host plant resistance (Rajasekaran *et al.*, 2002, 2005*a*, 2006; Punja, 2004). However, several laboratories are experimenting with potential antifungal gene constructs that offer resistance *in vitro*, *in situ* or *in planta* to *A. flavus*, often pyramided with insect-resistant genes. For example, bollworm or insect injury to cotton bolls is thought to provide entry for *A. flavus* (Zipf and Rajasekaran, 2003), which means that direct insect control may provide indirect control of infection by *Aspergillus* spp.

A number of potentially useful antifungal enzymes/proteins are produced either constitutively or in response to fungal attack in plants, including chitinases and β-1,3-glucanases (Broglie *et al.*, 1991; Meins *et al.*, 1992; Cornelissen and Melchers, 1993; Thomma *et al.*, 2002), osmotins (Singh *et al.*, 1989), protease inhibitors (Ryan, 1990), thionins (Bohlmann, 1994), and polygalacturonase inhibiting proteins (Toubart *et al.*, 1992). However, compounds active against other fungal species often are not effective against *A. flavus*.

Several transgenes are effective to at least some extent against *A. flavus*. For example, trypsin inhibitor (Chen *et al.*, 1998) is correlated with maize kernel resistance to *A. flavus* and enhances resistance to foliar pathogens when expressed in transgenic tobacco (Rajasekaran *et al.*, 2002). Trypsin inhibitor activity also is important to the resistance phenotype seen in progeny resulting from the IITA-USDA collaborative crosses. Cotton transformed with the trypsin inhibitor gene under the control of a constitutive promoter was resistant to *Verticillium dahliae* but not to *A. flavus* (Rajasekaran *et al.*, 2002). Higher seed-specific expression of the trypsin inhibitor gene in cotton might increase the resistance phenotype.

Other potentially useful transgenes include maize PR-10, a pathogenesis-related protein with antifungal and RNase activity (Chen *et al.*, 2006), and glyoxalase I (Chen *et al.*, 2004). A synthetic version of a maize ribosome inhibiting protein gene, *mod1*, controls *A. flavus* in maize and peanuts (Weissinger *et al.*, 2003). Resistance of a maize hybrid to *A. flavus* infection was correlated with elevated α-1-3-glucanase levels in transgenic cells (Lozovaya *et al.*, 1998).

Table 2. Inhibitory concentrations of D4E1 towards cotton pathogens.

Pathogen	Disease	IC_{50} (μM)[a]	MIC (μM)[b]
Aspergillus flavus	Boll rot	7.8	25
Aspergillus flavus 70-GFP	Boll rot	11	25
Cercospora spp.	Cercospora leaf spot	8.7	25
Colletotrichum spp.	Anthracnose and boll rot	13	25
Fusarium graminearum	Seedling disease	2.1	25
Fusarium oxysporum	Fusarium wilt and boll rot	2.1	12.5
Fusarium verticillioides	Seedling disease and boll rot	0.88	12.5
Penicillium spp.	Boll rot	5.9	25
Phytophthora spp.	Boll rot	nd[c]	4.6
Pythium ultimum	Seedling disease and root rot	nd	13
Rhizoctonia solani	Soreshin and Leaf spot	nd	27
Thielaviopsis basicola	Black root or Collar rot	0.52	6.0
Verticillium dahliae	Verticillium wilt	0.60	5.2
Xanthomonas campestris pv. *malvacearum*	Bacterial blight	0.19	1.2

[a]IC_{50} – the concentration of D4E1 (μM) that inhibits the growth of 50% of the spores or bacterial cells relative to a control treatment (0 μM D4E1).
[b]MIC – the minimum concentration of D4E1 required to completely inhibit the growth of fungal or bacterial cells.
[c]Not determined.

Genes encoding haloperoxidases (Wolffram *et al.*, 1988) increase the resistance to microbial pathogens in transgenic tobacco plants (Rajasekaran *et al.*, 2000). In bioassays with *A. flavus* as the test organism, addition of a myeloperoxidase, one of the haloperoxidases, increased the lethality of H_2O_2 90-fold by catalyzing its conversion to sodium hypochlorite (Jacks *et al.*, 1991). Bacterial chloroperoxidases can reduce the viability of *A. flavus* conidia *in vitro* (Jacks *et al*,. 1999; Rajasekaran *et al.*, 2000) and can increase resistance to *Aspergillus* spp. and other plant pathogenic fungi in transgenic tobacco, cotton and peanut (Rajasekaran *et al.*, 2000; Jacks *et al.*, 2004; Chu *et al.*, 2006). H_2O_2 is produced by plants under pest attack and often serves as the substrate for these unique peroxidases in host plant tissues (Jacks and Hinojosa, 1993; Rajasekaran *et al.*, 2000).

Small molecular weight peptides isolated from organisms other than plants also show promise as antifungal agents (Rao, 1995; Broekart *et al.*, 1997; Hancock and Chapple, 1999; Reddy *et al.*, 2004). For example, the cecropins (Bowman and Hultmark, 1987) and magainins (Zasloff, 1987) of insect and amphibian origins, respectively, and their synthetic analogs (Cary *et al.*, 2000; Rajasekaran *et al.*, 2001; DeGray *et al.*, 2001; Li *et al.*, 2001; Chakrabarti *et al.*, 2003) also have significant antifungal activity against *Aspergillus* spp. and can lyse fungal mycelium at concentrations < 10 μM. A synthetic lytic peptide (D4E1) gene, when transformed into tobacco and cotton, increased resistance *in planta* to several fungal and bacterial plant pathogens (Table 2; Cary *et al.*, 2000; Rajasekaran *et al.*, 2001). In cottonseed expressing the D4E1 gene, seed coats and cotyledons resisted penetration and colonization by a virulent *A. flavus* strain carrying a green fluorescent protein (GFP) reporter gene (Rajasekaran *et al.*, 2005*a*,*b*).

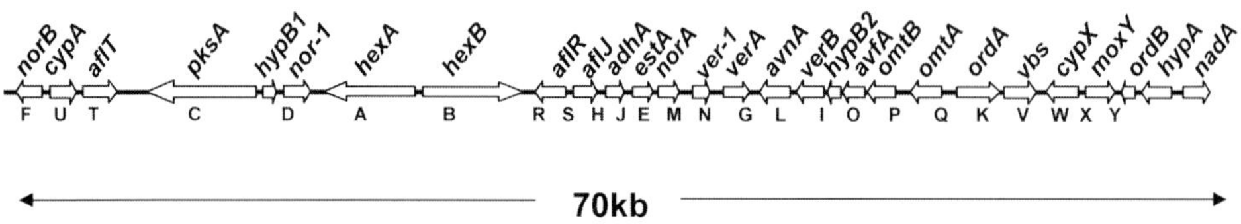

Figure 1. Schematic of the aflatoxin biosynthetic pathway gene cluster. The gene names are on the side of the cluster and arrows indicate the direction of gene transcription (for more details see Yu *et al.*, 2004*a*).

D4E1 also inhibited the germination of spores of *A. flavus* and caused severe abnormal lytic effects to the mycelial wall and within the cytoplasm and nuclei (Rajasekaran *et al.*, 2005*a*).

Genetics of aflatoxin formation

The only feature common to aflatoxin contamination of various crops is the causative fungus. Therefore, it is extremely important to understand how the fungus makes the toxin, how it responds to environmental signals, and the genetic mechanism(s) that allows it to survive in the field and to invade a very diverse set of plant hosts, *i.e.*, the plant-fungal interaction. The biosynthesis of aflatoxins has been studied extensively (reviewed in Payne and Brown, 1998; Yu *et al.*, 2004*a*; Bhatnagar *et al.*, 2003, 2006*a*; Chang *et al.*, 2007) and requires > 25 enzymatic steps (Fig. 1). Genes involved in aflatoxin biosynthesis are clustered in a 70-kb region in both the *A. parasiticus* and the *A. flavus* genomes.

Regulatory elements in aflatoxin biosynthesis

Aflatoxin biosynthesis in fungi is responsive to a number of nutritional and environmental cues such as carbon and nitrogen source, *e.g.*, amino acids, plant metabolites, *e.g.*, volatiles and tannins, stress, and physical factors such as pH, moisture and temperature. These stimuli are transduced via complex signaling cascades that control the expression of both global-acting and aflatoxin biosynthetic pathway-specific transcription factors. Two aflatoxin pathway-specific regulator genes, *aflR* and *aflJ*, are located in the aflatoxin biosynthetic gene cluster in both *A. flavus* and *A. parasiticus*, and in the sterigmatocystin cluster of *Aspergillus nidulans* (Matsushima *et al.*, 2001; Chang *et al.*, 1993, 2000).

AflJ is another regulatory gene in the aflatoxin biosynthetic pathway gene cluster and also has a role in the regulation of aflatoxin biosynthesis (Meyers *et al.*, 1998). *AflJ* is adjacent to *aflR* and the two genes are divergently transcribed (Fig. 1). The AFLJ protein has no significant sequence similarity with proteins of known function in the available databases. *AflJ* mutants produce no more than minute amounts of aflatoxin, but express all of the aflatoxin biosynthetic pathway genes at reduced levels. Thus, AFLJ does not function as does AFLR as a transcriptional activator of aflatoxin biosynthetic pathway genes. AFLJ can bind to the C-terminal region of AFLR and has been proposed to function as either a transcriptional enhancer or co-activator of AFLR (Chang, 2003). The *aflR* gene was first characterized in *A. flavus* by Payne *et al.* (1993) and in *A. parasiticus* by Chang *et al.* (1993). The AFLR protein is a positive-acting transcription factor that is required for expression of all known aflatoxin biosynthetic genes. AFLR activates transcription of aflatoxin pathway genes following binding to a specific site in the promoter regions of most of the pathway genes. Fungal isolates in which *aflR* has been inactivated no longer produce aflatoxin and do not express pathway genes. When the fungus over-expresses *aflR*, pathway genes also are overexpressed and toxin production increases (reviewed in Payne and Brown, 1998; Yu *et al.*, 2004*a*).

A number of globally-acting transcription factors also are involved in the regulation of aflatoxin biosynthesis as evidenced by the presence of proven and putative binding sites for these factors in the promoter regions of genes in the aflatoxin cluster, including the *aflR-aflJ* intergenic region. Binding sites for both PACC, the pH-responsive global regulator, and AREA, the nitrogen catabolism regulator, occur in the *aflR-aflJ* intergenic region (Chang *et al*,. 2000; Ehrlich *et al.*, 2003). *Aspergillus flavus* synthesizes aflatoxin in acidic media, but not in alkaline media, which is consistent with PACC repression of acid-expressed genes under alkaline conditions (Tilburn *et al.*, 1995). Nitrogen source regulation of aflatoxin biosynthesis is mediated by AREA binding to GATA sites in the *aflR-aflJ* intergenic region, which controls their expression (Chang *et al.*, 2000). Other global regulatory proteins, *e.g.*, BRLA – regulation of development, and CREA – regulation of carbon catabolism, also have putative binding sites in the *aflR-aflJ* intergenic region of a number of aflatoxin-producing *Aspergillus* species, including *A. nidulans* (Ehrlich *et al.*, 2003). The variability in the number and location of these transcription factor binding sites within the aflatoxin gene cluster may reflect the ecological niches of these species and the environmental stimuli that they encounter.

"Strain degeneration" in *A. parasiticus* following repeated subculturing of macerated mycelia resulted in morphological variants with "fan" and "fluff" phenotypes due to increased mycelial growth and reduced conidiation and aflatoxin production and provided the first link between aflatoxin production and fungal development (Bennett *et al.*, 1981). These variants, termed *sec-* strains, for secondary metabolite negative, are stable and do not revert back to wild-type levels of aflatoxin production or conidiation (Kale *et al.*, 1994, 2003). *Sec-* strains do not express most of the aflatoxin biosynthetic genes, including *aflJ*, and *aflR* is expressed at only 10-20% of the level observed in the parental strain. The mechanism linking sterigmatocystin and aflatoxin production to fungal development involves a G-protein/cAMP/protein kinase A (PKA) signaling pathway (Hicks *et al.*, 1997). The AFLR protein is inactivated by PKAA-mediated phosphorylation and the transcription of *aflR* is repressed by PKAA activity (Shimizu *et al.*, 2003).

The velvet gene, *veA*, mediates a developmental light-response and also is essential for the development of cleistothecia (sexual structures) and sclerotia (Calvo *et al.*, 2002, 2004). Deletion of *veA* blocks aflatoxin/sterigmatocystin production by inhibiting *aflR* and *aflJ* transcription (Kato *et al.*, 2003) and reduces the levels of the mycotoxins, cyclopiazonic acid and aflatrem (Cary *et al.*, 2007). VEA homologs have been identified in a number of filamentous fungi, *e.g.*, *Fusarium*, *Aspergillus*, *Magnaporthe*, *Sclerotinia*, and *Coccidioides*, but are not known in either yeasts, *e.g.*, *Saccharomyces cerevisiae* or *Schizosaccharomyces pombe*, or higher eukaryotes.

The *laeA* gene also encodes a global regulator of secondary metabolite production in *Aspergillus* species (Bok and Keller, 2004). Loss of LAEA prevents sterigmatocystin and penicillin production by *A. nidulans* and gliotoxin production in *Aspergillus fumigatus*, while overexpression of *laeA* increases penicillin and lovastatin production in *A. nidulans* and *A. terreus*, respectively. LAEA has limited homology to histone and arginine methyltransferases involved in chromatin modification (reviewed in Keller *et al.*, 2005).

Both the aflatoxin and sterigmatocystin biosynthetic gene clusters are located near the end of their respective chromosomes in a subtelomeric region enriched in transcriptionally-inactive heterochromatin. One model for LAEA function requires that secondary metabolite gene clusters, *e.g.*, the aflatoxin gene cluster, are maintained in silenced heterochromatin that is surrounded by transcriptionally active euchromatin. LAEA initiates chromatin re-modeling and converts heterochromatin to euchromatin perhaps by interfering with methylases or deacetylases that are associated with the heterochromatin. Thus, LAEA regulation of secondary metabolism may represent a form of epigenetic control of gene transcription and may affect the evolution and maintenance of clustering of secondary metabolic genes. *LaeA* expression in *A. flavus* is not affected by inactivation of *veA*, but it has not yet been determined if *veA* expression is linked to LaeA function (Cary *et al.*, unpublished).

As for VEA, LAEA homologs are known in a number of filamentous fungi but not in yeasts or higher eukaryotes. Thus, both *veA* and *laeA* are targets for aflatoxin control strategies as they appear to be unique to filamentous fungi and their inactivation is not expected to result in any adverse effects to higher eukaryotes. Both Δ*veA* and Δ*laeA* mutants are being used to identify genes whose expression depends on one of these two genes, with 136 genes identified as potentially *veA*-dependent in their expression (Cary *et al.*, 2007) and a gene cluster responsible for the production of the antitumor compound terrequinone A apparently regulated by *laeA* (Bok *et al.*, 2006).

Aspergillus flavus genomics

The two objectives of *A. flavus* genomics are: (i) to understand the genetic control and regulation of toxin production, and (ii) to understand the evolutionary process in *Aspergillus* section *Flavi* (Bhatnagar *et al.*, 2006*a*,*b*). *Aspergillus flavus* strain NRRL 3357 (ATCC 200026; SRRC 167), a wild type *A. flavus* strain widely used in laboratory and field studies, has been sequenced with 5× coverage (Bhatnagar *et al.*, 2006*b*; Payne *et al.*, 2006). The genome has been assembled into 79 scaffolds ranging in size from 1.0 kb to 4.5 Mb. Over 75% of the genome is represented in the 10 largest scaffolds. The estimated genome size of 36.3 Mb is similar to that for *Aspergillus oryzae* (36.8 Mb), but larger than that for *A. nidulans* or *A. fumigatus*. Links to other information on the sequencing project can be found at: www.Aspergillusflavus.org.

Functional genomics

An EST strategy provided rapid identification of genes potentially involved in aflatoxin contamination of crops by *A. flavus* (O'Brian *et al.*, 2003; Yu *et al.*, 2004*b*, 2007). There were 7,218 unique genes identified from > 26,000 cDNA clones in an *A. flavus* cDNA expression library with average cDNA insert size ~ 1.2-1.5 kb in length (Yu *et al.*, 2004*b*). However, 34% of the unique genes do not have homologs in existing databases, so two, or more, non-overlapping ESTs may be transcribed from the same gene but be counted as two, or more, unique genes. The 7,218 unique genes account for ~ 60% of the total functional genes in the *A. flavus* genome, but without additional biological evidence it is very difficult to predict whether these genes are involved in primary or secondary metabolism. Further characterization of these genes requires additional genomic scale studies such as gene expression profiling in microarray experiments followed by analysis of targeted deletion mutants.

DNA microarrays can be powerful tools for identifying previously uncharacterized genes that modulate aflatoxin biosynthesis. The production of aflatoxin is affected by both nutritional and environmental factors. In maize, *A. flavus* colonization is favored during plant stress, with aflatoxin levels increasing with drought, heat, nitrogen deficiency, and increased pressure from insects and weeds. Therefore, the effect of several biotic and abiotic factors on toxin production were evaluated, with carbohydrates, nitrogen source, metals, lipids, amino acids, temperature, and pH found to have the most effect (*cf.* Wilkinson *et al.*, 2007*a*,*b*).

Initially a small, targeted DNA microarray composed of 753 elements was used to identify genes involved in aflatoxin production. The elements in this array were chosen because they previously were identified as differentially expressed during aflatoxin biosynthesis in macroarrays (O'Brian *et al.*, 2003). The hypothesis was that genes with expression profiles similar to that of genes for aflatoxin biosynthesis under several cultural and environmental conditions might be involved in aflatoxin biosynthesis. Gene expression profiles in four conditions conducive for aflatoxin biosynthesis and four that were not conducive for aflatoxin biosynthesis were compared. Twenty genes were identified that were differentially expressed in the conducive nitrogen and pH treatments and in the non-conducive carbon and temperature treatments (Price *et al.*, 2005), with 17 genes exhibiting consistent induction or repression across all treatments.

An EST-based genomic DNA amplicon microarray was constructed with 9,445 pairs of sequence-specific primers for known unique ESTs. *Aspergillus flavus* genomic DNA was used as the template for the PCR amplification. The successful amplicons plus 31 aflatoxin pathway genes, representing 5,002 unique ESTs, were arrayed in triplicate onto Telechem Superamine aminosilane coated microscope slides and profiled under aflatoxin-producing and aflatoxin-non-producing conditions (Price *et al.*, 2006; O'Brian *et al.*, 2007; Wilkinson *et al.*, 2007*a*,*b*). In addition to the genes directly involved in aflatoxin formation, hundreds of additional genes were significantly up- or down-regulated. These genes encode hypothetical proteins with unknown function(s) and potential regulatory factors.

Comparative genomics

In addition to *A. flavus*, several other genomes of *Aspergillus* species have been sequenced, including the non-aflatoxin-producing species *A. fumigatus*, which is a human pathogen; *A. oryzae*, which is used in food fermentation, and *A. niger*, which is used in industrial fermentation. All of these species are close relatives of aflatoxin-producing species *A. flavus* and *A. parasiticus*. Sterigmatocystin, an aflatoxin precursor, is produced by some strains of *A. nidulans*, which is widely considered a model fungus for biological studies. Comparisons between the genomes of *A. flavus*, *A. fumigatus*, *A. oryzae* and *A. nidulans* are underway (reviewed in Bhatnagar *et al.*, 2006*b*; Payne *et al.*, 2006) and should provide insights into how *A. flavus* survives in the field, and what is unique about this fungus, as it is the only one of the *Aspergillus* species with a sequenced genome that infects crops.

Conclusions

Studies of economically important fungi at the molecular level are an innovative strategy to unravel the mystery of aflatoxin biosynthesis and to understand the interaction of the fungus with the host crops. Genomic studies of the fungus will increase our understanding of the biology, evolution, biochemical function and genetic regulation of the genes in the fungal system. EST technology allows rapid identification of most, if not all, of the genes expressed from the fungal genome and helps define their functions and regulation, the coordination of gene expression in response to internal and external factors, the relationship between primary and secondary metabolism, plant-fungal interactions and fungal pathogenicity and evolutionary biology. A microarray composed of EST sequences can be used to detect the genes expressed under specific environmental conditions. This technology enables the simultaneous study of a complete set of fungal genes that are responsible for or related to aflatoxin production. The identification of specific biochemical factors linked to resistance against *A. flavus* may contribute not only to the reduction or elimination of aflatoxin contamination, but also may contribute to the control of other fungal diseases. Biocontrol of aflatoxin contamination has significant promise, and may be a viable strategy for crops in which there are insufficient genetic diversity to develop resistant germplasm, *e.g.*, cotton.

References

Azziz-Baumgartner, E., Lindblade, K., Gieseker, K. Rogers, H.S., Kieszak, S., Njapau, H., Schleicher, R., McCoy, L.F., Misore, A., DeCock, K., Rubin, C., Slutsker, L., and the Aflatoxin Investigative Group. (2005) Case-control study of an acute aflatoxicosis outbreak, Kenya, 2004. *Environmental Health Perspectives* 113, 1779-1783.

Bennett, J.W., Silverstein, R.B. and Kruger, S.J. (1981) Isolation and characterization of two nonaflatoxigenic classes of morphological variants of *Aspergillus parasiticus*. *Journal of the American Oil Chemists Society* 58, 952-955.

Bhatnagar, D., Cleveland, T.E. and Cotty, P.J. (1994) Mycological aspects of aflatoxin formation. In: Eaton, D.L. and Groopman, J.D. (eds.) *The Toxicology of Aflatoxins*. Academic Press, San Diego, California, pp. 327-345.

Bhatnagar, D., Yu, J. and Ehrlich, K.C. (2002) Toxins of filamentous fungi. *Chemical Immunology* 81, 167-206.

Bhatnagar, D., Ehrlich, K.C. and Cleveland, T.E. (2003) Molecular genetic analysis and regulation of aflatoxin biosynthesis. *Applied Microbiology and Biotechnology* 61, 83-93.

Bhatnagar, D., Cary, J. W., Ehrlich, K., Yu, J. and Cleveland, T. E. (2006*a*) Understanding the genetics of regulation of aflatoxin production and *Aspergillus flavus* development. *Mycopathologia* 162, 155-166.

Bhatnagar, D., Proctor, R., Payne, G. A., Wilkinson, J., Yu, J., Cleveland, T. E. and Nierman, W. C. (2006*b*) Genomics of mycotoxigenic fungi. In: Barug, D., Bhatnagar, D., van Egmond, H.P., van der Kamp, J.W., van Osenbruggen, W.A., and Visconti, A. (eds.) *The Mycotoxin Fact Book*. Wageningen Academic Publishers, Wageningen, The Netherlands, pp 157-177.

Bohlmann, H. (1994) The role of thionins in plant protection. *CRC Critical Reviews in Plant Science* 13, 1-16.

Bok, J.W. and Keller, N.P. (2004) LaeA, a regulator of secondary metabolism in *Aspergillus* spp. *Eukaryotic Cell* 3, 527-535.

Bok, J.W., Hoffmeister, D., Maggio-Hall, L.A., Murillo, R., Glasner, J.D. and Keller, N.P. (2006) Genomic mining for *Aspergillus* natural products. *Chemical Biology* 13, 31-37.

Bowman, H.G. and Hultmark, D. (1987) Cell-free immunity in insects. *Annual Review of Microbiology* 31, 103-126.

Broekaert, W.F., Cammue, B.P.A., Debolle, M.F.C., Thevissen, K., Desamblanx, G.W. and Osborn, R.W. (1997) Antimicrobial peptides from plants. *CRC Critical Reviews in Plant Science* 16, 297-323.

Broglie, K., Chet, I., Holliday, M., Cressman, R., Biddle, P., Knowlton, S., Mauvais, C.J. and Broglie, R. (1991) Transgenic plants with enhanced resistance to the fungal pathogen *Rhizoctonia solani*. *Science* 254, 1194-1197.

Brown, R.L., Cleveland, T.E., Payne, G.A., Woloshuk, C.P., Campbell, K.W. and White, D.G. (1995) Determination of resistance to aflatoxin production in maize kernels and detection of fungal colonization using an *Aspergillus flavus* transformant expressing Escherichia coli β-glucuronidase. *Phytopathology* 85, 983-989.

Brown, R.L., Bhatnagar, D., Cleveland, T.E. and Cary, J.W. (1998) Recent advances in preharvest prevention of mycotoxin contamination. In: Sinha, K.K. and D. Bhatnagar (eds.) *Mycotoxins in Agriculture and Food Safety*. Marcel Dekker, New York, pp. 351-379.

Brown, R.L., Chen, Z.-Y., Menkir, A., Cleveland, T.E., Cardwell, K., Kling, J. and White, D.G. (2001) Resistance to aflatoxin accumulation in kernels of maize inbreds selected for ear rot resistance in West and Central Africa. *Journal of Food Protection* 64, 396-400.

Brown, R.L., Chen, Z.-Y., Menkir, A. and Cleveland, T.E. (2003) Using biotechnology to enhance host resistance to aflatoxin contamination of maize – A mini-review. *African Journal of Biotechnology* 2, 557-562.

Brown, R.L., Chen, Z.-Y. and Cleveland, T.E. (2004) Molecular biology for control of mycotoxigenic fungi. In: Arora, D., Bridge, P. and Bhatnagar, D. (eds.) *Fungal Biotechnology in Agricultural, Food, and Environmental Applications*. Marcel Dekker, New York, pp. 69-77.

Brown, R.L., Chen, Z.-Y., Menkir, A., Bandyopadhyay, R. and Cleveland, T.E. (2005) Development of aflatoxin-resistant maize inbreds and identification of potential resistance markers through USA-Africa collaborative research. In: Robens, J. (ed.) *Proceedings of the USDA-ARS Aflatoxin and Fumonisin Elimination and Fungal Genomics Workshops (October 2005, Raleigh, North Carolina)*. USDA, ARS, Beltsville, Maryland, p. 78.

Calvo, A.M., Wilson, R.A., Bok, J.-W. and Keller, N.P. (2002) Relationship between secondary metabolism and fungal development. *Microbiology and Molecular Biology Reviews* 66, 447-459.

Calvo, A.M., Bok, J.-W., Brooks, W. and Keller, N.P. (2004) VeA is required for toxin and sclerotial production in *Aspergillus parasiticus*. *Applied and Environmental Microbiology* 70, 4733-4739.

Campbell, K.W. and White, D.G. (1995) Evaluation of corn genotypes for resistance to *Aspergillus* ear rot, kernel infection, and aflatoxin production. *Plant Disease* 79, 1039-1045.

Cary, J.W., Rajasekaran, K., Jaynes, J.M. and Cleveland, T.E. (2000) Transgenic expression of a gene encoding a synthetic antimicrobial peptide results in inhibition of fungal growth *in vitro* and in plants. *Plant Science* 153, 171-180.

Cary, J.W., O'Brian, G.R., Nielsen, D.M., Nierman, W., Harris-Coward, P., Yu, J., Bhatnagar, D., Cleveland, T.E., Payne, G.A. and Calvo, A.M. (2007) Elucidation of *veA*-dependent genes associated with aflatoxin and sclerotial production in *Aspergillus flavus* by functional genomics. *Applied Microbiology and Biotechnology* 76, 1107-1118.

CDC (Center for Disease Control). (2004) Outbreak of aflatoxin poisoning – Eastern and Central Provinces, Kenya, January-July 2004. *Morbidity and Mortality Weekly Report* 53, 790-793.

Chakrabarti, A., Ganapathi, T.R., Mukherjee, P.K. and Bapat, V.A. (2003) MSI-99, a magainin analogue, imparts enhanced disease resistance in transgenic tobacco and banana. *Planta* 216, 587-596.

Chang, P.-K. (2003) The *Aspergillus parasiticus* protein AFLJ interacts with the aflatoxin pathway-specific regulator AFLR. *Molecular Genetics and Genomics* 268, 711-719.

Chang, P.-K., Cary, J.W., Bhatnagar, D., Cleveland, T.E., Bennett, J.W., Linz, J.E., Woloshuk, C.P. and Payne, G.A. (1993) Cloning of the *Aspergillus parasiticus apa-2* gene associated with the regulation of aflatoxin biosynthesis. *Applied and Environmental Microbiology* 59, 3273-3279.

Chang, P.-K., Yu, J., Bhatnagar, D. and Cleveland, T.E. (2000) Characterization of the *Aspergillus parasiticus* major nitrogen regulatory gene, *areA*. *Biochimica et Biophysica Acta* 1491, 263-266.

Chang, P.-K., Horn, B.W. and Dorner, J.W. (2005) Sequence breakpoints in the aflatoxin biosynthesis gene cluster and flanking regions in non-aflatoxigenic *Aspergillus flavus* isolates. *Fungal Genetics and Biology* 42, 914-923.

Chang, P.-K., Matsushima, K., Takahashi, T., Yu, J., Abe, K., Bhatnagar, D., Yuan, G.-F., Koyama, Y. and Cleveland, T. (2007) Understanding nonaflatoxigenicity of *Aspergillus sojae:* A windfall of aflatoxin biosynthesis research. *Applied Microbiology and Biotechnology* 76, 977-984.

Chen, Z.-Y., Brown, R.L., Russin, J.S., Lax, A.R. and Cleveland, T.E. (1998) Resistance to *Aspergillus flavus* in corn kernels is associated with a 14-kDa protein. *Phytopathology* 88, 276-281.

Chen, Z.-Y., Brown, R.L., Damann, K.E. and Cleveland, T.E. (2000) Proteomics analysis of kernel embryo and endosperm proteins of corn genotypes resistant or susceptible to *Aspergillus flavus* infection. In: Robens, J., (ed.) *Proceedings of the USDA-ARS Aflatoxin Elimination Workshop (October, 2000, Yosemite, California)*. USDA, ARS, Beltsville, Maryland, p. 88.

Chen, Z.-Y., Brown, R.L., Cleveland, T.E., Damann, K.E. and Russin, J.S. (2001) Comparison of constitutive and inducible maize kernel proteins of genotypes resistant or susceptible to aflatoxin production. *Journal of Food Protection* 64, 1785-1792.

Chen, Z.-Y., Brown, R.L., Damann, K.E. and Cleveland, T.E. (2002) Identification of unique or elevated levels of kernel proteins in aflatoxin-resistant maize genotypes through proteome analysis. *Phytopathology* 92, 1084-1094.

Chen, Z.-Y., Brown, R.L., Damann, K.E. and Cleveland, T.E. (2004) Identification of a maize kernel stress-related protein and its effect on aflatoxin accumulation. *Phytopathology* 94, 938-945.

Chen, Z.-Y., Brown, R.L., Rajasekaran, K., Damann, K.E. and Cleveland, T.E. (2006) Identification of a maize kernel pathogenesis-related protein and evidence for its involvement in resistance to *Aspergillus flavus* infection and aflatoxin production. *Phytopathology* 96, 87-95.

Chu, Y., Ramos, M.L., Faustinelli, P., Holbrook, C., Rajasekaran, K., Cary, J. and Ozias-Akins, P. (2006) Host genetic components for aflatoxin reduction strategies in peanut. In: Robens, J. (eds.) *Proceedings of the Annual Multicrop Aflatoxin/Fumonisin Elimination and Fungal Genomics Workshop (October 2006, Ft. Worth, Texas)*. USDA, ARS, Beltsville, Maryland, p. 53.

Cornelissen, B.J.C. and Melchers, L.S. (1993) Strategies for the control of fungal diseases with transgenic plants. *Plant Physiology* 101, 709-712.

Cotty, P.J. (1994) Influence of field application of an atoxigenic strain of *Aspergillus flavus* on the population of *A. flavus* infecting cotton bolls and on the aflatoxin content of cottonseed. *Phytopathology* 84, 1270-1277.

Cotty, P.J., and Bayman, P. (1993) Competitive exclusion of a toxigenic strain of *Aspergillus flavus* by an atoxigenic strain. *Phytopathology* 83, 1283-1287.

Daniell, H., Ruiz, O.N. and Dhingra, A. (2005) Chloroplast genetic engineering to improve agronomic traits. *Methods in Molecular Biology* 286, 111-138.

Davis, G.L., Windham, G.L. and Williams, W.P. (1999) QTL for aflatoxin reduction in maize. *Maize Genetics Conference Abstracts (September 1999, Lake Geneva, Wisconsin)*. 41(T8).139.

DeGray, G., Rajasekaran, K., Smith, F., Sanford, J. and Daniell, H. (2001) Expression of an antimicrobial peptide via the chloroplast genome to control phytopathogenic bacteria and fungi. *Plant Physiology* 127, 852-862.

Dorner, J.W. (2004) Biological control of aflatoxin contamination of crops. *Journal of Toxicology – Toxin Reviews* 23, 425-450.

Eaton, D.L and Groopman, J.D., eds. (1994) *The Toxicology of Aflatoxins: Human Health, Veterinary, and Agricultural Significance*. Academic Press, San Diego, California.

Ehrlich, K.C. and Cotty, P. J. (2004) An isolate of *Aspergillus flavus* used to reduce aflatoxin contamination in cottonseed has a defective polyketide synthase gene. *Applied Microbiology and Biotechnology* 65, 473-478.

Ehrlich, K.C., Montalbano, B.G. and Cotty, P.J. (2003) Sequence comparison of *aflR* from different *Aspergillus* species provides evidence for variability in regulation of aflatoxin production. *Fungal Genetics and Biology* 38, 63-74.

FAO. (2004) *Worldwide Regulations for Mycotoxins in Food and Feed in 2003. Food and Nutrition Paper* 81. Rome, Italy.

Gorg, A., Postel, W. and Gunther, S. (1988) The current state of two- dimensional electrophoresis with immobilized pH gradients. *Electrophoresis* 9, 531-546.

Greene-McDowelle, D.M., Ingber, B., Wright, M.S., Zeringue, H.J., Jr., Bhatnagar, D. and Cleveland, T.E. (1999) Effect of selected cotton leaf-life volatiles on growth, development and aflatoxin production of *Aspergillus parasiticus*. *Toxicon* 37, 883-893.

Guo, B.Z., Widstrom, N.W., Cleveland, T.E. and Lynch, R.E. (2000) Control of preharvest aflatoxin contamination in corn: Fungus-plant-insect interactions and control strategies. *Recent Research Developments in Agricultural and Food Chemistry* 4, 165-176.

Hamblin, A.M. and White, D.G. (2000) Inheritance of resistance to *Aspergillus* ear rot and aflatoxin production of corn from Tex6. *Phytopathology* 90, 292-296.

Hancock, R.E.W. and Chapple, D.S. (1999) Peptide antibiotics. *Antimicrobial Agents and Chemotherapy* 43, 1317-1323.

Henry, S.H., Bosch, F.X. and Bowers J.C. (2002) Aflatoxin, hepatitis and worldwide liver cancer risks. *Experimental Medicine and Biology* 504, 229-233

Hicks, J.K., Yu, J.H., Keller, N.P. and Adams, T.H. (1997) *Aspergillus* sporulation and mycotoxin production both require inactivation of the FadA Gα protein-dependent signaling pathway. *EMBO Journal* 16, 4916-4923.

Jacks, T.J. and Hinojosa, O. (1993) Superoxide radicals in intact tissues and in dimethyl sulfoxide-based extracts. *Phytochemistry* 33, 563-568.

Jacks, T.J., Cotty, P.J. and Hinojosa, O. (1991) Potential of animal myeloperoxidase to protect plants from pathogens. *Biochemical and Biophysical Research Communications* 178, 1202-1204.

Jacks, T.J., Delucca, A.J. and Morris, N.M. (1999) Effects of chloroperoxidase and hydrogen peroxide on the viabilities of *Aspergillus flavus* conidiospores. *Molecular and Cellular Biochemistry* 195, 169-172.

Jacks, T.J., Cary, J.W., Rajasekaran, K., Cleveland, T.E. and van Pée, K.-H. (2004) Transformation of plants with a chloroperoxidase gene to enhance disease resistance. U.S. Patent 6,703,540.

Jelinek, C.F., Pohland, A.E. and Wood, G.E. (1989) Worldwide occurrence of mycotoxins in foods and feeds – An update. *Journal of the Association of Official Analytical Chemists* 72, 223-230

Kale, S.P., Cary, J.W., Bhatnagar, D. and Bennett, J.W. (1994) Isolation and characterization of morphological variants of *Aspergillus parasiticus* deficient in secondary metabolite production. *Mycological Research* 98, 645-652.

Kale, S.P., Cary, J.W., Baker, C., Walker, D., Bhatnagar, D. and Bennett, J.W. (2003) Genetic analysis of morphological variants of *Aspergillus parasiticus* deficient in secondary metabolite production. *Mycological Research* 107, 831-840.

Kato, N., Brooks, W. and Calvo, A.M. (2003) The expression of sterigmatocystin and penicillin genes in *Aspergillus nidulans* is controlled by *veA,* a gene required for sexual development. *Eukaryotic Cell* 2, 1178-1186.

Keller, N.P., Calvo, A. and Gardner, H. (1999) Linoleic acid and linoleic acid derivatives regulate *Aspergillus* development and mycotoxin production. In: Robens, J. (ed.) *Proceedings of the USDA-ARS Aflatoxin Elimination Workshop (October 1999, Atlanta, Georgia.)*. USDA, ARS, Beltsville, Maryland p.43.

Keller, N.P., Turner, G. and Bennett, J.W. (2005) Fungal secondary metabolism – From biochemistry to genomics. *Nature Reviews* 3, 937-947.

Li, Q.S., Lawrence, C.B., Xing, H.Y., Babbitt, R.A., Bass, W.T., Maiti, I.B. and Everett, N.P. (2001) Enhanced disease resistance conferred by expression of an antimicrobial magainin analog in transgenic tobacco. *Planta* 212, 635-639.

Lozovaya, V.V., Waranyuwat, A. and Widholm, J.M. (1998) α-1-3-glucanse and resistance to *Aspergillus flavus* infection in maize. *Crop Science* 38, 1255-1260.

Mahoney, N. and Molyneux, R.J. (2004) Phytochemical inhibition of aflatoxigenicity in *Aspergillus flavus* by constituents of walnut, *Juglans regia*. *Journal of Agricultural and Food Chemistry* 52, 1882-1889.

Matsushima, K., Chang, P.-K., Yu, J., Abe, K., Bhatnagar, D. and Cleveland, T.E. (2001) Pretermination in AFLR of *Aspergillus sojae* inhibits aflatoxin biosynthesis. *Applied Microbiology and Biotechnology* 55, 585-589.

Maupin, L.M., Clements, M.J. and White, D.G. (2003) Evaluation of the MI82 corn line as a source of resistance to aflatoxin in grain and use of BGYF as a selection tool. *Plant Disease* 87, 1059-1066.

Meins, F., Neuhaus, J.-M., Sperisen, C. and Ryals, J. (1992) The primary structure of plant pathogenesis-related glucanohydrolases and their genes. In: Boller, T. and Meins, F. (eds.) *Genes Involved in Plant Defense*. Springer-Verlag, Berlin., pp. 245-282.

Menkir, A., Brown, R.L., Bandyopadhyay, R., Chen, Z.-Y. and Cleveland, T.E. (2006) A U.S.A. – Africa collaborative strategy for identifying, characterizing, and developing maize germplasm with resistance to aflatoxin contamination. *Mycopathologia* 162, 225-232 .

Meyers, D.M., O'Brian, G., Du, W.L., Bhatnagar, D. and Payne, G.A. (1998) Characterization of *aflJ*, a gene required for conversion of pathway intermediates to aflatoxin. *Applied and Environmental Microbiology* 64, 3713-3717.

O'Brian, G.R., Fakhoury, A.M. and Payne, G.A. (2003) Identification of genes differentially expressed during aflatoxin biosynthesis in *Aspergillus flavus* and *Aspergillus parasiticus*. *Fungal Genetics and Biology* 39, 118-127.

O'Brian, G. R., Georgianna, D. R., Wilkinson, J. R., Yu, J., Abbas, H. K., Bhatnagar, D., Cleveland, T. E., Nierman, W. and Payne, G.A. (2007) The effect of elevated temperature on gene transcription and aflatoxin biosynthesis. *Mycologia* 99, 232-239.

Payne, G.A. (1992) Aflatoxin in maize. *CRC Critical Reviews in Plant Sciences* 10, 423-440.

Payne, G.A. (1998) Process of contamination by aflatoxin-producing fungi and their impact on crops. In: Sinha, K.K. and Bhatnagar, D. (eds.) *Mycotoxins in Agriculture and Food Safety*. Marcel Dekker, New York, pp. 279-306.

Payne, G.A. and Brown, M.P. (1998) Genetics and physiology of aflatoxin biosynthesis. *Annual Review of Phytopathology* 36, 329-362.

Payne G.A., Nierman, W.C., Wortman, J.R., Pritchard, B.L., Brown, D., Dean, R.A., Bhatnagar, D., Cleveland, T.E., Machida, M. and Yu, J. (2006) Whole genome comparison of *Aspergillus flavus* and *A. oryzae*. *Medical Mycology* 44 Suppl, 9-11.

Payne, G.A., Nystrom, G.J., Bhatnagar, D.,Cleveland, T.E. and Woloshuk,C.P. (1993) Cloning of the *afl-2* gene involved in aflatoxin biosynthesis from *Aspergillus flavus*. *Applied and Environmental Microbiology* 59, 156-162.

Pennington, S.R., Wilkins, M.R., Hochstrasser, D.F. and Dunn, M.J. (1997) Proteome analysis: From protein characterization to biological function. *Trends in Cell Biology* 7, 168-173.

Pitt, J.I. (2000) Toxigenic fungi and mycotoxins. *British Medical Bulletin* 56, 184-192.

Price, M.S., Conners, S.B., Tachdjian, S., Kelly, R.M. and Payne, G.A. (2005) Aflatoxin conducive and non-conducive growth conditions reveal new gene associations with aflatoxin production. *Fungal Genetics and Biology* 42, 506-518.

Price, M.S., Yu, J., Nierman, W.C., Kim, H.S., Pritchard, B., Jacobus, C.A., Bhatnagar, D., Cleveland, T. E. and Payne, G. A. (2006) The aflatoxin pathway regulator AflR induces gene transcription inside and outside of the aflatoxin biosynthetic cluster. *FEMS Microbiology Letters* 255, 275-279.

Punja, Z.K. (2004) *Fungal Disease Resistance in Plants – Biochemistry, Molecular Biology, and Genetic Engineering*. Food Products Press, NewYork.

Rajasekaran, K., Cary, J.W., Jacks, T.J., Stromberg, K. and Cleveland, T.E. (2000) Inhibition of fungal growth *in planta* and *in vitro* by transgenic tobacco expressing a bacterial nonheme chloroperoxidase gene. *Plant Cell Reports* 19, 333-338.

Rajasekaran, K., Stromberg, K., Cary, J.W. and Cleveland, T.E. (2001) Broad-spectrum antimicrobial activity *in vitro* of the synthetic peptide D4E1. *Journal of Agricultural and Food Chemistry* 49, 2799-2803.

Rajasekaran, K., Cary, J.W., Jacks, T.J. and Cleveland, T.E. (2002) Genetic engineering for resistance to phytopathogens. *American Chemical Society Symposium Series* 829, 97-117.

Rajasekaran, K., Bhatnagar, D., Brown, R.L., Chen, Z.-Y., Cary, J.W. and Cleveland, T.E. (2005*a*) Enhancing food safety: Prevention of preharvest aflatoxin contamination. In: Ramasamy, C., Ramanathan, S. and Dhakshinamoorthy, M. (eds.) *Perspectives of Agricultural Research and Development.* Tamil Nadu Agricultural University, Coimbatore, India, pp. 434-467.

Rajasekaran, K., Cary, J.W., Jaynes, J.M. and Cleveland, T.E. (2005*b*) Disease resistance conferred by the expression of a gene encoding a synthetic peptide in transgenic cotton, *Gossypium hirsutum* L., plants. *Plant Biotechnology Journal* 3, 545-554.

Rajasekaran, K., Ulloa, M., Hutmacher, R., Cary, J.W. and Cleveland, T.E. (2006) Disease resistance in transgenic cottons. In: Dugger, P. and Richter, D.A. (eds.) *Proceedings of 2006 Beltwide Cotton Conferences (January 2006, San Antonio, Texas).* National Cotton Council, Memphis, Tennessee, pp. 895-903.

Rao, A.G. (1995) Antimicrobial peptides. *Molecular Plant-Microbe Interactions* 8, 6-13.

Reddy, K.V., Yedery, R.D. and Aranha, C. (2004) Antimicrobial peptides: Premises and promises. *International Journal of Antimicrobial Agents* 24, 536-547.

Richard, J. and Payne, G.A. (2003) *Mycotoxins: Risk in Plant, Animal, and Human Systems.* CAST Report 139. CAST, Ames, Iowa.

Ryan, C.A. (1990) Protease inhibitors in plants: Genes for improving defenses against insects and pathogens. *Annual Review of Phytopathology* 28, 425-449.

Shimizu, K., Hicks, J., Huang, T.-P. and Keller, N.P. (2003) Pka, Ras and RGS protein interactions regulate activity of AflR, a Zn(II).2Cys6 transcription factor in *Aspergillus nidulans. Genetics* 165, 1095-1104.

Singh, N.K., Nelson, D.E., Kuhn, D., Hasegawa, P.M. and Bressan, R.A. (1989) Molecular cloning of osmotin and regulation of its expression by ABA and adaptation to low water potential. *Plant Physiology* 90, 1096-1101.

Thomma, B.P.H.J., Cammue, B.P.A. and Thevissen, K. (2002) Plant defensins. *Planta* 216, 193-202.

Tilburn, J., Sarkar, S., Widdick, D.A., Espeso, E.A., Orejas, M., Mungroo, J., Penalva, M.A. and Arst, H.N., Jr. (1995) The *Aspergillus* PacC zinc finger transcription factor mediates regulation of both acid- and alkaline-expressed genes by ambient pH. *EMBO Journal* 14, 779-790.

Toubart, P., Desiderio, A., Salvi, G., Cervone, F., Daroda, L., DeLorenzo, G., Bergmann, C., Darvill, A.G. and Albersheim, P. (1992) Cloning and characterization of the gene encoding the endopolygalacturonase-inhibiting protein, PGIP, of *Phaseolus vulgaris* L. *Plant Journal* 2, 367-373.

van Egmond, H.P., Schothorst, R.C. and Jonker, M.A. (2007) Regulations relating to mycotoxins in food: Perspectives in a global and European context. *Analytical and Bioanalytical Chemistry* 389, 147-157.

Vardon, P. McLaughlin, C. and Nardinelli, C. (2003) Potential economic costs of mycotoxins in the United States. In: Richard, J. and Payne, G.A. (eds.) *Mycotoxins: Risks in Plant, Animal, and Human Systems*; CAST Task Force Report No. 139. CAST, Ames, Iowa, pp. 136-142.

Walker, R.D. and White, D.G. (2001) Inheritance of resistance to *Aspergillus* ear rot and aflatoxin production of corn from CI2. *Plant Disease* 85, 322-327.

Weissinger, A., Wu, M. and Cleveland, T.E. (2003) Expression in transgenic peanut of maize RIP 1, a protein with activity against *Aspergillus* spp. In: Robens, J. (ed.) *Proceedings of the USDA-ARS Aflatoxin Elimination Workshop (October 2003, Savannah, Georgia).* USDA, ARS, Beltsville, Maryland, p. 100.

White, D.G., Rocheford, T.R., Kaufman, B. and Hamblin, A.M. (1995*a*) Chromosome regions associated with resistance to *Aspergillus flavus* and inhibition of aflatoxin production in maize. In: Robens, J. (ed.) *Proceedings of the USDA-ARS Aflatoxin Elimination Workshop (October, 1995, Atlanta, Georgia).* USDA, ARS, Beltsville, Maryland, p 8.

White, D., Rocheford, T.R., Kaufman, B. and Hamblin, A.M. (1995*b*) Further genetic studies and progress on resistance to aflatoxin production in corn. In: Robens, J. (ed.) *Proceedings of the USDA-ARS Aflatoxin Elimination Workshop (October 1995, Atlanta, Georgia).* USDA, ARS, Beltsville, Maryland, p. 7.

White, D.G., Rocheford, T.R., Naidoo, G., Paul, C., Hamblin, A.M. and Forbes, A.M. (1998) Inheritance of molecular markers associated with, and breeding for resistance to *Aspergillus* ear rot and aflatoxin production in corn using Tex6. In: Robens, J. (ed.) *Proceedings of the USDA-ARS Aflatoxin Elimination Workshop (October 1998, St. Louis, Missouri).* USDA, ARS, Beltsville, Maryland, pp. 4-6.

White, D.G., Rocheford, T.R., Naidoo, G., Paul, C., Rozzi, R.D., Severns, D.E. and Forbes, A.M. (1999) Inheritance of molecular markers associated with and breeding for resistance to *Aspergillus* ear rot and aflatoxin production in corn. In: Robens, J. (ed.) *Proceedings of the USDA-ARS Aflatoxin Elimination Workshop (October 1999, Atlanta, Georgia).* USDA, ARS, Beltsville, Maryland, pp. 7-8.

Wilkins, M.R., Pasquali, C., Appel, R.D., Ou, K., Golaz, O., Sanchez, J.C., Yan, J.X., Gooley, A.A., Hughes, G., Humphrey-Smith, I., Williams, K.L. and Hochstrasser, D.F. (1996) From proteins to proteomes: Large-scale protein identification by two-dimensional electrophoresis and amino acid analysis. *Biotechnology* 14, 61-65.

Wilkins, T.A., Rajasekaran, K. and Anderson, D.M. (2000) Cotton biotechnology. *Critical Reviews in Plant Science* 19, 511-550.

Wilkinson, J.R., Yu, J., Abbas, H.K., Scheffler, B.E., Kim, H.S., Nierman, W.C., Bhatnagar, D. and Cleveland, T.E. (2007*a*) Aflatoxin formation and gene expression in response to carbon source media shift in *Aspergillus parasiticus. Food Additives and Contaminants* 24, 1-10.

Wilkinson, J. R., Yu, J., Bland, J. M., Nierman, W. C., Bhatnagar, D. and Cleveland, T. E. (2007*b*) Amino acid supplementation reveals differential regulation of aflatoxin biosynthesis in *Aspergillus flavus* NRRL 3357 and *Aspergillus parasiticus* SRRC 143. *Applied Microbiology and Biotechnology* 74, 1308-1319.

Wolffram, C., van Pee, K.-H. and Lingens, F. (1988) Cloning and high-level expression of a chloroperoxidase gene from *Pseudomonas pyrrocinia. FEBS Letters* 238, 325-328.

Wright, M. S., Greene-McDowelle, D.M., Zeringue, H.J., Jr., Bhatnagar, D. and Cleveland, T.E. (2000) Effects of volatile aldehydes from *Aspergillus*-resistant varieties of corn on *Aspergillus parasiticus* growth and aflatoxin biosynthesis. *Toxicon* 38, 1215-1223.

Wu, F. (2004) Mycotoxin risk assessment for the purpose of setting international regulatory standards. *Environmental Science and Technology* 38, 4049-4055.

Yu, J., Chang, P.-K., Ehrlich, K.C., Cary, J.W., Bhatnagar, D., Cleveland, T.E., Payne, G.A., Linz, J.E., Woloshuk, C.P. and Bennett, J.W. (2004*a*) Clustered pathway genes in aflatoxin biosynthesis. *Applied and Environmental Microbiology* 70, 1253-1262.

Yu, J., Whitelaw, C.A., Nierman, W.C., Bhatnagar, D. and Cleveland, T.E. (2004*b*) *Aspergillus flavus* expressed sequence tags for identification of genes with putative roles in aflatoxin contamination of crops. *FEMS Microbiology Letters* 237, 333-340.

Yu, J., Ronning, C.M., Wilkinson, J.R., Campbell, B.C., Payne, G.A., Bhatnagar, D., Cleveland, T.E. and Nierman, W.C. (2007) Gene profiling for studying the mechanism of aflatoxin biosynthesis in *Aspergillus flavus* and *A. parasiticus. Food Additives and Contaminants* 24, 1035-1042.

Yu, J.-H. and Keller, N. (2005) Regulation of secondary metabolism in filamentous fungi. *Annual Review of Phytopathology* 43, 437-458.

Zasloff, M. (1987) Magainins, a class of antimicrobial peptides from *Xenopus* skin: Isolation, characterization of two active forms, and partial cDNA sequence of a precursor. *Proceedings of the National Academy of Sciences USA* 84, 5449-5453.

Zipf, A.E. and Rajasekaran, K. (2003) Ecological impact of Bt cotton. *Journal of New Seeds* 5, 115-135.

23

Breeding Maize for Resistance to Mycotoxins at IITA

Abebe Menkir*, Robert L. Brown, Ranajit Bandyopadhyay, Zhi-yuan Chen and Thomas E. Cleveland

Abstract

Ear-rot-causing fungi, including *Aspergillus* and *Fusarium* spp., are common in maize in West and Central Africa. These fungi contaminate maize with mycotoxins that pose serious potential health hazards to humans in these areas. A collaborative germplasm screening project was initiated between the International Institute of Tropical Agriculture (IITA) and the Southern Regional Research Center (SRRC) of the USDA's Agricultural Research Service in 1998 to develop maize germplasm with resistance to aflatoxin contamination. In a laboratory screen, some IITA inbred lines had potential levels of resistance to aflatoxin production as high as or higher than the best lines from the United States. These results prompted the initiation of a breeding project to combine resistance factors from the IITA lines with resistance factors from the US inbred lines. Several crosses and backcross populations were made from selected resistant or potentially resistant inbred lines from the US and IITA. Sixty-five S_5 lines were developed from the backcross populations and 144 S_5 lines were derived from the F_1 crosses. Kernels from these lines were screened in a laboratory assay. Significant differences in aflatoxin accumulation were detected amongst the lines within each group. Several S5 lines in which aflatoxin contamination was significantly less than in either parent were selected for resistance-confirmation tests. We found pairs of S_5 lines with 88-97% common genetic backgrounds that differed significantly in aflatoxin accumulation. These pairs of lines are being used for proteomic analyses to identify the proteins and the corresponding genes that limit aflatoxin accumulation. We also found significant differences in fumonisin accumulation amongst 58 elite maize inbred lines in which variation in aflatoxin accumulation was found. Both the new inbreds and the elite lines can be exploited as new genetic sources in breeding programs in which the objective is to develop maize cultivars/hybrids that accumulate lesser amounts of mycotoxins.

Introduction

Maize is a major staple food crop widely grown from the wet climate of the coast to the dry Sudanese savannas of West and Central Africa. The most common ear and kernel rot causing fungal species in maize in this area include *Aspergillus*, *Botryodiplodia*, *Diplodia* and *Fusarium* (Fajemisin *et al.*, 1985). Some of these fungi are widespread in different tropical maize growing environments and can cause considerable economic losses by reducing grain quality in maize (Miller, 1996). When two of these fungal species, *Aspergillus flavus* and *Fusarium verticillioides*, infect maize plants, the developing kernels can be contaminated with aflatoxins and fumonisins, respectively, that render the harvested grain unfit for use as human food or animal feed.

Contamination with aflatoxins and fumonisins occur in warm, humid, tropical and subtropical maize-growing environments that are conducive to growth and development of the two fungi (Widstrom, 1996; Kpodo and Bankole, Chapter 9; Siame and Nawa; Chapter 10). Maize contaminated with mycotoxins is a serious problem in Sub-Saharan Africa because most of the maize grain in many countries is used for human consumption and the capacity to monitor the mycotoxin levels in the grain is limited (Widstrom, 1996). The widespread exposure to aflatoxin in Africa has been implicated in the increased incidence of acute toxicosis, liver cancer, and morbidity in children suffering from kwashiorkor (Miller, 1996; Widstrom, 1996; Gong *et al.*, Chapter 6). The consumption of fumonisin-contaminated maize foods also has been associated with outbreaks of esophageal cancer (Rheeder *et al.*, 1992; Thiel *et al.*, 1992) and neural tube birth defects in humans (Stack, 1998). Some reports suggest that various processing methods including roasting, boiling, frying, baking and fermenting, may not effectively eliminate either aflatoxin (Widstrom, 1996) or fumonisin (Hendrich *et al.*, 1993; Voss *et al.*, 1996; Fandohan *et al.*, Chapter 26) from contaminated maize food products.

Several approaches have been proposed for reducing aflatoxin (Widstrom, 1996) and fumonisin (Norred *et al.*, 1991; Hendrich *et al.*, 1993; Voss *et al.*, 1996; Katta *et al.*, 1997) levels in unprocessed grain and processed maize-based food products. One promising strategy is to grow maize cultivars that are resistant to *Aspergillus* and *Fusarium* ear rot and accumulate less mycotoxin in the grain (Widstrom, 1996; Kleinschmidt *et al.*, 2005). Moderate to high levels of resistance to *A. flavus* (Gorman and Kang, 1991; Brown *et al.*, 1999, 2001) and *F. verticillioides* (Widstrom, 1996; Clements *et al.*, 2004) are known in maize. The use of these types of resistance in combination with appropriate cultural practices can reduce the total amount of mycotoxin accumulation. Preharvest host resistance to *A. flavus and F. verticillioides* is a simple and economical technology that leaves no harmful residue in food or the environment, and can be applied over a broad range of environmental and socioeconomic conditions. Although several maize genotypes that accumulate only low levels of aflatoxin and fumonisin have been identified, most of these lines lack desirable agronomic backgrounds, with adaptation problems and the relatively high levels of toxin that can still accumulate sufficing to prevent commercial deployment (Gorman and Kang, 1991; Brown *et al.*, 1999; Clements *et al.*, 2004; Brooks *et al.*, 2005; Kleinschmidt *et al.*, 2005). Thus, there is a need to develop maize germplasm with desirable agronomic traits and reduced levels of mycotoxin contamination. This chapter evaluates the breeding strategies currently used at IITA to develop germplasm resistant to infection by *A. flavus and F. verticillioides*, which includes: (i) field selection for reduced ear rot infection, (ii) breeding for reduced aflatoxin accumulation, and (iii) screening elite germplasm for resistance to specific fungi.

Field selection for reduced ear rot infection

Selecting germplasm resistant to ear rots has a crucial role in IITA's strategy for developing maize germplasm targeted to the forest zones and mid-altitude regions of Central and West Africa. Every year, early generation and advanced breeding lines developed from diverse sources of germplasm, as well as varieties and hybrids, are screened in "hot spot" locations representative of the production zones where ear rot occurs regularly at high levels. Promising materials are selected following visual assessment for reduced levels of ear rot. Environmental factors affect naturally occurring ear rot severity, so the breeding materials are evaluated repeatedly at the "hot spot" locations and at different stages of development to identify materials with reasonable levels of resistance to kernel and ear rot infection. The breeding materials also have been evaluated for tight husk cover, which is the first line of defense against infection by fungi causing ear rot (Widstrom, 1996). The detection of high concentrations of mycotoxins in maize genotypes exhibiting minimal ear rot infection symptoms (Clements *et al.*, 2004; Kleinschmidt *et al.*, 2005) emphasizes the need to select directly for maize germplasm with reduced mycotoxin accumulation.

Breeding for reduced aflatoxin accumulation

Screening for reduced aflatoxin accumulation in parental materials

Effective, reliable, rapid screening techniques are indispensable for breeding for lower levels of aflatoxin accumulation in maize (Gorman and Kang, 1991; Brown *et al.,* 1999). Brown *et al.* (1995) developed a rapid laboratory-based kernel-screening assay that creates high uniform levels of infection and aflatoxin production and enables the differentiation of maize genotypes that accumulate low and high levels of aflatoxins. This assay provides a consistent ranking of maize genotypes in different tests and the results are correlated with results obtained in field trials (Brown *et al.*, 1995).

Seventy-six inbred lines from IITA with moderate to high levels of field resistance to ear rots in the forest zone and the mid-altitudes were evaluated at the Southern Regional Research Center (SRRC)-USDA-ARS laboratory with the kernel screening assay (Brown *et al.*, 2001). Eighteen of these inbred lines had aflatoxin levels that were as low as or lower than those of the best lines from the United States. Further studies with some lines found that the protein profiles of the IITA lines were different from those developed in the United States, suggesting that the IITA lines and those from the United States carry different alleles for the reduction of aflatoxin accumulation. By assessing fungal growth on selected lines, a unique line with low aflatoxin accumulation, but a high level of fungal growth was identified (Brown *et al.*, 2001). This result suggests that toxin accumulation may be inhibited directly in addition to being related to the amount of fungal infection present. The lowest toxin accumulating lines from IITA were crossed with similar genotypes from the United States in a collaborative breeding project (Brown *et al.*, 2003). This strategy increases the probability of developing inbred lines with good agronomic traits that accumulates less toxin than do the currently available commercial lines.

Genetics of resistance to aflatoxin accumulation

The mode of inheritance of resistance to *Aspergillus* ear rot and reduced aflatoxin accumulation in maize grain is not settled. In some studies resistance to ear rot and lower aflatoxin accumulation levels are quantitatively inherited (Walker and White, 2001), with additive gene effects playing a major role the inheritance of resistance (Norred *et al.*, 1991; Miller, 1996; Naidoo *et al.*, 2002). In other studies, dominance has a greater effect on the reduction of aflatoxin accumulation than does additive gene action (Campbell *et al.*, 1997; Campbell and White, 1995; Maupin *et al.*, 2003; Busboom and White, 2004). Broad-sense herita-

Table 1. Mean and range of values (ng/g) for each group of inbred lines evaluated with the kernel screening assay in 2003 and 2004.

Group	Number of lines	Range	Mean ± Standard error
Lines derived from backcrosses			
I	14	6 – 6,000	1,700 ± 420
II	6	0 – 5,200	2,000 ± 670
III	10	360 – 5,100	1,500 ± 530
IV	13	75 – 10,200	1,500 ± 740
V	15	40 – 2,100	930 ± 170
VI	12	150 – 7,900	1,600 ± 660
VII	8	140 – 4,800	2,700 ± 740
Lines derived from tropical × temperate crosses – Set-1			
I	12	890 – 15,000	6,500 ± 1,300
II	10	740 – 14,000	6,500 ± 1,400
III	11	3,000 – 9,100	6,000 ± 530
IV	9	0 – 4,400	1,900 ± 450
V	9	80 – 3,700	2,200 ± 380
VI	9	500 – 9,300	5,300 ± 880
VII	9	390 – 3,600	1,500 ± 310
VIII	7	250 – 5,600	3,300 ± 740
IX	11	230 – 4,600	1,300 ± 430
Lines derived from tropical × temperate crosses – Set-2			
I	9	990 – 14,000	7,900 ± 1,500
II	11	130 – 7,400	2,100 ± 790
III	11	460 – 6,600	2,500 ± 660
IV	11	650 – 9,500	5,000 ± 830
V	12	700 – 12,000	4,200 ± 1,300

bility estimates for both ear rot resistance and lowered aflatoxin levels are moderate to high (Norred *et al.*, 1991; Maupin *et al.*, 2003), suggesting that selection for resistance should be feasible. Significant progress has been made in identifying sources of resistance and understanding their genetic basis, but neither the germplasm nor the genetic information has been used to breed commercially useful maize that accumulates less aflatoxin. The trait's complex of inheritance, the erratic nature of field infection by *A. flavus*, and the year-to-year variability in aflatoxin levels have limited transfer of these traits to elite maize inbred lines since selections made under field conditions often could not be relied on (Gorman and Kang, 1991; Brooks *et al.*, 2005). The development of new efficient tools for screening maize genotypes in both field and laboratory settings increased the number of breeding strategies available for developing resistant maize germplasm that accumulates less toxin (Brown *et al.*, 2003).

Creating populations and developing lines

Both pedigree and backcross breeding methods have been used to develop maize lines with new combinations of agronomic traits and resistance to diseases. Five elite tropical inbred lines from IITA

Table 2. Mean aflatoxin values (ng/g) for a group of inbred lines derived from backcrosses evaluated with the kernel screening assay.

Line	Pedigree	Aflatoxin
L01	P3142 (US Susceptible check)	2,100 a[a]
L02	(MP420 × 4001 × MP420)-2-2-3-1-B	1,700 a
L03	(MP420 × 4001 × MP420)-2-2-3-3-B	1,600 a
L04	MP420 (Recurrent parent)	1,500 a
L05	MP420 × 9450 × MP420-3-1-1-2-B	1,400 ab
L06	MP420 × 9450 × MP420-3-1-1-3-B	1,400 ab
L07	(MP420 × 4001 × MP420)-2-2-3-2-B	1,200 ab
L08	(MP420 × 4001 × MP420)-3-1-3-2-B	800 abc
L09	(MP420 × 4001 × MP420)-3-1-3-1-B	770 abc
L10	(MP420 × 4001 × MP420)-2-2-3-4-B	640 bcd
L11	(MP420 × 4001 × MP420)-3-1-2-1-B	320 cde
L12	MP420 × 9450 × MP420-3-1-1-4-B	300 cde
L13	MI82 (US resistant check)	110 def
L14	(MP420 × 4001 × MP420)-2-1-1-1-B	63 def
L15	(MP420 × 4001 × MP420)-3-1-2-2-B	43 fg

[a]Means followed by the same letter are not significantly different based on the least significant difference test (p = 0.05).

(Babangoyo, KU1414-SR, 1368, 4001, and 9450) were crossed to three or four selected genotypes from the United States (B73×Tex6, C12, GT-MAS:gk, MI82, MO 17×Tex6, MP420, OH516, and T115) that accumulated low levels of aflatoxin (Brown *et al.*, 1995) to form 16 F_1 crosses. Each F_1 cross was crossed with the parental genotype from the United States as a recurrent parent to generate 16 backcross (BC_1) populations. In addition, seven elite IITA inbred lines (Babangoyo, KU1414-SR, 1368, 4001, 5012, 9071, and 9450) were crossed to two or three of the same set of US genotypes to develop 16 F_1 crosses.

Measuring the amount of aflatoxin produced by *A. flavus* in maize is both tedious and expensive, so we could not evaluate aflatoxin production in a large number of individual plants derived from each of the many segregating populations. Aflatoxin production was not assessed until homozygous lines (S_5) were developed following selection for agronomic traits and resistance to foliar diseases during the earlier stages of inbreeding. From 2000 to 2002, self-pollinated ears were selected from each row to develop lines from each BC_1 or F_1 cross. At each stage of inbreeding, visual selection within and among lines was made on the basis of synchrony between pollen shed and silking, low ear placement, well-filled ears and resistance to lodging and foliar diseases, including *Puccinia polysora*, *Bipolaris maydis* and *Curvularia lunata*, under naturally occurring disease pressure at Ibadan, Nigeria. Sixty-five S_5 lines were developed from the backcross populations and 144 S_5 lines were derived from F_1 crosses to be screened with the kernel screening assay (Brown *et al.*, 1995).

Screening lines derived from populations with the kernel screening assay

The 57 S_5 lines derived from the backcross populations were divided into seven groups, each containing three to eleven S_5 lines, the recurrent parent and resistant and a susceptible inbred checks. These groups were screened for reduced aflatoxin accumulation (Brown *et al.*, 1995). The maize inbred lines within each group exhibited a broad range in aflatoxin

Table 3. Mean aflatoxin values (ng/g) for a group of inbred lines derived from tropical × temperate crosses evaluated with the kernel screening assay.

Line	Pedigree	Aflatoxin
TL01	1368	7,400 a[a]
TL02	P3142 (US Susceptible check)	7,200 a
TL03	1368 × MI82-13-1-1-1-B	2,100 ab
TL04	1368 × MI82-23-1-1-2-B	1,500 b
TL05	1368 × MI82-19-4-1-1-B	1,200 bc
TL06	1368 ×MI82-11-2-1-1-B	1,200 bc
TL07	1368 × MI82-23-1-1-1-B	770 cd
TL08	MI82 (US resistant check)	670 cd
TL09	1368 × MI82-23-1-1-3-B	580 cd
TL10	1368 × MI82-11-1-1-1-B	560 cd
TL11	1368 × MI82-17-1-1-1-B	130 d

[a]Means followed by the same letter are not significantly different based on the least significant difference test ($p = 0.05$).

accumulation (Table 1). Inbred lines derived from backcrosses included in Group V and those derived from tropical × temperate crosses included in Group II, which were reported in Table 1, were chosen as examples to provide highlights of the results of the screening assay presented in Tables 2 and 3, respectively. Among the lines included in Group V, five S_5 lines (L10 – L15) accumulated significantly ($p < 0.05$) less aflatoxin than did the recurrent parent from the United States, MP420. L11, L12, L14, and L15 did not differ significantly in aflatoxin accumulation from the resistant US inbred check, MI82. Two pairs of S_5 lines (L02 and L10, and L05 and L12), which were advanced to the S_3 stage of inbreeding from the same single plant, differed significantly ($p < 0.05$) in aflatoxin accumulation (Table 2). Of the 57 S_5 lines evaluated, 23 accumulated significantly ($p < 0.05$) less aflatoxin than did their respective recurrent parent. Some of these lines also had aflatoxin contamination levels similar to or lower than the resistant inbred check from the United States, MI82.

The S_5 lines derived from the F_1 crosses also were divided into groups and screened for aflatoxin accumulation with the kernel screening assay. Significant ($p < 0.05$) differences in aflatoxin production again were detected amongst the lines within each group (Table 1). All of the S_5 lines (TL04 to TL011) differed significantly ($p < 0.05$) in aflatoxin accumulation from the elite tropical parental line, 1368, but not from the resistant inbred check, MI82 (Table 3). We found two pairs of inbred lines (TL04 and TL09 and TL04 and TZ07) that originated from the same single plant at the S_3 stage that had contrasting aflatoxin accumulation levels (Table 3). Thirty-two of the 102 S_5 lines evaluated had significantly lower aflatoxin levels than the elite tropical inbred parent. About half of these lines did not differ significantly in aflatoxin accumulation from the inbred check, MI82.

Selection of genetically similar inbred lines for proteome analysis

We defined a pair of inbred lines from the same backcross or F_1 cross, expected to share at least 88% common genetic background, as genetically similar. Seven pairs of S_5 lines derived from backcrosses and ten pairs of lines extracted from F_1 crosses had 88-97% genetic identity but differed significantly in aflatoxin accumulation. Proteomic analyses of kernel, embryo and

Table 4. Eigenvectors of the first three principal component axes (PC1, PC2 and PC3) for the various traits of 54 maize inbred lines grown at Saminaka and Ikenne in Nigeria in 2004.

Trait	PC1	PC2	PC3
Days to silking	0.11	0.10	0.82****
Plant height (cm)	-0.05	0.68****	-0.03
Ear height (cm)	-0.17*	0.65****	0.01
Husk cover (1-5)[a]	0.33****	-0.10	-0.32**
Plant aspect (1-5)[b]	0.41****	0.24*	-0.33**
Ear aspect (1-5)[c]	0.47****	0.06	0.25*
Ear rot (1-5)[d]	0.48****	-0.05	0.18
Lowland leaf blight (1-5)[e]	0.41****	0.17	-0.13
Lowland leaf rust (1-5)[e]	0.25***	0.00	0.00
Variance	0.39	0.21	0.14

*, ***, **** Significantly different from zero at $p < 0.05$, $p < 0.001$ and $p < 0.0001$ levels, respectively
[a]Husk cover: A scale of 1 to 5, where 1 = very tight husk extending well beyond the ear tip and 5 = exposed ear tip.
[b]Plant aspect (1-5): 1 = excellent overall phenotypic appeal and 5 = poor overall phenotypic appeal.
[c]Ear aspect (1-5): 1 = clean, uniform, large, and well-filled ears and 5 = rotten, variable, small and partially filled ears.
[d]Ear rot (1-5): 1 = little or no visible ear rot and 5 = extensive visible ear rot.
[e]Disease scores recorded at 26 days after mid-silking on a 1-5 scale, where 1 = no visible infection and 5 = severe infection on all leaves.

endosperm proteins associated with lower levels of accumulated aflatoxin has relied on side-by-side comparisons of lines with genotypes with different genetic backgrounds (Chen *et al.,* 2004, *a,b*). The identification of genetically similar lines differing significantly in aflatoxin accumulation should enable the identification of candidate genes underlying resistance to *A. flavus* infection and/or reduction of aflatoxin production without the confounding effects that result when lines of diverse genetic background are compared (Brown *et al.*, 2003). Comparing pairs of genetically similar lines in a proteome analysis, has thus far identified several resistance-associated proteins, categorized as stress-related, and a putative regulatory protein (Brown *et al.*, 2003). The expression of stress-related proteins may enable a plant to defend against fungal invasion under stress conditions. Extensive analysis of kernel endosperm proteins of several pairs of genetically similar lines emanating from our collaborative breeding project are described elsewhere (Brown *et al.*, 2007). This may facilitate identification of potential markers for rapid screening of genetic materials in a breeding program. Proteins identified in such a screen may be useful markers for rapidly screening genetic materials in a breeding program.

Evaluating agronomic performance of selected lines

Inbred lines that accumulate less aflatoxin need to be evaluated for agronomic performance as inbred lines *per se* and as parents of hybrids to determine their usefulness in a breeding program. We screened 54 inbred lines for aflatoxin accumulation under field conditions at Saminaka and Ikenne, Nigeria in 2004. Principal component analysis was used to evaluate the field data and to assess the performance of the new inbred lines on the basis of their agronomic traits. The first principal component axis (PC1) accounted for nearly 40% of the total variation in the data set (Table 4). A high PC1 score was associated with significant reduction in ear height, poor husk cover, plant aspect and ear aspect scores, and increased ear rot, leaf blight and leaf rust infec-

Table 5. Mean aflatoxin values (ng/g) of inbred maize lines selected for low aflatoxin production in the kernel screening assay (KSA) and evaluated under artificial inoculation in the field at Ibadan, Nigeria in 2003 and 2004.

Inbred line	Field trials			KSA
	2003	2004	Mean	
1368	1240	260	750	78
1823	370	96	230	39
TZMI102	1700	130	930	21
TZM104	310	310	310	270
TZMI502	240	93	170	70
Mean ± S.E.	780 ± 290	180 ± 71	440 ± 130	95 ± 43

tions. The second component axis (PC2) explained 21% of the total variation in agronomic traits recorded in this trial and its large scores were associated with taller plants, high ear placement and poor ear aspect scores. The third component axis (PC3) accounted for 14% of the total variation in agronomic traits recorded in this trial and its large scores were associated with a delay in silking, good husk cover and plant aspect scores but with poor ear aspect score. Grain yield of the inbred lines was negatively correlated with PC1 (r = -0.73, $p < 0.0001$) scores but not with PC2 (r = -0.24, p = 0.08) and PC3 (r = -0.12, p = 0.40) scores of the inbred lines. We found some inbred lines with high grain yields and negative PC1 scores, indicating that some of them accumulated less aflatoxin but also had good husk cover, plant aspect and ear aspect scores as well as lower levels of ear rot, leaf blight and leaf rust infections.

Screening elite germplasm for resistance to specific fungi

Evaluating resistant lines under artificial field infection

Field tests of inbred lines identified in the kernel screening assay are critical for identifying the best lines with consistently low levels of aflatoxin accumulation. The ultimate phase in the identification of lines that consistently accumulate less aflatoxins is the exposure of such lines to many populations of *A. flavus* under as wide a range of environmental conditions as possible. Nine inbred lines selected for low levels of aflatoxin production based on the kernel-screening assay were evaluated with artificial inoculation in the field in Nigeria in 2003. Among these lines, 1823, TZMI104 and TZMI502 had low levels of aflatoxin contamination under field conditions (Table 5). These three lines along with 1368 and TZMI102 also were evaluated in 2004. Two of the three inbred lines (1823 and TZMI502) that had low aflatoxin levels in 2003 also had low levels of aflatoxin in 2004, indicating that the laboratory-based kernel screening assay can be used to reliably screen breeding material prior to testing under field conditions (Brown *et al.*, 1999).

Screening elite germplasm for reduced fumonisin production

Mixed infections with *A. flavus* and *F. verticillioides* occur in maize under field conditions. Thus, any strategy to reduce mycotoxin accumulation also should identify sources of resistance to fumonisin accumulation and incorporate them into adapted germplasm. IITA is screening elite maize germplasm for reduced fumonisin accumulation. A replicated field trial of 58 elite inbred lines was conducted at Ibadan, Nigeria in 2003 and 2004, in which emerging silks were artificially inoculated with a spore suspension of an isolate of *F. verticillioides*. At Ikenne in 2003 the response to natural infection was evaluated. The inbred lines differed markedly in fumonisin accumulation at both locations and in both seasons. At Ibadan, mean fumonisin concentration in the grain ranged from 1.1 to 130 μg/g in 2003 and from 0.2 to 99 μg/g in 2004. At Ikenne, mean fumonisin concentration in the grain ranged from 0 to 120 μg/g in 2003. The number of lines at Ibadan with ≤ 5.0 μg/g fumonisin was nine in 2003 and 21 in 2004. Thirty-five inbred lines had < 5 μg/g fumonisin in the grain at Ikenne in 2003. Three inbred lines [(1368/S.A. Pub Lines36/1368)-2-2-2-B, KU1414×ICAL 36-1×KU1414-6-1-B and (CIM 116 × TZMi 302 × CIM 116)-2-2-B] and a commercial hybrid (Oba Super I) had < 5 μg/g fumonisin in the grain in both seasons at Ibadan and in the one season at Ikenne. These results re-emphasize the need to conduct multi-location and multi-season evaluations of genetic materials to identify sources of resistance with consistently low levels of fumonisin accumulation.

Conclusions

Maize inbred lines with consistently low aflatoxin levels after repeated evaluation in the laboratory and in the field could be used in the development of hybrids and synthetic varieties that can be deployed in farmers' fields to help reduce mycotoxin contamination. The new inbred lines also could be used to broaden and to diversify the genetic base of resistant germplasm in maize breeding programs. The advances made in identifying genetically similar lines with contrasting aflatoxin levels may enable the characterization of mechanisms responsible for lower levels of aflatoxin accumulation and the identification of candidate genes that underlie resistance to *A. flavus* infection and/or aflatoxin accumulation, access to novel variants, and the development of markers for rapid screening of breeding materials. Promising lines with low fumonisin accumulation also could be used in crosses with inbred lines that accumulate lower levels of aflatoxin to develop new lines with combined resistance to both fungal species and to lower the contamination levels of both of these very important mycotoxins.

References

Brooks, T.D., Williams, W.P., Windham, G.L., Willcox, M.C. and Abbas, H.K. (2005) Quantitative trait loci contributing resistance to aflatoxin accumulation in the maize inbred Mp313E. *Crop Science* 45,171-174.

Brown, R.L., Chen, Z.Y., Cleveland, T.E. and Russin, J.S. (1999) Advances in the development of host resistance in corn to aflatoxin contamination by *Aspergillus flavus*. *Phytopathology* 89,113-117.

Brown, R.L., Chen, Z.Y., Menkir, A., Cleveland, T.E., Cardwell, K., Kling, J. and White, D.G. (2001) Resistance to aflatoxin accumulation in kernels of maize Inbreds selected for ear rot resistance in West and Central Africa. *Journal of Food Protection* 64, 396-400.

Brown, R.L., Chen, Z.Y., Menkir, A. and Cleveland, T.E. (2003) Using biotechnology to enhance host resistance to aflatoxin contamination of corn. *African Journal of Biotechnology* 2, 557-562.

Brown, R.L., Cleveland, T.E., Payne, G.A., Woloshuk, C.P., Campbell, K.W. and White, D.G. (1995) Determination of resistance to aflatoxin production in maize kernels and detection of fungal colonization using an *Aspergillus flavus* transformant expressing *Escherichia coli* β-glucuronidase. *Phytopathology* 85, 983-989.

Brown, R.L., Menkir, A., Bandyopadhyay, R., Cleveland, T.E., and Chen, Z. (2007) Comparative proteomics of near-isogenic maize inbred lines to identify potential aflatoxin-resistance markers. *Phytopathology* 97, S14.

Busboom, K.N. and White, D.G. (2004) Inheritance of resistance to aflatoxin production and *Aspergillus* ear rot of corn from the cross of inbreds B73 and Oh516. *Phytopathology* 94, 1101-1115.

Cambell, K.W., Hamblin, A.M. and White, D.G. (1997) Inheritance of resistance to aflatoxin production in the cross between corn inbreds B73 and L31. *Phytopathology* 87, 1144-1147.

Campbell, K.W. and White, D.G. (1995) Evaluation of corn genotypes for resistance to *Aspergillus* ear rot, kernel infection and aflataxion production. *Plant Disease* 79, 1039-1045.

Chen, Z.Y., Brown, R.L., Cleveland, T.E. and Damann, K.E. (2004*a*) Investigating the roles of an aflatoxin resistance-associated protein in maize using RNAi. *Phytopathology* 94, S18

Chen, Z.Y., Brown, R.L., Damann, K.E. and Cleveland, T.E. (2004*b*) Identification of a maize kernel stress-related protein and its effect on aflatoxin accumulation. *Phytopathology* 94, 938-945.

Clements, M.J., Maragos, C.M., Pataky, J.K. and. White, D.G. (2004) Sources of resistance to fumonisin accumulation in grain and *Fusarium* ear and kernel rot of corn. *Phytopathology* 94, 251-280.

Fajemisin, J., Efron, Y., Kim, S.K., Khadr, F.H., Dabrowski, Z.T., Mareck, J.H., Bjarnason, M., Parkinson, V., Everett, L.A. and Diallo, A. (1985) Population and varietal development in maize for tropical Africa through resistance breeding approach. In: Brandolini, A. and Salamini, F. (eds.). *Breeding Strategies for Maize Production Improvement in the Tropics.* FAO and Istituto Agronomico per l'Oltremare, Florence, Italy, pp. 385-407.

Gorman, D.P. and Kang, M.S. (1991) Preharvest aflatoxin contamination in maize: Resistance and genetics. *Plant Breeding* 107, 1-10.

Hendrich, S., Miller, K.A., Wilson, T.M. and Murphy, P.A. (1993) Toxicity of *Fusarium proliferatum*-fermented nixtamalized corn-based diets fed to rats: Effect of nutritional status. *Journal of Agricultural and Food Chemistry* 41, 1649-1654.

Katta, S.K., Cagampang, A.E., Jackson, L.S and Bullerman, L.B. (1997) Distribution of *Fusarium* molds and fumonisins in dry-milled corn fractions. *Cereal Chemistry* 74, 858-863.

Kleinschmidt, C.E., Clements, M.J., Maragos, C.M., Pataky, J.K. and White, D.G. (2005) Evaluation of food-grade dent corn hybrids for severity of *Fusarium* ear rot and fumonisin accumulation in grain. *Plant Disease* 89, 291-297.

Maupin, L.M., Clements, M.J. and White, D.G. (2003) Evaluation of the M182 corn line as a source of resistance to aflatoxin in grain and use of BGYF as a selection tool. *Plant Disease* 87, 1059-1066.

Miller, J.O. (1996) Mycotoxins. In: Cardwell, K.F. (ed.) *Proceedings of the Workshop on Mycotoxins in Foods in Africa (6-10 November 1995, Cotonou, Benin)*. IITA, Ibadan, Nigeria, pp. 18-22.

Naidoo, G., Forbes, A.M., Paul, C., White, D.G. and Rocheford, T.R. (2002). Resistance to *Aspergillus* ear rot and aflatoxin accumulation in maize F_1 hybrids. *Crop Science* 42, 360-364.

Norred, W.P., Voss, K.A, Bacon, C.W. and Riley, R.T. (1991) Effectiveness of ammonia treatment in detoxification of fumonisin-contaminated corn. *Food and Chemical Toxicology* 29, 815-819.

Rheeder, J.P., Marasas, W.F.O, Thiel, P.G., Sydenham, E.W., Shephard, G. S. and Schalkwyk, D.J.V. (1992) *Fusarium moniliforme* and fumonisins in corn in relation to human esophageal cancer in Transkei. *Phytopathology* 82, 353-357.

Stack, M.E. (1998) Analysis of fumonisin B_1 and its hydrolysis product in tortillas. *Journal of the Association of Official Analytical Chemists International* 81, 737-740.

Thiel, P.G., Marasas, W.F.O., Sydenham, E.W., Shephard, G.S. and Gelderblom, W.C.A. (1992) The implications of naturally occurring levels of fumonisins in corn for human and animal health. *Mycopathologia* 117, 3-9.

Voss, K.A., Bacon, C.W., Meredith, F.I. and Norred, W.P. (1996) Comparative subchronic toxicity studies of nixtamalized and water-extracted *Fusarium moniliforme* culture material. *Food and Chemical Toxicology* 34, 623-632.

Walker, R.D. and White, D.G. (2001) Inheritance of resistance to *Aspergillus* ear rot and aflatoxin production of corn from C12. *Plant Disease* 85, 322-327.

Widstrom, N.W. (1996) The aflatoxin problem with corn grain. *Advances in Agronomy* 56, 219-280.

24

Etiology and Management of Aflatoxin Contamination

Peter J. Cotty*, Claudia Probst and Ramon Jaime-Garcia

Abstract

Aflatoxins are potent poisons that contaminate crops in warm regions worldwide and reduce health and economic welfare in several portions of Africa. Crops are contaminated in two phases: (i) *Aspergillus* species infect crops during development; and (ii) after maturation contamination builds during exposure to warm humid conditions. Identification of the exact fungi causing contamination can provide clues to management strategies. Crops usually are infected by complex mixtures of aflatoxin-producing and closely related fungi. Among these are atoxigenic strains that produce no aflatoxins. In the United States atoxigenic strains are used to reduce contamination. Such technologies also may have value in Africa.

Introduction

Aflatoxins are a group of highly toxic, cancer-causing chemicals produced by several members of the fungal genus *Aspergillus*. The presence of these mycotoxins in human foods can cause acute and chronic health effects (aflatoxicoses) including immune-system suppression, growth retardation, cancer, and death (Wild and Turner, 2002; Gong *et al.*, 2004; Williams *et al.*, 2004; Azziz-Baumgartner *et al.*, 2005). Aflatoxins are carcinogens and genotoxins that directly influence the structure of DNA (Williams *et al.*, 2004) and, as a result, occurrence of aflatoxins in human foods is strictly regulated to very low concentrations in developed countries. Indeed, in developed countries the exposure of domestic animals, even pets, is of both regulatory and economic concern. Deaths of pets due to aflatoxins in U.S. pet foods has had international economic impact in terms of both trade and litigation (Anonymous, 2006). Thus, in developed countries, the drive to abate aflatoxin contamination is due to loss in crop value resulting from stringent government regulations on maximum permissible levels in crops and crop related products used as foods or feeds. In crops intended for human consumption, maximum permitted aflatoxin levels range from 2 ng/g in the European Union to 20 ng/g in the United States. Aflatoxins are readily transferred from feed to milk resulting in similarly stringent regulations on feed intended for dairies (van Egmond, 2004, Wu, 2004). Maximum permissible levels of aflatoxins in milk are 0.05 ng/g in the European Union and 0.5 ng/g in the United States.

The requisite destruction of highly contaminated agricultural products combined with reduced value for products with lower contamination levels makes aflatoxin economically expensive in developed countries. Contamination may limit the economic viability of agriculture in some regions and, in others, reduces the acreage on which susceptible crops, *e.g.*, maize and peanuts, may be grown (Wu, 2004). In the United States, areas with severe contamination may yield crops with > 500 ng/g total aflatoxins (Jaime-Garcia and Cotty, 2003). However, in developing countries the contamination of crops with aflatoxin leads not only to economic losses, but also has a tremendous impact on human health. In Africa, a continent that relies on vulnerable crops such as peanuts and maize as dietary staples, aflatoxin contamination causes major health problems (Shephard, 2003). People in rural areas may have no option but to consume contaminated crops on a daily basis. This moderate, chronic intake of aflatoxin via food can lead to severe pathological conditions, including liver cancer, immune system deficiency and impaired development of children (Wild *et al.*, 1992; Wild *et al.*, 1993; Gong *et al.*, 2004; Williams *et al.*, 2004). Malnutrition, a common condition in rural Africa, increases disease prevalence and further reduces the ability of the human body to cope with aflatoxin exposure. Chronic aflatoxin poisoning reduces life expectancy.

Acute aflatoxin poisoning is caused by ingestion of high levels of the toxin. Immediate consequences are severe liver damage, acute jaundice and hepatitis, which subsequently may result in death (Bennett and Klich, 2003; Shephard, 2003; Williams *et al.*, 2004). Although on a global basis deaths from acute aflatoxin poisoning are rare, Kenya has experienced such episodes repeatedly for at least 25 years. In 2004, 317 cases of acute aflatoxicosis were reported, resulting in 125 deaths (case fatality rate = 39%), with additional cases probably unreported. This epidemic was caused by ingestion of maize with aflatoxin concentrations of up to 4,400 ng/g (Anonymous, 2004; Azziz-Baumgartner *et al.*, 2005).

Unfortunately, in an African setting, crops from small scale farmers frequently pass from field to storage to consumption with no regulatory oversight and without a test of the extent of aflatoxin contamination. Aflatoxin contamination often is mysterious to farmers because the extent of contamination is not readily evident and because it appears unrelated to crop yield or quality. Indeed, complex and expensive sampling and analyses often are required to estimate the extent of contamination (Whitaker and Johansson, 2005). The fungi that produce aflatoxins grow best under warm conditions and therefore, aflatoxins are of greatest concern in warm agricultural production areas especially during dry periods (Cotty *et al.*, 1994). Such areas of high vulnerability are common in parts of Africa where subsistence farmers frequently rely on contaminated maize and peanuts as life-sustaining staples (Egal *et al.*, 2005; Hell *et al.*, 2003). Aflatoxin contamination varies in most areas and crops (Wilson and Payne, 1994). This variation has been attributed to climatic factors, especially drought and high temperature, in maize (Cole *et al.*, 1982; Wilson and Payne, 1994; Widstrom, 1996) and peanuts (Cole *et al.*, 1982, 1989; Wilson and Payne, 1994) with increased contamination being associated with reduced rainfall. However, in areas like Arizona and South Texas in the United States, increased contamination also is associated with exposure of the mature crop to warm temperatures and increased humidity provided by irrigation and/or rain (Bock and Cotty, 1999; Cotty, 2001; Jaime-Garcia and Cotty, 2003).

Management

Prevention or management of aflatoxin contamination may be directed at both the process of contamination and the fungi causing contamination. The contamination process can be divided into two phases based on crop maturity (Cotty, 2001). The first phase occurs during crop development and is generally associated with physical damage to the crop typically by either physiologic stress or insect activity (Russell, 1982; Cotty, 2001). Crop components contaminated during the first phase often fluoresce a bright green-yellow as a result of kojic acid production in crop tissue by the aflatoxin-producing fungi (Zeringue *et al.*, 1999).

After maturation, the crops remain vulnerable to contamination, providing a window during which a second phase of contamination may occur (Bock and Cotty, 1999; Cotty, 2001). Exposure of the mature crop to both high humidity and temperatures conducive to aflatoxin producing fungi can result in both new crop infections and increases in the aflatoxin content of crop components already infected (Russell *et al.*, 1976; Cotty, 1991). The second phase may occur prior to harvest in the field or after harvest during transportation, storage, or at any point until the crop is consumed.

Hot dry conditions during crop development favor the first phase of contamination, whereas rain and high humidity with warm temperatures after crop maturation favor the second phase. Reliable management practices must address both phases. Improving the resistance of cultivars to contamination is one method of simultaneously addressing both phases of contamination. Although proper cultivar selection and crop management can limit vulnerability to both phases, environmental changes can better even the best management practices and result in a highly contaminated crop (Wilson and Payne, 1994; Cotty *et al.*, 2001).

When management procedures fail to prevent accumulation of unacceptable levels of aflatoxins, there are still options for the utilization of the contaminated crops. These options include detoxification. Chemical detoxification is a viable option for even very highly contaminated crops, with ammoniation the detoxification method currently in the widest use. Ammoniation inactivates aflatoxins by hydrolysis of the lactone ring, which is followed by further breakdown. Ammoniation has been used in North America, Europe, and Africa on crops including maize, cottonseed, and peanut meal (Park *et al.*, 1988; Bailey *et al.*, 1994). Following detoxification by ammoniation, the treated crop products are nutritionally valuable for domestic animals, but are not suitable for human consumption.

Etiology

Plant pathologists generally consider establishing the etiology, or cause, of a plant disease problem an initial step in developing management strategies for the problem. Since the establishment by Anton de Bary that potato late blight was caused by *Phytophthora infestans* and the formulation by Robert Koch of rules for establishing the cause of infectious disease, plant pathologists have drawn insight from improved knowledge of disease etiology to establish and improve disease management (Agrios, 2004). A clear understanding of disease etiology enables efficient screening for improved host resistance, identification of chemical pesticides toxic to causative agents, and development of biological control strategies that utilize less problematic organisms, *e.g.*, saprophytes, epiphytes, endophytes, and even less damaging pathogens, to minimize the impact of disease through a variety of mechanisms. Management procedures for prevention of aflatoxin-contamination frequently are directed at either controlling the environment, *i.e.*, either storage conditions or crop management are altered (Turner *et al.*, 2005) or reducing host susceptibility, *i.e.*, insect damage is reduced or crop barriers to infection are increased (Draughon and Ayres, 1981; Dowd, 1992). To direct management at the etiologic agent(s), *i.e.*, the fungus(i) producing aflatoxin, the contaminating fungi present must be characterized. The process of identifying the most important aflatoxin producers can be complex. Members of the species that produce aflatoxins vary widely in their aflatoxin producing ability, with some aflatoxin producers being of little or no concern while others are of vital interest (Schroeder and Boller, 1973; Lisker *et al.*, 1993; Cotty, 1997).

The communities of aflatoxin-producing fungi resident in agricultural and native ecosystems have a complexity that reflects the diverse geography and numerous substrates and hosts in which these species are found both across Africa and elsewhere (Cotty *et al.*, 1994). Aflatoxin-producing fungi occur in many regions of the world, but they are most commonly associated with agriculture in warm production areas. A few aflatoxin-producing fungi outside of *Aspergillus* section *Flavi* have been described (Cary *et al.*, 2005); however, the role of these species in the contamination of crops is not clear. Similarly, the extent to which some of the species within *Aspergillus* section *Flavi* contribute to aflatoxin contamination of crops also is unclear, as few episodes of contamination are attributed to either *Aspergillus nomius* (Cotty *et al.*, 1994), *Aspergillus bombycis* or *Aspergillus pseudotamarii*. This lack of attribution may be due to the relatively recent description of the latter two species and to their apparently low incidences in some crop environments (Ito *et al.*, 2001; Peterson *et al.*, 2001).

Aspergillus flavus and *A. parasiticus* are the most commonly implicated causal agents of aflatoxin contamination, with *A. flavus* by far the most common (Cotty *et al.*, 1994). *Aspergillus flavus* may be divided into two distinct morphotypes, the S and L strains (Cotty, 1989). Each morphotype is composed of many clonal lineages (called vegetative compatibility groups or VCGs) defined by a vegetative compatibility system that limits gene flow between dissimilar individuals (Papa, 1986; Bayman and Cotty, 1991*a*,*b*). Both morphotypes and VCGs differ in many characteristics; the most frequently studied of which is aflatoxin-producing ability. The S strain, on average, produces much higher concentrations of aflatoxins than does the L strain (Cotty, 1989, 1997; Garber and Cotty, 1997). Consequently, if the S strain commonly infects a vulnerable crop, this morphotype is a primary target for management of aflatoxin contamination. Members of different L-strain VCGs vary widely in their ability to produce aflatoxins. Members of some L-strain VCGs produce very large amounts of aflatoxins, while the members of many other L-strain VCGs produce very little, if any, aflatoxins. Isolates that produce no aflatoxins at all are termed "atoxigenic".

Communities of aflatoxin-producing fungi are complex and composed of multiple strains and VCGs. Although, these fungal communities are complex, the proportion of strains and VCGs with different aflatoxin-producing abilities varies widely among communities resident in different fields, valleys, and regions (Cotty, 1997). Consequently, the average aflatoxin-producing ability of those communities varies as well. Contamination events are not caused by individual specific fungi but by complex communities of aflatoxin-producing fungi that may contain several species, multiple morphotypes, and strains that belong to numerous VCGs (Bayman and Cotty, 1991*b*; Horne and Green, 1995; Doster *et al.*, 1996).

If all organisms that produce aflatoxins are fungi in the genus *Aspergillus*, how much more specificity is needed to manage aflatoxin contamination? Understanding the interaction between the diverse aflatoxin-producing fungi resident in soils, on plants, and throughout the environment with crop contamination can result in improved management strategies. Specific knowledge of the etiology of contamination is a first step, but the etiology of specific contamination events often is difficult to determine, since contamination events are not caused by individual fungi, but rather by complex communities of fungi that partially

Table 1. Percent of infection and aflatoxin contamination caused by the S and L strains of *Aspergillus flavus* in commercial cotton seed in Arizona, *i.e.* in 1990 11% of the crop was infected by only the S strain, but this 11% of the crop contained 81% of the aflatoxin present in the entire crop. Measurements were made on a seed by seed basis. Data from Cotty (1996).

Infecting Strain	% Infections				% Aflatoxin Content			
	1990	1991	1993	Mean	1990	1991	1993	Mean
L only	71	69	57	66	8	14	13	12
S only	11	10	17	13	81	27	63	57
S and L	18	20	25	21	11	58	24	31

reflect those distributed throughout the crop's extended environment. Within these communities not all fungi are equal. Individual strains may vary in crop preference and the ability to ramify through and rot crop tissues (Cotty, 1989; Brown *et al.*, 1992; Shieh *et al.*, 1997). The divergence in aflatoxin-producing ability is striking, as some isolates produce much higher levels of aflatoxins than others. It is not unusual to find crop components that range in aflatoxin content from < 10 ng/g to several million ng/g (Lee *et al.*, 1990; Mellon and Cotty, 2004). This range partially reflects the diversity present amongst the infecting fungi. Thus relatively minor, in number, components of the infecting fungal community may have major roles in determining the ultimate quantity of aflatoxins in the crop.

Work with the S strain of *A. flavus* illustrates both the potential importance of etiological knowledge and how minor components of infecting fungal communities can be the most important etiological agents of contamination. The S strain is of great importance both in North America and Africa (Cotty, 1989, 1997; Jaime-Garcia and Cotty, 2006*a*,*b*; Probst *et al.*, 2007). The apparent importance of the S strain as a causative agent can be erroneously minimized when researchers either overlook S strain isolates or when they preferentially select "typical", or L strain, isolates during primary isolations from crops. S strain isolates can appear either fluffy white, without sclerotia or spores, early in isolation or primarily black from abundant small sclerotia late in isolation. In Arizona, the S strain infects a relatively low proportion of cotton seed and yet causes the vast majority of the aflatoxin contamination (Table 1; Cotty, 1996). Thus, although the L strain of *A. flavus* is the most common strain infecting the cotton seed, it is not the most important cause of contamination. The S strain also is the most important etiological agent of aflatoxin contamination in South Texas, even though S strain isolates make up only a minor portion of the *Aspergillus* section *Flavi* propagules on harvested crops (Jaime-Garcia and Cotty, 2006*a*). The S strain morphotype also was the primary cause of the aflatoxin contamination events that resulted in hundreds of deaths in Kenya in 2004 (Probst *et al.*, 2007).

Development of cultivars with reduced susceptibility to aflatoxin contamination is a strategy for limiting contamination that is applicable to many crops. However, selection of the fungi used in the resistance screens does not typically include evaluation of the most important causal agents as they may vary by region (Jaime-Garcia and Cotty, 2006*b*), and it is not uncommon for researchers to request isolates from other regions, crops, and even continents for such screens. Differential virulence to hosts is well characterized within many fungal species (Agrios, 2004), and it is reasonable to question the assumption that resistance to one strain of *A. flavus* implies resistance to all strains of *A. flavus*. Characterization of

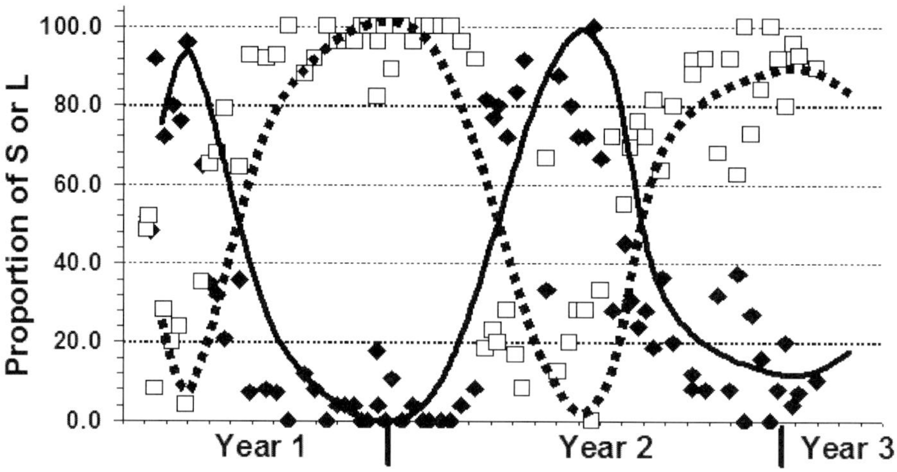

Figure 1. Proportion of the overall *A. flavus* community composed of the S and L strains in the air over agricultural fields in the Sonoran Desert. The most competitive strain varies with season. Redrawn from Bock *et al.* (2004). ♦ – % S strain observed, □ – % L strain observed.

the most important causal agents might result in cultivar screens that increase host resistance levels. S strains of *A. flavus* are an important cause of contamination in several areas where contamination is a perennial problem (Cotty, 1989, 1997; Jaime-Garcia and Cotty, 2006*a,b*), yet there are relatively few host resistance screens that incorporate S strain isolates.

The S and L strains of *A. flavus* are adapted to distinct ecological niches. This adaptation can be seen in the size and production habit of sclerotia, the variation in hydrolase production (Cotty *et al.*, 1990; Mellon and Cotty, 2004), the distribution by soil type (Jaime-Garcia and Cotty, 2006*b*), and even in seasonality (Fig. 1; Orum *et al.*, 1997; Bock *et al.*, 2004). In agricultural regions of the Sonoran desert conditions favoring the S and L strains differ, with S strains dominating during the warmest periods and L strains dominating during the winter and spring (Fig. 1). Similar responses to the environment may exist in African deserts including regions bordering the Sahara (Cardwell and Cotty, 2002). Thus, control measures optimized for one environment may not work in others, with management strategies utilizing cultural, chemical, and/or biological control methods particularly sensitive to their environments.

Under some conditions crop rotations and crop mixtures can influence the composition of *A. flavus* communities in soils, the average aflatoxin producing potential of resident fungi, and, the vulnerability of crops to contamination (Jaime-Garcia and Cotty, 2006*b*). It is not clear how cropping systems influence fungal communities and whether such influences relate to crop infection, crop debris, or other aspects of the cropping process. However, the results do suggest that by identifying the most important etiologic agents and the influences of agronomic practices on the prevalence of these agents, cropping systems can be altered to favor less toxigenic but similarly adapted fungi, and, in so doing, reduce the aflatoxin-producing potential of fungal communities to which susceptible crops are exposed.

Use of atoxigenic strains of *A. flavus* to limit contamination

Chemical examination of seeds infected by either the S or the L strains indicated that seed infected by the S strain alone had much higher aflatoxin content than seed coinfected by both S and L strain isolates. This phenomenon has been observed repeatedly: co-infecting fungi modulate aflatoxin production by each other (Cotty, 1990; Cotty and Bhatnagar, 1994; Garber and Cotty, 1997). Infection with a highly toxigenic strain of *A. flavus* alone results in much more toxin than co-infection by both a high toxin producer and a low toxin producer. Sometimes even a relatively low incidence of the low toxin producer can greatly reduce the extent of contamination. This phenomenon was an important factor in determining which Kenyan maize became contaminated with toxic levels of aflatoxins during the 2004 aflatoxicosis epidemic (Probst *et al.*, 2007). Modulation of aflatoxin biosynthesis by coinfecting strains has led to the development of techniques that exploit this phenomenon to reduce aflatoxin contamination. These procedures utilize the application of atoxigenic strains of *A. flavus* as biological control agents directed at minimizing aflatoxin contamination. Thus, by detailed exploration of the etiology of aflatoxin contamination, some members of the most important species causing contamination were shown not only not to make aflatoxins but to be useful as tools for limiting aflatoxin contamination.

Two observations were central to the initial development of atoxigenic strains as tools for limiting aflatoxin contamination. First, in the 1980s, investigations of the etiology of contamination resulted in the discovery that neither isolate pathogenicity nor the ability of a fungal isolate to ramify through crop tissues were related to aflatoxin-producing ability (Cotty, 1989). Thus, the aflatoxin-producing ability of an *A. flavus* strain is unrelated to its success during crop colonization, and, in theory, isolates that do not produce aflatoxins might be effective competitors of aflatoxin producers during crop infection. Some atoxigenic strains can competitively exclude aflatoxin producers during the infection of crop tissues (Cotty and Bayman, 1993), and, in so doing, markedly reduce or eliminate aflatoxin production by highly toxigenic strains during co-infection (Cotty and Bhatnagar, 1994; Cotty, 1990; Garber and Cotty, 1997). Increasing the frequency of this natural interference with seed contamination by atoxigenic strains would reduce the extent to which crops become contaminated. Second, the fungal community resident at one location may differ considerably in aflatoxin-producing potential from the fungal community at second location (Joffe, 1969; Schroeder and Boller, 1973; Cotty, 1997). Therefore, communities with lower aflatoxin-producing potentials exist and reductions in aflatoxin-producing potential might be induced by certain agronomic practices. Crops infected by fungal communities with relatively low aflatoxin-producing potential should be less contaminated with aflatoxin than crops infected by communities with high aflatoxin-producing abilities. If the aflatoxin-producing potential of fungal communities resident in an area could be reduced, then the extent to which crops in that area become contaminated also should be reduced. Breeders and other researchers of aflatoxin contamination routinely practice this principle when they spread substrates colonized by highly toxigenic isolates of *A. flavus* or *A. parasiticus* throughout test plots to increase the extent and uniformity of crop aflatoxin content (Batson *et al.*, 1997; Holbrook *et al.*, 2000; Betran *et al.*, 2002). Field tests have shown that application of atoxigenic strains in a similar manner could be used to reduce the aflatoxin-producing potential of fungal communities resident on crops and that these applications could have far reaching influences on the fungi resident in the environment (Fig. 2: Antilla and Cotty, 2002; Cotty and Antilla, 2003).

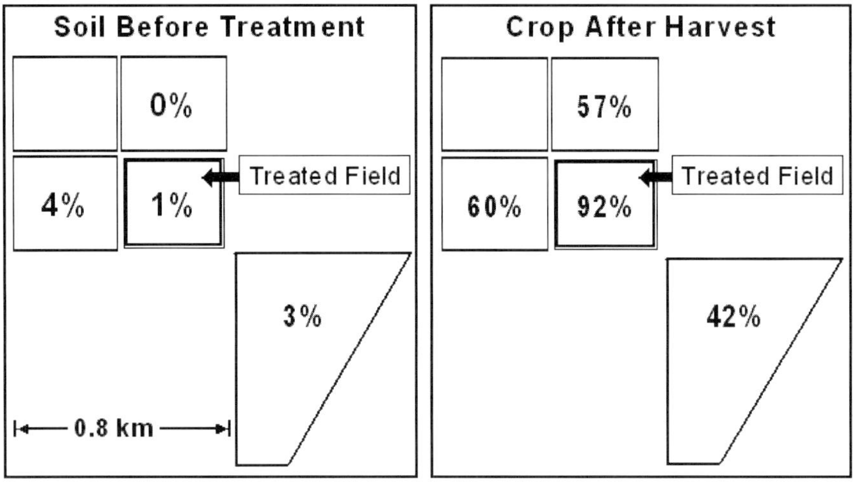

Figure 2. Proportion of the overall *A. flavus* community composed of the applied atoxigenic strain (AF36) in the soil prior to application and on the harvested crop. Application (9 kg of colonized wheat seed per hectare) was made to a single 16 hectare field. AF36 is native and occurred in soil prior to treatment. Application influences extended to untreated adjacent fields. Data are the result of 564 vegetative compatibility analyses. Redrawn from Cotty and Antilla (2003).

Strategies that use atoxigenic strains of *Aspergillus flavus* as biological control agents directed at limiting the production of aflatoxins exploit both the ability of certain atoxigenic strains to modulate aflatoxin biosynthesis during crop infection and the ability of atoxigenic strain applications to displace aflatoxin producers throughout the crop environment and in so doing, reduce the frequency and extent of crop infection by aflatoxin producers (Cole and Cotty, 1990; Cotty *et al.*, 1994; Dorner, 2004). These strategies seek to competitively exclude aflatoxin producers from crops and thus reduce both the incidence of aflatoxin producers in the environment and the level of aflatoxin contamination (Cotty and Antilla, 2003).

Aflatoxin prevention technologies based on atoxigenic strains of *A. flavus* and *A. parasiticus* are being developed for several crops in diverse agricultural systems (Doster *et al.*, 2002; Cotty and Antilla, 2003; Dorner, 2004). Most strategies apply relatively low amounts of *A. flavus* to a food source on which the fungus reproduces and from which distribution to secondary food sources and the crop occurs. Besides innate competitive ability, the applied atoxigenic strains usually have advantages provided by management practices that allow improved competition over aflatoxin-producers resident in the field. These management practices include having atoxigenic strains arrive with a food source formulated for utilization only by the applied strain, and applying the strains to the top of the soil, which eliminates the need for atoxigenics to escape the soil matrix to colonize above-ground crops. Applications are timed to ensure that the environment will support growth and reproduction by the atoxigenic strain and that the resident aflatoxin-producing strains have not previously multiplied and colonized the crop to such an extent that applications would be futile. In theory, application of an atoxigenic strain when overall *A. flavus* levels are low provides preferential exposure to the crop and an advantage in competing for crop resources. A collaboration between the International Institute of Tropical Agriculture, the University of Bonn, and the U.S. Department of Agriculture's Agricultural Research Service (USDA-ARS) is selecting useful atoxigenic strains from crops grown in Africa and developing strategies to apply atoxigenic strain technology in Africa.

Atoxigenic strains are considered biopesticides and, as such, the use of atoxigenic strains must comply with applicable pesticide laws. Two atoxigenic strains currently have pesticide registrations in the United States. Atoxigenic strains have been used most extensively on cotton crops in Arizona. Development of atoxigenic strains in Arizona has been through a collaborative partnership between the Arizona Cotton Research and Protection Council (ACRPC), a farmer run organization, and the USDA-ARS (Cotty and Antilla, 2003, Antilla and Cotty, 2002). This collaboration has resulted in the construction of a facility for the production of an atoxigenic strain product that is run by ACRPC. Commercial cotton in Arizona has been treated since 1996, first under an experimental use registration and, since 2002, under a full section 3 pesticide registration. All material used to treat fields in Arizona, California, and Texas since 1999 has been produced at the ARS-ACRPC facility with over 20,000 acres per year treated since 2002 (Cotty and Antilla, 2003). In the United States, atoxigenic strains are applied mechanically either by airplane or tractor; however, these procedures should be readily adaptable to smaller scale operations. Treatments increase the incidence of the applied atoxigenic strain in *A. flavus* communities associated with the crop and the soil. This increase reduces both the average aflatoxin-producing potential of the communities and the quantity of aflatoxins in the crop (Cotty and Antilla, 2003). Treatments do not increase the overall quantity of *A. flavus* on the crop at harvest or after ginning. There is an inverse relationship between the incidence of the applied strain and the concentration of aflatoxin in the crop (Cotty, 1994).

Changes to the average aflatoxin producing potential of *A. flavus* communities induced by atoxigenic strain applications typically last for several years. Cumulative benefits are expected due to effects on untreated fields adjacent to treated fields (Fig. 2) and to repeated treatments through sequences of crop rotation. Atoxigenic strain applications may provide area-wide benefits. Many farmers hope that by reducing the aflatoxin-producing potential of *A. flavus* communities throughout an area, the vulnerability of all of the crops grown in the area may be reduced, as it is common to cultivate multiple susceptible crops in the same region, *e.g.*, peanuts and maize. Atoxigenic strain technologies provide the hope of addressing all of these contamination issues with a single technology and in so doing reduce the burden of contamination over entire areas not just one crop within that area.

Atoxigenic strains, like aflatoxin-producing fungi, become associated with crops in the field during crop production. These fungi remain with the harvested crops after harvest and in storage. Since crop contamination with aflatoxins may occur in the field, in storage or anytime until the crop is consumed, if conditions are conducive for fungal growth, *e.g.*, high humidity and high temperature, then crop infection and contamination will continue as well. Like their aflatoxin-producing relatives, atoxigenic strains also move into storage with the crop and provide residual protection in transport, storage, and processing until consumption (Brown *et al.*, 1991; Dorner and Cole, 2002). Crops infected in the field and already contaminated at harvest accumulate less aflatoxin in storage when treated with an atoxigenic strain just prior to entering storage (Brown *et al.*, 1991). However, postharvest applications lack many advantages of field applications when the atoxigenic strains multiply in competition with the aflatoxin producers and become associated with the crop before extensive infection by aflatoxin producers.

Atoxigenic strain technology provides an opportunity to reduce the overall risk of contamination during all phases of aflatoxin contamination including in the field during crop development, in storage or at any other time after harvest until the mature crop is eventually utilized. Atoxigenic strains are but one example of how improved knowledge of both the contamination process and the etiologic agents can result in improved methods for limiting human exposure to aflatoxins.

References

Agrios, G.N. (2004) *Plant Pathology*, 5th ed. Academic Press, New York.

Anonymous. (2004) Outbreak of aflatoxin poisoning – eastern and central provinces, Kenya, January-July 2004. *Morbidity and Mortality Weekly Report (MMWR)* 53, 790-793.

Anonymous (2006) Diamond Pet Foods narrows recall after aflatoxin tests. *Journal of the American Veterinary Medical Association* 228, 490.

Antilla, L. and Cotty, P.J. (2002) The ARS-ACRPC partnership to control aflatoxin in Arizona: Current status. *Mycopathologia* 155, 64.

Azziz-Baumgartner, E., Lindblade, K.,Gieseker, K., Schurz Rogers, H., Kieszak, S., Njapau, H., Schleicher, R., McCoy, L.F., Misore, A., DeCock, K., Rubin, C., Slutsker, L. and the Aflatoxin Investigative Group. (2005) Case-control study of an acute aflatoxicosis outbreak – Kenya-2004. *Environmental Health Perspectives* 113, 1779-1783.

Bailey, G.S., Price, R.L., Park, D.L. and Hendricks, J.D. (1994) Effect of ammoniation of aflatoxin B_1-contaminated cottonseed feedstock on the aflatoxin M_1 content of cows' milk and hepatocarcinogenicity in the trout bioassay. *Food and Chemical Toxicology* 32, 707-715.

Batson, W.E., Caceres, J., Cotty, P.J. and Isakeit, T. (1997) Aflatoxin levels in cottonseed at weekly intervals in Arizona, Mississippi and Texas modules. *Proceedings of the Beltwide Cotton Production Research Conferences. (January, 1997 New Orleans, LA)*, pp. 116-118.

Bayman, P. and Cotty, P.J. (1991*a*) Improved media for selecting nitrate-nonutilizing mutants in *Aspergillus flavus*. *Mycologia* 83, 311-316.

Bayman, P. and Cotty, P.J. (1991*b*) Vegetative compatibility and genetic diversity in the *Aspergillus flavus* population of a single field. *Canadian Journal of Botany* 69, 1707-1711.

Bennett, J.W. and Klich, M. (2003) Mycotoxins. *Clinical Microbiology Reviews* 16, 497-516

Betran, F.J., Isakeit, T. and Odvody, G. (2002) Aflatoxin accumulation of white and yellow maize inbreds in diallel crosses. *Crop Science* 42, 1894-1901.

Bock, C.H., and Cotty, P.J. (1999) The relationship of gin date to aflatoxin contamination of cottonseed in Arizona. *Plant Disease* 83, 279-285.

Bock, C.H., Mackey, B. and Cotty, P.J. (2004) Population dynamics of *Aspergillus flavus* in the air of an intensively cultivated region of southwest Arizona. *Plant Pathology* 53, 422-433.

Brown, R.L., Cotty, P.J. and Cleveland, T.E. (1991) Reduction in aflatoxin content of maize by atoxigenic strains of *Aspergillus flavus*. *Journal of Food Protection* 54, 623-626.

Brown, R.L., Cleveland, T.E., Cotty, P.J. and Mellon, J.E. (1992) Spread of *Aspergillus flavus* in cotton bolls, decay of intercarpellary membranes, and production of fungal pectinases. *Phytopathology* 82, 462-467.

Cardwell, K.F. and Cotty, P.J. (2002) Distribution of *Aspergillus flavus* Section *Flavi* among soils from the four agroecological zones of the Republic of Benin, West Africa. *Plant Disease* 86, 434-439.

Cary, J.W., Klich, M.A. and Beltz, S.B. (2005) Characterization of aflatoxin-producing fungi outside of *Aspergillus* section *Flavi*. *Mycologia* 97, 425-432.

Cole, R.J. and Cotty, P.J. (1990) Biocontrol of aflatoxin production by using biocompetitive agents. In Robens, J., Huff, W. and Richard, J. (eds.) *A Perspective on Aflatoxin in Field Crops and Animal Food Products in the United States: A Symposium; ARS-83*. U.S. Department of Agriculture, Agricultural Research Service, Washington, D.C., pp. 62-66.

Cole, R.J., Hill, R.A., Blankenship, P.D., Sanders, T.H. and Garren, K.H. (1982) Influence of irrigation and drought stress on invasion by *Aspergillus flavus* of corn kernels and peanut pods. *Developments in Industrial Microbiology* 23, 229-236.

Cole, R.J., Sanders, T.H., Dorner, J.W. and Blankenship, P.D. (1989) Environmental conditions required to induce preharvest aflatoxin contamination of groundnut: Summary of six years research. In McDonald, D. and Mehan, V.K. (eds.) *Aflatoxin Contamination of Groundnut: Proceedings of an International Workshop.* ICRISAT, Patancheru, India, ICRISAT, pp. 279-287.

Cotty, P.J. (1989) Virulence and cultural characteristics of two *Aspergillus flavus* strains pathogenic on cotton. *Phytopathology* 79, 808-814.

Cotty, P.J. (1990) Effect of atoxigenic strains of *Aspergillus flavus* on aflatoxin contamination of developing cottonseed. *Plant Disease* 74, 233-235.

Cotty, P.J. (1991) Effect of harvest date on aflatoxin contamination of cottonseed. *Plant Disease* 75, 312-314.

Cotty, P.J. (1994) Influence of field application of an atoxigenic strain of *Aspergillus flavus* on the population of *A. flavus* infecting cotton bolls and on the aflatoxin content of cottonseed. *Phytopathology* 84, 1270-1277.

Cotty, P.J. (1996) Aflatoxin contamination of commercial cottonseed caused by the S strain of *Aspergillus flavus*. *Phytopathology* 86, S71.

Cotty, P.J. (1997) Aflatoxin-producing potential of communities of *Aspergillus* section *Flavi* from cotton producing areas in the United States. *Mycological Research* 101, 698-704.

Cotty, P.J. (2001) Cotton seed losses and mycotoxins. In: Kirkpatrick, T.L. and Rothrock, C.S. (eds.) *Compendium of Cotton Diseases. Part 1. Infectious Diseases.* 2nd ed. APS Press, St. Paul, Minnesota, pp. 9-13.

Cotty, P.J. and Antilla, L. (2003) *Managing Aflatoxins in Arizona.* United States Department of Agriculture, Agricultural Research Service, New Orleans, Louisiana.

Cotty, P.J. and Bayman, P. (1993) Competitive exclusion of a toxigenic strain of *Aspergillus flavus* by an atoxigenic strain. *Phytopathology* 83, 1283-1287.

Cotty, P.J. and Bhatnagar, D. (1994) Variability among atoxigenic *Aspergillus flavus* strains in ability to prevent aflatoxin contamination and production of aflatoxin biosynthetic pathway enzymes. *Applied and Environmental Microbiology* 60, 2248-2251.

Cotty, P.J., Cleveland, T.E., Brown, R.L. and Mellon, J.E. (1990) Variation in polygalacturonase production among *Aspergillus flavus* isolates. *Applied and Environmental Microbiology* 56, 3885-3887.

Cotty, P.J., Bayman, P., Egel, D.S. and Elias, K.S. (1994) Agriculture, aflatoxins and *Aspergillus*. In Powell K.A., Renwick, A. and Peberdy, J.F. (eds.) *The Genus* Aspergillus*: From Taxonomy and Genetics to Industrial Application.* Plenum Press, New York, pp. 1-27.

Cotty, P.J., Jaime-Garcia, R. and Kobbeman, K. (2001) The S strain of *A. flavus* in South Texas. *Phytopathology* 91, S19.

Dorner, J.W. (2004) Biological control of aflatoxin contamination of crops. *Journal of Toxicology - Toxin Reviews* 23, 425-450.

Dorner, J.W. and Cole, R.J. (2002) Effect of application of nontoxigenic strains of *Aspergillus flavus* and *A. parasiticus* on subsequent aflatoxin contamination of peanuts in storage. *Journal of Stored Products Research* 38, 329-339.

Doster, M.A., Michailides, T.J. and Morgan, D.P. (1996) *Aspergillus* species and mycotoxins in figs from California orchards. *Plant Disease* 80, 484-489.

Doster, M., Michailides, T., Cotty, P.J., Holtz, B., Bentley, W., Morgan, D. and Boeckler, L. (2002) Aflatoxin control in pistachios: Removal of contaminated nuts, ecological relationships, and biocontrol. *Mycopathologia* 155, 44.

Dowd, P.F. (1992) Insect interaction with mycotoxin-producing fungi and their hosts. In: Bhatnagar, D., Lillehoj, E.B. and Arora, D.K. (eds.) *Mycotoxins in Ecological Systems.* Marcel Dekker, New York, pp. 137-155.

Draughon, F.A. and Ayres, J.C. (1981) Inhibition of aflatoxin production by selected insecticides. *Applied and Environmental Microbiology* 41, 972-976.

Egal, S., Hounsa, A., Gong, Y.Y., Turner, P.C., Wild, C.P., Hall, A.J., Hell, K. and Cardwell, K.F. (2005) Dietary exposure to aflatoxin from maize and groundnut in young children from Benin and Togo, West Africa. *International Journal of Food Microbiology* 104, 215-224.

Garber, R.K. and Cotty, P.J. (1997) Formation of sclerotia and aflatoxins in developing cotton balls infected by the S strain of *Aspergillus flavus* and potential for biocontrol with an atoxigenic strain. *Phytopathology,* 87, 940-945.

Gong, Y.Y., Hounsa, A., Egal, S., Turner, P.C., Sutcliffe, A.E., Hall, A.J., Cardwell, K. and Wild, C.P. (2004) Postweaning exposure to aflatoxin results in impaired child growth: A longitudinal study in Benin, West Africa. *Environmental Health Perspectives* 112, 1334-1338.

Hell, K., Cardwell, K.F. and Poehling, H.M. (2003) Relationship between management practices, fungal infection and aflatoxin for stored maize in Benin. *Journal of Phytopathology* 151, 690-698.

Holbrook, C.C., Kvien, C.K., Rucker, K.S., Wilson, D.M., Hook, J.E. and Matheron, M.E. (2000) Preharvest aflatoxin contamination in drought-tolerant and drought-intolerant peanut genotypes. *Peanut Science* 27, 45-48.

Horne, B.W. and Green, R.L. (1995) Vegetative compatibility within populations of *Aspergillus flavus, A. parasiticus*, and *A. tamarii* from a peanut field *Mycologia* 87, 324-332.

Ito, Y., Peterson, S.W., Wicklow, D.T. and Goto, T. (2001) *Aspergillus pseudotamarii*, a new aflatoxin producing species in *Aspergillus* section *Flavi*. *Mycological Research* 105, 233-239.

Jaime-Garcia, R. and Cotty, P.J. (2003) Aflatoxin contamination in commercial cottonseed in South Texas. *Phytopathology* 93, 1190-1200.

Jaime-Garcia, R. and Cotty, P.J. (2006*a*) Spatial distribution of *Aspergillus flavus* and its toxigenic strains on commercial cottonseed from South Texas and its relationship to aflatoxin contamination. *Plant Pathology* 55, 358–366.

Jaime-Garcia, R. and Cotty, P.J. (2006*b*) Spatial relationships of soil texture and crop rotation to *Aspergillus flavus* community structure in South Texas. *Phytopathology* 96, 599-607.

Joffe, A.Z. (1969) Aflatoxin produced by 1,626 isolates of *Aspergillus flavus* from groundnut kernels and soils in Israel. *Nature* 221, 492.

Lee, L.S., Wall, J.H., Cotty, P.J. and Bayman, P. (1990) Integration of enzyme-linked immunosorbent assay with conventional chromatographic procedures for quantitation of aflatoxin in individual cotton bolls, seeds, and seed sections. *Journal of AOAC* 73, 581-584.

Lisker, N., Michaeli, R. and Frank, Z.R. (1993) Mycotoxigenic potential of *Aspergillus flavus* strains isolated from groundnuts growing in Israel. *Mycopathologia* 122, 177-183.

Mellon, J.E. and Cotty, P.J. (2004) Expression of pectinase activity among *Aspergillus flavus* isolates from southwestern and southeastern United States. *Mycopathologia* 157, 333-338.

Orum, T.V., Bigelow, D.M., Nelson, M.R., Howell, D.R. and Cotty, P.J. (1997) Spatial and temporal patterns of *Aspergillus flavus* strain composition and propagules density in Yuma County, Arizona, soils. *Plant Disease* 81, 911-916.

Papa, K.E. (1986) Heterokaryon incompatibility in *Aspergillus flavus*. *Mycologia* 78, 98-101.

Park, D.L., Lee, L.S., Price, R.L. and Pohland, A.E. (1988) Review of the decontamination of aflatoxin by ammoniation: Current status and regulation. *Journal of AOAC* 71, 685-703.

Peterson, S.W., Ito, Y., Horn, B.W. and Goto, T. (2001) *Aspergillus bombycis*, a new aflatoxigenic species and genetic variation in its sibling species, *A. nomius*. *Mycologia* 93, 689-703.

Probst, C., Njapau, H. and P. J. Cotty, P.J. (2007) Outbreak of an acute aflatoxicosis in Kenya 2004: Identification of the causal agent. *Applied and Environmental Microbiology* 73, 2762-2764.

Russell, T.E. (1982) *Aflatoxins in Cotton Seed*. Publication Q422, University of Arizona Cooperative Extension Service, Tucson, Arizona.

Russell, T.E., Watson, T.F. and Ryan, G.F. (1976) Field accumulation of aflatoxin in cottonseed as influenced by irrigation termination dates and pink bollworm infestation. *Applied and Environmental Microbiology* 31, 711-713.

Schroeder, H.W. and Boller, R.A. (1973) Aflatoxin production by species and strains of the *Aspergillus flavus* group isolated from field crops. *Applied Microbiology* 25, 885-889.

Shephard, G.S. (2003) Aflatoxin and food safety: Recent African perspective. *Journal of Toxicology* 22, 267-286.

Shieh, M.T., Brown, R.L., Whitehead, M.P., Cary, J.W., Cotty, P.J., Cleveland, T.E. and Dean, R.A. (1997) Molecular genetic evidence for the involvement of a specific polygalacturonase, P2c, in the invasion and spread of *Aspergillus flavus* in cotton bolls. *Applied and Environmental Microbiology* 63, 3548-3552.

Turner, P.C., Sylla, A., Gong, Y.Y., Diallo, M.S., Sutcliffe, A.E., Hall, A.J. and Wild, C.P. (2005) Reduction in exposure to carcinogenic aflatoxins by postharvest intervention measures in west Africa: A community-based intervention study. *The Lancet* 365, 1950-1956.

van Egmond, H.P. (2004) Natural toxins: risks, regulations and the analytical situation in Europe. *Analytical and Bioanalytical Chemistry* 378, 1152-1160.

Whitaker, T.B. and Johansson, A.S. (2005) Sampling uncertainties for the detection of chemical agents in complex food matrixes. *Journal of Food Protection* 68, 1306-1313.

Widstrom, N.W. (1996) The aflatoxin problem with corn grain. *Advances in Agronomy* 56, 219-280.

Wild, C.P., Hudson, G.J., Sabbioni, G. Chapot, B., Hall, A.J., Wogan, G.N., Whittle, H., Montesano, R. and Groopman, J.D. (1992) Dietary-intake of aflatoxins and the level of albumin-bound aflatoxin in peripheral-blood in the Gambia, West Africa. *Cancer Epidemiology, Biomarkers and Prevention* 1, 229-234.

Wild, C.P., Jansen, L.A., Cova, L. and Montesano, R. (1993) Molecular dosimetry of aflatoxin exposure: Contribution to understanding the multifactorial etiopathogenesis of primary hepatocellular carcinoma with particular reference to hepatitis B virus. *Environmental Health Perspectives* 99, 115-122.

Wild, C.P. and Turner, P.C. (2002) The toxicology of aflatoxins as a basis for public health decisions. *Mutagenesis* 17, 471-481.

Williams, J.H., Phillips, T.D., Jolly, P.E., Stiles, J.K., Jolly, C.M. and Aggarwal, D. (2004) Human aflatoxicosis in developing countries: a review of toxicology, exposure, potential health consequences, and interventions. *American Journal of Clinical Nutrition* 80, 1106-1122.

Wilson, D.M. and Payne, G.A. (1994) Factors affecting *Aspergillus flavus* group infection and aflatoxin contamination of crops. In Eaton, D.L. and Groopman, J.D. (eds.) *The Toxicology of Aflatoxins. Human Health, Veterinary, and Agricultural Significance.* Academic Press, San Diego, California, pp. 309-329.

Wu, F. (2004) Mycotoxin risk assessment for the purpose of setting international regulatory standards. *Environmental Science and Technology* 38, 4049-4055.

Zeringue, H.J.J., Shih, B.J., Maskos, K., and Grimm, D. (1999) Identification of the bright-greenish-yellow-fluorescence (BGY-F) compound on cotton lint associated with aflatoxin contamination in cottonseed. *Phytochemistry* 52, 1391-1397.

25

NovaSil™ Clay for the Management of Dietary Aflatoxins in Human Populations

Evans Afriyie-Gyawu, Jonathan H. Williams, Henry J. Huebner, Nii-Ayi Ankrah, David Ofori-Adjei, Pauline E. Jolly, Jia-Sheng Wang and Timothy D. Phillips*

Abstract

Cost-effective interventions that reduce human diseases and death from aflatoxin exposure are high priorities particularly for developing nations. A calcium montmorillonite clay, NovaSil™, when mixed at low inclusion rates with animal feeds can act as a selective enterosorbent for aflatoxins. NovaSil™ sequesters aflatoxins in the gastrointestinal tract and neutralizes their toxic effects by reducing the bioavailability of the toxins in the blood and other organs. These findings are especially significant for developing countries since clay-based strategies for managing aflatoxicosis would be practical, culturally acceptable, sustainable and unique in reducing external exposure and risk. When included in the diet of animals at levels up to 2.0% (w/w), NovaSil™ clay did not interfere with vitamin or micronutrient utilization. These results also support the hypothesis that NovaSil™ clay interventions in developing countries could have a major impact on health and well-being of susceptible humans who are highly and frequently exposed to these dietary toxins without compromising their nutritional status. Recent work has confirmed the relative safety of NovaSil™ in a long-term rodent study at and in a Phase I Adverse Events trial. No significant NovaSil™-related effects were observed in either study, and the conclusions were that NovaSil™ should be tested in a Phase II human trial in Ghana. A study population in Ghana has been identified that is exposed to high levels of dietary aflatoxins based on biomarkers in their blood and urine samples. This study will evaluate the consequences of NovaSil™ treatment on aflatoxin exposure over a 3-month period. In summary, enterosorption strategies, based on dietary NovaSil™ clay, hold great promise for the management of aflatoxins in high-risk human populations especially in developing countries. The remedy is novel, inexpensive and easily disseminated.

Introduction

Aflatoxins are a group of carcinogenic mycotoxins produced primarily by the fungi *Aspergillus flavus* and *Aspergillus parasiticus*. These toxins occur frequently in agricultural commodities including cereal grains, cottonseed, peanuts and tree nuts (CAST, 2003). These compounds are heat stable and can survive a variety of food processing procedures; thus, aflatoxins can occur as "unavoidable" contaminants in many foods and feeds, particularly those derived from maize and peanuts. Of the four naturally occurring aflatoxins (B_1, B_2, G_1, and G_2), aflatoxin B_1 is the most toxic and has been shown to interact with genes involved in carcinogenesis (McMahon *et al.*, 1986; 1987; Smela *et al.*, 2001) and tumor suppression (Aguilar *et al.*, 1993). Several studies also suggest that aflatoxins impair the cellular and humoral immune system, and that chronic low level exposure to these food contaminants may result in suppression of the immune system and increased susceptibility to disease (Rodricks and Stoloff, 1977; Miller *et al.*, 1978; Richards *et al.*, 1978; Pestka and Bondy, 1994; Hinton *et al.*, 2003; Turner *et al.*, 2003). In addition, dietary aflatoxins also reduce significantly hepatic vitamin A in chickens (Pimpukdee *et al.*, 2004). In the United States, the action level of aflatoxins in foods intended for human use is 20 ng/g. Although many countries have regulatory limits for aflatoxin in foods/feeds, outbreaks of poisoning still occur. For instance, Kenya (with an aflatoxin regulatory level of 20 ng/g) recently experienced an outbreak of aflatoxin poisoning in humans with a case-fatality rate of 39%, which was reportedly linked to consumption of maize containing aflatoxin levels as high as 8,000 ng/g (CDC, 2004).

Intervention strategies for the aflatoxin problem

Due to the significantly negative impact of aflatoxins on human and animal health, particularly in developing countries, intervention strategies that are effective yet economically feasible are needed. Several intervention strategies are available, including diverse physical, chemical, and biological methods, for managing aflatoxins in agricultural commodities, the simplest of which requires isolation and destruction of the contaminated source. This approach, however, often is not practical since alternative food supplies may not be available, or replacement supplies may not be affordable. Thus, the use of detoxification and decontamination procedures has become a viable option for remediating aflatoxin-contaminated foodstuffs. Optimally, remediation procedures should: (i) remove, inactivate or effectively destroy the mycotoxin from the commodity; (ii) not result in the deposition of toxic substances, metabolites, or by-products in the commodity; (iii) retain nutritional value and food/feed acceptability of the commodity; (iv) not cause significant alterations in the physical properties of the commodity; and (v) destroy fungal spores, if possible (Park *et al.*, 1988). Importantly, detoxification or decontamination procedures should be readily available, easily utilized and cost-effective (CAST, 2003).

A practical approach of current interest for the prevention of aflatoxicosis is the incorporation of non-nutritive clay minerals in contaminated food/feed to sorb aflatoxins in the gastrointestinal tract, thus reducing toxin bioavailability and distribution in the blood and other target organs (Phillips *et al.*, 1995; Phillips, 1999). This dietary consumption of clay, *i.e.*, geophagy, has been observed for centuries in both human and animal populations worldwide (Carretero *et al.*, 2002). Potential health benefits of geophagy presumably include enterosorption, *i.e.*, binding the toxic agent in the gastrointestinal tract, which decreases the bioavailability of toxic chemicals and hazardous microbes from both contaminated foods and water.

Animal studies with mycotoxins and calcium montmorillonite clay

In previous enterosorbent studies with aflatoxins, NovaSil™, a processed calcium montmorillonite clay, bound aflatoxin B_1 with high affinity and high capacity both *in vitro* and *in vivo*. When added to the diet, NovaSil™ significantly protected broiler and Leghorn chicks from the toxic effects of dietary aflatoxin exposure (Phillips *et al.*, 1988). Following these initial findings, the efficacy of NovaSil™ and similar calcium montmorillonite clays for aflatoxins was confirmed in multiple animal species including pregnant rodents (Mayura *et al.*, 1998), chickens (Phillips *et al.*, 1988; Kubena *et al.*, 1990; Pimpukdee *et al.*, 2004), turkeys (Kubena *et al.*, 1991), swine (Lindemann *et al.*, 1993), and lambs (Harvey *et al.*, 1991*a*). NovaSil™ also decreases the bioavailability of radiolabeled aflatoxins and reduces aflatoxin residues in poultry (Davidson *et al.*, 1987), rats (Sarr *et al.*, 1995), and pigs (Beaver *et al.*, 1990). Aflatoxin M_1, an oxidative metabolite of aflatoxin B_1, formed in the milk of lactating dairy cattle and goats was reduced significantly when NovaSil™ was incorporated into contaminated diets for the animals (Harvey *et al.*, 1991*b*; Smith *et al.*, 1994).

NovaSil™ is selective for aflatoxins since, when included in the diet of animals, NovaSil™ did not significantly reduce the toxicity of other common (and structurally diverse) mycotoxins, *i.e.*, zearalenone, deoxynivalenol, T-2 toxin, ochratoxin A, cyclopiazonic acid, ergotamine, and fumonisins. For example, the use of clay in mink fed zearalenone helped to reduce fetotoxicity, but did not alter the hyperestrogenic effects (Bursian *et al.*, 1992). Also, supplementation of swine diets with NovaSil™ clay at 0.5 and 1.0% w/w did not influence the average daily gain of pigs exposed to deoxynivalenol. In poultry studies, the inclusion of NovaSil™ clay in the diet as an enterosorbent did not significantly prevent the adverse effects of T-2 toxin (Kubena *et al.*, 1990), ochratoxin A (Huff *et al.*, 1992), cyclopiazonic acid (Dwyer *et al.*, 1997), and fumonisins (Lemke, 2000).

In these short-term animal studies, no observable adverse effects were reported following ingestion of the dietary NovaSil™ clay. Prior to chronic animal and short-term human studies, NovaSil™ was analyzed for concentrations of various environmental contaminants, including dioxins/furans and heavy metals to insure compliance with federal and international standards. A more recent study in which Sprague-Dawley rats ingested NovaSil™ clay at dietary concentrations as high as 2% throughout pregnancy showed neither maternal nor fetal toxicity, and did not significantly alter trace metal bioavailability in a variety of tissues (Wiles *et al.*, 2004). These findings suggest that NovaSil™ clay may be useful in enterosorbent therapies for the management of aflatoxicosis in high-risk human populations.

Prior to human feasibility and/or intervention trials utilizing NovaSil™, the potential adverse effects of NovaSil™ clay following chronic dietary ingestion need to be established in animals. We used a rodent model to evaluate the relative safety of chronic exposure to NovaSil™ clay via the diet. Male and female Sprague-Dawley rats were fed rations containing 0, 0.25, 0.5, 1.0 and 2.0% NovaSil™ clay *ad libitum* for 6.5 months. Rats treated with 0.25-2% NovaSil™ clay in the diet had neither dose-dependent nor NovaSil™-related adverse effects on body weight gains, feed conversion ratios, relative organ weights, gross anatomy, histological appearance of major organs, hematology or serum biochemistry parameters. Levels of selected essential nutrients, including Fe, Zn and vitamins A and E, were unaffected (Table 1; Afriyie-Gyawu *et al.*, 2005).

Table 1. Vitamin and mineral concentrations in serum and liver samples from Sprague-Dawley rats following dietary ingestion of 0-2.0% NovaSil™ for 6.5 mo; after Afriyie-Gyawu *et al.* (2005).

Parameter	Treatment Group				
	Control	0.25%	0.5%	1.0%	2.0%
Male					
[a]Serum VA (µg/l)	570 ± 25	570 ± 17	530 ± 18	540 ± 22	560 ± 219
[a]Serum VE (mg/l)	13.4 ± 1.2	13.2 ± 1.3	11.5 ± 0.5	11.8 ± 1.2	12.0 ± 1.3
[a]Serum Fe (µg/dl)	137 ± 8.9	153 ± 8.7	131 ± 12	180 ± 12*	151 ± 4.9
[a]Serum Zn (mg/l)	1.29 ± 0.04	1.37 ± 0.04	1.35 ± 0.05	1.38 ± 0.06	1.22 ± 0.05
[b]Hepatic VA (µg/g)	29 ± 3.2	28 ± 2.6	27 ± 0.9	27 ± 1.3	29 ± 1.7
[b]Hepatic VE (µg/g)	395 ± 2.8	393 ± 3.7	385 ± 5.0	407 ± 3.4	393 ± 2.4
Female					
[a]Serum VA (µg/l)	250 ± 6.0	230 ± 8.3	250 ± 9.9	300 ± 11*	290 ± 13
[a]Serum VE (mg/l)	14.5 ± 1.2	13.6 ± 0.6	13.8 ± 0.6	14.7 ± 1.0	15.7 ± 0.7
[a]Serum Fe (µg/dl)	321 ± 14	309 ± 18	344 ± 24	337 ± 23	344 ± 27
[a]Serum Zn (mg/l)	1.05 ± 0.03	1.11 ± 0.03	1.16 ± 0.04	1.11 ± 0.04	1.11 ± 0.06
[b]Hepatic VA (µg/g)	60 ± 3.9	61 ± 4.6	62 ± 4.7	51 ± 5.1	54 ± 5.7
[b]Hepatic VE (µg/g)	482 ± 14	471 ± 1.7	469 ± 13	469 ± 16	481 ± 42

[a]Data are reported as mean values (± SEM).
[b]Data represent mean values (± SEM) of three randomly selected liver samples per treatment/sex .
*Indicates statistical difference compared to controls ($p \leq 0.05$).

Short-term human studies with NovaSil™

A two-week short-term safety evaluation of NovaSil™ was conducted with healthy human volunteers (Wang *et al.*, 2005). This phase I clinical study was designed to determine short-term safety and tolerance of NovaSil™ in normal human subjects. The NovaSil™ was sterilized at 121°C and packaged into capsules. The NovaSil™ capsules were produced under sterile conditions according to U.S. Good Manufacturing Practices.

The overall design of the human study followed the guidelines for a randomized, double-blind phase I clinical trial. Fifty adults who met the recruiting standards were voluntarily enrolled in the study and divided into two groups. The low-dose group took three capsules of NovaSil™ (0.5 g) three times a day for two weeks. The high-dose group took three capsules of NovaSil™ (1.0 g) three times a day for two weeks. All capsules were the same color and size. The two dose levels were extrapolated from dosimetry data based on animal studies (Phillips, 1999; Phillips *et al.*, 2002; Afriyie-Gyawu *et al.*, 2005). This dosimetry protocol seems suitable for future long-term human intervention studies in larger populations.

Both doses of NovaSil™ used in this study were tolerated by all study participants. Gastrointestinal adverse effects were noticed in 6/25 subjects, in the 1.5 g/day group and by 7/25 subjects in the 3.0 g/day group. Symptoms included bloating, constipation, diarrhea, flatulence, and abdominal discomfort. Two participants in the 1.5 g/day group reported some dizziness, an effect that was not seen in the 3.0 g/day group. All symptoms described were recorded within the first two days after taking the NovaSil™ capsules and no symp-

Table 2. Analysis of minerals and vitamins in serum obtained from human subjects following ingestion of NovaSil™ for 14 days; after Wang *et al.* (2005).

Mineral or Vitamin	Group			
	Low-dose (1.5 g/day)		High-dose (3.0 g/day)	
	Before	After	Before	After
Ca (mg/l)	91 ± 4.8[a]	87 ± 11	88 ± 13	91 ± 3.6
Cu (mg/l)	1.01 ± 0.24	0.93 ± 0.21	1.18 ± 0.57	1.17 ± 0.52
Fe (mg/l)	1.05 ± 0.31	1.06 ± 0.44	1.05 ± 0.46	1.10 ± 0.40
K (mg/l)	168 ± 16	164 ± 22	160 ± 25	167 ± 11
Mg (mg/l)	19.6 ± 1.4	18.6 ± 2.6	19.3 ± 2.7	19.6 ± 1.3
Mn (ng/l)	3.9 ± 2.1	3.2 ± 0.9	4.4 ± 5.4	3.1 ± 0.4
Na (mg/ml)	3.12 ± 0.10	3.03 ± 0.35	3.03 ± 0.40	3.15 ± 0.08
Ni (ng/l)	6.8 ± 3	6.8 ± 3	9.3 ± 17	8.2 ± 5.5
P (mg/l)	122 ± 14	111 ± 18	121 ± 25	128 ± 18
Pb (ng/l)	3.4 ± 5.1	3.8 ± 0.33	1.8 ± 1.4	1.9 ± 1.3
S (mg/ml)	1.10 ± 0.08	1.01 ± 0.14*	1.03 ± 0.15	1.05 ± 0.05
Se (mg/l)	0.11 ± 0.01	0.10 ± 0.02	0.11 ± 0.01	0.11 ± 0.01
Si (mg/l)	2.74 ± 0.31	2.83 ± 0.57	3.26 ± 0.84	3.38 ± 0.75
Sr (ng/l)	55 ± 18	74 ± 17 **	54 ± 24	96 ± 24 **
Zn (mg/l)	1.2 ± 0.2	1.3 ± 0.5	1.1 ± 0.3	1.2 ± 0.3
Vitamin A (μmol/l)	2.33 ± 0.42[a]	2.35 ± 0.47	2.48 ± 0.81	2.59 ± 0.63
Vitamin E (μmol/l)	20.7 ± 6.7	19.2 ± 5.9	22.0 ± 6.8	22.5 ± 5.9

[a]mean ± SD; * $p < 0.05$, ** $p < 0.01$ as compared to values before the study.

toms (or complaints) were recorded thereafter. Statistically, no significant difference was observed for these adverse symptoms between the low- and the high-dose groups. All reported side-effects, with one exception, were assessed to be mild, and there was no significant difference between the two treatment groups (Wang *et al.*, 2005).

Thus, the consumption of NovaSil™ capsules at 1.5-3.0 g/day by healthy human subjects for 14 days was relatively safe based on biochemical and hematological parameters and physical examinations. Several parameters, such as red blood cell counts, hemoglobin, total protein, albumin, ALT, and sulfur, decreased significantly in blood samples collected after treatment in the low-dose group, however, none of the parameters were significantly different from normal in blood samples from the high-dose group. All of the observed changes were within the normal range of clinical references. Some clay minerals are postulated to sorb vitamins. In this study there were no statistical differences in serum levels of vitamins A and E after treatment with either dose of NovaSil™ (Table 2). Thus, NovaSil™ can bind aflatoxins but does not interact with vitamins A or E. There were no significant differences in the levels of the minerals analyzed, with two exceptions: lower inorganic sulfur concentration in the low-dose group and higher strontium concentrations in both groups (Table 2). The clinical significance of these findings is not yet known (Wang *et al.* 2005).

Conclusions

In conclusion, clay-based enterosorbents that are selective for aflatoxins, *e.g.*, NovaSil™, offer a practical, beneficial, and economically feasible solution for the aflatoxin problem, particularly in developing countries. Notably, NovaSil™ is unique in that it diminishes the bioavailability of aflatoxins from the gastrointestinal tract, thus minimizing exposure and health risks to the consumer. Both the chronic animal and short-term human feasibility studies are consistent with the hypothesis that NovaSil™ clay in human diets could prevent or significantly reduce exposure to aflatoxins and reduce or eliminate the adverse effects known to occur in humans that chronically consume aflatoxin-contaminated foods. The relative safety and efficacy of dietary NovaSil™ clay in the animal models, coupled with the results of the short-term human feasibility trial, served as the basis for initiating a three-month phase II intervention trial of dietary NovaSil™ clay in human subjects. Clay-based enterosorbents, and other sorbent materials that may be added to the diet, must be rigorously tested in appropriate animal models to determine their potential for nutrient interactions and toxicity. Importantly, all enterosorbents should be routinely evaluated to confirm their efficacy and safety before being considered for possible human applications.

Acknowledgments

Supported in part by USAID LAG-G-00-96-90013-00 and NIEHS P42-ES04917.

References

Afriyie-Gyawu, E., Mackie, J., Dash, B., Wiles, M., Taylor, J., Huebner, H., Tang, L., Guan, H., Wang, J.-S., and Phillips, T.D. (2005) Chronic toxicological evaluation of dietary NovaSil clay in Sprague-Dawley rats. *Food Additives and Contaminants* 22, 259-269.

Aguilar, F., Hussain, S.P. and Cerutti, P. (1993) Aflatoxin B_1 induces the transversion of G $\rightarrow$ T in codon 249 of the *p53* tumor suppressor gene in human hepatocytes. *Proceedings of the National Academy of Sciences (USA)* 90, 8586-8590.

Beaver, R.W., Wilson, D.M., James, M.A., Haydon, K.D., Colvin, B.M., Sangster, L.T., Pikul, A.H. and Groopman, J.D. (1990) Distribution of aflatoxins in tissues of growing pigs fed an aflatoxin-contaminated diet amended with a high affinity aluminosilicate sorbent. *Veterinary and Human Toxicology* 32, 16-18.

Bursian, S.J., Aulerich, R.J., Cameron, J.K., Ames, N.K. and Steficek, B.A. (1992) Efficacy of hydrated sodium calcium aluminosilicate in reducing the toxicity of dietary zearalenone to mink. *Journal of Applied Toxicology* 12, 85-90.

Carretero, M.I. (2002) Clay minerals and their beneficial effects on human health. *Applied Clay Science* 21, 155–163.

CAST (Council for Agricultural Science and Technology). (2003) *Mycotoxins: Risks in plant, animal, and human systems*. Task Force Report No. 139. CAST, Ames, Iowa.

Centers for Disease Control and Prevention (CDC). (2004) Outbreak of aflatoxin poisoning – Eastern and Central Provinces, Kenya. *Morbidity and Mortality Weekly Report* 53, 790-793.

Davidson, J.N., Babish, J.G., Delaney, K.A., Taylor, D.R. and Phillips, T.D. (1987) Hydrated sodium calcium aluminosilicate decreases the bioavailability of aflatoxin in the chicken. *Poultry Science* 66, 89.

Dwyer, M.R., Kubena, L.F., Harvey, R.B., Mayura, K., Sarr, A.B., Buckley, S., Bailey, R.H. and Phillips, T.D. (1997) Effects of inorganic adsorbents and cyclopiazonic acid in broiler chickens. *Poultry Science* 76, 1141-1149.

Harvey, R.B., Kubena, L.F., Phillips, T.D., Corrier, D.E., Elissalde, M.H. and Huff, W.E. (1991*a*) Diminution of aflatoxin toxicity to growing lambs by dietary supplementation with hydrated sodium calcium aluminosilicate. *American Journal of Veterinary Research* 52, 152-156.

Harvey, R.B., Phillips, T.D., Ellis, J.A., Kubena, L.F., Huff, W.E. and Petersen, H.D. (1991*b*) Effects of aflatoxin M_1 residues in milk by addition of hydrated sodium calcium aluminosilicate to aflatoxin-contaminated diets of dairy cows. *American Journal of Veterinary Research* 52, 1556-1559.

Hinton, D.M., Myers, M.J., Raybourne, R.A., Francke Carroll, S., Sotomayor, R.E., Shaddock, J., Warbritton, A. and Chou, M.W. (2003) Immunotoxicity of aflatoxins in rats: Effects on lymphocytes and the inflammatory response in a chronic intermittent dosing study. *Toxin Science* 73, 362-377.

Huff, W.E., Kubena, L.F., Harvey, R.B. and Phillips, T.D. (1992) Efficacy of a hydrated sodium calcium aluminosilicate to reduce the individual and combined toxicity of aflatoxin and ochratoxin A. *Poultry Science* 71, 64-69.

Kubena, L.F., Harvey, R.B., Corrier, D.E., Phillips, T.D. and Huff, W.E. (1990) Diminution of aflatoxicosis in growing chickens by the dietary addition of a hydrated sodium calcium aluminosilicate. *Poultry Science* 69, 727-735.

Kubena, L.F., Huff, W.E., Harvey, R.B., Yersin, A.G., Elissalde, M.H., Witzel, D.A., Giroir, L.E. and Phillips, T.D. (1991) Effects of hydrated sodium calcium aluminosilicate on growing turkey poults during aflatoxicosis. *Poultry Science* 70, 1823-1830.

Lemke, S.L. (2000) *Investigation of clay-based strategies for the protection of animals from the toxic effects of selected mycotoxins*. Ph.D. Dissertation, Department of Veterinary Anatomy and Public Health, Texas A&M University, College Station.

Lindemann, M.D., Blodgett, D.J., Kornegay, E.T. and Schurig, G.G. (1993) Potential ameliorators of aflatoxicosis in weanling/growing swine. *Journal of Animal Science* 71, 171-178.

Mayura, K., Abdel-Wahhab, M.A., McKenzie, K.S., Sarr, A.B., Edwards, J.F., Naguib, K., Phillips, T.D. (1998) Prevention of maternal and developmental toxicity in rats via dietary inclusion of common aflatoxin sorbents: Potential for hidden risks. *Toxicological Sciences* 41, 175-182.

McMahon, G., Hanson, L., Lee, J.J. and Wogan, G.N. (1986) Identification of an activated *c-Ki-ras* oncogene in rat liver tumors induced by aflatoxin B_1. *Proceedings of the National Academy of Sciences (USA)* 83, 9418-9422.

McMahon, G., Davis, E. and Wogan, G.N. (1987) Characterization of *c-Ki-ras* oncogene alleles by directing sequencing of enzymatically amplified DNA from carcinogen-induced tumors. *Proceedings of the National Academy of Sciences (USA)* 84, 4974-4978.

Miller, D.M., Stuart, B.P., Crowell, W.A., Cole, J.R., Goven, A.J. and Brown, J. (1978) Aflatoxicosis in swine: Its effect on immunity and relationship to salmonellosis. *American Association of Veterinary Laboratory Diagnosticians* 21, 135-146.

Park, D.L., Lee, L.S., Price, R.L. and Pohland, A.E. (1988) Review of the decontamination of aflatoxin by ammoniation: Current status and regulation. *Journal of the Association of Official Analytical Chemists* 71, 685-703.

Pestka, J.J. and Bondy, G.S. (1994) Mycotoxin-induced immunomodulation. In: Dean, J.H., Luster, M.I., Munson, A.E. and Kimber, I. (eds.) *Immunotoxicology and Immunopharmacology*. Raven Press, New York, pp. 163-182.

Phillips, T.D. (1999) Dietary clay in the chemoprevention of aflatoxin-induced disease. *Toxin Science* 52, 118-126.

Phillips, T.D., Kubena, L.F., Harvey, R.B., Taylor, D.R. and Heidelbaugh, N.D. (1988) Hydrated sodium calcium aluminosilicate: A high affinity sorbent for aflatoxin. *Poultry Science* 67, 243-247.

Phillips, T.D., Sarr, A.B. and Grant, P.G. (1995) Selective chemisorption and detoxification of aflatoxins by phyllosilicate clay. *Natural Toxins* 3, 204-213.

Phillips, T.D., Lemke, S.L., and Grant, P.G. (2002) Characterization of clay-based enterosorbents for the prevention of aflatoxicosis. *Advances in Experimental Medicine and Biology* 504, 157-171.

Pimpukdee, K., Kubena, L.F., Bailey, C.A., Huebner, H.J., Afriyie-Gyawu, E. and Phillips, T.D. (2004) Aflatoxin-induced toxicity and depletion of hepatic vitamin A in young broiler chicks: Protection of chicks in the presence of low levels of NovaSil PLUS in the diet. *Poultry Science* 83, 737-744.

Richard, J.L., Thurston, J.R. and Pier, A.C. (1978) Effects of mycotoxins on immunity. In: Rosenberg, P. (ed.) *Toxins: Animal, Plant and Microbial*. Pergamon Press, New York, pp. 801-817.

Rodricks, J.V. and Stoloff, L. (1977) Aflatoxin residues in contaminated feed in edible tissues of food-producing animals. In: Rodricks, J.V., Hesseltine, C.W. and Mehlman, M.A. (eds.), *Mycotoxins in Human and Animal Health*. Pathotox, Park Forest South, Illinois, pp. 67-69.

Sarr, A.B., Mayura, K., Kubena, L.F., Harvey, R.B., and Phillips, T.D. (1995) Effects of phyllosilicate clay on the metabolic profile of aflatoxin B_1 in Fisher-344 rats. *Toxicology Letters* 75, 145-151.

Smela, M.E., Currier, S.S., Bailey, E.A. and Essigmann, J.M. (2001) The chemistry and biology of aflatoxin B_1: From mutational spectrometry to carcinogenesis. *Carcinogenesis* 22, 535-545.

Smith, E.E., Phillips, T.D., Ellis, J.A., Harvey, R.B., Kubena, L.F., Thompson, J. and Newton, G. (1994) Dietary hydrated sodium calcium aluminosilicate reduction of aflatoxin M_1 residue in dairy goat milk and effects on milk production and components. *Journal of Animal Science* 72, 677-682.

Turner, P.C., Moore, S.E., Hall, A.J., Prentice, A.M. and Wild, C.P. (2003) Modification of immune function through exposure to dietary aflatoxin in Gambian children. *Environmental Health Perspectives* 111, 217-220.

Wang, J.-S., Luo, H., Bilam, M., Wang, Z., Guan, H., Tang, L., Goldston, T., Afriyie-Gyawu, E., Lovett, C., Griswold, J., Brattin, B., Taylor, R.J., Huebner, H.J. and Phillips, T.D. (2005) Short-term safety evaluation of processed calcium montmorillonite clay (NovaSil) in humans. *Food Additives and Contaminants* 22, 270-279.

Wiles, M.C., Huebner, H.J., Afriyie-Gyawu, E., Taylor, R.J., Bratton, G.R. and Phillips, T.D. (2004) Toxicological evaluation and metal bioavailability in pregnant rats following exposure to clay minerals in the diet. *Journal of Toxicology and Environmental Health* (Part A) 67, 863–874.

26

Food Processing to Reduce Mycotoxins in Africa

Pascal Fandohan*, Kerstin Hell and Walter F. O. Marasas

Abstract

Mycotoxins have both economic and health impacts because they contaminate human food and animal feeds. A central question is whether there are food processing strategies that reduce mycotoxin levels in food products. In this chapter we critically review food processing methods tested in Africa for their efficacy in reducing mycotoxin contamination in maize and peanut. These methods include cleaning, separation of screenings, washing, steeping, aqueous extraction, dehulling, milling, fermentation, cooking and roasting. Some methods reduced mycotoxin levels significantly while others were less effective. Encouraging the widespread use of the toxin-reducing processing techniques would lower consumption of contaminated food products in Africa, and improve food quality and human health.

Introduction

Mycotoxins are noxious secondary metabolites produced by fungi that are found in food products worldwide. In Africa, mycotoxins, particularly aflatoxins and fumonisins, are receiving increasing attention because of their impact on human health and international trade. Contamination of food commodities by these toxins results in reduction in quality and market value, with significant economic losses for farmers and food processors, and serious health implications for consumers (Cardwell and Miller, 1996). As much as 40% of the human productivity lost to diseases in developing countries may be due to diseases exacerbated by aflatoxin contamination (Miller, 1996). Aflatoxicosis outbreaks occurred in Kenya in 2004 causing 125 deaths (Azziz-Baumgartner *et al.*, 2005; Okioma, Chapter 11), and reoccurred in 2005. Fumonisins have been associated with esophageal cancer in South Africa (Rheeder *et al.*, 1992) and in China (Chu and Li, 1994). An outbreak of abdominal pain and diarrhea has been reported in India due to consumption of products contaminated with fumonisins (Bhat *et al.*, 1997).

Food products that are particularly susceptible to mycotoxin contamination in Africa include maize and peanuts. These products are widely cultivated and consumed across the continent and are very susceptible to contamination with mycotoxins. For example, 5-56% of maize collected in a nationwide survey in Benin were contaminated with aflatoxin (Hell *et al.*, 2003). Various interventions to minimize human exposure to mycotoxins and thereby prevent economic loss have been proposed or are being tested with appropriate integrated management from field to fork. In this respect, recent research has focused on the development of food processing methods as a strategy to lower mycotoxin content in foods. The use of methods such as cleaning, separation of screenings, washing, aqueous extraction, dehulling and milling, and alkaline cooking all are effective, at least to some extent, in reducing mycotoxins in food commodities (Charmley and Prelusky, 1995; Shetty and Bhat, 1999; Voss *et al.*, 2001). In Africa, diverse traditional processing procedures for food products are used, some of which reduce consumers' exposure to mycotoxins. The present chapter critically reviews these methods and analyzes their efficacy in reducing mycotoxin contamination, with a focus on maize- and peanut-based foods.

Processing procedures for maize and peanut in Africa

In Africa, maize and peanut may undergo short or long processing protocols that lead to various derived fermented and non-fermented food products. For instance, *owo*, a non-fermented food product, is obtained from maize in Benin by a short processing procedure involving sorting/winnowing of raw maize, followed by milling and cooking of maize meal (Fandohan *et al.*, 2005). *Makume*, a fermented food product, is prepared by a long processing protocol comprising several steps including sorting, winnowing, washing, crushing, dehulling, dry screening, milling, fermentation and cooking (Fandohan *et al.*, 2005). Another example of a long processing procedure is that of the peanut-based *kuli-kuli* found in West Africa (Fig. 1). In Benin, ~ 40 different derived products are obtained from maize (Nago, 1997). Most of the processing procedures involve simple but critical operations, *e.g.*, cleaning, dehulling, milling, fermentation, cooking, *etc*. Some of these operations are very effective in reducing mycotoxin content in foods while others are less so.

Cleaning

Cleaning is generally the first step when processing maize and peanuts. Initially the grains/seeds are sorted, winnowed and/or washed. Cleaning can be done manually at the household level or by using specific machines at an industrial level. During this process, visibly moldy and damaged grains as well as chaff and other undesirable particles are systematically discarded. The remaining apparently healthy grains may undergo a washing process during which grains that float are collected and removed from the bulk.

Sorting out the moldy, damaged and shriveled grains and impurities is a very useful decontamination operation. In South Africa, Sydenham *et al.* (1994) reduced fumonisin levels 29-69% by removing screenings from maize bulk. Sorting followed by winnowing of naturally contaminated maize grains resulted in a mean reduction of 59% and 69% in aflatoxin and fumonisin levels, respectively (Fandohan *et al.*, 2005). Aflatoxins generally are associated with the smaller, shriveled peanut seeds (Davidson *et al.*, 1982), and with seeds or grains that often are different in color (IPCS, 1979). Martin *et al.* (1999) recommend sorting as the only remedial technique available for reducing aflatoxin contamination of peanut in Senegal. Sorting may begin by subsistence farmers in the field at harvest, before storage and processing, with the removal of damaged cobs, pods and grains, and cobs with poor husk cover (Hell *et al.*, 2000).

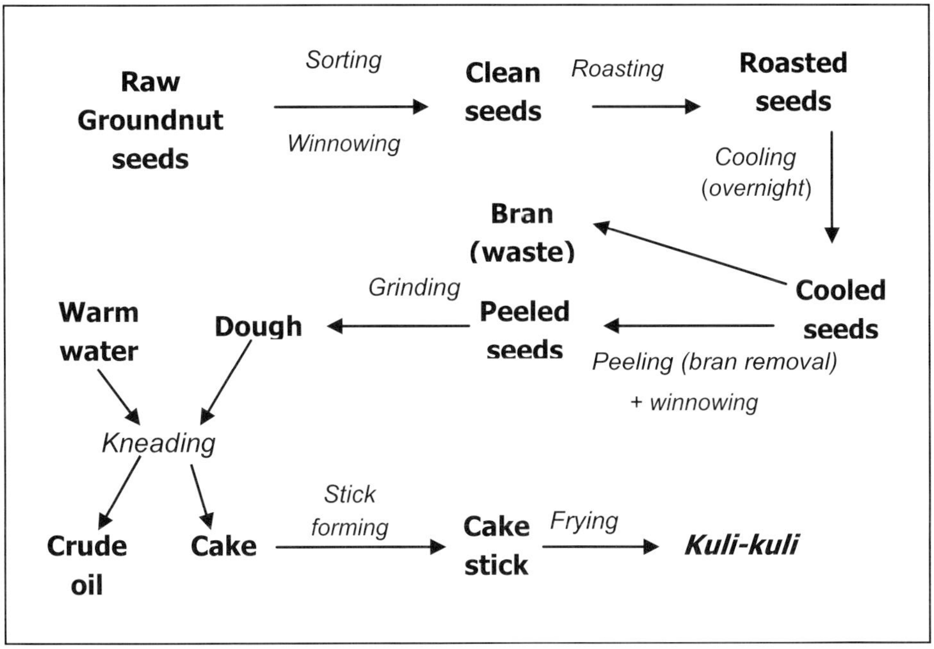

Figure 1. A long processing procedure for the preparation of *kuli-kuli*, a peanut-based food in West Africa.

The efficacy of sorting depends on the extent of grain contamination (Martin *et al.*, 1999), and the ability of the people who do it (Fandohan *et al.*, 2005). People trained to easily recognize diseased grains execute this process more efficiently (Desjardins *et al.*, 2000). Sorting also depends on the goodwill of the people in charge and their willingness to spend enough time on this operation for it to be worthwhile. As a large segment of the African population is unaware of the human health problem posed by mycotoxins, they have no incentive to sort out poor quality grains or seeds (Bankole *et al.*, 2004). Even if they do sort their grain, they are likely to feed the discarded grain to their animals reducing their health and productivity (Williams *et al.*, 2004). Sorting is unlikely to be a practical solution in Africa as long as poverty, hunger and food insecurity are major issues. In such situations, having something to eat today is more important than the usually long-term effects that accompany mycotoxin toxicoses.

Following sorting, simply washing or steeping the grains or seeds in water also can contribute significantly to mycotoxin decontamination. Due to their relative water solubility, aflatoxins and fumonisins often are leached into washing or steeping water during food processing. For example, Se (2002) recorded a 48% reduction in aflatoxin levels due to loss in steeping water during an experiment on maize wet-milling in Egypt. Similar losses of aflatoxins (37%) and fumonisins (51%) to wash water have been reported when maize was processed into derived products in Benin (Fandohan *et al.*, 2005). The leaching of mycotoxins into the wash water may be even more substantial in the presence of additives, *e.g.*, salt, in the wash water (Shetty and Bhat, 1999). Njapau *et al.* (1998) recovered 31% of the aflatoxin B_1 in the steeping water that resulted from the preparation of *nshima*, the main edible maize-based food in Zambia.

Removing the floating grain fraction when washing also can reduce substantially the amount of mycotoxin present, as contaminated grains usually are less dense than the healthy grains (Shetty and Bhat, 1999). The high mycotoxin content of steeping water or supernatant of fermented foods is a cause for concern in many parts of Africa. These liquids often are not discarded, but instead are used to prepare porridges, beverages and traditional herbal medicines consumed by both adults and children (Njapau *et al.*, 1998; Fandohan *et al.*, 2005).

Dehulling and crushing

Dehulling is one of the oldest processing operations in Africa. This process removes the outer parts of grains or seeds, either mechanically or chemically, and often is followed by crushing or milling. In maize, mechanical dehulling often is accompanied by degerming of the grains, *i.e.*, removal of the embryo.

Dehulling or crushing the grains can reduce mycotoxin contamination significantly as mycotoxins are concentrated in the outer layers of the contaminated grains (Sydenham *et al.*, 1995). Dehulling maize grain can reduce aflatoxin contamination by 92% (Siwela *et al.*, 2005). Crushing or grinding peanut seeds generally results in oil free of aflatoxin with the toxin remaining in the cake (Mbaye, 2004), which is used to prepare foods such as *kuli-kuli*, which is consumed by adults and children in West Africa as a snack or as a food supplement.

The efficacy of the dehulling process depends on the degree of fungal penetration of the grain, the degree of contamination and the distribution of the toxin in the grain as mycotoxins are unevenly distributed in the contaminated unprocessed commodity (Charmley and Prelusky, 1995; Bolger *et al.*, 2001).

Fermentation

In many African countries, traditional food preparation is based on fermentation. These fermented foods have numerous advantages including: reduced spoilage, longer shelf-life, increased nutritional value, enhanced flavor and acceptability, and reduction in the toxic and antinutritive compounds present (Holzapfel *et al.*, 1998).

Fermentation can increase the safety of some food products contaminated with mycotoxins (Westby *et al.*, 1997). However, the available reports are contradictory, with some showing very efficient reductions in mycotoxins associated with fermentation, whereas others find lesser or no effects. Lactic acid fermentation can reduce the amount of aflatoxin present in some soybean-based foods (Ogunsanwo *et al.*, 1989) and by up to 70% when contaminated maize and sorghum were processed into *ogi*, a popular West African food (Adegoke *et al.*, 1994). Nout (1994) explained the significant effect of lactic fermentation on aflatoxin B_1 reduction as the result of complete degradation of the molecule following the opening of the lactone ring. Significant decreases, 68-75%, in the levels of fumonisin and zearalenone in fermented maize meal also occur after four days of lactic fermentation (Mokoena *et al.*, 2005). However, these authors suggest that this reduction may not suffice to significantly alter the toxic effects of the two toxins.

In contrast to these rather positive reports, Dada and Muller (1983) found that lactic fermentation during the preparation of *ogi* reduce aflatoxin levels by only 12-16% and the lactic fermentation involved in the preparation of *kenkey* from maize does not reduce aflatoxin levels at all (Jespersen *et al.*, 1994; Kpodo *et al.*, 1996). Indeed Kpodo *et al.* (1996) observed that aflatoxin levels increased initially and then persisted during the fermentation. The authors argue that a reduction of aflatoxin levels is unlikely under acid conditions, and that chemical transformation of the toxin is more likely, since reduction in aflatoxin content usually is accompanied by opening the lactone ring in the presence of alkali. Small reductions in mycotoxin levels (18% for aflatoxins and 13% for fumonisins) also have been observed following lactic fermentation when preparing *ogi* (Fandohan *et al.*, 2005). The ability of ethanol fermentations to reduce mycotoxin content during preparation of beer at traditional and industrial levels also has been evaluated, but no significant reduction of aflatoxin or fumonisin contamination was observed (Desjardins *et al.*, 2000; Shephard *et al.*, 2005).

The mixed results with lactic acid fermentations reported to date suggest that further in-depth studies are needed to clarify the effects of fermentation on mycotoxin, primarily aflatoxins and fumonisins, contamination in foods. Some researchers who collect survey data evaluate the mycotoxin content of the surveyed foods at fermentation stage. These surveys do not always take into account the other production steps in preparing fermented foods, *e.g.*, cleaning, steeping, dehulling, milling and cooking, that also can contribute to the overall reduction in mycotoxin levels observed in the final fermented products and should not be ignored (Westby *et al.*, 1997).

Thermal processing

Cooking and roasting are the most common thermal food-processing treatments at the household level in Africa. Mycotoxins such as aflatoxins and fumonisins are relatively heat-stable and cannot be easily destroyed by ordinary cooking. Significant reductions in toxin levels are more likely at higher cooking temperatures. Temperatures > 150°C are required for a reduction in fumonisins, and >195°C for a reduction in aflatoxins (Bolger *et al.*, 2001). Such temperatures are difficult to reach during ordinary household cooking.

Cooking of food at the household level generally lasts no more than 30 min and is unlikely to result in a significant reduction of the mycotoxins present. After 20 min of cooking of a South African stiff porridge the fumonisin content was reduced by 23% (Shephard *et al.*, 2002). A three hour cooking time could reduce aflatoxin levels by up to 80% in *kenkey*, a Ghanaian maize-based food (Kpodo *et al.*, 1996). If *nshima* was cooked for 15 min, then aflatoxin B_1 was reduced by 13% and no further reduction occurred if the cooking time was extended to 2 hours (Njapau *et al.*, 1998). Conversely, Adegoke *et al.* (1994) observed a 68% reduction in aflatoxin B_1 content after 30 min of cooking *tuwo*, a Nigerian maize-based product, and this reduction increased to 81% after 60 min. In general, moist conditions are more favorable for degradation of mycotoxins during cooking (Rehana and Basappa, 1990; Kpodo *et al.*, 1996).

Relative to ordinary cooking or boiling, roasting is more likely to result in a substantial reduction of mycotoxin content in food. Roasting peanuts gives a highly significant reduction (85%) in the level of aflatoxin B_1 whereas boiling results in a mean reduction in toxin of only 19% (Njapau *et al.*, 1998). A complete loss of fumonisin B_1 was observed when artificially contaminated maize meal was roasted at 218°C for 15 min (Castelo *et al.*, 1998).

Further studies are needed on the effects of cooking on mycotoxin contamination, especially with respect to moisture conditions during cooking and pH, as many food products in Africa initially are fermented before being cooked. Of particular importance is the need to determine if the mycotoxin apparently lost during cooking and roasting has been degraded or if it is instead simply bound to a food matrix or is transformed into another product(s) toxic to humans during the cooking process.

Conclusions

Specific processing techniques in Africa clearly can be used to reduce human exposure to mycotoxins. Of these techniques, cleaning by removing poor quality products, washing the product before processing, and mechanically dehulling the grain are by far the most important and can be accomplished with simple tools. Cleaning and mechanical dehulling also are the least likely to produce other toxic residues in the food. Processes such as fermentation and cooking also need further investigations to determine their influence on mycotoxin decontamination in food. Interestingly, in Africa several unit operations, *e.g.*, cleaning, dehulling, milling, fermentation and cooking, all take place during the processing of food products and provide multiple opportunities for high reductions of mycotoxin levels. However, there is no single method that is likely to result in considerable food decontamination.

References

Adegoke, G.O., Otumu, E.J. and Akanni, A.O. (1994) Influence of grain quality, heat and processing time on the reduction of aflatoxin B_1 levels in *tuwo* and *ogi*: Two cereal-based products. *Plant Foods for Human Nutrition* 45, 113-117.

Azziz-Baumgartner, E., Lindblade, K., Gieseker, K., Rogers, H.S., Kieszak, S., Njapau, H., Schleicher, R., McCoy, L.F., Misore, A., DeCock, K., Rubin, C. and Slutsker, L., and The Aflatoxin Investigative Group. (2005) Case-control study of an acute aflatoxicosis outbreak, Kenya, 2004. *Environmental Health Perspectives* 113, 1779-1783.

Bankole, S.A., Ogunsanwo, B.M. and Mabekoje, O.O. (2004) Natural occurrence of molds and aflatoxins in melon seeds from markets in Nigeria. *Food and Chemical Toxicology* 44, 1209-1214.

Bhat, R.V., Shetty, P.H., Amruth, R.P. and Sudershan, R.V.(1997) A foodborne disease outbreak due to the consumption of moldy sorghum and maize containing fumonisin mycotoxins. *Clinical Toxicology* 35, 249-255.

Bolger, M., Coker, R.D., DiNovi, M., Gaylor, D., Gelderblom, W., Olsen, M., Paster, N., Riley, R.T., Shephard, G. and Speijers, G.J.A. (2001) Fumonisins. In: *Safety Evaluation of Certain Mycotoxins in Food*, WHO Food Additives Series 47, FAO Food and Nutrition Paper 74, Prepared by the 56th Meeting of the Joint FAO/WHO Expert Committee on Food Additives (JECFA), WHO, Geneva, Switzerland, pp. 103-279.

Cardwell, K.F. and Miller, J.D. (1996) Mycotoxins in foods in Africa. *Natural Toxins* 4, 103-107.

Castelo, M.M., Summer, S.S. and Bullerman, L.B. (1998) Stability of fumonisins in thermally processed corn products. *Journal of Food Protection* 61, 1030-1033.

Charmley, L.L. and Prelusky, D.B. (1995) Decontamination of *Fusarium* mycotoxins. *Applied and Environmental Microbiology* 61, 421-435.

Chu, F.S. and Li, G.Y. (1994) Simultaneous occurrence of fumonisin B_1 and other mycotoxins in moldy corn collected from the People's Republic of China in region with high incidence of esophageal cancer. *Applied and Environmental Microbiology* 60, 847-852.

Dada, L.O. and Muller, H.G. (1983) The fate of aflatoxin B_1 in the production of *ogi*, a Nigerian fermented sorghum porridge. *Journal of Cereal Science* 1, 63-70.

Davidson, J.I., Whitaker, T.B. and Dickens, J.W. (1982) Grading, cleaning, storage, shelling, and marketing of peanuts in the United States. In: Pattee, H.E. and Young, C.T. (eds.) *Peanut Science and Technology*. American Peanut Research and Education Society, Yoakum, Texas, USA, pp. 571-623.

Desjardins, A.E., Manandhar, G., Plattner, R.D., Maragos, C.M., Shrestha, K. and McCormick, S.P. (2000) Occurrence of *Fusarium* species and mycotoxins in Nepalese maize and wheat and the effect of traditional processing methods on mycotoxin levels. *Journal of Agricultural and Food Chemistry* 48, 1377-1383.

Fandohan, P., Zoumenou, D., Hounhouigan, D.J., Marasas, W.F.O., Wingfield, M.J. and Hell, K. (2005) Fate of aflatoxin and fumonisin during the processing of maize into food products in Benin, West Africa. *International Journal of Food Microbiology* 98, 249-259.

Hell, K., Cardwell, K.F., Setamou, M. and Poehling, H.M. (2000) The influence of storage practices on aflatoxin contamination in maize in four agroecological zones of Benin, West Africa. *Journal of Stored Products Research* 36, 365-382.

Hell, K., Cardwell, K.F. and Poehling, H.M. (2003) Distribution of fungal species and aflatoxin contamination in stored maize in four agroecological zones in Benin, West-Africa. *Journal of Phytopathology* 151, 690-698.

Holzapfel, W., Olasupo, N. and Haberer, P. (1998) Lactobacilli in a healthy diet. In: Gaukel, V. and Spieß, W.E.L. (eds.), *Proceedings of the 3rd Karlsruhe Nutrition Symposium, European Research towards Safer and Better Food*, Karlsruhe, Germany, Part 1, pp. 20–32.

IPCS. (1979) Environmental Health Criteria 11. *Mycotoxins*. International Program on Chemical Safety. World Health Organization, Geneva, Switzerland, 138 pp.

Jespersen, L., Halm, M., Kpodo, K. and Jakobsen, M. (1994) Significance of yeasts and molds occurring in maize dough fermentation for "kenkey" production. *International Journal of Food Microbiology* 24, 239-248.

Kpodo, K., Sorensen, A.K. and Jakobsen, M. (1996) The occurrence of mycotoxins in fermented maize products. *Food Chemistry* 56, 147-153.

Martin, J., Ba, A., Dimanche, P. and Schilling, R. (1999) How groundnut contamination can be controlled? Work in Senegal. *Agriculture et Développement* 23, 58-67.

Mbaye, A.A. (2004) *Sanitary and Phytosanitary Requirements and Developing-country Agro-food Exports. An Assessment of the Senegalese Groundnut Sub-sector.* Agriculture and Rural Development Discussion Paper. Agriculture and Rural Development (ARD), The International Bank for Reconstruction and Development/The World Bank, Washington, D.C. 41 pp.

Miller, J.D. (1996) Mycotoxins. In: Cardwell, K.F. (ed.) *Proceedings of the Workshop on Mycotoxins in Food in Africa (6-10 November 1995, Cotonou, Republic of Benin)*. International Institute of Tropical Agriculture, Ibadan, Nigeria, pp. 18-22.

Mokoena, M.P., Chelule, P.K. and Gqaleni, N. (2005) Reduction of fumonisin B_1 and zearalenone by lactic acid bacteria in fermented maize meal. *Journal of Food Protection* 68, 2095-2099.

Nago, C.M. (1997) La transformation alimentaire traditionnelle du maïs au Bénin: Détermination des caractéristiques physico-chimiques des variétés en usage. Relation avec l'obtention et la qualité des principaux produits dérivés. Thèse de Doctorat d'Etat. Université de Paris 7.

Njapau, H., Muzungaile, E.M. and Changa, R.C. (1998) The effect of village processing techniques on the content of aflatoxins in corn and peanuts in Zambia. *Journal of the Science of Food and Agriculture* 76, 450-456.

Nout, M.J.R. (1994) Fermented foods and food safety. *Food Research International* 27, 291-298.

Ogunsanwo, B.M., Faboya, O.O., Ikotun, T. and Idowu, R. (1989) Fate of aflatoxins in soybeans during the preparation of "soyogi". *Nahrung* 33, 485-487.

Rehana, F. and Basappa, S. (1990) Detoxification of aflatoxin B_1 in maize by different cooking methods. *Journal of Food and Science Technology* 27, 379-399.

Rheeder, J.P., Marasas, W.F.O., Thiel, P.G., Sydenham, E.W., Shephard, G.S. and van Schalkwyk, D.J. (1992) *Fusarium moniliforme* and fumonisins in corn in relation to human esophageal cancer in Transkei. *Phytopathology* 82, 353-357.

Se, A. (2002) Distribution of aflatoxins in product and by-products during glucose production from contaminated corn. *Nahrung* 46, 341-344.

Shephard, G.S., Leggott, N.L., Stockenstrom, S., Somdyala, N.I.M. and Marasas, W.F.O. (2002) Preparation of South African maize porridge: Effect on fumonisin mycotoxin levels. *South African Journal of Science* 98, 393-396.

Shephard, G.S., van der Westhuizen L., Gatyeni, P.M., Somdyala, N.I.M., Burger, H.-M. and Marasas, W.F.O. (2005) Fumonisin mycotoxins in traditional Xhosa maize beer in South Africa. *Journal of Agricultural and Food Chemistry* 53, 9634-9637.

Shetty, P.H. and Bhat, R.V. (1999) A physical method for segregation of fumonisin-contaminated maize. *Food Chemistry* 66, 371-374.

Siwela, A.H., Siwela, M., Matindi, G., Dube, S. and Nziramasanga, N. (2005) Decontamination of aflatoxin-contaminated maize by dehulling. *Journal of the Science of Food and Agriculture* 85, 2535-2538.

Sydenham, E.W., Stockenstrom, S., Thiel, P.G., Shephard, G.S., Koch, K.R. and Marasas, W.F.O. (1995) Potential of alkaline hydrolysis for the removal of fumonisins from contaminated corn. *Journal of Agricultural and Food Chemistry* 43, 1198-1201.

Sydenham, E.W., van der Westhuizen, L., Stockenstrom, S., Shephard, G.S. and Thiel, P.G. (1994) Fumonisin contaminated maize: Physical treatment for the decontamination of bulk shipments. *Food Additives and Contaminants* 11, 25-32.

Voss, K.A., Poling, S.M., Meredith, F.I., Bacon, C.W. and Saunders, D.S. (2001) Fate of fumonisins during the production of fried tortilla chips. *Journal of Agricultural and Food Chemistry* 49, 3120-3126.

Westby, A., Reilly, A. and Bainbridge, Z. (1997) Review of the effect of fermentation on naturally occurring toxins. *Food Control* 8, 329-339.

Williams, J.H., Phillips, T.D., Jolly, P.E., Stiles, J.K., Jolly, C.M. and Aggarwal, D. (2004) Human aflatoxicosis in developing countries: A review of toxicology exposure, potential health consequences, and interventions. *American Journal of Clinical Nutrition* 80, 1106-1122.

27

Indoor Airborne Exposure to Molds and Mycotoxins

Brian G. Shelton* and John F. Leslie

Introduction

The potential for a wide variety of fungal toxins to cause adverse health effects from the ingestion of mycotoxin-contaminated foods and feeds is a well-documented and widely accepted health problem of global importance. Some of these fungal metabolites are extremely toxic even at very low concentrations and are included as possible bioterrorism agents. Consequently, detecting the presence of mycotoxins in food, and knowledge of its (their) concentration(s) is important in assessing the disease risk of ingesting fungal-contaminated foodstuffs. The documented hazard posed by mycotoxins ingested has caused some to postulate that other potential routes of exposure, *e.g.*, inhalation of mold spores and mycotoxins, may lead to mycotoxicoses.

In the past decade, a number of studies have been published (Croft *et al.*, 1986; Etzel *et al.*, 1998; Hodgson *et al.*, 1998; Johanning *et al.*, 1996; Vesper *et al.*, 2007*a*) that purport to describe the hazards of exposure to airborne fungi in indoor environments of homes and offices. In these studies, a wide variety of occupant symptoms, *e.g.*, headaches, chronic fatigue syndrome, focal alopecia, neural and respiratory disorders and even deaths, have been attributed to exposure to airborne mycotoxins. These studies also have resulted in widely used and undefined lay terms, such as "toxic mold," to describe the fungi that produce mycotoxins and also appear commonly in indoor settings. These studies commonly are quoted by the media, and used to generate hypotheses advanced in court cases that leave scientific laymen with the impression that "toxic mold" is a scientifically well-documented and important public health problem. This is not the case.

A fundamental problem is that the studies evaluating "toxic mold" hypotheses are few in number and beset with major epidemiological shortcomings that undermine the reliability of their conclusions. The scientific literature for the past several years yields a nearly yearly review, *e.g.*, Kuhn and Ghannoum (2003), Kelman *et al.* (2004) and Bush *et al.* (2006), that conclude that the widespread concern about indoor exposures to so-called "toxic mold" is unproven and not backed by convincing scientific evidence. In this chapter, we summarize our understanding of problems with "toxic mold" and whether these types of concerns are important in developing countries.

Possible health effects of airborne fungi

Typically, concentrations of fungi in outdoor air are higher than in indoor air (Shelton *et al.*, 2002). Mold-related problems arise indoors when, because of water intrusions and dampness, environmental conditions favor fungal growth, reproduction and sporulation. In general, complaints related to mold occur, as expected, whenever visible fungal growth and sporulation has occurred indoors whether or not the indoor exposure levels exceed those typically encountered out of doors.

Several types of adverse outcomes can result from exposure to airborne fungi including allergic reactions, such as allergic rhinitis and sinusitis, *e.g.*, Bush *et al.* (2006), that are well accepted by the medical community. (Note: Fungal sinusitis typically is rare and secondary to bacterial sinusitis; rhinitis by definition is not toxin mediated). There also is some evidence that asthma in asthmatics may be exacerbated following exposure to indoor airborne fungi. [Note: There are many more common triggers for asthma than indoor fungi including exposure to pets, insects, cold air and exercise (American Academy of Allergy, Asthma and Immunology, 2006). The fungal involvement in the exacerbation of asthma is not known to be toxin mediated.] Other potential outcomes also may occur, *e.g.*, fungal infections and systemic mycoses, but serious fungal infections usually are rare and generally are limited to people with immune deficiencies.

Airborne mycotoxins

A major question is whether mycotoxins occur in fungal spores that are the etiologic cause of adverse health effects from inhalation in indoor environments. Available studies usually are limited in scope and do not tie cultures from fungi purportedly responsible for an outbreak to toxic spores or fungal strains that produce spores carrying mycotoxins to any particular disease symptoms, *e.g.*, Panaccione and Coyle (2005). Case studies and anecdotal reports are relatively common (see above), but well-designed epidemiological studies of this issue are needed. If mycotoxins are not carried in sufficient quantities by fungal spores, then any problems caused by airborne mycotoxins must result from their absorption from airborne dust or other particles into which the toxins were secreted by the fungi that produced them.

Airborne fungi as allergens

There is no known direct "dose-response" relationship between the concentrations of fungi and allergic health effects as there is with many industrial chemical exposures (NIOSH, 2005*b*). This lack of a direct relationship is especially true of allergic reactions, as the idiosyncrasies of the individual's hypersensitivity to the fungal allergens is the primary determinant of whether an allergenic interaction would occur. In hypersensitive individuals, a very small amount of the right allergen can trigger an allergic reaction. If an illness is mediated by a mycotoxin, then similar interaction patterns to those seen for other chemical toxins should occur, making the presence of viable toxin-producing organisms less critical.

Organic Dust Toxic Syndrome (ODTS)

The only "toxic" condition reported to result from exposure to airborne fungi, is the "Organic Dust Toxic Syndrome" a transient occupational disorder in agricultural workers resulting from intense exposures to extremely high levels of both fungal spores and bacteria from decaying organic material (Musgrave *et al.*, 1994). The exposures that produce this syndrome are much higher than those found in indoor settings. ODTS has numerous names, including: (i) farmer's lung disease (Warren, 1977), (ii) pulmonary mycotoxicosis (Emanuel *et al.*, 1975), (iii) grain fever in grain elevator workers (doPico *et al.*, 1982), (iv) silo unloader's syndrome (Pratt and May, 1984), (v) mill fever in cotton textile workers (Rylander *et al.*, 1987), and (vi) inhalation fever (Rask-Andersen and Pratt, 1992).

The classic "toxic mold" case study

During the 1990s concerns emerged that exposure to toxigenic molds, particularly *Stachybotrys*, in indoor air could cause severe illness.

Sick infants in Cleveland

Prior to 1993 investigators promoting the hypothesis that "toxic mold" was a health problem in indoor air were almost universally discounted. The hypothesis that airborne toxins associated with indoor mold cause serious illness got a major boost from preliminary reports of an investigation in Cleveland, Ohio in 1993 of infants who developed acute pulmonary hemorrhage/hemosiderosis and one of whom died. Investigators, including some from the Centers for Disease Control (CDC) in Atlanta, Georgia, concluded that *Stachybotrys atra (chartarum)* in the homes of victims was the cause (Etzel *et al.*, 1998; Montana *et al.*, 1997). As some of the investigators were from CDC, these findings attracted attention and had instant credibility. These investigators hypothesized that toxins produced by *Stachybotrys* caused pulmonary hemorrhage. The resulting media coverage was incendiary and phrases such as "killer mold," "black mold" and "toxic mold" entered the vernacular. However, other epidemiologists at CDC were skeptical of the alleged association with *Stachybotrys*. Their skepticism increased when a similar "outbreak" of acute pulmonary hemorrhage occurred in infants in Chicago, but no *Stachybotrys* was found in any of the homes of the affected infants, even though the fungus was found in one of the control homes. This result prompted CDC to launch an internal review of its Cleveland study.

Internal and external reviews by CDC

CDC conducted a rigorous internal scientific review of the Cleveland studies followed by an external review by a distinguished panel of experts. In June, 1999, CDC's internal scientific Work Group produced its report detailing methodological and statistical flaws (CDC, 1999*a*). The CDC External Panel on Acute Pulmonary Hemorrhage in Infants first met in August, 1999. They reviewed the report of the internal Work Group, the records from Cleveland, the relevant scientific literature, and interviewed the investigators. The report from the internal Work Group and the External panel (CDC, 1999*b*) caused CDC to back away from the initial conclusions of the Cleveland studies. This repudiation was published in March, 2000 and included the statement, "Conclusions regarding the possible association between cases of pulmonary hemorrhage/hemosiderosis in infants in Cleveland and household water damage or exposure to *S. chartarum*, are not substantiated adequately by the scientific evidence ... the association should be considered not proven ..." (DHHS/CDC, 2000). CDC followed up these reviews by forming expert advisory groups to establish a case definition for Acute Pulmonary Hemorrhage/Hemosiderosis in Infants (APHI), to facilitate nationwide surveillance of APHI and to identify opportunities for better studies.

Review by the American Industrial Hygiene Association (AIHA)

In 2000, one of us (BS) convened a panel of experts at the request of the Indoor Environmental Quality Committee of AIHA to review the epidemiologic literature most often cited to support the hypothesis that "toxic mold" causes toxic human illness. Four papers were selected for review: Croft *et al.* (1986), Johanning *et al.* (1996), Hodgson *et al.* (1998) and Etzel *et al.* (1998, the Cleveland studies). The question asked was, "Do toxins associated with these fungi in indoor settings result in increased frequency, severity, or a change in spectrum, of illness?" The panel presented their results at the 2000 annual AIHA meeting (Kirkland, 2001). Reported flaws in these studies include: (i) absence of clear and consistent case definitions, (ii) lack of consistent medical examinations, (iii) ignoring and not excluding other risk factors, (iv) non-blinded and inconsistent environmental assessment including biased sampling and flawed statistical analyses, (v) sample sizes insufficient to warrant conclusions, and (vi) the inability to combine and evaluate separate studies because they addressed different outcomes or different definitions of illness. Based on these shortcomings, the evaluators concluded that "… the evidence was insufficient to support the hypothesis that toxigenic fungi caused health complaints different from or worse than those associated with other common molds."

Reviews by the Institute of Medicine (IOM)

IOM (2000) analyzed the evidence for a causal association between asthma and indoor exposures, including fungi. IOM concluded that the assertion that mold causes asthma must be qualified. In particular, the data available suffice only for conclusions that fungi can trigger *exacerbation* of asthma in asthmatics. The available data were not sufficient to support the conclusion that fungi cause the *development* of asthma. IOM added further that, "… fungal exposure ... is complex ... (which has) … led to confusion, poorly constructed studies on the role of fungi, ... inconclusive results, and avoidance of the field by the best investigators."

In 2004, another IOM committee produced another report, *Damp Indoor Spaces and Health* (IOM, 2004). This committee reached very similar conclusions regarding asthma, and also concluded that there was "inadequate or insufficient evidence to determine whether an association exists" between the "presence of mold" and 15 other health effects as claimed by proponents of problems due to "toxic" fungal exposures including APHI, chronic fatigue, neuropsychiatric symptoms, cancer, reproductive effects, and rheumatologic and other immune disorders.

Independent scientific reviews

In most years since 2000 at least one major review of the scientific literature concerning the relationship of "toxic mold" to human illness has appeared. Page and Trout (2001) reviewed the literature indexed under "mycotoxins" and cross-referenced with "indoor air pollution" and/or "sick building syndrome." They concluded that "… there is inadequate evidence to support the conclusion that exposure to mycotoxins in the indoor (nonindustrial) environment is causally related to symptoms or illness among building occupants." These conclusions are consistent with those of CDC's internal and external reviews, and also with those of the AIHA Forum. They further advised that, "… there is inadequate evidence to support recommendations for greater urgency in cases where mycotoxin-producing fungi have been isolated," as suggested in the AIHA Bioaerosol manual (Dillon *et al.*, 1996, pp. 58-59).

Since then reviews have appeared approximately annually that reach the same consistent conclusion that there is a lack of scientific evidence for a relationship between toxigenic molds in indoor environments and human health problems, *e.g.*, Hardin *et al.* (2002), Texas Medical Association (2002), Fung and Hughson (2003), Kuhn and Ghannoum

(2003), Lees-Haley (2003), Harbinson and Hillman (2004), Kelman *et al.* (2004) and Bush *et al.* (2006). In NIOSH (2005*a*) there is a description of a Health Hazard Investigation of an elementary school in California where the teachers had been diagnosed with "toxic encephalopathy" attributed to "toxic mold." NIOSH concluded that "none had evidence of toxic encephalopathy." While not a review article *per se*, this report restated the NIOSH conclusions that, "… there is currently no evidence of a link between mycotoxin exposure in the indoor environment and human illness" and specifically noted "insufficient evidence that mold, mycotoxins, or damp environments cause neuropsychiatric disease." Based on this steady accumulation of peer-reviewed scientific data, there is no scientifically valid reason to view "toxic mold," *i.e.*, fungi that might be capable of producing mycotoxins, as a greater hazard than any other fungi in terms of nonindustrial indoor exposures.

Lack of causation criteria

The anecdotal studies of airborne exposure to "toxic mold" usually share the same limitations and violate basic epidemiologic principles. For example, the disease outcome usually is not pre-defined and medical ascertainment usually is not performed. Since no pre-defined disease definition is adhered to, subjects may be included (improperly) in the study based solely on symptoms that can vary to include whatever people complain about at the time of the study. Therefore the disease outcome can range from symptoms that are difficult to verify, *e.g.*, headaches and chronic fatigue, to serious pulmonary hemorrhage. In short, the disease outcome is a moving target. The term "toxic mold" also often is poorly defined and may mean only a catch-all group of fungi and associated chemicals. There is no definitive list of "toxic molds" of concern in indoor air settings and no defined concentration threshold at which an adverse response would occur. The closest thing to such a list is that developed by Vesper *et al.* (2004), but this list includes many species regarded as non-toxigenic, *e.g.*, species of *Aureobasidium*, *Chaetomium* and *Cladosporium*, and excludes some important toxin producers, *e.g.*, all *Fusarium* species. Without such a list, the specific fungi and mycotoxins of concern can be almost never-ending, with as yet unidentified toxins implicated as necessary. The "relative moldiness index" proposed by Vesper *et al.* (2007*b*) is similarly ill-defined both in terms of fungi analyzed and in the total number of fungal spores present. In short, without a pre-defined exposure of interest, the exposure variable is not only undefined but also can take on an effectively infinite number of possible outcomes. Studies in which both the exposure variable and/or the outcome variable are moving targets, are flawed. At their worst these studies document a building with complaining occupants in which "toxic" fungi are found and assume that the association of the two is meaningful. It is not surprising that these associations are found, but it is surprising that they are published as conclusions from "epidemiological" studies.

Evidence needed to substantiate the claim that mycotoxins in indoor air cause disease(s) include: (i) defining the airborne inhalation dose needed to elicit defined symptoms for each mycotoxin, (ii) data showing that the specific mycotoxin was present at or above the dose level by inhalation in the indoor air in the disease setting, and (iii) a link between criteria (i) and (ii) in terms of sick people with specific pre-defined symptoms. Without such supporting evidence, the associations may be coincidental rather than causal, and those who publicize them may appear to be on a crusade rather than providing a public service.

A typical problematic study of "toxic mold" is that of Auger *et al.* (1994). This short two-page report summarizes three different case studies and concludes that toxic mold is linked to both upper respiratory infections and a chronic fatigue-like syndrome. The basic epidemiological parameters of exposure and medical ascertainment are inadequately defined in this study. Hence, both the disease outcomes and the exposures of interest are moving targets in this study. That "toxic mold" might cause a chronic fatigue-like syndrome is perhaps the ultimate irony for this study, as it concludes that an epidemiologically undefined exposure agent can cause a vague and medically undefined disease with a sample size of $n \leq 3$. Current studies of "toxic mold" are at the anecdotal stage. Going beyond the anecdotes requires rigorously defined studies to determine what risks, if any, airborne fungal mycotoxins pose in a nonindustrial setting.

Potential problems in developing countries

In developing countries, the situation will not be the same as that found in the developed countries of Europe and North America. As detailed elsewhere in this book, mycotoxin exposure by ingestion may be high and all but unavoidable in many developing countries. In most parts of Africa, the major threats will be the aflatoxins and fumonisins associated with maize and peanuts (Marasas *et al.*, Chapter 4). African diets consisting primarily of more traditional cereals, *e.g.* sorghum and millet, should be safer in terms of lower levels of mycotoxin exposure (Bandyopadhyay *et al.*, 2007). Airborne exposure to mycotoxins is probably much less than that encountered in foods. Fungal growth and degradation of thatch roofing material, bedding and grain stored in houses could result in spores and dust carrying mycotoxins that could potentially be absorbed following inhalation.

In terms of exposure to airborne fungi, problems that parallel ODTS observed in agrarian portions of the United States and Western Europe seems a much more likely problem. ODTS results from exposure to large numbers of spores and high levels of dust. Such conditions are common in agrarian settings in developing countries, where there might be no paved surface in an entire village. Hand-harvesting of crops and drying and storing the harvest in the indoor family living areas would increase the exposure to dust and fungal spores even further.

Health problems associated with ODTS are readily detected in Western Europe and the United States. Reports to a physician in a developed country of recurring symptoms such as shortness of breath, coughing, wheezing and fever, although not life-threatening in-and-of-themselves, would elicit follow-up examinations and treatment to reduce symptom severity. In a rural agrarian setting in a developing country, an individual with such symptoms, if they contacted a physician at all, would be unlikely to receive more than a cursory examination and the symptoms probably would be attributed to one or more tropical diseases or to general poor health. Thus, a problem viewed as a significant occupational hazard for farmers in Western Europe and the United States is probably ignored and untreated in most developing countries.

Conclusions

Problems attributable to airborne fungal spores, especially in an agrarian setting, are well known and well documented. Dangers due to airborne exposure to mycotoxins and to "toxic mold" in indoor air in developed countries have been claimed, but the claims have not been substantiated through epidemiological studies. Good scientific process demands evidence, proof, verification and reproducibility, all of which are available for ODTS, but almost none of which are available for the currently claimed "toxic mold" cases. Without the establishment of causation, *e.g.*, a defined toxic dose of a specific toxin, confirmation of the toxin in the environment at or above the threshold, and reliable, verified symptoms in exposed individuals consistent with the exposure chemical and level, problems with "toxic mold" are questionable in developed countries and are probably negligible in developing countries. Instead the lung and other health problems clearly associated with agriculture in Western Europe as far back as at least the early 1700s (Warren, 1977) need to be given more attention as regards their role in reducing economic output and quality of life in current developing countries.

References

American Academy of Allergy, Asthma and Immunology (AAAAI). (2006) Tips to remember: Asthma triggers and management. www.aaaai.org/patients/publicedmat/tips/asthmatriggersandmgmt.stm.

Auger, P.L., Gourdeau, P. and Miller, J.D. (1994) Clinical experience with patients suffering from a chronic fatigue-like syndrome and repeated upper respiratory infections in relation to airborne molds. *American Journal of Industrial Medicine* 25, 41-42.

Bandyopadhyay, R., Kumar, M. and Leslie, J.F. (2007) Relative severity of aflatoxin contamination of cereal crops in West Africa. *Food Additives and Contaminants* 24, 1109-1114.

Bush, R.K., Portnoy, J.M., Saxon, A., Terr, A.I. and Wood, R.A. (2006) The medical effects of mold exposure. *Journal of Allergy and Clinical Immunology* 117, 326-333.

Center for Disease Control (CDC). (1999*a*) *Report of the CDC Working Group on Pulmonary Hemorrhage/Hemosiderosis*. CDC, Atlanta, Georgia. www.cdc.gov/mold/pdfs/hemorrhage_report.pdf.

CDC. (1999*b*) *Reports of Members of the CDC External Expert Panel on Acute Idiopathic Pulmonary Hemorrhage in Infants: A Synthesis*. CDC, Atlanta, Georgia. www.cdc.gov/mold/pdfs/aiphi_report.pdf.

Croft, W.A., Jarvis, B.B. and Yatawara, C.S. (1986) Airborne outbreak of trichothecene toxicosis. *Atmospheric Environment* 20, 549-552.

DHHS/CDC. (2000) Update: Pulmonary hemorrhage/hemosiderosis among infants – Cleveland, Ohio, 1993-1996. *Morbidity and Mortality Weekly Report* 49: 180-184.

Dillon, H.K., Heinsohn, P.A. and Miller, J.D. (1996) *Field Guide for the Determination of Biological Contaminants in Environmental Samples*, 1st ed. AIHA Publications, Fairfax, Virginia.

doPico, G.A., Flaherty, D., Bhansali, P. and Chavaje, N. (1982) Grain fever syndrome induced by inhalation of airborne grain dust. *Journal of Allergy and Clinical Immunology* 69, 435-443.

Emanuel D.A., Wenzel, F.J. and Lawton B.R. (1975). Pulmonary mycotoxicosis. *Chest* 67, 293-297.

Etzel, R.A., Montana, E., Sorenson, W.G., Kullman, G.J., Allan, T.M. and Dearborn, D.G. (1998) Acute pulmonary hemorrhage in infants associated with exposure to *Stachybotrys atra* and other fungi. *Archives of Pediatrics and Adolescent Medicine* 152, 757-762.

Fung, F. and Hughson, W.G. (2003) Health effects of indoor fungal bioaerosol exposure. *Applied Occupational and Environmental Hygiene* 18, 535-544.

Harbinson, R. and Hillman, J.V. (2004) Evaluation of mold-induced adverse health effects. *Harris Martin's Columns - Mold.* 3(3): 6-7, 59-61.

Hardin, B.D., Kelman, B.J. and Saxon, A. (2002) Adverse human health effects associated with molds in indoor environments. American College of Occupational and Environmental Medicine, Arlington Heights, IL. Web Site: www.acoem.org/guidelines.aspx?id=850;_Adopted: October 27, 2002.

Hodgson, M.J., Morey, P., Leung, W.-Y., Morrow, L., Miller, J.D., Jarvis, B.B., Robbins, H., Halsey, J.F. and Storey, E. (1998) Building associated pulmonary disease from exposure to *Stachybotrys chartarum* and *Aspergillus versicolor*. *Journal of Occupational and Environmental Medicine* 40, 241-249.

Institute of Medicine of the National Academies (IOM). (2000) *Clearing the Air: Asthma and Indoor Air Exposures*. National Academy Press, Washington, D.C.
IOM. (2004) *Damp Indoor Spaces and Health*. National Academy Press, Washington, D.C.
Johanning, E., Biagini, R., Hull, D., Morey, P., Jarvis B.B. and Landsbergis, P. (1996) Health and immunology study following exposure to toxigenic fungi (*Stachybotrys chartarum*) in a water-damaged office environment. *International Archives of Occupational and Environmental Health* 68, 207-218.
Kelman, B.J., Robbins, C.A., Swenson, L.J. and Hardin, B.D. (2004) Risk from inhaled mycotoxins in indoor office and residential environments. *International Journal of Toxicology* 23, 3-10.
Kirkland, K. (2001) Health hazards from exposure to mycotoxic fungi in indoor environments. *The Synergist* 12 (4), 25-26.
Kuhn, D.M. and Ghannoum, M.A. (2003) Indoor mold, toxigenic fungi, and *Stachybotrys chartarum*: Infectious disease perspective. *Clinical Microbiology Reviews* 16, 144-172.
Lees-Haley, P.R. (2003) Toxic mold and mycotoxins in neurotoxicity cases: *Stachybotrys, Fusarium, Trichoderma, Aspergillus, Penicillium, Cladosporium, Alternaria* and trichothecenes. *Psychological Reports* 93, 561-584.
Montana, E., Etzel, R.A., Allan, T., Horgan, T.E. and Dearborn, D.G. (1997) Environmental risk factors associated with pediatric idiopathic pulmonary hemorrhage and hemosiderosis in a Cleveland community. *Pediatrics* 99, 117-124.
Musgrave, K.J., Parker, J.E., Olenchock, S.A. and Castellan, R.M. (1994) *Request for Assistance in Preventing Organic Dust Toxic Syndrome*. NIOSH publication 94-102, NIOSH, Dept of Health and Human Services, Washington, D.C. http://www.cdc.gov/nasd/docs/d001001/d001100/d001027/d001027.html.
National Institute for Occupational Safety and Health (NIOSH). (2005*a*) Health hazard evaluation report No. 2005-0122. CDC, Atlanta, Georgia.
NIOSH. (2005*b*) NIOSH Pocket Guide to Chemical Hazards, NIOSH publication no. 2005-149. CDC, Atlanta, Georgia.
Page, E. H. and Trout, D.B. (2001) The role of *Stachybotrys* mycotoxins in building-related illness. *AIHA Journal* 62, 644-648.
Panaccione, D. and Coyle, C.M. (2005) Abundant respirable ergot alkaloids from the common airborne fungus *Aspergillus fumigatus*. *Applied and Environmental Microbiology* 71, 3106-3111.
Pratt, D.S. and May, J.J. (1984). Feed-associated respiratory illness in farmers. *Archives of Environmental Health* 39, 43-48.
Rask-Andersen, A. and Pratt D.S. (1992) Inhalation fever: A proposed unifying term for febrile reactions to inhalation of noxious substances. *British Journal of Industrial Medicine* 49, 40.
Rylander, R., Schilling, R.S.F., Pickering, C.A.C., Rooke, G.B., Dempsey, A.N. and Jacobs, R.R. (1987) Effects after acute and chronic exposure to cotton dust: The Manchester criteria (editorial). *British Journal of Industrial Medicine* 44, 577-579.
Shelton, B.G., Kirkland, K.H., Flanders, W.D. and Morris, G.K. (2002) Profiles of airborne fungi in buildings and outdoor environments in the United States. *Applied and Environmental Microbiology* 68, 1743-1753.
Texas Medical Association. (2002) Black mold and human illness. Report of Council on Scientific Affairs Report 1-1-02.
Vesper, S.J., Varma, M., Wymer, L.J., Dearborn, D.G., Sobolewski, J. and Haugland, R.A. (2004) Quantitative polymerase chain reaction analysis of fungi in dust from homes of infants who developed idiopathic pulmonary hemorrhaging. *Journal of Occupational and Environmental Medicine* 46, 596-601.
Vesper, S. J., Mckinstry, C., Haugland, R. A., Iossifovva, Y., Lemasters, G., Levin, L., Hershey, G.K.K., Villareal, M., Bernstein, D.I., Lockey, J. and Reponen, T. (2007*a*) Relative moldiness index as predictor of childhood respiratory illness. *Journal of Exposure Science and Environmental Epidemiology* 17, 88-94.
Vesper, S., McKinstry, C., Haugland, R., Wymer, L., Bradham, K., Ashley, P., Cox, D., Dewalt, G. and Friedman, W. (2007*b*) Development of an environmental moldiness index for US homes. *Journal of Occupational and Environmental Medicine* 49, 829-833.
Warren, C.P. (1977) Lung disease in farmers. *Canadian Medical Association Journal* 116, 391-394.

Institutional Issues in Mycotoxin Management

28

Are Ghanaians Aware of the Aflatoxin Menace?

Richard T. Awuah*, Kwaku O. Agyemang,
Simon C. Fialor and Curtis M. Jolly

Abstract

We conducted surveys in Ghana in 1999 and 2003 to assess the level of awareness of aflatoxins and their relevant harmful effects. The first survey was of peanut farmers, poultry farmers, retailers, processors, feed millers and consumers. It found very poor awareness of the problem, with only 8% of the 1983 respondents indicating knowledge of the word "aflatoxin". The second survey was of medical doctors, nurses, agriculturists, pharmacists, biologists and other professionals. This survey found a much higher knowledge of the words "mycotoxin" or "aflatoxin," $\geq$ 70% for most categories of interviewed persons, and, to a lesser extent, $\leq$ 40%, of the relevant adverse effects. The level of awareness depended on education, profession and position. Thus, awareness of aflatoxins among health, agricultural and some other professionals in Ghana is fair, but the general Ghanaian population is mostly unaware of the problem or its implications. The general lack of awareness, together with the inability of most of the responding professionals to accurately identify the harmful effects of aflatoxins suggests that the menace posed by the toxin is not well appreciated in Ghana.

Introduction

Aflatoxins are potent mycotoxins produced by the fungi, *Aspergillus flavus* and *Aspergillus parasiticus*. These toxins are involved in several human and animal health conditions (Dimitri and Gabal, 1996; Obasi *et al.*, 1996; Ramos *et al.*, 1996; Sahoo *et al.*, 1996; Oyelami *et al.*, 1997). In acute doses, aflatoxin is toxic and can be lethal (Chen, 1992). In Ghana, aflatoxins occur in peanut (Beardwood, 1964; Mintah and Hunter, 1979; Kpodo, 1995; Awuah and Kpodo, 1996) and some of its processed products (Kpodo, 1995), and in maize and kenkey, a maize-based food (Kpodo *et al.*, 1996).

In 1998, a news item with the sensational banner headline, "Shocking Scientific Report: Kenkey Causes Cancer" appeared in the August 17-18 edition of the *Ghanaian Chronicle*, a Ghanaian newspaper. The author of the article, who at the time also was the editor of the newspaper wrote, "Scientific studies from major processing sites and markets in Accra have concluded comprehensively that there is widespread occurrence of the *strain* (*italics* mine) that causes cancer, particularly liver cancer in Ga kenkey" (Koomson, 1998*a*). The writer based his report on a publication by Kpodo *et al.* (1996), that documented the occurrence of mycotoxins in fermented maize products. Koomson (1998*a*) showed some lack of understanding of the matter when, for example, he wrote "According to the report, maize which is a major dietary staple has been identified as the 'villain'. The molds which develop on the maize particles, because of delayed or poor drying process, contain a *bacteria* called *aflatoxin* (*italics* mine), a proven cause of cancer." In a related report (Koomson, 1998*b*), confused the issue further by referring to aflatoxin as a *fungi* (*italics* mine) that can cause certain types of cancer.

Many Ghanaians eat kenkey and the *Ghanaian Chronicle*, at that time, was one of the most widely read newspapers in Ghana. The article on the carcinogenicity of kenkey was, therefore, read by many Ghanaians. Naturally the publication attracted a lot of rejoinders and counter rejoinders some with headlines such as "The Kenkey Palava" (Odei, 1998), "The Kenkey Debate: Must We Tell the People" (Koomson, 1998*c*), and "The Battle of the Kenkeys" (Season, 1998). In the weeks following publication of the initial news item (Koomson, 1998*a*), panelist after panelist discussed the issue on several radio stations in Ghana. Phone-in callers, mostly lay persons, contributed to the discussions. The impression from most of these discussions and from some of the newspaper rejoinders pointed to a lack of understanding by Ghanaians of aflatoxin and its harmful effects. Thus, we decided to determine the level of awareness of aflatoxin and its associated problems in Ghana. Such a study would be useful in the formulation of aflatoxin management strategies.

Methodology

In 1999, 331 peanut farmers, 727 peanut consumers, 372 peanut retailers, 400 peanut processors, 18 feed millers and 135 poultry farmers from three broad geographical zones in Ghana were surveyed with a pre-tested, structured questionnaire. This survey was part of a USAID Peanut Collaborative Research Support Program (Peanut CRSP) project titled "Economic effect of aflatoxins on the well being of Ghanaian peanut producers and consumers". The Upper East region, Upper West region and the Northern region comprised the Northern Zone. The middle Zone was comprised of the Ashanti and Brong Ahafo regions. The Central, Greater Accra, Western, Volta and Eastern regions constituted the southern zone. Participants were interviewed individually, usually in the relevant local language. Information on socio-economic variables, awareness of the word "aflatoxin", health effects of aflatoxin, and on several issues linked to aflatoxin awareness was obtained from respondents.

In 2003, a similar study was conducted, but with a focus on people involved in the delivery of agricultural and health services in the country. The second study involved 385 agriculturists, 103 health professionals, 17 biologists and 69 other professionals.

The Statistical Package for Social Sciences (SPSS) for Windows XP version 11.0 was used to analyze the data. Analytical tools were descriptive, *e.g.*, means, frequencies and percentages, as well as inferential, *e.g.*, χ^2 tests.

Results and discussion

In the 1999 study, 331 peanut farmers, 727 consumers, 372 peanut retailers, 400 peanut processors, 18 feed millers and 135 poultry farmers participated. Feed millers and poultry farmers were not included in the northern zone because the feed milling and commercial poultry industries are generally absent from these areas. Eighty-three percent of the respondents were male. Other than the feed millers, 50% of whom indicated knowledge of the word "aflatoxin", knowledge of the word by farmers, poultry farmers, retailers, processors and consumers, respectively, were 6.0, 18.5, 6.5, 2.8 and 6.9% (average 8.1%). For the various groups, indication of knowledge of the word "aflatoxin", except for peanut farmers, was significantly dependent ($p < 0.01$) on education. Thus, feed millers in Ghana, most of whom are well educated, have a better appreciation of aflatoxins than do the other groups in this survey. Indication of awareness of the word "aflatoxin" by geographic zone was 10.4, 6.7 and 8.0%, respectively, for the northern, middle and southern zones.

The overall low awareness of the word "aflatoxin", among the general Ghanaian population probably results, in part, from the highly technical nature of the subject and, in part, because aflatoxin has never been considered a serious enough issue in Ghana to merit an urgent awareness campaign. This low level of awareness may result in recommended management not being seriously implemented.

A large number of farmers and food processors in Ghana became aware of aflatoxin through interactions with personnel from the Ghana Ministry of Food and Agriculture. This education was not surprising since some of the Ministry of Food and Agriculture staff had previously attended a workshop on aflatoxin in peanuts held in Kumasi, Ghana and had gone back to their communities and were disseminating the knowledge they had acquired. Amongst all of the survey participants, only 2.5% of the respondents became aware of aflatoxin through workshops, which is probably because workshops on aflatoxin are not common in Ghana. The low level of education resulting from these workshops is undesirable, as they are important fora for the dissemination of technical and scientific information.

Ninety-one percent of the total respondents indicated that they sorted their nuts prior to consumption, with 53% indicating that the sorting was thorough. The fate of the sorted bad nuts depended on the geographic zone. In the northern zone, half of the respondents used the "bad" nuts while 16% and 29% of respondents in the middle and southern zones, respectively, did so. Uses for the bad nuts included feeding to backyard poultry and processing into peanut butter after adding some good nuts. More respondents in the northern zone use the bad nuts because of the chronic food insecurity and poverty problems that occur there.

In 2003, we surveyed personnel involved in the delivery of agricultural and health services in Ghana because aflatoxin is, among others, both an agricultural and a medical issue. Thus, personnel in the agricultural and health services delivery sectors and policy makers in both sectors need to play lead roles in the dissemination of aflatoxin information in Ghana. To do so effectively, they must be well apprised of the subject. More agriculturists (67% of the respondents) participated in the study, while relatively few health personnel (18% of the respondents) did. Of the medical personnel, physicians were the most under-represented, with an ~ 20% response. The other personnel were mainly social scientists, physical scientists, *etc.*, associated with agricultural and medical establishments and they constituted about 15% of the sample. Seventy-eight percent of the respondents were male.

Amongst the professionals (Table 1), physicians, with 61% positive responses, were the most aware of aflatoxins/mycotoxins and their potential complications, followed by pharmacists (49%), biologists (47%), agriculturists (40%), nurses (34%) and the other physical/social scientists (32%). Although many of the health and agricultural professionals said they knew of the term "mycotoxins" (Issue 1, Table 1), relatively few of them could correctly identify the organisms that produce them (Issue 4, Table 1). Similarly, most professionals were aware of the term "aflatoxin" (Issue 2, Table 1), but only a few of them could

Table 1. Positive responses (%) by professionals in Ghana to mycotoxin/aflatoxin issues (2003 survey).

Issue[a]	Profession						Mean[b]
	Agriculturist	Physician	Nurse	Pharmacist	Biological Scientist	Other	
1	62	100	53	88	65	44	68
2	83	100	54	97	71	68	79
3	39	100	29	71	71	68	63
4	36	67	12	28	24	26	32
5	38	42	22	69	71	17	43
6	33	42	19	44	47	20	34
7	47	100	63	73	41	30	59
8	43	75	59	46	47	39	52
9	27	50	25	18	53	15	31
10	71	58	46	63	77	44	60
11	68	92	50	97	70	49	71
12	7	17	7	23	18	10	14
13	9	33	2	0	11	13	11
14	16	33	12	13	12	1	14
15	20	33	27	28	41	23	29
16	16	17	14	30	18	7	17
17	43	50	31	56	59	48	48
18	36	75	48	13	24	19	36
19	73	72	71	66	56	62	67
Mean	40	61	34	49	46	32	44

[a]Issues Issue: **1** – Awareness of the term "mycotoxin"; **2** – Awareness of term "aflatoxin"; **3** – Awareness of what produces mycotoxins; **4** – Correct indication of what produces mycotoxins; **5** – Awareness of effects of aflatoxins on animals; **6** – Correct indication of the effects of aflatoxins on animals; **7** – Awareness of the effects of aflatoxins on humans; **8** – Correct indication of the effects of aflatoxins on humans; **9** – Awareness of aflatoxin accumulation in animal products; **10** – Awareness of aflatoxin contamination of groundnut; **11** – Awareness of aflatoxin contamination of food items; **12** – Attended an aflatoxin workshop(s); **13** – Availability of documents on aflatoxins at workplace; **14** – Formal training on aflatoxins; **15** – Have discussed aflatoxins with colleagues and/or superiors; **16** – Have discussed aflatoxins with subordinates; **17** – A few discolored nuts will not sicken me; **18** – I sort groundnuts before eating them so that I won't be sick; and **19** – Aflatoxin contamination can be minimized.

[b]Mean of the means of each of the six professions, *i.e.*, not weighted for the size of the sample for each profession.

correctly indicate its harmful effects on humans (Issue 8, Table 1). With only 44% of the professionals responding positively to the various aflatoxin/mycotoxin-related issues, awareness of the toxins by professionals can at best be described as fair. This result is worrisome as these key personnel, who should lead the effort to disseminate information on aflatoxin in Ghana, are not sufficiently knowledgeable of the subject to be able to provide leadership.

Professionals with higher academic qualifications such as MB.Ch., B./M.D., Ph.D. *etc.*, were significantly more likely to indicate positive knowledge of the world "aflatoxin" than those with lesser academic credentials, with more than 80% of those with a diploma or more training recognizing the word. Of those with only a certificate, the lowest academic

Table 2. Positive responses (%) by agriculturists in Ghana to mycotoxin/aflatoxin issues (2003 survey).

Issue[a]	Agriculturist Category					Mean[b]
	Extension Agents	District/ Regional Directors	Production Officers	Lecturers	Researchers	
1	73	100	89	89	79	86
2	64	91	87	85	71	80
3	58	89	80	65	71	72
4	56	77	81	77	71	72
5	66	91	79	73	64	75
6	29	59	46	42	61	48
7	13	38	27	23	36	27
Mean	51	78	70	65	65	66

[a]Issues were: **1** – Awareness of word "aflatoxin"; **2** – Awareness of crop contamination by aflatoxins; **3** – Awareness of peanut contamination by aflatoxins; **4** – Awareness of food/food item contamination by aflatoxins; **5** – Correct indication of foods that could be contaminated with aflatoxins; **6** – Awareness of effects of aflatoxins on animals; and **7** – Correct indication of harmful effects of aflatoxins on animals.

[b]Mean of the means of each of the five categories of agriculturists, *i.e.*, not weighted for the size of the sample for each category.

qualification for most professionals in Ghana, somewhat < 60% recognized the word "aflatoxin". Profession was not significantly associated with the recognition of the term "mycotoxin", although it was significantly associated with general awareness of the effects of aflatoxins ($p < 0.05$) and with knowledge of what produced them ($p < 0.01$). Sex was neither associated with knowledge of the word "aflatoxin" nor with awareness of the effects of mycotoxins on humans. Those within a profession were more likely to discuss information on mycotoxins with their colleagues or their superiors ($p < 0.05$) than they were with their subordinates.

With respect to the agriculturists, the District and Regional Directors were more aware of aflatoxins (78% positive responses) while Extension Agents were the least aware (51% positive responses; Table 2). The difference in recognition of the term "aflatoxin" varied by their rank and was statistically significant ($p < 0.001$). Most of the agriculturists, however, did not know of the effects of aflatoxins on animals (Issue 7, Table 2) even though the majority of them had heard of the term "aflatoxin" (Issue 1, Table 2). The reason for these gaps in knowledge regarding aflatoxins in Ghana probably is related to the highly technical nature of the subject and to inadequate dissemination of information on aflatoxins.

In the present study only 14% of the responding professionals have had any formal training on aflatoxins (Issue 14, Table 1) or had attended a workshop on the subject (Issue 12, Table 1). More awareness of the aflatoxin problem is, therefore, needed in Ghana. This process could include workshops during which the subject is properly explained to participants. The participants would then become secondary disseminators of aflatoxin-related information in their community, and would be encouraged to share information on aflatoxins with others, as aflatoxin information dissemination in Ghana even by agricultural/health and related personnel is poor (Issues 15 and 16, Table 1).

Conclusions

In conclusion, we found that there is some awareness of aflatoxins among health, agricultural and other professionals in Ghana, but that the same cannot be said of the general Ghanaian population. The ignorance of the general populace, together with the inability of most of the responding professionals to accurately identify the harmful effects resulting from aflatoxin exposure, suggests that the problems associated with aflatoxins are not well appreciated in Ghana. Of the professionals, the physicians were the most knowledgeable of aflatoxins and its associated problems, followed in descending order by pharmacists, biologists, agriculturists, nurses and the other social/physical scientists. Aflatoxin awareness creation workshops should have a central role in aflatoxin management interventions in Ghana. Such workshops should involve all stakeholders, particularly the agricultural extension agents who are the frontline officers in agricultural information dissemination in Ghana and, nurses, who by the nature of their work, associate more closely with patients. These categories of professionals also were the least knowledgeable of aflatoxins and the problems it poses. The relatively low rate of response to the 2003 survey means that a follow-up survey might be worthwhile, especially for physicians, to confirm the accuracy of the present findings.

Acknowledgements

This study was funded by the USAID Peanut CRSP.

References

Awuah, R.T. and Kpodo, K.A. (1996) High incidence of *Aspergillus flavus* and aflatoxins in stored groundnut in Ghana and the use of a microbial assay to assess the inhibitory effects of plant extracts on aflatoxin synthesis. *Mycopathologia* 134, 109-114.

Beardwood, C.J. (1964) Detection of aflatoxin in groundnuts in Accra. *Ghana Medical Journal* 3, 87-88.

Chen, C.T. (1992) Perak, Malaysia, mass poisoning. Tale of the nine emperor gods and rat tail noodles. *American Journal of Forensic Medicine and Pathology* 13, 261-263.

Dimitri, R.A. and Gabal, M.A. (1996) Immunosuppressant activity of aflatoxin ingestion in rabbits measured by response to *Mycobacterium bovis* antigen. 1. Cell mediated immune response measured by skin test reaction. *Veterinary and Human Toxicology* 38, 333-336.

Koomson, K. (1998*a*) Shock scientific report: Kenkey causes cancer. *The Ghanaian Chronicle* (Accra, Ghana). August 17-18 edition, p. 1.

Koomson, K. (1998*b*) Kenkey: The risk is real, says Danish professor. *The Ghanaian Chronicle* (Accra, Ghana). August 31-September 1 edition, p. 1.

Koomson, K. (1998*c*) The kenkey debate: Must we tell the people? (1). *The Ghanaian Chronicle* (Accra, Ghana). September 2-3 edition, p. 3.

Kpodo, K.A. (1995) Present status of research on the aflatoxin problem in groundnut in Ghana. *Arachide* 6, 5.

Kpodo, K.A., Sorensen, A.K. and Jakobsen, M. (1996) The occurrence of mycotoxins in fermented maize products. *Food Chemistry* 56, 147-153.

Mintah, S. and Hunter, R.B. (1979) The incidence of aflatoxin found in groundnut (*Arachis hypogaea* L.) purchased from markets in and around Accra, Ghana. *Peanut Science* 5, 13-16.

Obasi, S.C., Njoku, O.U., Obidoa, O. and Ononogbu, I.C. (1996) Effect of single oral doses of scopoletin and aflatoxin B_1 on the bleeding time, serum cholesterol and phospholipid levels of Guinea pigs. *Nutrition Research* 16, 667-672.

Odei, V. (1998) The kenkey palava. *Daily Graphic* (Accra, Ghana). August 28 edition, p. 7.

Oyelami, O.A., Maxwell, S.M., Adelusola, K.A., Aladekoma, T.A. and Oyelese, A.O. (1997) Aflatoxins in the lungs of children with Kwashiorkor and children with miscellaneous diseases in Nigeria. *Journal of Toxicology and Environmental Health* 51, 623-628.

Ramos, J.J., Fernandez, A., Saez, T., Sanz, M.C., and Marca, M.C. (1996) Effect of aflatoxicosis on blood mineral constituent of growing lambs. *Small Ruminant Research* 21, 233-238.

Sahoo, P.K., Chattopadhyay, S.K. and Sidkar, A. (1996) Immunosuppressive effects of induced aflatoxicosis in rabbits. *Journal of Applied Animal Research* 9, 17-26.

Season, A. (1998) The battle of the kenkeys. *The Mirror* (Accra, Ghana). September 12 edition, p. 7.

29

Institutional Aspects of Sanitary and Phytosanitary Issues in ECOWAS Trade

Jacqueline d'A Hughes, Ranajit Bandyopadhyay*, Kehinde Makinde and Sarah Olembo

Abstract

The Economic Community of West African States (ECOWAS) has accepted trade liberalization and globalization as important policy directions. West African trade with Europe and the United States is already much greater than trade with other developed countries or intra-regional trade, although trading with developed countries may entail considerable difficulties due to trade regulations and the need to conform to Sanitary and Phytosanitary (SPS) standards. There is generally a low level of awareness of quality standards among produce exporters in West Africa. SPS focal points are not established in all countries, which makes it difficult for exporters to check on standards and requirements. Frequent changes to standards, excessive procedural requirements, high costs for testing and certification, and a lack of transparency in the application of standards combine to compromise the ability of many countries to comply effectively with SPS. Many West African countries have not upgraded their national SPS systems in response to the introduction of the SPS Agreement, leading to differences between local and international standards that makes meeting standards difficult for firms that do business in multiple markets. There also is insufficient testing capability to meet the needs for international trade and a lack of regional coordination. Standards application is not enforced in a number of countries, but others have set up institutions for testing, certification, and quality control of both domestic products and imported goods. The effectiveness of these agencies often is weak due to inadequate equipment, a dearth of skilled technical personnel, inability to assess risks, inadequate laboratory accreditation, and a lack of enforcement.

Introduction

The introduction of trade liberalization through the World Trade Organization (WTO) and the dismantling of traditional trade barriers were expected to provide considerable export growth opportunities for developing countries to satisfy emerging markets in developed countries. However, in less than a decade of operation, it is clear that developing countries are having problems realizing this potential. Technical, non-tariff regulations in developed countries are constraining the growth of agricultural food and produce exports of many developing countries, including those in Africa (Townsend, 1999; Henson and Loader, 2001; Otsuki *et al.*, 2001; Kandiero and Randa, 2004; Jha, 2005).

West African countries face enormous and rather daunting challenges to implement and comply with international standards. The negative impact of standards and technical regulations on the trade positions of West African countries cannot be overemphasized. These countries seek to increase production for export, but there are substantial difficulties because of the influence of trade regulations on trade patterns and the ability of producers to enter new markets. Evidence suggests that the losses associated with divergent national regulations, both in exporting and importing countries, may keep producers in developing countries from entering new markets. An improved understanding of the positive effects of Sanitary and Phytosanitary (SPS) measures could alleviate the situation in which developing countries claim that their access to markets in developed countries is constrained by arbitrary, unreasonable technical requirements. A better understanding of both the positive and the negative effects of the technical issues might increase the willingness of developing countries to participate fully in new rounds of trade negotiations.

All West African countries except Liberia are members of WTO. The obligations under the agreement apply equally to all member countries. Thus, compliance with national and international standards is binding and becoming a prominent issue in successful export promotions. The economies of most West African states are predominantly agricultural and these countries rely on the export of primary products for more than half of their export earnings. To increase their access to global markets these economies must increase product competitiveness, strengthen their ability to promote trade, and meet global demands and standards. The current emphasis on expanding intra-regional trade, and regional integration and market opportunities, shows that the Economic Community of West African States (ECOWAS) member states are aware of this requirement. Increased emphasis on standards and quality would enhance these efforts to boost participation of ECOWAS countries in regional and global trade.

Data and their analysis

To provide evidence of the importance and scope of SPS issues in West Africa, data were collected from primary and secondary sources. Primary data were collected from export, trade, standards, and quarantine agencies in ECOWAS countries by using questionnaires either during personal visits to these agencies or distributed by mail to key contacts in the ECOWAS region. Secondary data were collated from websites of national, regional and international data sources. Some of the available data must be interpreted with caution as most West African countries have inadequate reporting practices. Trade barriers and restrictive exchange controls in Africa provide incentives to falsify the customs vouchers that are used to compile trade statistics, and goods in some countries flow through unofficial channels that may not be included in the available statistics. Twenty to 35% of the total trade amongst the ECOWAS countries may be unrecorded (Hardy, 1992). For example, the unrecorded trade between Togo and Ghana is believed to be several times the amount of the official trade.

Two questionnaires were designed to complement the data from secondary sources. A country level questionnaire was designed to elicit information on the pervasiveness of SPS regulations and the effects of these standards on exporting countries. The second questionnaire was intended as an industrial survey of exporters and was given to selected firms in chosen sectors. These firms were requested to provide information on cost structure, production and exports, and impediments to domestic sales, exports and operations resulting from difficulties in complying with SPS regulations. Public agencies and standard setting bodies also were surveyed to elicit information on important standards and perceived trade barriers.

Survey methods have been used in the past, particularly when other sources of information were lacking, to assess the barriers to trade faced by developing countries willing to export to the United States and the European Union. Questionnaires were sent to contact points by e-mail, and in-depth interviews were conducted in three countries: Nigeria, Mali and Ghana. This approach (Henson *et al.*, 2002) was adopted to identify the most relevant issues affecting the ECOWAS countries. A weakness of this approach is that exporting companies and governmental agencies may provide biased data if they perceive that the survey is to be used for policy purposes, or that the results could be used for politically motivated purposes or could contribute to dispute settlements.

Trade data were compiled from the websites of major trade institutions and organizations worldwide, including the United States Food and Drug Administration (USFDA), the United Nations Commodity Trade Statistics Database (UN Comtrade), the Food and Agriculture Organization of the United Nations (FAO), the Organization for the Advancement of Structured Information Standards (OASIS), and ECOWAS. Some data on rejections at destinations were obtained from reviewed documents and information from the questionnaires, as well as the international trade websites. To facilitate empirical analysis of the survey results, supplemental and anecdotal information also was collected.

This study was initiated to address the constraints faced by all fifteen ECOWAS member countries in West Africa. Three countries, Nigeria, Ghana and Mali, were selected as representative countries for further studies. The largest trade partner for West Africa is the European Union (EU). The largest EU partners within ECOWAS were Côte d'Ivoire and Nigeria. These two countries accounted for 49% of all ECOWAS imports from the EU and almost 68% of the exports from ECOWAS were to the EU. Senegal and Ghana occupy the third and fourth positions, respectively.

The European Union is gradually introducing reciprocal trade between European Union member countries and sub-regions within the African-Caribbean-Pacific (ACP) group of developing nations since the Lomé Convention provisions expired in February 2000. Agricultural commodities were recommended because they come from sectors that produce goods similar to those found in European Union countries and that are important to the ECOWAS countries due to their role in industrial development and poverty reduction. The agricultural sector and local industries are the ECOWAS sectors with the most problems due to reciprocity in access between the European Union and ECOWAS.

Applicable SPS measures, conformity assessment procedures and enforcement procedures in the major export markets were identified and evaluated for their impact on production costs, export revenues and profit margins in ECOWAS member states. Where possible, macro compliance costs were identified, and cost estimates made for conformity assessment, product quality monitoring and control services by public and quasi-public agencies for each product or product group.

Discussion

Economic and trade profile of West Africa

This section broadly describes the overall economic performance of ECOWAS.

Table 1. Per capita GDP (US$) in 2003 and growth rate during 2001-2002 of West African countries and some other countries (based on the ranking of 231 countries). Compiled from: http/www.worldfactsandfigures.com/gdp_country_desc.php for per capita GDP and world ranking, and World Bank (2004) for growth rate.

Country	Per capita GDP	World ranking	Growth rate
Benin	1,100	202	3.3
Burkina Faso	1,100	204	2.1
Cape Verde	1,400	191	-
Cote d'Ivoire	1,400	194	-3.8
The Gambia	1,700	181	-5.7
Ghana	2,200	164	2.7
Guinea	2,100	168	2.0
Guinea Bissau	900	211	-9.8
Liberia	1,000	209	0.8
Mali	900	210	1.9
Mauritania	1,800	177	0.8
Niger	800	218	-0.1
Nigeria	800	214	-3.1
Senegal	1,600	188	-1.2
Sierra Leone	500	229	4.2
Togo	1,500	190	2.4
Luxembourg	55,100	1	-
United States	37,800	2	1.4
Canada	29,700	12	2.3
Japan	28,000	18	0.2
United Kingdom	27,700	19	1.5
South Korea	17,700	49	5.7
Argentina	11,200	72	-12.0
South Africa	10,700	77	1.8
Malaysia	9,000	84	1.9
East Timor	500	231	-
World average	8,200	91	0.7

Macroeconomic indicators

The 3.7% annual growth of the Gross Domestic Product (GDP) was higher in 2003 than in any of the years from 1999-2002. The growth rate was positive in all 15 countries except Côte d'Ivoire, which experienced severe civil disturbances. Seven of the 15 countries (Benin, Cape Verde, Gambia, Ghana, Nigeria, Senegal and Sierra Leone) had growth rates that exceeded the regional average (Table 1). This growth occurred despite the weak growth of the global economy and continued structural and political constraints to improved performance in a number of countries.

Trade performance

Accurate information on the level and composition of trade flows is essential to formulate trade policy reforms or to design regional trade arrangements. Africa has fared very poorly in its international trade performance over the last two decades (Orden and Roberts, 1997;

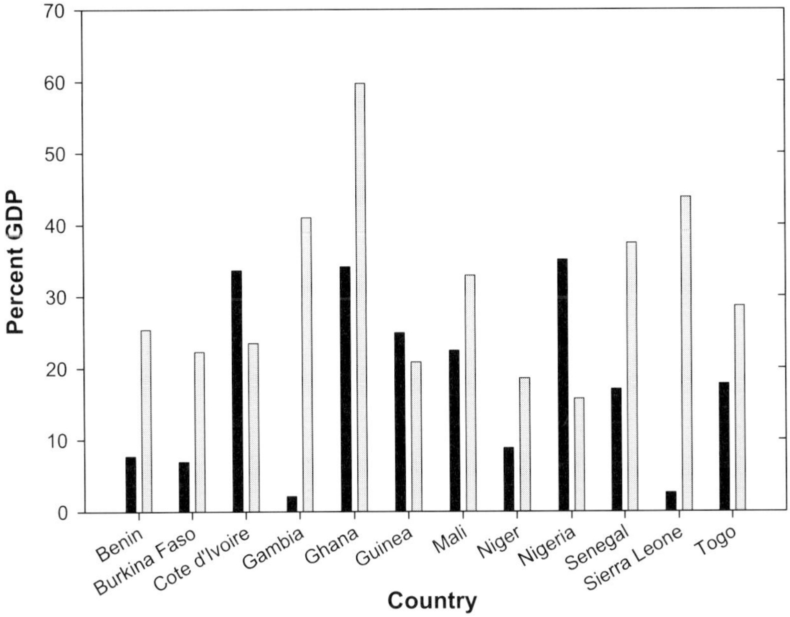

Figure 1. Imports (░) and exports (■) of goods by West African countries as % 2001 GDP. Compiled from: National Accounts of ECOWAS, ECOWAS Handbook of International Trade, ECOWAS website http://www.sec.ecowas.int/.

African Development Bank, 2004; World Bank, 2004), as its share of world exports has declined from ~ 5.5% in 1975 to ~ 2.5% in 2002 (Bora *et al.*, 2007).

Primary products dominate West Africa's exports. All ECOWAS countries except Côte d'Ivoire, Nigeria and Guinea experienced a trade deficit in 2001 (Fig. 1) with Gambia running the highest deficit. The primary products orientation of the exports means that processing within the country could increase export value and that these economies are all quite vulnerable to external shocks. ECOWAS trade with Europe and the United States is considerably greater than trade with other developed countries or intra-regional trade (Table 2), with 58% of exports from West African countries going to the EU and the United States. Trade between West African countries not only is low, but the volume of trade has fluctuated over time. In general, intra-regional trade increased steadily from \$1.8 billion in 1998 to \$2.6 billion in 2001 and dropped by 1.4 % in 2002.

Issues related to trade effects of SPS in ECOWAS

Lack of transparency and clarity

A key principle underlying the SPS Agreement is that countries have the right to decide on measures they deem necessary to protect human, animal or plant life or health in their own country. To prevent abuse, however, these measures should be based on scientific principles, should not be maintained without scientific justification and should not be applied in an arbitrary or unjustifiable manner (World Bank, 2005).

Table 2. Geographic destination of West African exports (1996-2001). Compiled from National Accounts of ECOWAS, ECOWAS Handbook of International Trade, ECOWAS website http://www. sec.ecowas.int/.

Country	Total exports (Million US$)	% of total export trade with			
		Japan	USA	Europe	Africa & the rest
Benin	304	0.2	0.2	14.6	85.1
Burkina Faso	171	1.8	0.5	54.5	43.2
Cape Verde	10	0.1	17.7	11.7	70.5
Cote d'Ivoire	5,406	0.2	7.1	53.8	38.9
Gambia	16	0.2	1.8	10.8	87.2
Ghana	1,671	1.4	5.9	45.6	47.2
Guinea	525	0	8.8	53.4	37.8
Mali	519	0	0.5	13.5	86.1
Niger	206	14.9	4.6	43.8	36.8
Nigeria	27,055	0.4	42.6	18.0	39.0
Senegal	696	0.1	0.2	14.4	85.4
Sierra Leone	41	0.1	1.0	0.2	98.7
Togo	251	0.1	0.4	6.4	93.3
Total	36,872	0.5	32.7	25.1	41.7

Despite these provisions, the SPS Agreement still has major weaknesses in the lack of transparency in the definition and application of standards, which often results in requests for clarification. For example, the 16th Technical Consultation between Regional Plant Protection Organizations held in Nairobi (30 August to 3 September 2004) highlighted the need for explanatory documentation to accompany International Standard for Phytosanitary Measures (ISPM) 15, the international phytosanitary measures developed by the International Plant Protection Convention, to clarify many issues that cause disagreements and concern. This lack of clarity results in different rules in different countries regarding SPS restrictions such as inspection of imported products, specific treatment or processing of products, setting maximum allowable levels of pesticide residues or mycotoxins, and the permitted use of some food additives. These deliberate "flexibilities" in the SPS Agreement leave room for discretion, but also lead to trade disputes when products are treated differently in different markets. Several revisions have been made to ISPM 15, and these details are updated at the International Portal on Food Safety, Animal and Plant Health (http://www.ipfsaph.org/En/default.jsp).

Lack of authoritative source from which information on standards can be obtained

At the onset of the SPS Agreement (available through the International Portal on Food Safety, Animal and Plant Health) in January 1995, all WTO Members assumed specific obligations that effectively prohibit the use of SPS measures as arbitrary or unjustifiable restrictions of trade. The obligation of each member includes the establishment of a "national enquiry point" and the designation of a national notification authority to ensure transparency on SPS matters. Familiarity with the WTO and other international standard bodies generally is low in West Africa and produce exporters often are unaware of relevant quality standards due to the absence or lack of recognition of such "enquiry points".

Complexity of SPS measures and issues

SPS measures and issues are becoming increasingly complex. In general, the SPS measures adopted by developed countries are considered incompatible with the traditional systems of production and marketing in developing countries. Developing countries view the costs of compliance to be high and sometimes prohibitive. For example, slight differences in sampling methods for the aflatoxin standard significantly influence the risk of rejection and illustrate both the increasing stringency of SPS measures and the complexity of testing methods (Jha, 2005). Many of these testing methods are relatively expensive and may be difficult to implement routinely in developing countries. Similarly, risk assessment methodologies are becoming increasingly complicated and cases with conflicts in the scientific data have yet to be resolved by WTO panels.

Capacity to challenge and defend positions on exports

The capacity to challenge and defend positions on exports regarding SPS issues often is very weak. This weakness results from inadequate human and capital resources, little information and inexperience. Private sector operators in West Africa have little or no capacity to influence the content and development of international standards even for their most important products. This inability to articulate and defend their interests also renders them vulnerable to changes in standards that may lack clarity in definition. Increased transparency in standards formulation requires that the views of the African countries be taken into account from their initial drafting to their ultimate implementation.

Lack of coherence in standards

There is a big difference between international standards and local or regional standards in many countries, with the harmonization within the EU as a notable exception (FAO, 2004), which can make exporting to multiple markets difficult. Discriminatory standards may limit trade by raising the costs of market entry, which eventually undermines global competition and imposes severe costs on consumers.

Costs of compliance

Other prominent issues include the changes in production costs, export revenues and profit margins as result of the SPS Agreement. The compliance costs for these international standards are prohibitive at both the level of the agencies responsible for standards development and enforcement as well as for the exporters and the producers. Thus, it is increasingly difficult to ensure compliance with the international standards. In some cases, standards agencies have found it difficult to develop, revise and implement relevant domestic standards and regulations in compliance with international standards and obligations. The major factors affecting compliance are insufficient financial resources and inadequate facilities, personnel, standards and technologies for African countries in general. There is a lack of adequate testing capability and much of that available is not internationally accredited (Waliyar *et al.*, Chapter 31).

Table 3. Number of contraventions cited for US Food and Drug Administration import detentions, June 1996 to June 1997 [after Henson and Loader (2001)].

Reason for contravention	Africa	Latin America & Caribbean	Europe	Asia	Total
Food additives	2	57	69	426	554
Pesticide residues	0	821	20	23	864
Heavy metals	1	426	26	84	537
Mold	19	475	27	49	570
Microbial contamination	125	246	159	895	1,425
Decomposition	9	206	7	668	890
Filth	54	1,253	175	2,037	3,519
Low acid canned foods	4	142	425	829	1,400
Labeling	38	201	237	622	1,098
Other	51	68	39	151	309
Total	303	3,895	1,184	5,784	11,166

Inadequate regional coordination

Many West African countries have neither upgraded nor reorganized their national SPS systems in response to the introduction of the SPS Agreement. These efforts should be harmonized to ensure compliance with ISPMs and SPS standards of major markets. Harmonization of African national standards with currently accepted ISPMs and other international standards should expand Africa's market opportunities since products could then be exported without further intervention to all of the countries with the same standards.

Product rejection

The following cases of rejections and the accompanying rationales are indicative of the challenges that exporters face. These examples are principally cases of bans and restriction of export products in traditional markets, especially the European Union and the United States. Discussions with trade organizations in Nigeria, Ghana and Mali, indicated that cases of rejection are incontrovertible. In most cases, however, rejection results from the refusal of the exporters to supply evidence of certification by the regulatory agencies with the obvious implication that the rejected consignments may not have met national and international standards before they were exported.

There were significant rejections of imports from Africa, Asia, Latin America and the Caribbean due to microbiological contamination, filth and decomposition between June 1996 and June 1997 (Table 3) (Henson and Loader, 2001). These rejections were attributed to the inability of these countries to meet basic heath requirements for food safety rather than the inability to test for compliance with more sophisticated standards such as those for pesticide residues and heavy metals.

Further evidence on rejection of imports from African countries is provided by a study of exporters in Ghana (Oduro, 2003). Ghanaian exporters have experienced problems with quality and phytosanitary requirements. For example, a shipment of cassava leaf exports was rejected in the UK because of the presence of insects. Reactive training of inspectors who could then identify the insects helped the country to overcome the problem.

National capacities and institutional arrangements

The need to increase consumer protection from health hazards, and the expectation that increased livelihoods will lead to greater demands for food, drugs and other health-related commodities, have led to the establishment of product quality monitoring and control services by public and quasi-public agencies. On the public sector side, SPS capability includes legislation, standards enforcement mechanisms, inspection and certification systems, monitoring and surveillance systems, management structures, trained staff, adequate laboratories and equipment, and communication systems. On the part of the private sector, there are complementary capabilities in production supervision, plant and animal health monitoring, and pest management. These systems together manage national food safety and biosecurity risks, and are expected to operate in a manner consistent with the country's obligations under the SPS Agreement.

The Senegalese peanut sector utilizes 60% of the cultivated land and 80% of the rural labor force in the country. In 2000, 60% of household agricultural income was generated from peanuts, and this sector made up 5% of GDP. Oil-mill peanut (peanut oil and oilcake) and confectionary peanut are the two major commodities from the peanut sector. Senegalese edible peanut products exported to Europe have decreased sharply in recent years, falling from 10,000 tons per year in the 1990s to approximately 500 tons in 2004. A major reason for the decline in peanut exports is stringent aflatoxin regulatory levels (4 ng/g for confectionary peanuts and 20 ng/g for oil-mill peanuts) in Europe. Aflatoxin contamination of edible peanuts occurs mainly in the field, and there is no method of detoxifying edible peanuts during processing at the factory. The agricultural practices for managing aflatoxin are well known and simple (Turner *et al.*, 2005; Waliyar *et al.*, Chapter 18). To follow these practices in the field, growers require support from extension services and incentives to market superior quality products. The cost/benefit analysis of Mbaye (2004) estimated that compliance with international standards for oil-mill production, and confectionary peanuts would accrue net benefits of CFAF 138 billion and CFAF 92 billion, respectively (US$ 1 = CFAF 490). These benefits would result from higher prices for the higher-quality products and the potential for more sales if the products met the quality standards of increasingly demanding markets. Increasing the role of the peanut industry-wide private sector would promote exports through institutional innovations, *e.g.*, aflatoxin testing and certification protocols, training and adoption of good agricultural practices.

Development of trade regulations

The ability of a country to meet SPS requirements depends on several elements including the regulatory, institutional and technical frameworks. Appropriate national legislation for implementation of the SPS is at various stages of development in West Africa. Standards application is not enforced in a number of countries, but others have set up institutions for testing, certification and quality control that are applicable to both domestic products and imported goods. Countries that have evolved quality control and monitoring systems include Côte d'Ivoire, Benin, Ghana, Guinea, Nigeria, Senegal and Togo. In Côte d'Ivoire, Benin, Nigeria and Guinea, application of standards to health-related products is compulsory, and Nigeria's environmental standards are compulsory as well. Most African countries base their standards on international standards, which implies that those standards are not substantially different from the international standards (Oyejide *et al.*, 2004). Benin, Ghana, Guinea and Nigeria have standards based on guidelines issued by International Organization for Standards (ISO), The *Codex Alimentarius* Commission and the African Regional Standards Organization (ARSO). Ghana has a domestic quality standard for exports, while Guinean exporters adopt foreign standards when selling in foreign markets.

Nigeria and Ghana have a reasonably well established body of contemporary legislation and government officials who can administer these regulations. The acceptance by importing countries of the phytosanitary certificates issued by these countries is indicative of the competence of these countries in these fields. In both countries, regulatory agencies to set and enforce standards are in place and appropriate legislation has been enacted to back their operations. In fact, the local legislation relating to standards and technical regulations predates the SPS Agreement in both countries.

In relation to the health and safety of consumer goods in Ghana, technical regulations and standards for the imports are set by the Ghana Standards Board, which is a member of the ISO and the *Codex Alimentarius* Commission. The Ghana Standards Board was originally established in August 1967 as the National Standards Board and renamed in 1973. The standardization objectives of the Ghana Standards Board include the establishment and promulgation of standards with the objectives of ensuring that high quality goods are produced in Ghana for both domestic consumption and export, promoting standardization in industry and commerce, promoting industrial efficiency and development, and promoting standards to protect public and industrial welfare, health and safety. The Ghana Standards Board also is a member of International Electrotechnical Commission (IEC), the International Organization of Legal Metrology (IOLM), and ARSO. It is the officially designated WTO–TBT (World Trade Organization – Technical Barriers to Trade) National Enquiry Point. The Ghana Standards Board is currently supported by the World Bank to seek accreditation of selected tests being conducted at Ghana Standards Board laboratories with the goal of improving the technical competence of the laboratories and improving trade.

The Food and Drugs Law of 1992 that established the Ghana Food and Drug Board invests it with the regulation and control of the manufacture, importation, exportation, distribution, use and advertisements of food, drugs, cosmetics, medical devices and household chemical substances with respect to ensuring their safety, quality and efficacy. The Board licenses and registers all manufacturers and their products and issues export certifications in accordance with international mandatory requirements. The Ghana Food and Drug Board's Quality Control Laboratory provides laboratory services in the form of quality evaluation of foods, drugs, cosmetics and chemical substances.

Enforcement mechanisms

The effectiveness of these standards agencies is weak due to constraints of inadequate equipment, availability of highly skilled technical persons, inadequate capacity in risk assessment, and a limited number of accredited testing laboratories (Waliyar *et al.*, Chapter 31).

Inspection and certification systems

The application and enforcement of SPS by Nigerian standards agencies are based on a monitoring procedure covering food products and consumption. The Hazard Analysis and Critical Control Point (HACCP) principle and a code of practices and general principles of food hygiene to manufacturing process are used in Nigeria. Similar activities also are conducted in Ghana, where there is no laboratory accredited by the International Organization for Standardization (ISO) to do the requisite tests. The Ghana Food and Drug Board also does laboratory tests. Some pre-shipment agencies operate alongside public standard agencies to effect the inspection and certification systems in the region. In Nigeria, some independent agencies are engaged by the government to verify the quality, quantity, price, financial terms and customs classification of goods imported into Nigeria.

Management structures

Although appropriate legislation is in place, the coordination and inter-agency cooperation mechanisms for the enforcement of these laws in ECOWAS are ineffective. Often the proper definition of roles and responsibilities among agencies is lacking. Public agencies responsible for standards belong to different supervising bodies, each with different targets. This separation may result in different actions for solving the same problem. Exporters also complain about cumbersome paperwork requirements and inefficient document handling.

Private sector participation

Appropriate recognition by the private sector of their role in effectively implementing standards and contributing to the enhancement of competitive production is lacking in many countries. Generally, the public and private sectors have not formed an effective partnership, which results in a lack of effective compliance with the standards, technical regulations and conformity assessment schemes required to improve both domestic and international trade. There are some examples of effective partnerships/cooperative agreements between public and private sector agencies in the region. Associations of manufacturers, traders and exporters also promote standards enforcement through self-regulatory practices. These associations are being used by National Agency for Food and Drugs Administration and Control (NAFDAC) in Nigeria to disseminate relevant information on SPS to their members. Such strong and healthy partnerships between standards agencies and stakeholders will enable the private sector to contribute to the standardization of the regulations they must operate under. Private sector organizations also could collaborate with public agencies to organize training and workshop activities that enhance public/private partnerships in food standard control strategies. Similarly the involvement of consumer organizations provides a strong lobby system that can balance pressure from industry.

Plant and animal health monitoring

There are complementary capabilities in production supervision, plant and animal health monitoring and pest management to monitor food safety and biosecurity risks in a manner consistent with the obligations under the SPS Agreement to facilitate the export of animal and plant products in conformity with the requirement of importing countries. Although legislation on sanitary standards appears to be firmly in place, the enforcement by regulatory authorities of phytosanitary issues is inadequate across the region. To promote the use of phytosanitary measures to ensure food safety standards, the 21st General Assembly of the

Inter-African Phytosanitary Council in Dakar, Senegal, drew attention to the enormous constraints and obstacles to the implementation of phytosanitary standards that exist in Africa, particularly in the areas of phytosanitary and plant protection regulation, capacity building in pest risk analysis, pest surveillance and phytosanitary inspection. Other areas of concern were certification as well as import control and transparency issues.

Conclusions

Food safety standards and the trade-off between these standards and agricultural export growth are currently at the forefront of the trade policy debate. How food safety is addressed in the world trade system is critical for the developing countries of West Africa that continue to rely on agricultural exports. In a fragmented system of conflicting national food safety standards, contentious globally accepted standards and limited capacity for certification, export prospects for the ECOWAS countries may be severely limited.

SPS regulations imposed on agricultural products affect trade patterns, the ability of exporters to enter new markets, and consumer costs. The number of SPS measures has steadily increased over time. These measures may be justified and appropriate compliance measures should be taken, but there are numerous inherent difficulties, detentions and complaints against food exporters from developing countries that reflect real SPS problems and that are not simply non-tariff barriers masquerading as SPS measures.

Significant donor resources are needed to strengthen national capacities. Large donors have a role in facilitating capacity-building programs. The Standards and Trade Development Facility (STDF) is a global program in capacity building and technical cooperation established by FAO, the World Organization for Animal Health (OIE), the World Bank, the World Health Organization (WHO) and the WTO. Countries such as Benin and Guinea have benefited from STDF funding. The similarities between the countries offer the opportunity for cooperation at a regional level for harmonization of capacity building strategies. For example, the Common Market for Eastern and Southern Africa (COMESA) has received funding from the African Development Bank in 2005 to strengthen SPS capacity and harmonization in member states of this Regional Economic Commission. Local institutions can enhance the effectiveness of development assistance programs by increasing their relevance, cost efficiency and sustainability.

The relevant institutions in these fields, however, face formidable problems. Some of these problems include: (i) the existence of significant numbers of plant pests and diseases, (ii) the threat of introduction of other serious pests and diseases, and (iii) food safety controls, especially in local food distribution chains and for street vendors, that are either rudimentary or completely lacking. It is difficult, if not impossible, to make land borders impervious to uncontrolled imports of plants and animal products, movement of wild animals and the seasonal incursions of pests and domesticated animals from one area to another.

The way forward

The major challenge for West African countries relates to capacity inadequacy at both the national and the regional levels. There is an acute need for appropriate policy responses to deal with capacity constraints. A number of trade and trade-related capacity building initiatives for Africa already exist or are in the planning stages. National training programs should be designed to educate technical personnel in laboratories in both the private and the public sectors. Since capacity constraints are a cross-cutting issue, regional training programs would provide multiple benefits including enhanced regional cooperation based on a common understanding of the growing concern about the problem of food safety

SPS Agreements should allow sufficient time for developing countries to adjust and implement new regulations and they also should be provided with appropriate technical assistance to enhance their expertise. West African countries also need to be pro-active in the enactment of legislation to adapt local standards to the requirements of external markets, *i.e.*, national standards should be based on recommendations from the National *Codex* Committee. Appropriate provision should be made for the periodic revision of these standards based on changes in the export markets particularly for significant exports. Efforts should be made to ensure that details of methodology, risk assessment and other factors are taken into consideration and shared with exporters.

Export development organizations in West African countries need to concentrate their efforts on building infrastructure and disseminating information to improve standards compliance in specific sectors. In some cases, exporters are simply unaware of the available support to enhance trade promotion in spite of awareness-raising programs promulgated by the standards agencies. Trade agencies also need to build partnerships to help them comply with export standards and related requirements and to use available sources of advice and support from within the importing country. Development of adequate capacity in SPS and other regulatory issues will allow West Africa to compete as an equal partner in global trade and by so doing address the economic constraints that are the foundation of many of the region's development needs.

Acknowledgments

We thank the Canadian International Development Agency for supporting this research.

References

African Development Bank. (2004) *African Development Report, Africa in the Global Trading System and Economic and Social Statistics of Africa.* Oxford University Press, Oxford, UK.

Bora, S., Bouët, A. and Roy, D. (2007) *The Marginalization of Africa in World Trade*. Research Brief No. 7. International Food Policy Research Institute, Washington, D.C., USA.

FAO. (2004) *Worldwide regulations for mycotoxins in food and feed in 2003. A compendium.* FAO Food and Nutrition Paper No 81. FAO, Rome, Italy.

Hardy, C. (1992) The prospects for intra-regional trade growth in Africa. In: Stewart, F., Lall, S. and Wangwe, S. (eds.) *Alternative Development Strategies in Sub-Saharan Africa.* MacMillan Press, London, pp. 426-444.

Henson, S. and Loader, R.J. (2001) Barriers to agricultural exports from developing countries: The role of sanitary and phytosanitary requirements. *World Development* 29, 85-102.

Henson, S., Jaffee, S., de Haan, C. and van der Meer, K. (2002) *Sanitary and Phytosanitary Requirements and Developing Country Agri-food Exports: Methodological Guidelines for Country and Product Assessments.* World Bank, Washington, D.C., USA.

Jha, V. (2005) South Asia. In: Jha, V. (ed) *Environmental Regulation and Food Safety: Studies of Protection and Protectionism.* Edward Elgar Publishing, Inc., Northampton, Massachusetts, USA, pp. 39-69.

Kandiero, T. and Randa, J. (2004) Agricultural Exports: Important issues for Sub-Saharan Africa. *African Development Review* 16, 1-35.

Mbaye, A.A. (2004) *Sanitary and Phytosanitary Requirements and Developing-Country Agro-Food Exports. An Assessment of the Senegalese Groundnut Subsector.* World Bank, Washington, D.C., USA.

Oduro, A.D. (2003) Agriculture and the new trade agenda in the WTO 2000 negotiations: Economic analysis of interests and options for Ghana. In: Ingco, M.D., Nash, J.D. and Njinkeu, D. (eds.) *Liberalizing Agricultural Trade: Issues and Options for Sub-Saharan Africa in the World Trade Organization (WTO).* MacMillan Nigerian Publishers Ltd, Lagos, Nigeria, pp. 229-263.

Orden, D. and Roberts, D., eds. (1997) *Understanding Technical Barriers to Agricultural Trade.* Department of Applied Economics, International Agricultural Trade Research Consortium, University of Minnesota, St. Paul, USA.

Otsuki, T., Wilson, J.S. and Sewadeh, M. (2001). What price precaution: European harmonization of aflatoxin regulation and African groundnut exports. *European Review of Agricultural Economics* 28, 263-283.

Oyejide, T.A., Ogunkola, E.O. and Bankole, A.S. (2004) Quantifying the trade impact of sanitary and phytosanitary standards: What is known and issues of importance for Sub-Saharan Africa. In: Maskus, K.E. and Wilson, J.S. (eds.) *Quantifying the Impact of Technical Barriers to Trade: Can it be Done?* The University of Michigan Press, Ann Arbor, pp. 137-154.

Townsend, R.F. (1999) *Agricultural Incentives in Sub-Saharan Africa: Policy challenges.* Technical paper No. 444. World Bank, Washington, D.C., USA.

Turner, P.C., Sylla, A., Gong, Y.Y., Diallo, M.S., Sutcliffe, A.E., Hall, A.J. and Wild, C.P. (2005) Reduction in exposure to carcinogenic aflatoxins by simple postharvest intervention measures in West Africa. *Lancet* 365, 1950-1956.

World Bank (2004) *World Development Report 2004: Making Services Work for Poor People.* World Bank, Washington, D.C., USA.

World Bank (2005) *Food Safety and Agricultural Health Standards: Challenges and Opportunities for Developing Country Exports.* Report No. 31207. World Bank, Washington, D.C., USA.

30

Institutional Stakeholders in Mycotoxin Issues – Past, Present and Future

Jonathan H. Williams*

Abstract

The generally accepted model for the management of mycotoxins is regulation and the institutions usually identified as having a stake in mycotoxin contamination represent either producers/suppliers or regulators/consumers at both the global and the national levels. Stakeholders also service the need for analysis and enforcement of the regulations in trade. Consumers, although paying more, are passive stakeholders. The regulatory approach is effective for developed country food systems, but is ineffective for most situations in developing countries. Food systems and scales of production, processing, and the lack of effective means of assuring food with safe levels of mycotoxins prevent regulation from being effective in developing countries where the model applied generally is one of denial or neglect. In these settings, a consumer management model is required to reduce mycotoxin exposure.

Introduction

Almost all commodities are at risk of contamination by one or more mycotoxins. Mycotoxins are common, and are considered by the Food and Drug Administration of the United States to be unavoidable. Thus everyone is a stakeholder in the mycotoxin problem as a consumer of potentially contaminated food. In developed countries, the popular knowledge of these toxins usually is minimal, since regulated food systems operate to provide safe food and such knowledge is not necessary for an individual's well being. In developing countries the poor people are the most at risk of exposure to mycotoxins since contamination commonly is associated with poorer quality and lower priced foods that can usually be avoided by the affluent.

Aflatoxins are the best known and most heavily regulated of the mycotoxins. There are individual stakeholders in aflatoxin contamination all along the value chain from producer to consumer and beyond, but this chapter is about the institutions that presently represent the individuals with a stake, and those who should have a role in the future. Since our perceptions of the problem change as our knowledge, economy and technology evolves, the stakeholders also change. Thus, we can identify a number of classes of institutional stakeholders: (i) those actively involved; (ii) those with mandated involvement; and (iii) those with a logical reason for future involvement. Rather than provide a detailed list, these institutional stakeholders are considered by their generic functions.

Two major classes of stakeholder can be defined – those on the supply side of the issue and those on the demand side (consumers), although some entities may be both a consumer of a commodity for processing and a supplier of a processed product. Thus, the institutional perspectives can be complicated. On the supply side of the equation there are the stakeholders that produce, trade, process and add value to the commodity. Their returns on investments are dependent upon the level of contamination in the commodities. The mycotoxin of concern depends on the commodity and the fungal species involved. A general model is provided by aflatoxin, which is perhaps the most extensively investigated mycotoxin, has numerous stakeholders, is the first identified and most extensively regulated mycotoxin; and is the toxin about which the author has the most knowledge.

Aflatoxins are a group of secondary metabolites produced primarily by *Aspergillus* spp. when they invade the tissue of many food products. These fungi grow most actively if the water content is between 10% and 18%, *i.e.*, partially dried produce, and when the temperatures are between 24 and 33°C. Such conditions often are found in crops exposed to drought or in storage, particularly in the tropics. Infection and contamination may occur preharvest, but most aflatoxin is synthesized postharvest. A number of staple tropical crops, *e.g.*, maize, peanut, rice and cassava can be contaminated and temperature conditions in the tropics and subtropics often are appropriate for contamination to occur. For example, aflatoxin contamination of maize is a common annual risk for farmers in Texas, but is an issue for corn belt farmers in the United States only in years of drought and abnormally high temperatures.

A large number of facts and intersecting forces must be described and then synthesized to make the present situation clear and to make logical predictions of the future for stakeholders in the issue.

Sequences of institutional involvement

There is a natural sequence of institutional involvement in any mycotoxin problem. Aflatoxin is a mature hazard in this regard so reviewing this sequence provides details about both the toxin and established management approaches. It also provides some insight into failures in the present paradigms for managing mycotoxins and enables us to predict that a new set of institutions should be involved in the issue. Maturity, after all, is a relative state.

Mycotoxin discovery

Discovery of a mycotoxin usually occurs in the animal science or veterinary fields because animals have a wide range of susceptibilities to mycotoxins and act as biological sentinels for problems. The institutions involved at this stage are the veterinary and toxicology research institutions. Once a possible risk to human health is defined the regulatory and food production agencies become involved as well.

Problem description

Description of the problem is the next stage, with the involved institutions seeking to establish a comprehensive description of the scale and scope of the problem, the difficulties of addressing it and the economic and health consequences. This description is needed for an economic assessment of the problem. Investigations now follow a standardized procedure for each newly identified hazard, and from these investigations a national or international set of regulations are proposed, adjusted to optimize the protection and cost considerations, and put into place.

Regulation

Regulation is established at both the national and the international level, but for different reasons. The first reason is to protect the health of citizens, while the second is to protect the assets of traders and industries. Economics plays an important role in the development of regulations. A critically important consequence of such an economic analysis is the development of dual standards for commodities that are used for both animal feeds and human foods. The limits to contamination allowed in animal feeds usually are guided by acute toxicities, while the limits applied to humans are, in the case of carcinogens, driven by cumulative toxicities, and thus may be very low. There usually is a significant difference between the levels set for animals and those set for humans with consequences to knowledge and management strategy.

Management

Management comes into effect as soon as people find that they have a problem, *i.e.*, that the mycotoxin is affecting their economic activity regardless of whether it is regulated. For example, if animals die or produce poorly, then farmers avoid the sources of contamination as soon as these sources are identified. As soon as there are market consequences, then the producers of a commodity usually respond with attempts to manage the problem that allow them to remain competitive and to continue selling their product. Management requirements usually are defined once research institutions identify the factors that are contributing to contamination and suggest ways to optimize the management response. For example, the water content range for aflatoxin production suggested that drying was essential in the American environment and has lead to the development of drying capacity for entire crops. Presently this capacity is located at buying points. Extension services and commercial laboratories are stakeholders from this phase of the process onwards as they, respectively, transfer research results to producers and monitor contamination.

Enforcement

Enforcement of regulations also requires that government institutions representing the consumer invest significantly in enforcement mechanisms, and that industries invest to minimize their risks in the given regulatory environment. Enforcement requires surveillance, laboratories, legal accountability and the people to implement them. The ability to enforce regulations is a feature of the mycotoxin management system that is a major challenge for most developing countries.

Toxicology

It is not the purpose of this chapter to detail the biochemical aspects of mycotoxin activity. However, some important toxicological concepts must be described as they condition how these toxins are perceived. Such perceptions help determine institutional interests and stakes. As in all toxicology the dose is critical and for aflatoxin three dose-related classes of intoxication can be defined.

Acute poisoning occurs when sufficient toxin has been ingested to cause the appearance of a set of characteristic symptoms, illness and/or death of the intoxicated person. For aflatoxin the characteristic symptoms manifest as severe and acute hepatotoxicity, anorexia, malaise, low-grade fever, vomiting, abdominal pain, jaundice, hepatitis, and death. A high mortality rate is observed in people with these symptoms. However, humans are one of the more "tolerant" species with a high LD_{50} relative to that of many domesticated animals, and outbreaks of acute human aflatoxicosis are rare enough to be newsworthy when they are correctly diagnosed.

Chronic asymptomatic poisoning occurs when exposure to the toxin is insufficient to provoke the characteristic symptoms of acute exposure. Often, people suffering from this problem don't even know that they are being exposed to the toxin. The symptoms vary with the toxin, but for aflatoxin the consequences of sustained chronic exposure include immune suppression, nutritional interference, teratogenic effects, and cancer promotion through genetic mutations.

Cumulative poisoning results from damage, normally genetic, from repeated exposures that enhance the risks of adverse events, particularly cancer. For regulatory purposes, tolerable dosages generally are very low for toxins with cumulative or genetic toxicities.

Species differences and these basic forms of toxicity interact in determining regulations and managing toxins in food. Humans may be exposed to a toxin over a long period of time, so cumulative poisoning is a serious consideration. In contrast, farm animals usually are not fed for extended periods of time so the thresholds for regulation often are set at levels that ignore the cumulative form of toxicity, but avoid the acute toxicities that would cause immediate economic losses to those engaged in animal feeding operations.

Management models for mycotoxins

We also must understand the currently available models for managing mycotoxin contamination problems. Three main approaches exist with the management strategy usually dependent upon governmental responses to the contamination problem.

Denial/neglect

This response is common and may result from limited institutional capacity, perceptions of health priorities, and economic considerations. Developing country health systems do not need to look for reasons to justify their existence. Instead they usually are trying to address serious, immediate problems and have neither the time nor the resources to invest in or provide the most cost-effective long-term interventions. The food systems in these countries are not easily regulated and they usually lack the resources to enforce existing regulations. Consequently, a wide range of degrees of neglect exist and it is only when there are significant economic losses, or the health consequences are clear that action is taken. Kenya responded to the outbreak of acute aflatoxicosis in 2004 (CDC, 2004; Okioma, Chapter 11) and 2005, but earlier reports of market contamination were largely unheeded, *e.g.*, Muriuki and Siboe (1995), since Kenya does not export these commodities. Denial also is a common response – particularly when the prevailing exposure occurs in the chronic range and other diagnoses, *e.g.*, infectious diseases are possible. Even when toxins are shown to be a component of the food being consumed, the response may be that the food is a traditional product and has not "harmed" people previously. Governments may consider either that a mycotoxin problem, if/when it is publicized, does not apply to their situation, or that it is an anomaly, outside of the normal set of events and not worthy of a response that will consume resources into the future. Many governments promulgate regulations and then neglect to provide the means to enforce them for obvious practical and economic reasons – it can be expensive to provide the services that quality assurance regulations demand. The challenge for science is to show that the cost of neglect is greater than the cost of enforcement.

Based on reported market food contamination, biomarkers for selected populations, and food system considerations (Williams *et al.*, 2004), some 4.5 billion people living in developing countries probably live with uncontrolled exposure to aflatoxins.

Regulation

This model is the present official policy of most governments since most have subscribed to *Codex Alimentarius*. Many countries have national regulations that define the allowable contamination levels in foods, but the success of this approach depends on empowered institutions being charged with protecting the public interest and health.

The intent of this model is to provide public health and food safety for the general population, but also to protect the interests of the supply side of the equation by making the regulations achievable. Balancing these interests has been the task of the WHO/FAO Joint Expert Committee on Food Additives (JECFA). Implementation of their recommendations has required authorities to promulgate laws, and establish mechanisms for their enforcement. In the United States, for example, independent inspectors sample each batch of peanuts delivered by farmers to buying points and make measurements to determine quality, while port authorities and their proxies (certifying laboratories) operate at the international level to protect trader interests.

The regulatory model clearly burdens the suppliers of commodities with the task of assuring the quality of their product and they bear the risks of failure. If the food industry is commercial, then mycotoxin contamination can largely be prevented through pre- and post-harvest management. For example, irrigation is a powerful factor influencing the risk of preharvest fungal invasion and mycotoxin contamination. Timely harvest and rapid drying regardless of the weather conditions also reduce the mycotoxins present. The cost of management and processes to meet a regulatory standard increase progressively as the allowable limits for the toxin decrease. Providing these services requires investment and increases the cost of production which leads to a more expensive product. In developed countries, safe food is a condition that consumers take for granted. These conditions often do not prevail in developing countries. Another consequence of regulation is that it creates market barriers for producers who lack the resources to reliably meet the required standards.

Voluntary consumer management

In this approach the consumer is responsible for managing the quality of the food consumed. Since the regulatory levels allowed in animal feeds do not protect against harmful chronic exposure and the cost of providing higher quality is a factor in profits, the self-management model is already used by large scale animal farmers in both developed and developing countries. This option is not widely recognized as a valid model for humans, but given the challenges of reducing exposure to mycotoxins in developing countries this model may be the most realistic for the countries in which 75% of the world's population live. If the scale of operation makes formal regulation uneconomic; if the producer and the consumer are the same entity; if the consumer needs greater quality than the regulations provide, or if more economic solutions than regulation to safe levels are available, then this type of regulation is the solution.

A number of approaches to assuring food quality are available. One particularly attractive method is to use enterosorbent feed additives to prevent exposure of farm animals to aflatoxin (Afriye-Gyawu *et al.*, Chapter 25). Here the regulations are defined as close to the limit for acute toxicity, but farmers can experience loss of productivity and greater disease incidence at sub-symptomatic chronic exposures. Under such circumstances, many farmers use additives that neutralize the contaminant to assure the full value of the feed. Additives usually are cheaper than methods of removing or degrading the contaminant.

Viewpoints and perspectives

Clearly, there are many different perceptions of mycotoxin problems. Each institution has a role and a perspective. The major determinants of perspective that I will consider are: (i) regulatory environment, (ii) health environment, (iii) living standards and food security, (iv) knowledge, and (v) trade and economics.

Regulated and unregulated environments

A major factor determining one's perspective on mycotoxin contamination is the presence of legally enforced regulations on the levels of a mycotoxin in foods. Safe food is considered a public good, but if one is on the supply side of the issue then the presence or absence of enforced regulations impacts management, market access and profitability. In response to these pressures, a farmer may change his choice of crops or may invest in technologies that can control and manage the contamination problem. To the protagonists of trade development, regulations may be viewed as protectionist trade barriers, but to laboratories providing analytical services they provide revenue opportunities.

Public health status

The general health situation and our knowledge of it also conditions perceptions of aflatoxin contamination. In countries with high medical and hygiene standards and where the benefits of good nutrition have added years to survival, cumulative toxicities assume greater significance than they do in countries where life expectancy is not as great due to the high incidence of infectious diseases, environmental hazards and unsafe water.

Statistics generally attribute to the presently recognized risks of aflatoxins (liver cancer) a small fraction of the burden of disease, and on this basis aflatoxin is rated a lower priority for Ministries of Health in developing countries than are many other more pressing problems. Hopefully this position will change, as it is based on untenable assumptions about the levels of exposure and toxicities occurring in developing countries.

Another factor that strongly influences the perception of aflatoxin is the presence of the hepatitis B virus, with which aflatoxin interacts synergistically. When these two agents are combined, the potency of aflatoxin as a carcinogen is increased 30 fold (Henry *et al.*, 2002). This synergism means that institutions concerned for people with hepatitis B should perceive aflatoxin as a much more dangerous toxin than they currently do. Although there are campaigns to inoculate people against hepatitis B, there currently are 80 million people with the disease for whom any aflatoxin exposure is an increased concern.

Living standards and food security

For the food insecure, the possible presence of mycotoxins in their food only gets attention when the consumption of that food leads to death sooner than the rejection of that food – and most mycotoxins do not kill their victims that quickly. Food insecurity, associated with a poor harvest, contributed to the 2004 and 2005 outbreaks of acute poisoning in Kenya (Okioma, Chapter 11). Economic stress also increases chronic exposure. In Ghana, AF-albumin levels increase with the number of children in high school (Jolly, unpublished data). The number of children in high school is an indicator of economic stress, since families must pay for school fees and books. Thus poverty is clearly a factor in determining the exposure of people to aflatoxin in the developing world. Lower quality grain with more risk of mycotoxin contamination at dangerous levels is purchased by those with limited incomes since their hunger overrides their preference for higher quality foods.

Knowledge bases

Knowledge of mycotoxins varies widely depending on the disciplinary or functional role of the institution. In most cases the driving force is the level of aflatoxin allowed by regulations in grain destined for human or animal consumption. For humans, where regulations are enforced, which usually also is where a medical research agenda has been established, the regulatory levels are set to address cumulative risks. In contrast, the levels for animal feeds are set just below the threshold for acute toxicities, so veterinary scientists usually have more experience and knowledge of the chronic toxicities of aflatoxin. In 2005, a literature search with Google Scholar™ using the keywords aflatoxin and cancer found 6,900 references, mostly in journals of human medicine, while a similar search on aflatoxin and immunity generates only 860 references, mostly in the veterinary literature. The importance of this difference in focus is critical for developing countries since an often unstated assumption is that exposure levels are consistent for humans in all locations worldwide. Unfortunately, the increased exposure in developing countries may mean that the animal data and knowledge may be more appropriate to the developing country situation than is the human-based but cancer-focused research. On a global scale, liver cancer is responsible for some 560,000 deaths each year and aflatoxin is not the only factor contributing to its incidence. Therefore aflatoxin does not make the WHO priority list (Rodgers *et al.*, 2002). This mortality is compared with some 50 million deaths attributed to infectious diseases, *e.g.*, HIV ~5 million and malaria ~3 million. If the animal-based reports of immunity and nutritional impact also apply to humans; however, then aflatoxin is contributing to suppressed immunity and poorer nutrition and is a modulating factor for infectious diseases thereby assuming a very different role in the world's health picture.

Trade and economics

The debate on the trade-off between health for developed country consumers and trade opportunities for developing countries is a powerful perception polarizer. Aflatoxin has been characterized as a useful tool to protect markets and to prevent market access by developing countries seeking to sell their products to the developed countries. Recent publications have estimated the trade-off between trade declines and the additional health benefits provided by the European Union regulations (Otsuki *et al.*, 2000; Coulibaly *et al.*, Chapter 7) adding to the perception that health-based regulations are being used to manipulate market access.

I think that a more important question for developing countries is, "What is the true health cost of neglecting aflatoxins?" I suspect that the health costs incurred are much higher than the trade benefits lost. If one stops at the point of sale one, then the conclusion that the economic loss through reduced market opportunity is the most important issue is easily reached. If the effects of aflatoxin include the full impact on health then a very different economic reality may become evident. Embracing this reality is an important challenge for developing nations. In developed countries we can estimate the cost of management actions to minimize contamination, and the costs of failing to meet the standards. But the cost of not addressing aflatoxin contamination must include both the lost exports and the loss of additional health costs incurred due to uncontrolled exposure.

Present stakeholders

In broad terms the present stakeholders are:

Trade related

The present concerns of developing countries in relation to aflatoxin are those associated with international trade. Countries and politicians are focused on the trade consequences of *Codex Alimentarius*, but neglect the fact that foods that are rejected in developed countries are consumed regularly by their own populations. The World Bank analyses have focused on this issue, but have not sought to quantify the health cost. For peanuts, the trade cost is based on only a very small portion of the total production. The FAO statistics database indicates that only 230,000 t of whole peanut from the 37 million tons of the peanut produced are involved in formal cross border trade (FAO, 2005). Our data show that there is little economic gain for investing in management practices that decrease the aflatoxin contamination in foods consumed in developing countries. Thus, there is little price incentive to change the contamination levels in peanuts.

Consumers

A major divide on the aflatoxin scene is the one separating human stakeholders in developed and developing countries. A reasonable representation is that the consumer in developed countries eats food with low levels of contamination and with the presence of a major divide between developed and developing countries (Table 1). This divide is the result of many of the factors discussed above and clearly demonstrates that the institutions that should protect populations in developing countries are failing to do so.

What are the consequences of this institutional failure? The animal data summarized by Williams *et al.* (2004) and the human epidemiological data (Gong *et al.*, 2002; Jiang *et al.*, 2005) that show that the changes in animal immunity and growth also occur in humans with chronic aflatoxicosis, suggest that aflatoxin is an important factor in the general health of human populations.

The possible connections between the risks to health defined by the WHO and aflatoxin have been reviewed and summarized by Williams *et al.* (2004). Based on published research these authors found logical, mechanistic grounds for aflatoxin modulation of risk factors that provide some 43% of the Disability Adjusted Life Years (DALYS) in the WHO risk analysis (Williams *et al.*, 2004). Since the above analysis, there has been further demonstration of the

Table 1. Aflatoxin exposure (ng/kg body wt/day) in some regions and countries.

Location	Aflatoxin exposure
Sub-Saharan Africa	
Gambia	4-115 (Hall and Wild, 1994)
Kenya	3.5-15 (Hall and Wild, 1994)
Mozambique	39-180 (Hall and Wild, 1994)
South Africa[a]	17 (Hall and Wild, 1994)
Swaziland	5.1-160 (Hall and Wild, 1994)
Zimbabwe	10-115 (Henry *et al.*, 1998)
Australia	0.15 (Henry *et al.*, 1998)
Asia	
South Korea	1.2-5.8 (Park *et al.*, 2004)
Thailand	6.5-53 (Hall and Wild, 1994)
Western Europe	0.3-1.3 (Henry *et al.*, 1998)
United States	18 ng/person/day (Henry *et al.*, 1998)

[a]Transkei province.

role of aflatoxin in these risk factors. Jolly *et al.* (Chapter 5) observed decreased cellular immunity in the most aflatoxin exposed half of the naturally exposed population in Ghana. The two most important risk factors; underweight (14.9% of DALYS) and risky sexual behavior and HIV (10.2% of DALYS) are correlated with aflatoxin exposure. The underweight status of children in West Africa has been related to their historic aflatoxin exposure as measured by AF-albumin adduct and interventions to decrease the exposure have been able to reverse this situation (Gong *et al.*, 2002, Chapter 6). In the case of HIV the connection remains statistical. Based on data in Table 1 and the data for HIV in those countries (UNAIDS, 2004) the relationship between aflatoxin exposure and HIV is statistically significant ($p < 0.057$). The statistical relationship is supported by mechanistic linkages (Williams *et al.*, 2006).

Future stakeholders

The demonstration of immune and nutritional effects of aflatoxin in the populations of developing countries implies that a new set of institutional stakeholders should be involved in the aflatoxin arena. The immune and nutritional implications of this toxin should make it a factor in most high priority issues for the Ministries of Health for governments in developing countries, who may need to reevaluate their priorities in the light of these emerging data.

The "cure" for aflatoxins though, will not be the provision of vaccines and nutritional supplements. Instead, the medical and agricultural extension services must educate people on the consequences of neglecting mycotoxin contamination. Since my assessment also is that food system considerations make enforced regulations a strategy unlikely to succeed in the near future, the model of consumer regulation probably is the best option. People need to be educated as to the effects of not paying attention to aflatoxin and encouraged to create the economic pressure that will result in better quality and self-regulating markets. Thus, the ultimate institution for the management of aflatoxin is educated consumers.

Acknowledgments

This chapter was prepared with support from USAID grant (LAG–G-00-96-90013-00) to the Peanut Collaborative Research Support Program.

References

CDC. (2004) Outbreak of aflatoxin poisoning – Eastern and central provinces, Kenya, January-July 2004. *Morbity and Mortality Weekly Report* 53, 790-793.

FAO (2005) http://faostat.fao.org/site/336/default.aspx (Databases interrogated 1 Nov 2005).

Gong, Y.Y., Cardwell, K., Hounsa, A., Egal, S. Turner, P.C., Hall, A.J. and Wild, C.P. (2002) Dietary aflatoxin exposure and impaired growth in young children from Benin and Togo: cross sectional study. *British Medical Journal* 325, 20-21.

Hall A.J and Wild C.P. (1994) Epidemiology of aflatoxin-related disease. In: Eaton, D.L. and Groopman, J.D. (eds.) *The Toxicology of Aflatoxins: Human Health, Veterinary, and Agricultural Significance*. Academic Press, London, pp. 233-258.

Henry, S.H., Bosch, F.X. and Bowers, J.C. (2002) Aflatoxin, hepatitis and worldwide liver cancer risks. *Advances in Experimental Medicine and Biology* 504, 229-233.

Henry, S., Bosch, F.X., Bowers, J.C., Portier, C.J., Petersen, B.J. and Barraj, L. (1998) Safety evaluation of certain food additives and contaminants. *WHO Food Additives Series 40*. WHO, Geneva, Switzerland, p. 359-468.

Jiang, Y., Jolly, P.E., Ellis, W.O., Wang, J.S., Phillips, T.D. and Williams, J.H. (2005) Aflatoxin B_1 albumin adduct levels and cellular immune status in Ghanaians. *International Immunology* 17, 807-814.

Muriuki, G.K. and Siboe, G.M. (1995) Maize flour contaminated with toxigenic fungi and mycotoxins in Kenya. *African Journal of Health Science* 2, 236-241.

Otsuki, T., Wilson, J.S. and Sewadeh, M. (2000) Saving two in a billion: Quantifying the trade effect of European food safety standards on African exports. *Food Policy* 26, 495-514.

Park, J.W., Kim, E.K. and Kim, Y.B. (2004) Estimation of the daily exposure of Koreans to aflatoxin B_1 through food consumption. *Food Additives and Contaminants* 21, 70-75.

Rodgers, A., Vaughan, P., Prentice, T., Edejer, T.T., Evans, D. and Lowe, J. (2002) Quantifying selected major risks to health. In: Campanini, B. and Haden, A. (eds.). *The World Health Report 2002: Reducing Risks, Promoting Healthy Life*. World Health Organization, Geneva, Switzerland, pp. 47-97.

UNAIDS. 2004 Report on the global AIDS epidemic. http://www.unaids.org/bangkok2004/GAR2004_html/GAR2004_00_en.htm.

Williams, J.H., Phillips, T.D., Jolly, P.E., Stiles, J.K., Jolly, C.M. and Aggarwal, D. (2004) Human aflatoxicosis in developing countries: A review of toxicology, exposure, potential health consequences and interventions. *American Journal of Clinical Nutrition* 80, 1106-1122.

Williams, J.H. (2006) Connecting the dots: The links between aflatoxin and the HIV epidemic. http://168.29.148.65/pdfs/af_hiv2005.pdf.

31

Institutionalizing Mycotoxin Testing in Africa

Farid Waliyar, Moses Siambi, Richard Jones, S. Veera Reddy, Dyborn Chibonga, P. Lava Kumar and Stella Denloye

Abstract

The aflatoxin class of mycotoxins is one of the most important food safety concerns at domestic and international levels. The widespread occurrence of aflatoxin contamination in Africa and other tropical countries is a major potential hazard to humans and domesticated animals, and causes severe health and economic problems. Africa is in a precarious position because the environmental conditions and cultural practices across the continent favor fungal attacks of crops and commodities. Other factors that exacerbate African problems include constraints in resources and infrastructure, a lack of adequate regulatory and control systems for monitoring mycotoxin contamination, and limited availability of food due to war, famine and other natural disasters. Several countries in Africa have established regulations on mycotoxins in food and feed to safeguard both health and trade interests. Due to inadequate facilities for monitoring mycotoxins, the institutionalization of food safety regulations in Africa has been difficult. The availability of accurate, cost-effective testing procedures for rapid mycotoxin analysis is a prerequisite for the enforcement of food safety regulations and to facilitate international trade. Recently, we established aflatoxin-testing facilities in Nigeria, Malawi, Mali and Mozambique. These facilities enable farmers and traders to make sales in the high-value international trade markets. In this chapter, we discuss the problems and opportunities for institutionalizing aflatoxin-testing facilities in Africa.

Introduction

Concerns about food safety are escalating globally, particularly in developed countries where food safety issues influence consumer perceptions and policies with respect to food production, processing, handling and trade. In developing countries, health risks from food and water-borne diseases are a major food-safety concern (Unnevehr, 2003). However, with the quest by developing countries for a larger share of the global food trade and the desire to earn the necessary income for development, attention is now being paid to food safety issues. The contamination of food and feed by mycotoxins (toxic metabolites of fungi) are recognized as significant sources of food-borne illnesses (FAO, 2004), and top priority has been given to their regulation in food and feed at the domestic and international levels.

Mycotoxins that pose major health risks include aflatoxins, deoxynivalenol, fumonisins, ochratoxins and ergot alkaloids (Detroy *et al.*, 1971; Table 1). Among the mycotoxins, aflatoxins raise the most concern. Aflatoxin B_1 is a potent mycotoxin that is distributed widely, of-

Table 1. Major toxigenic fungi, their toxic metabolites, target effects and methods used for mycotoxin detection.

Toxigenic fungi	Mycotoxin	Toxic effects	Commodities	Analytical Method
Aspergillus flavus *Aspergillus parasiticus*	Aflatoxins (B_1, B_2, G_1, G_2, M_1 and M_2)	Acute toxicity (especially to the liver), liver cancer, immune suppression, growth abnormalities, lowered productivity	Cereals, nuts, spices, dried fish, feed, tea, cocoa and cocoa products, milk and milk products	HPLC, ELISA, HPTLC
Aspergillus ochraceus *Penicillium* spp.	Ochratoxin A	Acute toxicity (especially to the kidney) and cancer	Cereals, beer, cocoa, oats, wheat, dried fruit and pulses	HPLC, ELISA
Fusarium verticillioides *Fusarium proliferatum* *Fusarium avenaceum*	Fumonisins	Leukoencephalomalacia, neural tube defects, pulmonary edema, esophageal cancer	Maize and maize products	HPLC, ELISA
Fusarium graminearum	Trichothecenes, Zearalenone	Acute toxicity, immune suppression, reproductive dysfunction	Wheat and maize	HPLC, ELISA

ten at high levels, in staple foods such as maize, peanuts and other commodities. The widespread occurrence of aflatoxin in Africa and other tropical countries make aflatoxin contamination a major potential hazard to human and animal health, and an important non-tariff barrier to international trade. Environmental conditions in Africa, *e.g.*, high temperatures, high humidity, terminal drought, unseasonable rains during harvest and flash floods, all favor fungal proliferation and the production of mycotoxins. In addition, constraints in resources and infrastructure for food storage, processing and preservation, the lack of adequate regulatory and control systems for monitoring mycotoxin contamination, and the limited availability of food due to war, famine or other natural disasters increase food safety problems.

Since the discovery of the aflatoxins in the 1960s, regulations have been established in many countries, and newer regulations are still being issued to control the level of aflatoxins and other mycotoxins in products meant for consumption by humans or domesticated animals. In most African countries, specific mycotoxin regulations do not exist, and in the 15 African countries where regulations do exist they primarily concern aflatoxins (FAO, 2004). Although all countries recognize the potential threat posed by aflatoxins to human health and trade, this threat has not been addressed due to various socio-economic and political reasons. For example, aflatoxin regulations have limited effects on health protection in African countries, because most farmers grow at most enough crops for their own consumption (Shephard, 2004). The increase in the number of deaths due to consumption of aflatoxin-contaminated food (Azziz-Baumgartner *et al.*, 2005), and in farmers' interest to gain revenue through export of agricultural commodities, however; are creating a demand for better regulation of aflatoxins in food and feed. Numerous African countries lack the capacity and facilities for testing for aflatoxins. Lack of well-equipped laboratories, trained personnel, strict legislation for inspection, surveillance and monitoring, poor communication and information systems, and weak government commitment to the regulation of mycotoxins in food and feed all contribute to the problem. This chapter discusses the problems and opportunities for establishing aflatoxin-testing facilities in Africa with specific examples from Nigeria and Malawi.

Need for mycotoxin testing and monitoring

Health effects of mycotoxin contamination

The effects of long term exposure to aflatoxins in humans are well documented and include damage to body organs such as the liver and suppression of the immune system, which increases the frequency of other health disorders (Williams *et al.*, 2004). Cases of acute toxicity have been reported and these often have been exacerbated by limited availability of food supplies, *e.g.*, in the drought-stricken areas of Kenya where in 2004 over 100 people died from consuming aflatoxin-contaminated maize meals, the major cereal in African diets (Azziz-Baumgartner *et al.*, 2005). Additional cases also were reported in 2005. High concentrations of aflatoxin in maize also were reported in Nigeria, Ghana and several other West African countries (Kpodo and Bankole, Chapter 9).

Millions of children in Africa suffer from acute or chronic aflatoxicity, which often results in lowered immunity, low birth weight, stunted growth and even death (Wojnowski *et al.*, 2004). Studies conducted in Togo and Benin showed that 99% of 480 children aged between 9 months and 5 years had high concentrations of an aflatoxin-adduct in the blood (Gong *et al.*, 2002). Growth of children in high aflatoxin exposure areas was stunted compared to that observed in low-exposure areas (Gong *et al.*, 2004). In all cases, white maize and/or peanuts were used as the weaning food. According to the United Nations Children's Fund, one in every five West and Central African children dies before reaching the age of five. The under-five mortality rate is 97/1000 in Ghana, 156/1000 in Benin, and 141/1000 in Togo, all countries where maize and/or peanuts are used as a weaning food (UNICEF, 2004). Small-scale peasant farmers in Africa face the problem of not producing enough food and therefore experience limited availability of food supply in terms of both quality and quantity. Due to market demand for better grain quality to fetch good prices, the best quality grain often is sold in the markets while shriveled, molded and damaged grains are retained for home consumption, incorporated into livestock feed, or sold to low-income groups in urban or rural markets. This practice increases the risk of aflatoxin exposure by local human and livestock populations with all of the accompanying health implications.

Export constraints

Aflatoxins have tremendous economic impact on international trade in the form of losses in African exports to the European Union and the United States. Aflatoxin contamination of peanuts is one of the most important constraints to their trade. Peanuts are the most important cash crop in most countries of western and southern Africa. Africa produces 20% of global production, 60% of which comes from West Africa, mostly from Senegal, Ghana and Nigeria

(Ntare *et al.,* 2005). In these three countries, the proportion of peanuts contaminated with aflatoxin beyond the maximum permissible limit for trade is high (Kpodo and Bankole, Chapter 9).

The sharp decline in international peanut trade has had severe economic consequences for development in many African countries, especially those that rely on agricultural commodities for export earnings. Rejection of export commodities by developed countries resulted in the loss of trade (Wu, 2004). The Nigerian Ministry of External Affairs reported that food export commodities were rejected by the European Union on several occasions due to high aflatoxin levels, especially in melon seeds. Such rejections have resulted in large, but unknown, losses in revenue and created poor images of the country in international food export markets.

The screening of agricultural commodities, processed food and animal feed for aflatoxins has become essential in international trans-border trade. African countries, in most cases, are unable to meet the stringent food safety requirements of importing countries, such as the European Union, which set the limit for aflatoxins at one tenth (2 ng/g) that of *Codex Alimentarius* (20 ng/g). Many African countries, due to their limited resources and infrastructure, are at a competitive disadvantage in guaranteeing/certifying the safety of their commodity exports.

Poor regulatory and control systems

Africa's inability to meet the regulatory standards set by many importing countries is caused by inadequate or nonexistent national regulatory systems for mycotoxin monitoring in the form of standards, regulations, analytical laboratories and qualified personnel. Seventy-six countries worldwide have regulations preventing the importation of aflatoxin-contaminated food commodities (FAO, 2004). Such widespread restrictions keep the situation for African countries bleak.

Regulatory standards worldwide are implemented to safeguard the health and rights of consumers, and to protect them from undue economic exploitation as a result of the supply of inferior quality goods. Satisfying these fairly ambitious standards requires the use of laboratory-generated data on consumer products, identifying quality parameters for regulating existing products and the enforcement of compliance with existing regulations.

Some African countries have attempted to implement regulatory measures to monitor and control the quality of food products meant for both local and international consumption. In Nigeria, the National Agency for Food, Drug, Administration and Control (NAFDAC) is responsible for the regulation and control of the importation, exportation, manufacture, advertisement, distribution, sale and use of food in the country. The Kenya Bureau of Standards (KEBS) formulates food standards and regulates food processing in Kenya, and Ghana has a similar regulatory and control body. All three countries have established national food standards and adopted international standards such as *Codex*, but none can obtain the necessary data on mycotoxin contamination levels necessary for the policy development. The problems associated with this lack of information reinforce the importance of surveillance and capacity building for mycotoxin testing in African countries. Testing laboratories form the backbone of global regulatory activities, as all other regulatory functions revolve around them.

The NAFDAC experience in Nigeria

Establishment of an analytical, regulatory laboratory is a formidable challenge, particularly in African countries. However, the Mycotoxins Laboratory initiated and completed by NAFDAC is today amongst the best equipped on the African continent.

Political commitment

The first challenge for NAFDAC was to establish the need for such a facility. As the sixth largest oil exporting country in the world, Nigeria derives over 90% of its foreign earnings from the sale of crude oil. Thus, agriculture is at best a secondary area of trade in the overall foreign trade context for the country. Recognizing the need for regulations against mycotoxins in exported agricultural commodities was a major hurdle. Yet policy makers were convinced that to meet the regulatory requirements of importing countries and to participate in the international trade in food, the quality of agricultural commodities must be monitored. Once convinced of this point, the government committed to support NAFDAC in its quest to ensure food safety.

NAFDAC began monitoring agricultural commodities for the aflatoxins in 1998 when the presence of aflatoxins was detected in a routine screen of Chinese green tea. This discovery provided the impetus to consider a dedicated laboratory for the analysis and monitoring of mycotoxins in foods for local consumption. In December 2000, NAFDAC secured a technical cooperation project with the International Atomic Energy Agency (IAEA) on "Regulatory Control and Monitoring of Contaminants and Residues in Fresh Produce". The project's objective was to increase the safety and quality of foods, to provide up-to-date information on methods of regulatory control and to build capacity and develop procedures for mycotoxin control testing. This project essentially led to the establishment of the Mycotoxin Laboratory, which now also monitors food products meant for export.

Infrastructure

An irregular electrical supply was a chronic hindrance for the performance of the laboratory. The electricity from the national power supply was both erratic and unreliable, which affected the performance of both the instruments and the personnel. The laboratory now maintains stand-by generators that provide uninterrupted electricity on a 24-hour basis throughout the year. Computerization and effective communication systems also were lacking in the early years of the laboratory, which made sourcing information difficult. However, with support from the NAFDAC management and the IAEA, the laboratory is functional and remains current in its scientific operations.

Personnel

Trained personnel are the most essential component in the establishment of laboratories such as this one. Through an on-going project supported by IAEA, six staff members were sent to Austria in August 2002 for six months of training on mycotoxin analysis in food. When they returned, they became the founding staff for the Mycotoxin Laboratory. Portions of the Mycotoxin Laboratory are being replicated in other NAFDAC area laboratories. The Mycotoxin Laboratory is equipped with modern analytical equipment, *e.g.*, high performance liquid chromatograph (HPLC) and high performance thin layer chromatograph (HPTLC), and meets all international standards. Methods of analysis available include HPLC, HPTLC and enzyme-linked immunosorbent assay (ELISA) techniques, all of which are used worldwide as standard methods for mycotoxin analyses. The availability of these facilities and methodologies have improved staff performance, increased the reliability of results and enabled the laboratory to apply for accreditation as a regional center of reference. The laboratory can test for all of the major mycotoxins (aflatoxins, ochratoxins and fumonisins) in cereals, oil seeds, spices, nuts and milk.

Sustainable laboratory supplies

Many African countries lack good quality laboratory supplies, which are essential for reliable analytical results, as these supplies usually are imported and expensive. It costs about US$ 30,000 to purchase a HPLC and an average of US$ 50 to analyze a sample by HPLC or US$ 5 for an ELISA analysis. Another reason for inadequate laboratory supplies is the lack of a planned procurement system. To address this problem, NAFDAC reached an agreement with VWR for sustainable provision of chemicals, media, reagents, glassware and other consumables through bulk purchases on a quarterly basis.

Instrument maintenance and repairs

A major constraint for laboratories in the developing countries is the maintenance and servicing of instruments and equipment. Due to the lack of technical expertise, many African countries rely on expatriate service engineers for maintenance and repairs of equipment. These laboratories often experience long down times for instruments, which reduces their output. Instrument, Maintenance and Engineering Units have been established to enable quick repair of at least minor problems with small equipment, *e.g.*, furnaces, ovens, refrigerators, freezers and air conditioners. When repairs cannot be handled by local units, engineers from suppliers are invited to make the repairs.

Quality assurance/quality control and laboratory accreditation

Reaching the standards for a quality analysis and quality control (QA/QC) system is another major hurdle for laboratories in Africa. International norms require that laboratories operate under a set quality system, *e.g.*, ISO 17025, which specifies quality criteria for operations in a laboratory prior to accreditation. The complex requirements often are underestimated. To ensure international acceptability of the analytical results that underpin export trade, NAFDAC has developed QA/QC requirements in the Mycotoxin Laboratory that are essential for accreditation.

The NASFAM experience in Malawi

Peanuts are subject to stringent food safety requirements usually presented as maximum permissible levels of aflatoxin contamination. These requirements are blocking access to high-value markets in the European Union by producers from African countries. Therefore, determining the amount of aflatoxin contamination of a peanut consignment at its source not only reduces the risks of loss of shipment associated with a food safety event, but also helps maintain the country's market share in an international context. Until 1990, Malawi was a major peanut exporter, but peanut exports declined due to aflatoxin contamination and stringent trade regulations. In the new millennium, the Department of Agriculture, Government of Malawi, initiated measures to recapture the export trade. However, the country lacked a laboratory and trained personnel for monitoring aflatoxin contamination. To overcome this problem, the National Smallholder Farmers' Association of Malawi (NASFAM) entered into a collaborative agreement with the International Crops Research Institute for the Semi-Arid Tropics (ICRISAT) centre based at Lilongwe, Malawi to test peanuts for aflatoxins.

Establishment of aflatoxin analysis laboratory

The ICRISAT-NASFAM partnership project initiated the establishment of an analytical laboratory for aflatoxins at ICRISAT-Lilongwe, Malawi in April 2004. The initial phase involved stocking the laboratory with equipment, glassware and consumables. A laboratory was already in place for doing pathology related research work, so some of the equipment was already available. During the season before the laboratory was functional, NASFAM and ICRISAT-Lilongwe collected 3,806 peanut samples from farmers. The samples were ground to a powder, labeled and shipped by courier to ICRISAT headquarters in Patancheru, India for aflatoxin analyses. At ICRISAT-Patancheru, aflatoxin was extracted from the samples and analyzed by ELISA to determine the aflatoxin concentration (Waliyar *et al.*, 2005). Analytical results were returned to ICRISAT-Lilongwe and provided to NASFAM for product differentiation and market targeting based on the maximum permissible limits of aflatoxin. Before the laboratory began operating, capacity building was done through a training course at ICRISAT-Lilongwe on aflatoxin management and aflatoxin levels determined by an ELISA method. Two scientists/technicians from NARS and NASFAM were trained at ICRISAT-Patancheru to estimate *Aspergillus flavus* seed infection and to use aflatoxin analytical techniques.

Marketing organizational structure

The NASFAM Associations are organized into units referred to as Chapters, which are subdivided into Marketing Action Centers (MACs), and the MACs subdivided into smaller units termed Clubs. For marketing purposes, several farmer groups or clubs sell their produce at designated marketing points (MACs). This structure allows for easy traceability because the producers are known and the peanut lots are labeled from this point.

Sampling procedure

ICRISAT designed and instituted a system for quality assurance, which involved sampling and estimation of aflatoxin contamination levels in peanuts purchased by NASFAM. A sampling auger was used to collect peanut samples from each 50 kg bag at various depths. A 200 g subsample was derived from the auger samples, placed in a plastic Ziploc bag and sent to the laboratory for processing.

From the 3,806 samples evaluated for NASFAM during the 2003/2004 season, 39% had no detectable aflatoxin, 29% had 1-2 ng/g aflatoxin, 20% had 2-4 ng/g aflatoxin and 12% had > 4 ng/g aflatoxin, with 1,500 ng/g the highest level of aflatoxin detected. NASFAM separated the peanut lots into those with < 4 ng/g aflatoxins and those with < 20 ng/g aflatoxins for export to Europe and South Africa, respectively. The markets of different countries have readily accepted the certified peanuts from Malawi. With this certification, Malawian farmers received better prices for their peanuts and increased their share of the export trade. The high numbers of samples with low levels of aflatoxin indicate that farmers have begun to apply the good agricultural practices and postharvest handling procedures advised by ICRISAT to reduce or eliminate aflatoxin contamination.

This improvement was due primarily to the aflatoxin analytical laboratory support and trained staff available at ICRISAT-Lilongwe. This laboratory is a simple one for testing peanuts for aflatoxins with an ELISA. The test is easy to perform, does not require expensive equipment and is suitable for laboratory conditions that exist in many developing countries.

Antibodies produced in-house at ICRISAT-Patancheru and standards necessary for the test are provided by ICRISAT. The cost of testing for each sample is expected to be less than US$ 1, which is several times cheaper than testing by HPLC or with a commercial ELISA kit. Moreover, this laboratory is easy to maintain and is not resource intensive. The facility functions as an autonomous unit and is seeking the accreditation necessary for certification purposes.

Levels of aflatoxin contamination in farmer samples and MACs

The number of farmers' peanut samples in each range of aflatoxin levels across the Chapters indicated that the Mikundi and Chiosya Chapters contributed a higher proportion of good quality peanut with aflatoxin levels between 0 and 4 ng/g (Fig. 1). Similarly, based on a maximum permissible level of 2 ng/g aflatoxin, the Mikundi chapter provided excellent quality nuts in terms of the total cumulative frequency of peanut samples that registered low aflatoxin contamination. The chapters of Mlonyeni and Msitu were not included in the determination of the best performing chapter as too few samples were collected from these Chapters. Of the eight MACs in the Chiosya Chapter, MACs 1, 4, 6 and 8 contributed the greater portion of good quality peanuts with low levels of aflatoxin contamination (Fig. 2).

By using ELISA for aflatoxin analyses, NASFAM could differentiate its products for different markets, and enable smallholder farmers to participate in markets that they would not otherwise have accessed. The collaboration between producers, traders and scientists in developing procedures for the management of aflatoxin requires resources and continuing

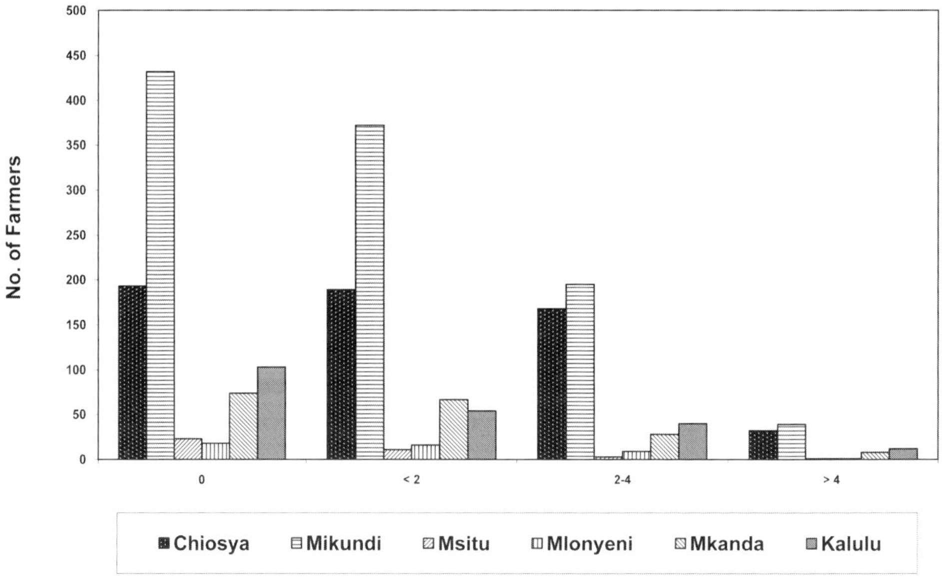

Figure 1. Levels of aflatoxin contamination (ng/g) in peanuts produced by NASFAM farmers from different chapters in Mchinji during the 2003-04 season.

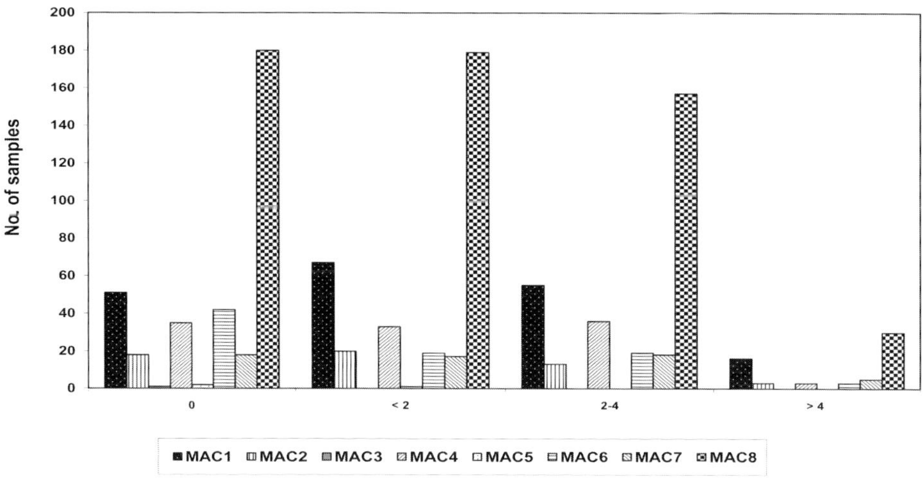

Figure 2. Levels of aflatoxin contamination (ng/g) in peanuts produced by NASFAM farmers from Chiosya Chapter in Mchinji during the 2003/2004 season.

support if sustained improvement in quality is the goal. As the results showed, even within Chapters there is variation in aflatoxin contamination levels. Thus, the use of this technique can provide information for better targeting of training of farmers and/or the handling of produce from specific areas.

Conclusions

African governments need to support food safety issues, such as mycotoxin contamination, because of their significance for public health and trade. To do so requires strengthening the capacity to effectively monitor mycotoxins, and to establish efficient food control systems through appropriate legislation. Establishment of mycotoxin facilities in Nigeria and Malawi is contributing to the increased international trade of agriculture commodities and augmenting local food safety standards. Studies in Nigeria and Malawi showed that government initiative is a key to the success of such initiatives. Simple aflatoxin detection methods, such as ELISA, are ideal for developing countries in Africa because of their ease of execution and relatively low operational costs.

Efforts to institutionalize mycotoxin testing in Africa need to be supported by international organizations such as FAO, WHO, IAEA, World Bank, USAID and EU through technical and financial assistance. Where such facilities are available, continued support is needed to upgrade laboratories and humans skills as new analytical methods are developed for food safety analyses. Developing competency in the risk assessment of mycotoxins will enable African countries to participate more effectively in international agricultural trade, strengthening Africa's position in the global food trade and indirectly contribute to the health of the local people.

References

Azziz-Baumgartner, E., Lindblade, K., Gieseker, K., Rogers, H.S., Kieszak, S., Njapu, H., Schleicher, R., McCoy, L.F., Misore, A., DeCock, K., Rubin, C., Slutsker, L. and the Aflatoxin Investigative Group. (2005) Case-control study of an acute aflatoxicosis outbreak, Kenya 2004. *Environmental Health Perspectives* 113, 1779-1783.

Detroy, R.W., Lillehoj, E.B. and Ciegler, A. (1971) Aflatoxin and related compounds. In: Ciegler, A., Kadis, S. and Ali, S.A. (eds.) *Microbial Toxins*, volume 6. Academic Press, New York, 3-178.

FAO. (2004) *Worldwide regulations for mycotoxins in food and feed in 2003. A compendium. FAO Food and Nutrition Paper No 81*. FAO, Rome, Italy.

Gong, Y.Y., Cardwell, K., Hounsa, A., Egal, S., Turner, P.C., Hall, A.J. and Wild, C.P. (2002) Dietary aflatoxin exposure and impaired growth in young children from Benin and Togo: Cross sectional study. *British Medical Journal* 325, 20-21.

Gong, Y.Y., Hounsa, A., Egal, S., Turner, P.C., Sutcliffe, A.E., Hall, A.J., Cardwell, K. and Wild, C.P. (2004) Post-weaning exposure to aflatoxin results in impaired child growth: A longitudinal study in Benin, West Africa. *Environmental Health Perspectives* 112, 1334-1338.

Ntare, B.R., Waliyar, F., Ramouch, M., Masters, E. and Ndjeunga, J. (2005) *Market Prospects for Groundnut in West Africa*. CFC Technical Paper No. 39. ICRISAT-CFC Publication, ICRISAT, Patancheru, India.

Shephard, G.S. (2004) Mycotoxins worldwide: Current issues in Africa. In: Barug, D., van Egmond, H.P., Lopez-Garcia, R., van Osenbruggen, W.A., and Visconti, A. (eds.) *Meeting the Mycotoxin Menace*. Wageningen Academic Publishers, Wageningen, The Netherlands, pp. 81-88.

UNICEF (2004) *Progress for Children: A Child Survival Report Card*. UNICEF, New York.

Unnevehr, L.J., ed. (2003) *Food Safety in Food Security and Food Trade*. International Food Policy Research Institute, Washington, D.C.

Waliyar, F., Reddy, S.V. and Kumar, P.L. (2005) *Estimation of* Aspergillus flavus *Infection and Aflatoxin Contamination in Seeds: Laboratory Manual*. International Crops Research Institute for the Semi-Arid Tropics, Patancheru, India.

Williams, J.H., Phillips, T.D., Jolly, P.E., Stiles, J.K., Jolly, C.M. and Aggarwal, D. (2004) Human aflatoxicosis in developing countries: A review of toxicology, exposure, potential health consequences, and interventions. *American Journal of Clinical Nutrition* 80, 1106-1122.

Wojnowski, L., Turner, P.C., Pedesen, B., Hustert, E., Brockmoller, J., Mendy, M., Whittle, H.C., Kirk, G. and Wild, C.P. (2004) Increased levels of aflatoxin-albumin adducts are associated with CYP3A5 polymorphisms in The Gambia, West Africa. *Pharmacogenetics* 14, 691-670.

Wu, F. (2004) Mycotoxin risk assessment for the purpose of setting international regulatory standards. *Environmental Science and Technology* 38, 4049-4055.

32

Prevention Strategies for Trichothecenes and Ochratoxin in Cereals

Naresh Magan*, Monica Olsen and David Aldred

Abstract

Contamination of cereal commodities with mycotoxins represents a significant hazard to consumer health and is receiving increasing attention from food safety authorities and legislators. The most important mycotoxins in small cereal grains in Europe are trichothecenes, including nivalenol, deoxynivalenol, T-2 toxin and HT-2 toxin in most regions, and ochratoxin, which is a major postharvest contaminant in northern Europe. Knowledge of the pre- and postharvest stages of the cereal production chain and in particular information on where prevention strategies can be implemented is critical to the development of quality assurance systems for improving food safety. In this chapter we review the key critical control points identified in the cereal production chain for trichothecenes and ochratoxin A that must be properly handled to minimize consumer exposure to these important mycotoxins. Our summary provides useful information for the cereal production and processing industries that can be used for developing appropriate specifications and management systems for preventing mycotoxin contamination in the food chain.

Introduction

Contamination of cereal commodities with mycotoxins represents a significant hazard to consumer health and is receiving increasing attention from food safety authorities and legislators. The most important species and mycotoxins in Europe are: (i) *Fusarium culmorum*, *Fusarium graminearum* and related species which produce trichothecenes including nivalenol, deoxynivalenol, T-2 toxin and HT-2 toxin in most regions, and for which EU legislative limits are being implemented, and (ii) *Penicillium verrucosum*, which is responsible for postharvest contamination of cereals with ochratoxin A in northern Europe. Knowledge of the pre- and postharvest stages in the cereal production chain and, in particular, information on where prevention strategies can be implemented is critical for developing quality assurance systems for improving food safety. These strategies have been developed in the framework of a HACCP program for managing ochratoxin A in cereals. When developing such strategies there are a number of key steps: (i) identifying the critical control points (CCP); (ii) establishing critical limits for the critical control points; (iii) developing rapid monitoring methods; and (iv) implementing corrective actions in the event of deviation

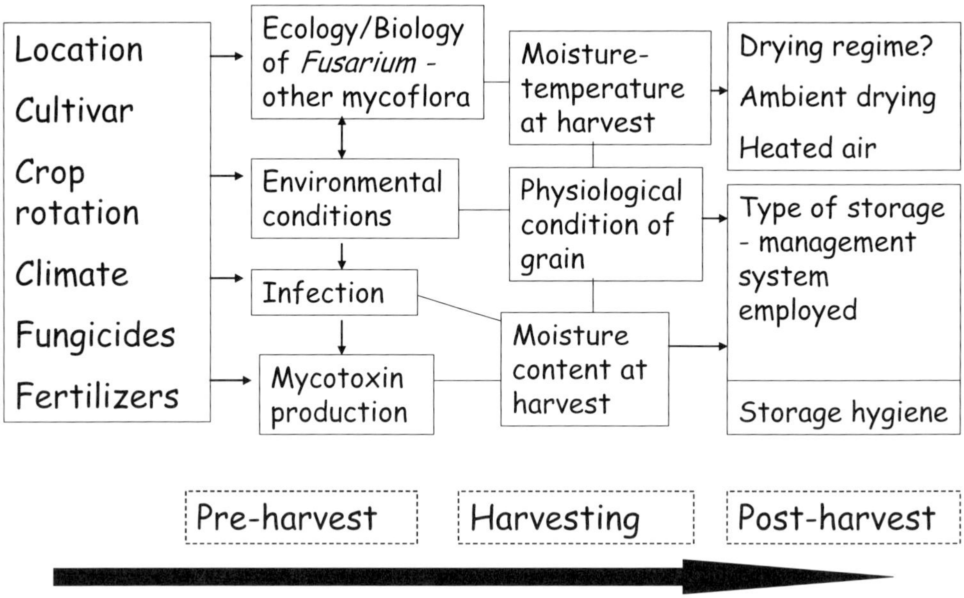

Figure 1. Interactions that impact trichothecene contamination of cereals pre- and post-harvest in Europe.

from a critical limit (Aldred *et al.*, 2004). In this chapter we review the key critical control points identified in the cereal production chain in relation to these two important classes of mycotoxins. Developing a HACCP program is essential to minimize consumer exposure to trichothecenes and ochratoxin A. The information used to develop the program also should be useful to the cereal production and processing industries as they develop specifications and management systems for preventing mycotoxin contamination in the food chain (Fig. 1).

Prevention strategies for trichothecenes

Since 2002, a concerted attempt has been made to evaluate the potential for effective control of trichothecenes, especially deoxynivalenol, by examining the whole production chain. For these mycotoxins the key components examined have included preharvest and postharvest factors. Weather conditions during flowering, type and amount/type of fungicides, ecology of the key species and moisture content at harvest are all key aspects that must be considered to reduce the risk of contamination with these mycotoxins.

Trichothecene production preharvest

The key preharvest considerations are the environmental conditions and the range of inputs into intensive monoculture cereal crop systems. The key hazard associated with natural contamination with trichothecenes, especially deoxynivalenol, in wheat preharvest in Europe is the presence of strains from one or more *Fusarium* spp. infecting the growing crop. *Fusarium* spp. are plant pathogens, some of which cause *Fusarium* head blight, also known as ear blight or head scab, which results in significant losses in grain quantity and quality (Jennings *et al.*, 2004). Measures taken against *Fusarium* development in the field are determined by the need to control *Fusarium* head blight rather than by the need to reduce the mycotoxin risk. However, with European Union legislation imminent and regulations already existing in many countries world-wide (FAO, 2004) the consideration of mycotoxin contamination has become more important. Although *Fusarium* infection generally is considered a preharvest problem, inadequate grain drying may result in colonization by *Fusarium* species in damp layers of drying grain and an increase in mycotoxin contamination.

The nature of Fusarium *head blight*

Fusarium head blight may be a disease complex, which means that it can be caused by individual species or by a combination of related species. In northern Europe the disease usually is caused by *F. culmorum*, and in southern Europe by *F. graminearum*, although recent studies suggest that the latter species is becoming more common across Europe, including the United Kingdom (Edwards and Ray, 2005; Waalwijk *et al.*, Chapter 17). Studies suggest that *F. graminearum* is very competitive and may be able to outcompete *F. culmorum* (Hope *et al.*, 2005*a*). Other species implicated are *F. avenaceum*, and *F. poae* (both mycotoxin producers) and some closely related *Microdochium nivale* varieties that are not known to produce toxins. The problem with *Fusarium* head blight seems to have been exacerbated by the increasing cultivation of bread wheats in Europe (Magan *et al.*, 2002). At present no durable, fully *Fusarium* head blight resistant wheat cultivars exist, so control relies on the use of cultivars with partial resistance together with field management and fungicide use (Diamond and Cooke, 2003). The development of resistant cultivars is an extremely important area, since wheat cultivars that are resistant to the most aggressive, high deoxynivalenol-producing strains of *F. graminearum* and *F. culmorum* have less disease progression and lower levels of toxin contamination (Mesterhazy, 2002).

Resistant cultivars

Recently, this research area has attracted a great deal of attention. Mesterhazy (2002) showed that the wheat varieties most resistant to Fusarium head blight also were contaminated by little, if any, deoxynivalenol. In fact, resistance appeared to depend on inhibition of toxin production directly, since the most aggressive disease causing fungal strains were those that produced the highest levels of deoxynivalenol. We think that increased availability of such resistant varieties, coupled with the use of appropriate fungicides are a critical part of any integrated approach to controlling *Fusarium* diseases and mycotoxin contamination. Such a strategy, with emphasis on mycotoxin control, could represent an important critical control point at the preharvest stage.

Field management

Appropriate field management and preparation is particularly important for *Fusarium* head blight and trichothecene control. An important part of the strategy is deep plowing to remove from the surface residual fungal material, which is present in debris from the previous crop. Crop rotation also is important and can break the chain of production of infectious material by reducing sporulation on crop residue. For example, wheat/legume rotations decrease the amount of inoculum present. The use of maize in such rotation should be avoided as this crop also is susceptible to infection by the same *Fusarium* strains that can infect wheat, and maize debris can be a very good source of carry-over onto wheat via stubble/crop residues. In Mediterranean climates it is a good practice to leave plowed land exposed to autumn sunshine to help destroy fungal material that could otherwise infect the following crop.

There is a narrow window (anthesis) during which ripening heads are susceptible to infection by splash dispersed soil-borne material onto the plants during periods of rain. If irrigation is used, then excessive splashing should be avoided and this practice may be an important control point in some instances. Effective management of the irrigation regime and the timing of irrigation applications to avoid infection at anthesis are very important. Lodging also is an important risk factor in trichothecene production. If lodging occurs then, deoxynivalenol production is very high irrespective of any fungicide treatment used as part of a control strategy (Nicholson *et al.*, 2003; Jennings *et al.*, 2004).

Environmental conditions

Environmental conditions such as relative humidity and temperature have an important effect on the onset of *Fusarium* head blight, with the moisture conditions during anthesis being particularly critical for *Fusarium* infection of the heads. Lacey *et al.* (1999) showed that *F. culmorum* infection in the UK increases if it is wet during the early flowering period in the summer, which is the optimum time for infection. Drought stress also can damage plants and make them more susceptible to infection. However, until recently, very little was known about the threshold limits for trichothecene production in the *Fusarium* head blight system. Agro-meteorological information for the period preceding and during ripening can be used to predict the risk of deoxynivalenol contamination of wheat by *F. graminearum* and *F. culmorum* [Hooker and Schaafsma (2003) in Canada; Detrixhe *et al.* (2003) in Belgium; Rossi *et al.* (2003) in Italy].

Hope and Magan (2003) compared two dimensional environmental profiles for growth and deoxynivalenol production by *F. culmorum* and *F. graminearum* to water availability (water activity, a_w) and temperature. They found that the range of conditions over which deoxynivalenol was produced was far more restricted than were the conditions under which growth could occur on wheat grain. Toxin production occurred in the relatively narrow a_w range 0.995-0.95 while growth persisted to 0.93-0.90 a_w. The optimal conditions for deoxynivalenol production (25°C, 0.995-0.98 a_w) were within the range that also was optimal for growth. These a_w levels correspond to water contents of approximately 30% and 26% respectively. In wet years, grain with > 20% moisture content can be present. Toxin production was significantly higher at 25°C than at 15°C. The significant difference between the species was temperature, with the a_w ranges being similar. Recent studies of *F. graminearum* strains from other regions of the world are consistent with these conclusions (Ramirez *et al.*, 2006).

Fungicide use

Part of the integrated control of *Fusarium* head blight in wheat involves the use of fungicides, but such a use also complicates predictions of trichothecene production, as under some conditions the fungicide may stimulate toxin production (Ramirez *et al.*, 2004). This situation is particularly dangerous since circumstances may occur in which the disease levels of *Fusarium* head blight are reduced or even eliminated, and yet high levels of mycotoxin contamination may still occur. Clearly grain affected in this way cannot be identified by visual inspection for signs of *Fusarium* head blight, *e.g.*, pink grain or tombstone kernels, and, in fact, cannot be identified as a problem at all until specific mycotoxin analyses are conducted.

Research on fungicide use in terms of *Fusarium* head blight and mycotoxin development has produced very interesting results. In particular, fungicides in common use have differential effects against toxin-forming *Fusarium* species and related non-toxin forming pathogens, *e.g.*, *Microdochium nivale* on wheat heads (Simpson *et al.*, 2001). The effectiveness of the use of a fungicide depends on the fungal species present, and the effect that a particular fungicide has on these species. For example, in work commissioned by the Home Grown Cereal Authority (HGCA), azoxystrobin significantly reduced disease levels but increased the levels of deoxynivalenol present in grain under experimental conditions when *F. culmorum* and *M. nivale* were both present (Nicholson *et al.*, 2003). These data were thought to result from the selective inhibition of *M. nivale* by azoxystrobin. *Microdochium nivale* is a natural competitor of toxin-forming *Fusarium* spp., particularly *F. culmorum*. Removal of *M. nivale* by the fungicide probably allowed the toxigenic *Fusarium* species to fill the niche with a concomitant increase in toxin formation. This result is an important finding because it indicates that the impact of the fungicide is only indirectly related to mycotoxin production. Other fungicides such as tebuconazole and metaconazole appeared to work in the opposite way, *i.e.*, they selectively inhibit *F. culmorum*, while having far less effect on *M. nivale*. The efficacy of the fungicides in these situations was directly correlated with the applied dose. Thus, whenever *Fusarium* head blight is caused by *Fusarium* species in the absence of *Microdochium*, disease severity can be associated with higher levels of toxin.

Simpson *et al.* (2001) also worked with azoxystrobin and reported similar findings, *i.e.*, mixtures of azoxystrobin with either prochloraz or fluquinconazole were less effective against *F. culmorum* than they were against *M. nivale*. Other fungicides that stimulate toxin production under some conditions include tridemorph, which stimulates T-2 toxin production by *F. sporotrichioides* (Moss and Frank, 1985), and tubiconazole and difenoconazole, which both stimulated production of monoacetyl deoxynivalenol (3-AcDON) by *F. culmorum*. *In-vitro* studies of several strains of *F. culmorum* from across Europe found that increased deoxynivalenol production occurred in the presence of low concentrations of epoxiconazole and propiconazole (Magan *et al.*, 2002; Ramirez *et al.*, 2004). Thus, it is important to use the recommended rate of fungicide to ensure full effects against *Fusarium* spp. Fungicides applied at early to mid-anthesis (mid-flowering) were the most effective at inhibiting both *Fusarium* head blight and deoxynivalenol/nivalenol contamination (Edwards and Ray, 2005).

Biological control agents

In a comprehensive screening of biocompetitive microorganisms to control *Fusarium* pathogens of cereal ears and maize, potential microorganisms were identified that decrease sporulation by *Fusarium* species on cereal stubble, and thus decrease the pool of inoculum for infection (Dawson *et al.*, 2002*a*,*b*; Luongo *et al.*, 2005). Studies to control *Fusarium* head blight and deoxynivalenol production have identified a number of microorganisms that can effectively decrease both toxin and disease levels. There is potential for further progress in this area, but there is a very narrow window of only 5-10 days during which protection is needed. Targeted spraying of biocontrol agents on flowering heads could provide the necessary protection. The potential for commercialization of some of these microbes is being evaluated (Dawson *et al.*, 2004).

Problems related to use of HACCP preharvest

The application of HACCP-type principles is most problematic during the preharvest period. There are several reasons for these problems:

- The field situation with respect to *Fusarium* head blight and mycotoxin development is complex, and some scientific information is lacking.
- At present, there are no control measures that can guarantee elimination of *Fusarium* head blight or the prevention of mycotoxin contamination.
- Economic factors also affect farming practices. In particular, field management, cropping systems and chemical input may reflect economic pressures more accurately than they do practices for disease control or for toxin reduction or elimination.
- In food processing plants, where HACCP was first developed, definitive control measures can be devised and implemented to close limits. The same level of control cannot be exercised in the field, *e.g.*, temperature and moisture depend on the weather conditions. Thus, some scientists think that attempts to apply HACCP to preharvest situations are fundamentally flawed.

Postharvest management

Grain at harvest and entering initial storage contains numerous microbes that could spoil the grain or produce toxins (Magan *et al.*, 2003*a*). The microbial population present depends largely on the field conditions and harvesting process, and may change during storage. Sometimes grain is kept for a short period of time on the farm in buffer storage before drying. During this time conditions conducive to *Fusarium* growth may occur and the mycotoxin contamination level increase. Thus, poor postharvest management can result in rapid loss of quality and increased mycotoxin contamination. These problems are exacerbated in wet harvest years.

At harvest, moisture management is the dominant control measure for preventing further mycotoxin production. Moisture management requires prompt accurate measurement of moisture content and procedures to efficiently dry the grain, if needed. Another equally important control measure at this stage is an effective assessment of the crop for the presence of diseases such as *Fusarium* head blight. If diseased kernels are present, then an efficient method for separating diseased material from healthy grain also is needed.

In general, if efficiently dried to the correct moisture content, *i.e.*, < 0.70 a_w (< 14.5% moisture by weight), then there will be no additional fungal spoilage or mycotoxin production. However grain often is harvested at moisture levels far in excess of this level and may even be traded on a wet weight basis. The bulk drying and storage of grain presents technological challenges in addition to those that result from poor practice or negligence. Thus, mycotoxin production is a hazard associated with both preharvest grain production and with postharvest grain handling and storage.

Overall, the most important control measures relevant to storage stages are:

- Regular and accurate moisture determination.
- Prompt, efficient drying of wet grain. This process includes consideration of holding time/temperature prior to drying as well as the actual drying conditions.

- Infrastructure for quick response, including provisions for segregation of contaminated grain and appropriate transportation conditions.
- Appropriate storage conditions at all stages in terms of moisture and temperature control, and general maintenance of facilities to prevent pest and water ingress.
- Ability to efficiently identify and reject material below specified standards in terms of fungal disease and mycotoxin level, especially when passing material to a third party.
- Operation of approved supplier systems, *i.e.*, setting the specifications for acceptance/rejection.

Although complex, postharvest stages of the wheat commodity chain, including drying, storage, transportation, milling and baking are far more conducive to a "classic" type HACCP analysis than are the preharvest stages. Unlike the preharvest stages, in the postharvest stages it is possible to apply definitive control measures, to set critical limits, and to initiate monitoring procedures. In particular, flour milling and baking can be viewed as straightforward food processing procedures that are immediately amenable to the HACCP approach.

Preventing ochratoxin from entering the cereal food chain

Background

The European Commission legislation and *Codex Alimentarius* currently are evaluating ochratoxin A contamination of food commodities and raw material. Ochratoxin A can be found in cereals, wine, grape juice, dried vine fruits, coffee, spices, cocoa, and animal derived products, *e.g.*, pork. The current European Commission legislation includes unprocessed cereals and cereal products, including baby food. JECFA (the Joint FAO/WHO expert committee on Food Additives) evaluated ochratoxin A at its 56th meeting in 2001 (FAO/WHO, 2001). They determined that ochratoxin A is nephrotoxic and causes renal cancer, but that the mechanism of action is not yet clear as both genotoxic and non-genotoxic mechanisms have been proposed.

JECFA retained the previously established provisional tolerable weekly intake (PTWI) of 100 ng/kg body weight (bw), corresponding to approximately 14 ng/kg bw/day. Estimates of tolerable daily intake for ochratoxin A, based on non-threshold mathematical modeling approaches or a safety factor/threshold approach, have ranged from 1.2 to 14 ng/kg bw/day. The Scientific Committee for Food of the European Commission (SCF, 1998) considered that, "It would be prudent to reduce exposure to ochratoxin A as much as possible, ensuring that exposures are towards the lower end of the range of tolerable daily intakes of 1.2-14 ng/kg bw/day that have been estimated by other bodies, *i.e.*, below 5 ng/kg bw/day". In the most recent assessment of ochratoxin A amongst European consumers (SCOOP, 2002), and in earlier investigations, cereals were the most important dietary source of ochratoxin A, contributing from 50 to 80% of the toxin intake. Thus, prevention of ochratoxin A formation by specific molds in cereals would have a significant impact on the consumer intake of ochratoxin A in Europe.

The food and brewing industries are increasingly demanding high quality cereals for food and drink products and require grain at least conforming to the statutory limits set for ochratoxin A. Thus, there is a major incentive for the European cereal industry to minimize ochratoxin A, and other mycotoxins, in grain to remain competitive worldwide and to reduce consumer risk as much as possible. There is an urgent need to understand the factors that enable or promote mycotoxin formation both pre- and postharvest stages so that strategies can be developed to minimize mycotoxin formation.

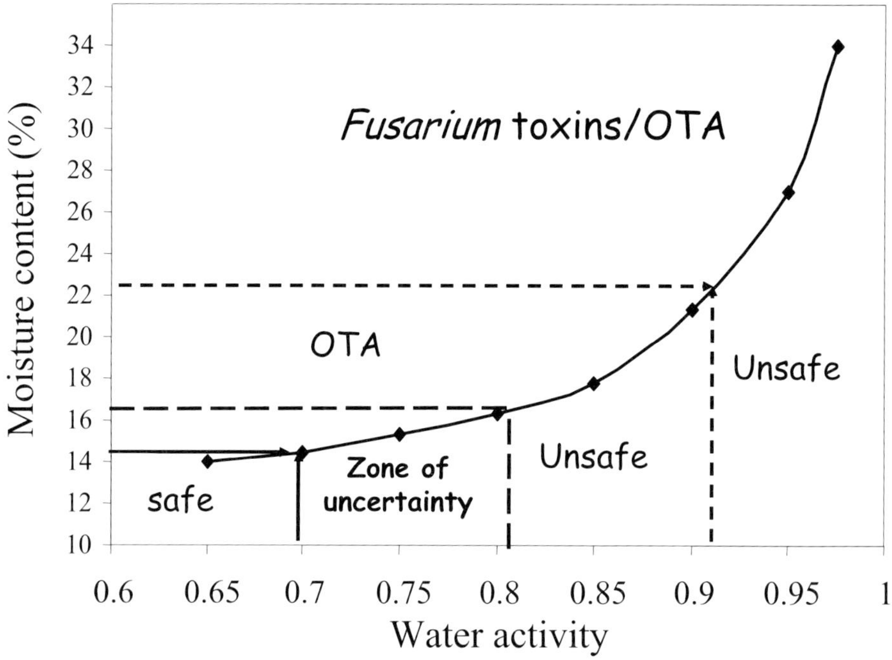

Figure 2. Postharvest grain moisture content ranges that are safe for storage of wheat, and those that are unsafe and more likely to lead to contamination with either *Fusarium* mycotoxins or ochratoxin A.

Application of HACCP-like assessment to the cereal food chain is consistent with the established knowledge that drying grain quickly at harvest is the most crucial factor in avoiding mold growth and mycotoxin formation during subsequent storage. Thus, instigation of a suitable monitoring check system to ensure that grain is dried rapidly to a safe moisture content, together with regular inspection and monitoring of grain moisture, should eliminate the risk of ochratoxin A formation during storage. In addition, following Good Agricultural Practices (GAPs) will increase the production of good quality grain. The continued occurrence of ochratoxin A in grain suggests that GAPs either are not or cannot always be fully followed, or that the factors involved in the formation of ochratoxin A are not completely understood. In other words, if grain cannot be dried quickly enough, then understanding all of the factors that affect the potential for ochratoxin A formation becomes important. Only by obtaining this information can sound and effective advice be developed on how to minimize this risk.

Cereals and ochratoxin A-producing fungi

Recent surveys of cereals from Europe, especially wheat and barley, have shown that *Penicillium verrucosum* is commonly isolated from cereals with *A. ochraceus* present only occasionally. Thus, contamination with ochratoxin A is primarily a postharvest problem. Lund and Frisvad (2003) showed that *P. verrucosum* contaminates grain during the harvesting process, drying and storage. In damp harvest years in northern Europe effective drying regimes after harvest are essential to prevent *P. verrucosum* from becoming established. Thus, the effective management of this phase is critical to the reduction or elimination of ochratoxin A contamination in the food chain. Ambient drying systems in damp autumns require longer times for drying which may result in layers of drying grain reabsorbing moisture and being colonized by mycotoxigenic molds. *P. verrucosum* also is a very competitive species and can dominate under conducive environmental conditions in stored grain (Magan *et al.*, 2003*b*). Some work has suggested that the level of contamination by *P. verrucosum* is a good indication of potential contamination by ochratoxin A (Ramakrishna *et al.*, 1996; Lund and Frisvad, 2003; Lindblad *et al.*, 2004). For example, Lund and Frisvad (2003) found that samples in which > 7% of the grain was contaminated with *P. verrucosum* also had significant ochratoxin A contamination as well, although fungal contamination and mycotoxin levels are not linearly correlated.

Sources of infection of grain by *P. verrucosum* include the contaminated environments of combines, dryers, and silos. Prompt and effective drying of cereals at harvest therefore is a major critical control point for preventing the formation of ochratoxin A. In the portions of Europe where the cereal harvest is at greatest risk, measures to avoid mold and toxin problems often are effective, while in areas normally at lower risk for ochratoxin A contamination may not be adequately prepared to avoid storage problems if unusual conditions occur. For example, it may not be economical to have expensive drying machinery idle in some years, while in others the supply of damp grain may exceed the drying capacity available even though delays in drying may put the grain at risk. Another problem arises if the infrastructure lacks sufficient funds and/or expertise to ensure best storage practices are followed (Scudamore, 2003).

Critical environmental limits, growth and ochratoxin A production

The most important abiotic factors influencing growth and ochratoxin A production include water availability, temperature and, when the grain is moist, gas composition (Magan *et al.*, 2003*a*). The interaction between these variables usually determines whether mold growth will occur and, if so, the composition and relative rate of development of the fungal community. An accurate determination of the marginal conditions for growth and ochratoxin A production by species such as *P. verrucosum* and *A. ochraceus* is important, since it can be used to develop guidelines and risk prediction tools for the entire food chain. Such studies must generate detailed information on the ability of numerous isolates of these species to colonize grain matrices over a relatively wide range of interacting conditions.

During respiration of damp grain, O_2 is utilized and CO_2 is produced. The O_2 and CO_2 in the inter-granular atmosphere are important factors in determining the pattern of fungal colonization of grain during storage. Most molds, including *P. verrucosum* and *A. ochraceus*, are obligate aerobes, so increasing the level of CO_2 should inhibit their growth. Yet, only a few studies have been conducted regarding the impact and interactions of CO_2 level, water activity (a_w) and temperature on fungal growth, particularly *P. verrucosum* and *A. ochraceus*, the main ochratoxin A producers on wheat during storage.

The occurrence of ochratoxin A in grain on farms usually is attributed to insufficient drying or to excessively long pre-drying storage (Jonsson and Pettersson, 1992). Studies of the safe storage period for cereal grains are few, and are based on visible molding (Kreyger, 1972), dry-matter loss (Steel *et al.*, 1969; White *et al.*, 1982) or loss of seed germination

(Kreyger, 1972; White *et al.*, 1982) rather than on ochratoxin A accumulation. For maize, the maximum allowable storage time has been estimated based on the time before dry matter loss exceeds 0.5% (Steel *et al.*, 1969). This loss is estimated to correspond to the loss of one US grade, which is based on visible inspection. Visible molding may be an unreliable criterion for ochratoxin A production, because considerable losses can occur before molding is visible, depending on whether the conditions favor fungal growth and sporulation (Seitz *et al.*, 1982).

Measurement of respiration is widely used to estimate fungal growth, biomass and dry-matter loss. The rate of CO_2 production also has been used to estimate total living microbial biomass in soil (Anderson and Domsch, 1975). CO_2 production was highly correlated with ergosterol content ($r = 0.98$) when *Eurotium repens* colonized maize (Martin *et al.*, 1989). Studies of respiration rates under different environmental conditions have enabled the development of a mathematical model, that describes the effect of a_w, *i.e.*, water availability, and temperature on safe storage time for cereal grain before obvious growth of *P. verrucosum* and formation of ochratoxin A. Thus, the logistic model of Lindblad *et al.* (2004) relates populations of *P. verrucosum*, measured as colony forming units (CFUs), to the probability of exceeding the European legislative limit of 5 ng/g of ochratoxin A in cereal grains under different a_w × temperature storage regimes. They suggested a threshold of 1000 CFUs of *P. verrucosum* per gram of grain as a threshold limit for the probability of risk of excessive ochratoxin A contamination. Cairns-Fuller *et al.* (2005) suggested that growth could occur under some conditions at $a_w = 0.80$, although ochratoxin A production may be limited, if it occurs at all, if $a_w < 0.83$. Thus, there is a "zone of uncertainty" (Fig. 2) for ochratoxin A contamination between 15 and 17.5% moisture content in terms of whether there is a high risk that ochratoxin A will be produced.

The general relationship between water availability, temperature, growth and ochratoxin A production has been discerned (Cairns-Fuller, 2004; Cairns-Fuller *et al.*, 2005). Rapid growth by *P. verrucosum* and *A. ochraceus* occurs at $a_w = 0.98$-0.99, *i.e.*, 27-30% moisture content, between 10 and 25°C, but is almost completely inhibited at $a_w = 0.80$-0.83, *i.e.*, 17.5-18% moisture content. No ochratoxin A is produced at $a_w = 0.80$, although some toxin may be produced at $a_w = 0.85$ (19% moisture content) at 15 and 20°C. Optimum conditions for ochratoxin A production were $a_w = 0.93$-0.98 (23.5-27.5% moisture content) at 10-25°C on wheat grain incubated for up to 56 days. On wheat grain, 7-14 days are required for ochratoxin A to begin to be produced at levels above the legislative limit (Cairns-Fuller *et al.*, 2005). Contour maps of the optimum and marginal conditions of water and temperature for growth and OTA production have been constructed. Approximately 17-18% moisture content, *i.e.*, $a_w = 0.80$-0.83, is the lower limit for growth and/or ochratoxin A production in wheat grain (Cairns-Fuller *et al.*, 2005). Thus, it is essential that grain is dried to lower moisture contents as quickly as possible regardless of the drying system employed. To avoid initiation of molding by xerophilic *Eurotium* species, drying to < 14.5% moisture content, *i.e.*, $a_w = 0.70$ (Magan *et al.*, 2003*b*) is essential. This level must be maintained during storage and transport to effectively prevent the accumulation of ochratoxin A.

Tests of controlled atmospheres to control ochratoxin A contamination in cereal grains have found that spore germination is not affected, but that germ tube elongation is significantly inhibited by 50% CO_2, especially at $a_w = 0.90$-0.995, for both *P. verrucosum* and *A. ochraceus* (Cairns-Fuller, 2004). Growth and ochratoxin A production were highest in air, followed by 25% and 50% CO_2 regardless of the a_w level tested. Generally, increased CO_2 and decreased a_w work together to reduce growth, although this interaction was not necessarily synergistic.

Table 1. The key control points, tools and potential effects on the entrance of ochratoxin A into the food production chain.

Control Point[a]	Control type	Tool	Impact
Harvest	GAP	Keep grain and machinery clean; remove old grain & dust	% toxin reduction not estimable, but significant
Storage before drying and during drying (in near-ambient dryers)	CCP	Model predicting safe storage time (critical limits)	Up to 100% toxin reduction
		Monitoring for toxin and producing fungi (LFDs and ELISAs)	
		Data on environmental conditions conducive to growth and ochratoxin A production	% toxin reduction not estimable, but useful tools in DSS
Storage	GSP/ CCP	Silo design and maintenance	% toxin reduction not estimable, but significant
		Critical limits for remoistening	Up to 100% toxin reduction
		Food grade antioxidants and natural control measures to prevent toxin formation in wet grain	> 80% toxin reduction possible, but not yet economically feasible
Cereal processing industry – Intake	CCP	Monitoring for toxin[b] and producing fungi (LFDs and ELISAs)	
		< 1000 cfu/g *P. verrucosum* in wheat	
		Monitoring fungus (DYSG, LFD, ELISA, and PCR)	
Cereal processing industry	GMP	Reductive measures during extrusion and baking	Baking up to 5-10%, extrusion up to 40%
Milling industry	GMP	Reductive measures during milling	Cleaning 2-3%, scouring 3-44%, milling up to 60%
Malting industry – Intake	CCP	< 3% internal infection or < 400 cfu/g of *P. verrucosum* in barley	Up to 100% toxin reduction
Malting industry	GMP	Control temperature during malting	Reducing temp. from 16-18°C to 12-14°C reduces toxin by 4×
Brewing industry – Intake	CCP	Monitoring toxin[b] in malt (LFD & ELISA)	
Brewing industry	GMP	Fate of ochratoxin A during brewing	Up to 80% toxin reduction
Regulatory control	CCPs	Monitoring toxin[b] (LFD & ELISA)	

[a]Abbreviations used: CCP, Critical Control Point; cfu, colony forming units; DSS, decision support systems; DYSG,Dichloran Yeast Extract Sucrose 18% Glycerol Agar; ELISA, Enzyme Linked Immunosorbent Assay; GAP, Good Agricultural Practice; GMP, Good Manufacturing Practice; GSP, Good Storage Practice; OTA, ochratoxin A; PCR, Polymerase Chain Reaction; LFD, lateral flow device.

[b]The critical limits at these points are the same as the legislative limits (currently 5 and 3 ng/g for unprocessed cereals and cereal products, respectively).

No other studies have been conducted of controlled atmosphere effects on growth and ochratoxin A production by *P. verrucosum*, although some information exists for some *Aspergillus* and *Fusarium* species. For example, Paster *et al.* (1983) reported that ochratoxin A production by *A. ochraceus* was completely inhibited by > 30% CO_2 on agar-based media after 14 days suggesting that there are differences between mycotoxigenic species. Thus, for the efficient storage of moist cereals > 50% CO_2 concentrations probably need to be achieved rapidly to prevent ochratoxin A contamination in storage or during transport.

Various postharvest treatments have been evaluated for their potential to prevent growth of *P. verrucosum* and to minimize ochratoxin A accumulation in stored grain. Essential oils, resveratrol and lactic acid bacteria all can limit ochratoxin A production by *P. verrucosum* and *A. ochraceus* on grain (Ricelli *et al.*, 2002; Fanelli *et al.*, 2003; Hope *et al.*, 2005*b*). However, while some essential oils, *e.g.*, thyme, cinnamon leaf and clove bud, and antioxidants, especially resveratrol, are quite effective, in pilot scale tests the economics of treatment prevents the implementation of these controls at the present time. In small scale storage experiments and experimental maltings, no inhibitory effect attributable to the selected lactic acid bacterial strain was observed.

Preventive actions required

Based on examination of the cereal production chain for both the milling and the malting industries a number of key points have been identified and must be effectively managed and monitored to minimize ochratoxin A accumulation in the food and feed chain (Table 1). With the identification of the critical control points, these points must be effectively and accurately monitored. Systems need to be available to enable accurate measurements to be made. This monitoring requires a mixture of techniques ranging from lab-based assays, with highly skilled personnel and/or sophisticated analytical equipment, to hand-held devices for on-farm and field-based measurements. Molecular-based tools for ochratoxin A detection (Geisen, 2004) have been developed in addition to immunoassays that can meet required legislative limits. More recently, lateral flow devices have been developed which may be relatively inexpensive and easy enough to use that natural levels of contamination could be routinely checked (Danks *et al.*, 2003).

Acknowledgements

This work was supported by the European Commission, Quality of Life and Management of Living Resources Program (Contracts QLK1-CT-1999-00433, QLK1-CT-1999- 00996).

References

Aldred, D., Olsen, M. and Magan, N. (2004) The use of HACCP in the control of mycotoxins: The case for cereals. In: Magan, N. and Olsen, M., (eds.) *Mycotoxin in food: Detection and control*. Woodhead Publishing, Ltd., Cambridge, U.K., pp. 139-173.

Anderson, J. P. E. and Domsch, K .H. (1975) Measurement of bacterial and fungal contributions to respiration of selected agricultural and forest soils. *Canadian Journal of Microbiology* 21, 314-322.

Cairns-Fuller, V. (2004) *Dynamics and control of ochratoxigenic strains of* Penicillium verrucosum *and* Aspergillus ochraceus *in the stored grain ecosystem*. Ph.D. Thesis. Cranfield University, Silsoe, Bedford, U.K.

Cairns-Fuller, V., Aldred, D. and Magan, N. (2005) Water, temperature and gas composition interactions affect growth and ochratoxin A production by isolates of *Penicillium verrucosum* on wheat grain. *Journal of Applied Microbiology* 99, 1215-1221.

Danks, C, Ostoja-Starzewska, S, Flint, J. and Banks, J. (2003). The development of a lateral flow device for the discrimination of ochratoxin A producing and non-producing fungi. *Aspects of Applied Biology* 68, 21-28.

Dawson, W.A.J.M., Bateman, G.L., Jestoi, M. and Rizzo, A. (2002*a*) Biological control of ear blight of wheat caused by *Fusarium culmorum*. *Journal of Applied Genetics* 43A, 217-222.

Dawson, W.A.J.M., Bateman, G.L., Kohl, J., Haas, B.H., de Lombaers-vander Plas, C.H., Corazza, L., Luongo, L., Galli, M., Jestoi, M. and Rizzo, A. (2002*b*) Controlling infection of cereal grain by toxigenic *Fusarium* species using fungal competitors. *Brighton Crop Protection Conference 2002 - Pests and Diseases* 1, 347-352.

Dawson, W.A.J.M., Jestoi, M., Rizzo, A., Nicholson, P. and Bateman, G.L. (2004) Field evaluation of fungal competitors of *Fusarium culmorum* and *F. graminearum*, causal agents of ear blight of winter wheat, for control of mycotoxin production in grain. *Biocontrol Science and Technology* 14, 783-799.

Detrixhe, P., Chandelier, A., Cavelier, M., Buffet, D. and Oger, R. (2003) Development of an agrometeorological model integrating leaf wetness duration estimation to assess the risk of head blight infection in wheat. *Aspects of Applied Biology* 68, 199-204.

Diamond, H. and Cooke, B.M. (2003) Preliminary studies on biological control of the *Fusarium* ear blight complex of wheat. *Crop Protection* 22, 99-107.

Edwards, S. and Ray, R. (2005) *Fusarium* mycotoxins in UK wheat production. *BCPC Crop Science and Technology* 5B, 395-402.

FAO (Food and Agriculture Organization of the United Nations). (2004) *Worldwide Regulations for Mycotoxins in Food and Feed in 2003. Food and Nutrition Paper* 81. Rome, Italy.

FAO/WHO (World Health Organization). (2001) Ochratoxin A. In: *Safety Evaluation of Certain Mycotoxins in Food. Fifty-sixth meeting of the Joint FAO/WHO Expert Committee on Food Additives (JECFA).* FAO Food and Nutrition Paper 74, FAO, Rome, Italy.

Fanelli, C., Taddei, F., Trionfetti Nisini, P., Jestoi, M., Ricelli, A., Visconti, A. and Fabbri, A.A. (2003) Use of resvaretol and BHA to control fungal growth and mycotoxin production in wheat and maize seeds. *Aspects of Applied Biology* 68, 63-72.

Geisen, R. (2004) Molecular monitoring of environmental conditions influencing the induction of ochratoxin A biosynthesis genes in *Penicillium nordicum*. *Molecular and Nutritional Food Research* 48, 532-540.

Hooker, D.C. and Schaafsma, A.W. (2003) The DONcast Model: Using weather variables pre- and post-heading to predict deoxynivalenol content in winter wheat. *Aspects of Applied Biology* 68, 117-122.

Hope, R and Magan, N. (2003) Two-dimensional environmental profiles of growth, Deoxynivalenol and nivalenol production by *Fusarium culmorum* on a wheat-based substrate. *Letters in Applied Microbiology,* 37, 70-74.

Hope, R., Aldred, D. and Magan, N. (2005*a*) Comparison of the effect of environmental factors on deoxynivalenol production by *F. culmorum* and *F. graminearum* on wheat grain. *Letters in Applied Microbiology* 40, 295-300.

Hope, R., Cairns, V., Aldred, D. and Magan, N. (2005*b*) Use of antioxidants and essential oils for controlling mycotoxins in grain. *BCPC Crop Science and Technology 2005* 5B, 429-436.

Jennings, P., Kohl, J. and Gosman, N. (2004) Control of mycotoxins: Raw material production. In: Magan, N. and Olsen, M. (eds.) *Mycotoxins in Food: Detection and Control.* Woodhead Publishing, Ltd., Cambridge, U.K., pp. 443-460.

Johnsson N. and Pettersson, H. (1992) Comparison of different preservation methods for grain. In: Johansson, H., ed. *Cereals in the Future Diet, Proceedings of the 24th Nordic Cereal Congress, (Stockholm 1990)*, Nordic Cereal Association, Lund, Sweden, pp. 357-364.

Kreyger, J. (1972) *Drying and Storing Grains, Seeds and Pulses in Temperate Climates.* Institute for Storage and Processing of Agricultural Produce, (IBVL), publication 205, Wageningen, Netherlands.

Lacey, J. Bateman, G.L. and Mirocha, C.J. (1999) Effects of infection time and moisture on the development of ear blight and deoxynivalenol production by *Fusarium* spp. in wheat. *Annals of Applied Biology* 134, 277-283.

Lindblad, M., Johnsson, P., Jonsson, N., Lindqvist, R. and Olsen, M. (2004) Predicting noncompliant levels of ochratoxin A in cereal grain from *Penicillium verrucosum* counts. *Journal of Applied Microbiology* 97, 609-616.

Lund, F. and Frisvad, J.C. (2003) *Penicillium verrucosum* in wheat and barley indicates presence of ochratoxin A. *Journal of Applied Microbiology* 95, 1117-1123.

Luongo, L., Galli, M., Corazza, L., Meekes, E., de Hass, L., van de Plas, C.L. and Kohl, J. (2005) Potential of fungal antagonists for control of *Fusarium* spp. on wheat and maize by competition on crop debris. *Biocontrol Science and Technology* 15, 229-242.

Magan, N., Hope, R., Colleate, A. and Baxter, E.S. (2002) Relationship between growth and mycotoxin production by *Fusarium* species, biocides and environment. *European Journal of Plant Pathology* 108, 685-690.

Magan, N., Hope, R., Cairns, V. and Aldred, D. (2003*a*) Post-harvest fungal ecology: Impact of fungal growth and mycotoxin accumulation in stored grain. *European Journal of Plant Pathology* 109, 723-730.

Magan N., Sanchis, V. and Aldred, D. (2003*b*) Role of spoilage fungi in seed deterioration. In: Aurora, D.K. (ed) *Fungal Biotechnology in Agricultural, Food and Environmetal Applications,* Marcel Dekker, New York, pp. 311-323.

Martin S., Tuite, J. and Diekman, M.A. (1989) Inhibition radioimmunoassay for *Aspergillus repens* compared with other indices of fungal growth in stored corn. *Cereal Chemistry* 66, 139-144.

Mesterhazy, A. (2002) Role of deoxynivalenol in aggressiveness of *Fusarium graminearum* and *F. culmorum* and in resistance to *Fusarium* head blight. *European Journal of Plant Pathology* 108, 675-684.

Moss, M.O., Frank, J.M., (1985) Influence of the fungicide tridemorph on T-2 toxin production by *Fusarium sporotrichoides. Transactions of the British Mycological Society* 84, 585-590.

Nicholson, P., Turner, J.A., Jenkinson, P., Jennings, P., Stonehouse, J., Nuttall, M., Dring, D., Weston, G. and Thomsett, M. (2003) Maximizing control with fungicides of *Fusarium* Ear Blight (FEB) in order to reduce toxin contamination of wheat. *Project Report No. 297, Home Grown Cereals Authority,* London.

Paster, N., Lisker, N. and Chet, I. (1983) Ochratoxin production by *Aspergillus flavus* Wilhem grown under controlled atmospheres. *Applied and Environmental Microbiology* 45, 1136-1139.

Ramakrishna, N., Lacey, J. and Smith, J.E. (1996) Colonization of barley grain by *Penicillium verrucosum* and ochratoxin A formation in the presence of competing fungi. *Journal of Food Protection* 59, 1311-1317.

Ramirez, M.L., Chulze, S and Magan, N (2004) Impact of environmental factors and fungicides on growth and deoxynivalenol production by *Fusarium graminearum* isolates from Argentinian wheat. *Crop Protection* 23, 117-125.

Ramirez, M.L., Chulze, S. and Magan, N. (2006) Interacting environmental factors, growth and temporal deoxynivalenol production by strains of *Fusarium graminearum* from Argentina on wheat grain. *International Journal of Food Microbiology* 106, 291-296.

Ricelli, A., Fabbri, A.A., Trionfetti-Nisini, P., Reverberi, M., Zjalic, S. and Fanelli, C. (2002) Inhibiting effect of different edible and medicinal mushrooms on the growth of two ochratoxigenic microfungi. *International Journal of Medicinal Mushrooms* 4, 173-179.

Rossi, V., Giosue, S. and Delogu, G. (2003) A model estimating risk for *Fusarium* mycotoxins in wheat kernels. *Aspects of Applied Biology* 68, 229-234.

SCF (Scientific Committee on Food) (1998) Opinion of the Scientific Committee on food on ochratoxin A (expressed on 17 September 1998). Directorate, Scientific Opinions. European Commission, Brussels, Belgium

SCOOP. (2002) Assessment of dietary intake of ochratoxin A by the population of EU member states. (European Commission web page: http://europa.eu.int/comm/food/fs/scoop/3.2.7_en.pdf

Scudamore, K. A., Banks, J., and MacDonald, S. J. (2003). Fate of ochratoxin A in the processing of whole wheat grains during milling and bread production. *Food Additives and Contaminants* 20, 1153-1163.
Seitz, L.M., Sauer, D.B. and Mohr, H.E. (1982) Storage of high-moisture corn: Fungal growth and dry matter loss. *Cereal Chemistry* 59, 100-105.
Simpson, D.R., Weston, G.E., Turner, J.A., Jennings, P. and Nicholson, P. (2001) Differential control of head blight pathogens of wheat by fungicides and consequences for mycotoxin contamination in grain. *European Journal of Plant Pathology* 107, 421-431
Steel, I.F., Saul, R.A. and Hukill, W.V. (1969) Deterioration rate of shelled corn as measured by carbon dioxide production. *Transactions of American Society of Agricultural Engineers* 12, 685-689.
White, N.D.G., Sinha, R.N. and Muir, W.E. (1982) Intergranular carbon dioxide as an indicator of biological activity associated with the spoilage of stored wheat. *Canadian Agricultural Engineering* 24, 35-42.

International Programs on Mycotoxins

33

FAO Program on Mycotoxin Management

Maya Piñeiro

The Food and Agriculture Organization of the United Nations (FAO) World Food Summit Plan of Action specifically states that "Food security exists when all people, at all times, have physical and economic access to sufficient, safe and nutritious food to meet their dietary needs and food preferences for an active and healthy life". FAO works to eliminate poverty and hunger by promoting sustainable agricultural development, improving nutrition and guiding the pursuit of food security. Food safety is an essential component of improved nutrition and food security.

There are two major approaches to food safety issues – the food chain approach and fulfilling obligations resulting from World Trade Organization (WTO) agreements. The food chain approach optimizes results, by sharing the responsibility for safe food with all who produce, process and trade in food from primary production to final consumption, *e.g.*, farmers, fishermen, slaughterhouse operators, food processors, storage and transport operators, processors and distributors. Information regarding food safety is provided to the next link in the food chain. This approach to food safety distributes the responsibility for food safety beyond the food processing sector.

The FAO Nutrition and Consumer Protection Division, which hosts the *Codex* Secretariat, the Joint Expert Committee on Food Additives and Contaminants (JECFA) and the food quality and safety capacity building work, was moved in 2006 from the Economic and Social Department to the Agriculture, Biosecurity, Nutrition and Consumer Protection Department. This move is consistently in line with the farm-to-table approach to nutrition, food safety and consumer protection along the food chain and provides new opportunities for cooperation between the units involved in the production, processing, handling, storage and distribution of food products as well as those in food safety control and standards development.

The Agreements on the Application of Sanitary and Phytosanitary Measures (the SPS Agreement) and on Technical Barriers to Trade (the TBT Agreement) of the WTO include food safety issues and balance the competing demands of domestic regulatory autonomy and trade facilitation. Both agreements are compromises that permit the adoption of measures within a member country that fulfill legitimate objectives and restrict trade as little as possible. Member countries also should balance human, animal and plant health with protection from arbitrary or unjustifiable discrimination resulting from different sanitary and phytosanitary standards. Countries also must abide by the provisions of the TBT Agreement that allow measures to prevent deceptive trade practices, to protect human health or safety and animal or plant life and health or the environment, and to meet national security requirements.

FAO's mandate calls for "raising levels of nutrition and standards of living … and contributing towards an expanding world economy and ensuring humanity's freedom from hunger" (FAO Constitution). Thus, the Food Quality and Standards Service engages in a variety of capacity building activities (publications, training, technical assistance projects, policy advice) to help national governments increase safe food supplies and become more competitive in international trading markets. Within FAO, the Nutrition and Consumer Protection Division integrates both the food chain and the international agreement approaches into three major working groups: setting international food standards (*Codex Alimentarius* Secretariat), food safety risk assessment and provision of scientific advice (JECFA, JEMRA and other committees and specialized meetings), and capacity building.

FAO capacity building activities in the mycotoxin area

Programs for prevention and control of mycotoxins are very important due to the impact of these contaminants on consumer safety (IARC, 1993), food availability and trade. Despite many years of research and the introduction of good practices in the food production, storage and distribution chain, mycotoxins continue to be a problem. In 1996, the UN Commission on Sustainable Development approved a work program on indicators of sustainable development that included mycotoxins in food as one component of an indicator related to protection and promotion of human health.

Regulations that limit the levels of mycotoxins in foods and feed because of their public health significance also can affect food security (food availability), especially in developing countries. Losses due to mycotoxins in developing countries extend beyond the losses in grain and animal production, as export markets may be lost to the stricter limits and possible non-tariff barriers that result from these mycotoxin regulations. FAO estimates of world losses of foodstuffs due to mycotoxins are around 1,000,000,000 t per year.

Losses can be minimized by a multi-factorial food chain mycotoxin management approach. Four principles underlie this approach: (i) establishing science-dependent, realistic and risk-based regulatory levels that are proportionate to the risk incurred and not overtly protective, (ii) systematically managing mycotoxin risks through country-specific practical recommendations and national action plans, (iii) implementing adequate tools, *e.g.*, good agricultural practices (GAP) and a HACCP system, along the food chain that prevents, controls or reduces the damage resulting from fungal contamination, and (iv) adopting early warning systems based on climatic models that predict when mycotoxin outbreaks are likely.

The most important challenge faced in mycotoxin research and management in developing countries is the development of sustainable prevention and control strategies based on appropriate and feasible methods within the context of making available food for one of the fastest growing populations in the globe. The risks posed by mycotoxins in food present enormous difficulties to these countries and the international organizations responsible for controlling mycotoxin levels. International consensus on tolerable levels for mycotoxins is very difficult to reach due to deficiencies in the risk database and inconsistencies in risk management decisions in spite of concerted action by international organizations and institutions such as FAO with the *Codex Alimentarius* and JECFA to address all of the components of the mycotoxin issue, including new methods for risk evaluation and control strategies.

FAO facilitates assistance in capacity building programs in six major areas: (i) assisting countries in implementing national standards and harmonizing them with *Codex*, (ii) risk assessment and scientific advice, (iii) training at both normative (guidelines, codes and tools) and operational levels (workshops and seminars); (iv) advice and strategies for prevention and control of mycotoxins, (v) technical assistance (field projects), and (vi) strategies for increasing information communication and dissemination (brochures, web pages and newsletters). FAO capacity building activities in these areas may be in conjunction with international organizations, national governments, international and regional financial institutions and Non-Governmental Organizations (NGOs). The activities and strategies differ by region and by country. Often, technical assistance is provided to assess and analyze the institutional set up for food control to identify the main weaknesses, and to formulate recommendations and proposals for establishing technically sound food control systems that are harmonized with international standards. Specific capacity building activities usually include training food control officials and technical staff through seminars, workshops and study tours, enhancing food control laboratory and inspection capabilities, preparing training manuals and guidelines, and establishing and strengthening policies, regulatory frameworks and National *Codex* Committees.

Key Areas in Mycotoxins

The areas of risk analysis/risk assessment, analytical quality assurance, and prevention and control provide the basic foundations of most technical assistance projects.

Risk analysis

Until 1997, a proposed tolerable limit, *e.g.*, level of consumption/kg body weight/day, was set for non-genotoxic substances or a recommendation of levels "as low as reasonably achievable" (ALARA) was made for genotoxic substances. Improvements in mycotoxin detection and the implications for health, trade and food availability lead to the development of risk analysis methodology that requires more detailed information on mycotoxin actions and interactions. Risk analysis strengthens the ability of traditional food safety systems by providing a framework to effectively manage, assess and communicate risks in cooperation with diverse stakeholders allowing effective decision-making.

Risk analysis, as defined by the *Codex Alimentarius* (FAO/WHO/UNEP, 1999; *Codex Alimentarius*, 2003*b*), has three components: risk assessment, risk management and risk communication. These components are interdependent and cannot occur separately. The process often begins with risk management, which defines the problem, articulates the goals of the risk analysis and defines the questions to be answered by the risk assessment. The science-based tasks of "measuring" and "describing" the nature of the risk being analyzed, *i.e.*, risk characterization, are part of the risk assessment.

Risk communication encompasses an interactive exchange amongst risk managers, risk assessors, the risk analysis team, consumers and stakeholders. The process often culminates with action being taken by the risk managers. This approach increases the ability of food safety regulators to identify hazards, characterize them, assess exposure, and estimate the likelihood of the resulting risks to and potential impact on human health.

A science-based approach strengthens traditional food safety systems, minimizes foodborne hazards, enables risk management, and improves decision-making processes. This concept is not new and is closely related to good practices (agricultural, hygienic and manufacturing) and to the Hazard Analysis and Critical Control Point system (HACCP) already in place in many countries. What is new is the use of risk analysis as a framework to view and to respond to food safety problems in a systematic, structured, scientific manner that increases the quality of the decision-making process. Mycotoxins are natural contaminants that are particularly difficult for risk assessors, as they usually are considered unavoidable and therefore must be regulated differently than synthetic chemicals, which are considered avoidable. Thus, small amounts of mycotoxins are allowed in foods if the permitted levels are not injurious to human or animal health (Miller, 1998).

Natural chemical contaminants produced by biological organisms usually are considered independently of the producing organism with respect to hazard identification and characterization. However, identification of a producing organism may be of use in hazard assessment before the chemical contaminant is characterized. The producing organism also is an important factor in risk management since control strategies that target the producing organism may be the most effective way to reduce the levels of the contaminant in food.

Understanding the environmental and physiological factors that control mycotoxin production is critical if models predicting contamination levels are to be developed and for the development of control strategies. Often, multiple toxigenic fungi may be present and each species may produce a different spectrum of mycotoxins. Deciding which toxins, or combinations, are important can be difficult for the risk assessor. Finally, because mycotoxins often occur in basic commodities (cereals, nuts and fruits) and often are found in important export products, the health risk posed by exposure to the contaminant must be considered concurrently with the potential adverse impact on food availability and other economic impacts.

Complete removal of a mycotoxin hazard usually is not possible, so a risk management scheme must be developed. Risk management requires the evaluation of management alternatives, followed by the implementation of the best alternative and its monitoring and enforcement. Risk management strategies applied to mycotoxins include: (i) establishment of science-based regulatory limits, (ii) monitoring of food products prior to and during harvest/processing through HACCP protocols, (iii) screening and testing of commercial products, (iv) development of decontamination procedures, and (v) diversion of products to uses that are more tolerant of the mycotoxins present.

Thus, it is essential to provide developing countries with the technical and financial resources required to obtain the necessary data and to support or conduct risk analysis (Rosner, 1998; Piñeiro, 2004*b*). Areas in which substantial need for assistance have been detected include: (i) infrastructure and technology transfer to enable international quality risk assessment, (ii) obtaining the data required to support mycotoxin risk assessments, *e.g.* exposure assessment data, (iii) bridging gaps in quantitative risk assessment, and (iv) incorporating data from developing countries into international risk assessments and the international standards and guidelines that result. FAO has helped developing countries to: (i) develop their own risk management strategy and prevention and control procedures for mycotoxins through workshops and the provision of expert advice through technical cooperation projects (TCPs), (ii) prepare a monograph on mycotoxin prevention in conjunction with IARCs and WHO, and (iii) finalize a CD-ROM training package on Food Safety Risk Analysis that includes a framework and overview manual, a training module presentation, case studies in risk analysis and access to FAO/WHO resources related to food safety risk analysis (FAO/WHO, 2006).

Analytical results and laboratory quality assurance

Reliable analytical results require the systematic application of quality assurance measures, including documentation, trained personnel, appropriate and calibrated instrumentation, validated methods and adequate laboratory infrastructure. Method performance is demonstrated through proficiency or interlaboratory studies, use of reference materials, and statistical evaluation (repeatability, reproducibility values and accuracy and precision control charts). Adequate quality assurance procedures enable both the identification of problems and their correction.

In June 1997 *Codex Alimentarius* (2003*d*) recommended that laboratories responsible for control of export and import foods comply with ISO 17025 "General Requirements for the Competence of Calibration and Testing Laboratories", and ultimately be accredited by a certified body. In addition, the only methods of analysis endorsed by the *Codex* Committee on Methods of Analysis and Sampling (CX/MAS) will only be validated methods for ISO 17025 accredited laboratories. Directive 93/99/EEC of the European Union (European Commission, 2002) also requires food control laboratories to be formally accredited to an internationally recognized standard such as ISO 17025. Thus, both governmental and industrial laboratories must have analytical capabilities with quality assurance systems in place.

The implementation of international standards/guidelines and recommendations under the SPS Agreement enforced by WTO, requires suitable laboratory facilities, adequately trained personnel, and, by implication, continuous training in quality assurance for mycotoxin analysis. National food control systems should improve food safety and stimulate international trade by employing efficient methods for detecting, quantifying and certifying mycotoxins in food. Strengthening the analytical capabilities of laboratories in developing countries to enable effective monitoring of the mycotoxin content of food should help overcome non-tariff trade barriers. Research data are needed on the effectiveness of various analytical methods used to monitor mycotoxin contamination to identify cost-effective validated procedures (Piñeiro, 2000).

Good, reliable data also are needed to perform risk assessments. Reliable, sensitive, accurate, precise methods of analysis and sampling, leading to internationally accepted analytical results are fundamental to international trade under the SPS and TBT agreements of the WTO. Mycotoxins often are regulated food contaminants that become non-tariff trade barriers. Laboratories in developing countries must implement quality assurance systems according to ISO 17025 and Good Laboratory Practices to demonstrate their analytical capability. FAO assists national food control laboratories in preparing for accreditation and in validating methods for detecting and quantifying mycotoxins. In addition to validated methods, internal quality control procedures must be implemented and document participation in proficiency testing and proper use of reference materials. Quality assurances programs, with their method validation procedures, interlaboratory schemes and accreditation awards provide evidence of consistent satisfactory performance.

Food safety is a shared responsibility among government, industry and consumers. Most governments have enacted legislation to assure the safety of the food supply, and over 100 countries regulate mycotoxin levels (FAO, 2004). Effective enforcement of mycotoxin legislation requires administration/policy and regulation, inspection and sampling, and analytical services. Analytical results submitted by an accredited laboratory are considered *prima facie* evidence of violations and also are necessary for health-oriented, population-based monitoring.

Risk analysis is dependent on adequate and reliable monitoring programs to provide reliable data for calculating exposure assessments (for IARC, JECFA). In the absence of reliable data and scientific consensus, there often is disagreement between importing and exporting countries, which each may set regulatory levels based on their perceptions of what is achievable. National limits for aflatoxins often are determined by a country's trading position (exporter/importer), by climate and by other non-risk factors. The resulting inconsistencies have incurred considerable economic costs, the rejection and downgrading of food shipments, increased expenses in quality assurance (QA) programs, and significant trade problems (Pitt, 1995; Rosner, 1998).

FAO's assistance in laboratory performance and analytical quality assurance are exemplified by two sub-regional projects to strengthen the InterAmerican Network of Food Analytical Laboratories (INFAL/RILAA) which began in 2004 and benefits nineteen countries from South and Central America and the Caribbean. The project's goal is to ensure that laboratories provide reliable analytical results for the national food control system and that they can certify food imports and exports. The participating laboratories strive for product conformity assessment through harmonization or the establishment of internationally equivalent analytical results. The main activities undertaken were training and the development of national action plans for strengthening the technical, analytical and administrative skills necessary for laboratory analyses of microbiological and chemical contaminants.

FAO also is assisting developing countries worldwide in the implementation of laboratory quality assurance measures and in generating data by using accepted method validation criteria through TCPs and training workshops, *e.g.*, Regional Training Courses on Development of Quality Assurance for Mycotoxin Analysis. These courses develop quality assurance systems for mycotoxin analysis of food and train participants in the establishment and implementation of an analytical quality system that complies with ISO Standard 17025, making accreditation possible. The FAO/IAEA CRP on mycotoxin methods, also provides support to countries worldwide, and has implemented a coordinated research project on mycotoxin methodology that involves development and validation of mycotoxin methods for use in developing countries.

Mycotoxin prevention and control practices

Mycotoxin accumulation in crops is increasing world-wide due to changes in climate and the widespread planting of inappropriate crops or genotypes of high yield but that are susceptible to mycotoxin accumulation. At recent international meetings (FAO/WHO/UNEP, IUPAC and AOAC; FAO/WHO/UNEP, 1999) the focus has shifted to the preventive side of the mycotoxin issue, with the HACCP approach integrated into agricultural management and postharvest control. This approach is potentially more economically rewarding than simply regulating mycotoxins by setting tolerance limits.

HACCP

Many food borne problems, including mycotoxin contamination, are increasing in incidence and emerging as major public health and economic issues because of reduced productivity, despite advances in control and prevention. Changing lifestyles, mass production of food, environmental contamination, travel, the globalization of the food trade and improved analytical techniques also influence food safety. The need for improved food safety occurs as regulatory resources for this purpose diminish and industry streamlines.

The HACCP approach appears to be a cost-effective food safety assurance system. It can be applied to all links in the food chain process and emphasizes prevention. Good Agricultural Practices (GAP) are the foundations on which HACCP is based. The application of HACCP to export trade has enormous benefits, in volume and in price, through opening markets and facilitating acceptance by buyers who place increased importance on food safety. The most important aspect of a HACCP program is the prevention of food associated diseases, including illnesses resulting from mycotoxins and other contaminants, which are particularly important in developing countries (FAO, 2001).

Whether mycotoxins are labeled biological or chemical hazards they fit in a HACCP program and appropriate Critical Control Points and their Critical Limits must be identified. Once the mycotoxin hazards are identified and the Critical Control Points determined, the Critical Limits can be set and the HACCP follows well-established steps. Associated relevant data are the established maximum limits, *i.e.*, national or international legislation, guidelines and veterinary recommendations, or based on the knowledge of the food processor of the safety risk associated with each mycotoxin. Within the *Codex* system, the hazard identification step of the risk analysis process for mycotoxins and other chemical contaminants (industrial and environmental contaminants and naturally occurring toxicants such as mycotoxins) is initiated by CCFAC or member countries. JECFA's toxicological evaluation results in a Tolerable Intake or, when no threshold is thought to exist, *e.g.*, for aflatoxins, identification of a level that cannot be eliminated from a food without discarding that food altogether or severely compromising the ultimate availability of major food supplies.

HACCP was developed in the early 1970s to assure the safety of food products. Its introduction signaled a shift in emphasis from end-product testing to preventive control of hazards at all stages of food production. HACCP passes the responsibility for food safety to food processors that must, through a hazard analysis, identify potential sources of food safety hazards impacting their products, assess the relative risks of these hazards and specify steps that can reduce the potential risks. The processors determine in advance how they will react if the preventive steps fail and products or processes deviate from prescribed parameters. Continual monitoring, documentation, and improvement of the plan are essential elements of the HACCP system and serve as the basis for governmentally sponsored audits.

Mycotoxins are one of the conditional hazards that must be assessed on a situational basis. For example, deoxynivalenol contamination in grain depends on climatic and storage conditions. Monitoring the growing area and storage sites may be control points (CP) in a HACCP program. If monitoring indicates that deoxynivalenol levels in the incoming grain are a significant concern, then the food processor reassesses the likelihood of occurrence of this hazard and initiates CCP level controls as necessary and applies a full HACCP plan. HACCP protocols for aflatoxins, patulin and other trichothecene mycotoxins also require field and storage components in addition to process monitoring.

What is most important is that the HACCP Team has access to sufficient expertise to remain abreast of newly arising mycotoxin issues, *e.g.*, toxicological evaluations, and that they can revise the HACCP protocols as necessary. Quantitative risk assessment is an area of particular concern since reliable data on mycotoxin no-observable effect levels (NOELs) or acceptable daily intake is scarce, especially for sensitive groups within both human and animal populations and for which active research is in progress.

CCFAC has developed a General Standard for Contaminants based on risk assessment and management that provides a process for incorporating exposure assessment into the standard setting process. In addition to its work on maximum levels, the *Codex Alimentarius* Commission also is developing Codes of Practice, so that Good or Improved Agricultural Practices can be used to decrease or eliminate mycotoxin contamination. HACCP principles are used to identify the major critical control points and to minimize the associated monitoring and remediation costs. The HACCP principles are introduced in the codes of practice to focus on major critical control points and provide a means of reducing the cost of such controls, while maintaining the required level of safety. FAO and WHO have developed guidelines for small and less-developed businesses (SLDBs).

The Food Quality and Standards Service (AGNS) of FAO has worked with government bodies and the food industry to implement HACCP. A major part of this program has been the development of tools and the conducting of training courses to strengthen national capacity in HACCP application and auditing. The FAO Manual "Food Quality and Safety

Systems – A training manual on food hygiene and the Hazard Analysis Critical Control Point (HACCP) system" serves as a reference for trainers and those with responsibility for HACCP plan development.

The FAO/IAEA Manual on the Application of the HACCP System in Mycotoxin Prevention and Control (FAO, 2001) provides examples in which the HACCP system is applied to mycotoxin contamination in food and feed in developing country scenarios. Workshops based on this manual are held in conjunction with on-site development of HACCP plans for mycotoxin prevention and control.

Integrated management

Mycotoxin contamination of food- and feedstuffs has become a top priority issue in human and animal health. In traded goods, limits on mycotoxins are becoming ever more restrictive and have become a major trade barrier for agricultural products from Lesser Developed Countries (LDCs) in international markets (Pitt, 1995; Rosner, 1998).

Mold and mycotoxin contamination may be detected in any point of the supply chain. The climate in tropical and subtropical countries offers ideal conditions for mycotoxin production that need to be controlled by postharvest processes, adequate equipment and sound handling practices. More data on these aspects of the agricultural supply chain in developing countries are needed to meet international trade requirements through an integrated agricultural management scheme to control mycotoxin contamination. Contamination associated with postharvest stages of cleaning, drying, storage and processing needs to be identified in terms of the type and amounts of mycotoxins present in each stage and in the by-products used to feed animals.

Managing mycotoxins also requires risk policy determination, *i.e.*, the determination of an "appropriate level of protection" and risk mitigation strategies. Risk mitigation requires an analysis and evaluation of the impact on production systems of mycotoxin levels and should include the setting of priorities based on risk and possible impact and advice to the private sector and individuals on prevention.

FAO provides technical assistance through projects for the prevention and control of mycotoxin contamination when introducing GAP/HACCP based management systems. All mycotoxin projects must overcome four obstacles in developing countries that increase the problem: (i) inadequate government and private resources, (ii) outdated food control systems, (iii) few, if any, action plans for prevention and control; and (iv) lack of awareness and communication. Some current projects include:

- A project on the prevention of ochratoxin A formation in coffee began in Ecuador in July 2004. The main objective was to create awareness of the effect of mold formation in coffee on public health and the economic and social consequences. The project's activities include the identification of cost-effective control systems and the implementation of a HACCP-based training program and a National Action Plan for the control of mold formation in coffee.
- A similar project on coffee quality and the reduction of ochratoxin A contamination began in Vietnam in September 2002 and was completed in 2007. The project defined mold formation mechanisms, identified critical control points, evaluated optimal drying conditions, and developed GAP, GMP and HACCP protocols to reduce and monitor ochratoxin A.
- A project in South Africa is focused on deoxynivalenol and ochratoxin A contamination in selected commodities and is designed to improve the government's ability to sample, monitor, prevent and control mycotoxins in foodstuffs. The activities centered on data gen-

eration and analysis to improve monitoring and control activities, enable scientific-based risk assessment with local data, and to provide relevant data to other SADC countries.
- In Iran the project assisted in the detection, control and management of mycotoxins in foodstuffs. The objective of the project was to enhance the government's ability to monitor and control mycotoxins in foodstuffs to ensure maximize consumer protection and to promote international trade.
- TCP/URU/2801 was an emergency technical assistance program in Uruguay to evaluate and control *Fusarium* contamination of wheat and wheat products and to establish adequate controls in the supply chain. The project included a study mission, extensive training activities in GAP, GMP and HACCP for mycotoxin prevention and control, an integrated national action plan, rapid detection methods and laboratory strengthening, the establishment of a deoxynivalenol rapid alert forecasting system and a crop information network.

Integrated management is a component of a control system composed of legislation and regulations, inspection, analysis and management (FAO/WHO, 2003). The need to develop and strengthen food control systems occurs repeatedly as such food control systems assure the health of the population, add value to domestic food products, protect against unlawful competitors, promote agro-industry development, increase access to international markets and prevent dumping of inferior quality or unsafe products. FAO's work in strengthening Food Control Systems provides advice to national authorities on legislation, infrastructure and enforcement mechanisms to strengthen food control systems to protect public health, prevent fraud, avoid food adulteration and facilitate trade. Over 40 TCPs provide assistance to national *Codex* Committees and for strengthening food control systems.

Mycotoxin Work at *Codex*

Standard setting

The *Codex Alimentarius* Commission (CAC) is an intergovernmental body of 174 member countries with a joint FAO/WHO secretariat (*Codex Alimentarius*, 2003*a*) that facilitates world trade and protects the health of the consumer by developing harmonized international standards for food. Within the CAC, the *Codex* Committee on Food Additives and Contaminants (CCFAC; *Codex Alimentarius*, 2003*c*) sets maximum limits (standards) for additives and contaminants in food, which are critical in trade conflicts. CCFAC develops standards through a risk analysis process that follows the general *Codex* Procedural Manual and the *Codex* General Standard for Contaminants and Toxins in Food. Discussion papers on all relevant aspects of a food contaminant are requested when there is reason to expect health concerns or trade problems, followed by proposals for maximum levels when all the necessary requirements for standard setting have been met.

Mycotoxin standards are set at the international level by the *Codex Alimentarius*. CAC through CCFAC and relevant commodity committees has established the first international guidelines for various mycotoxins based on risk assessments performed by the Joint FAO/WHO Expert Committee on Food Additives (JECFA). Recommended maximum levels for mycotoxins in various foods and feeds are determined by the CCFAC and eventually adopted by the CAC, both of whom base their work on the JECFA recommendations. The 56th meeting of JECFA (2001) was the last to address mycotoxins, and re-evaluated fumonisins B_1, B_2, B_3 and B_4, ochratoxin A, deoxynivalenol, T-2 and HT-2. JECFA also performed a quantitative risk assessment for ochratoxin A in cereal and cereal products at the levels of 5 and 20 ng/g, and for aflatoxin M_1 in milk at 0.5 and 0.05 ng/g. Thus far, the *Codex Alimentarius* (2005) has approved a maximum limit of 15 ng/g total aflatoxins in peanuts intended for further processing; a maximum limit of 0.5 ng/g for aflatoxin M_1 in milk, and a maximum limit of 50 ng/g for patulin in apple juice. The CCFAC also has devoted considerable attention to developing Codes of Practice that contain both principles and practical advice (*Codex Alimentarius*, 2003*a*,*b*). The CCFAC has adopted:

- A revised sampling plan and maximum level for aflatoxins in peanuts intended for further processing.
- A code of practice and a maximum level for patulin in apple juice and apple juice ingredients in other beverages.
- A maximum level for aflatoxin M_1 in milk.
- A code of practice for reduction of aflatoxin B_1 in raw materials and supplemental feedstuffs for milk-producing animals.
- A code of practice for the prevention of mycotoxin contamination in cereals including annexes on ochratoxin A, zearalenone, fumonisins and trichothecenes.

The CCFAC currently is working on:

- A discussion paper and draft code of practice on aflatoxin in tree nuts.
- A discussion paper on deoxynivalenol.
- A discussion paper on a code of practice for the reduction of aflatoxin contamination in peanuts.
- A draft maximum level for ochratoxin A in raw wheat, barley and rye and derived products.
- Sampling plans for aflatoxins in maize and peanuts and draft procedures for aflatoxins in almonds, Brazil nuts, hazelnuts and pistachios.
- Proposed draft maximum levels for aflatoxins in almonds, hazelnuts and pistachios.
- Proposed draft maximum levels for ochratoxin A in wine, coffee and cocoa.
- A general standard for contaminants based on risk assessment and management that will enable the incorporation of exposure assessment into the standard setting process.

The *Codex* Committee on Methods of Analysis and Sampling has work in progress on QA that is important for mycotoxin analysis including:

- Draft general guidelines on sampling.
- Draft guidelines on measurement uncertainty.
- Working instructions for the implementation of the criteria approach and the consequential amendment to the general criteria for the selection of methods of analysis using the criteria approach.
- A recommendation to adopt the IUPAC harmonized guidelines for single-laboratory validation of methods of analysis.

Strategies for communication and dissemination

Development and publication of training materials and tools for information exchange and capacity building are a significant portion of FAO's actions for mycotoxin prevention and control and for awareness creation of the mycotoxin problem. FAO/AGNS released in 2004 (FAO, 2004) the Worldwide Regulations for Mycotoxins in Food and Feed in 2003 (www.fao.org/es/ESN/food/quality_myco_en.stm; www.fao.org/es/ESN/food/quality_myco_fr.stm

or www.fao.org/es/ESN/food/quality_myco_es.stm), which updates the previous comprehensive review published by FAO in 1997. The number of countries with specific regulations for mycotoxins has increased, reflecting general concerns about the effects of mycotoxins on human and animal health and their implications for trade.

Technical assistance on prevention and control of mycotoxins

The importance assigned to actions in this area is growing with the perceived need of anticipating outbreaks and extracting as much information as possible from the, fortunately, scarce fatal outbreaks for which mycotoxins are responsible. The recent Kenyan aflatoxicosis offered an opportunity, to gather information on the past, present and future consequences of an outbreak of acute mycotoxicosis. The major areas of FAO action in technical assistance are: (i) provision of policy advice, awareness creation and knowledge/experience sharing/transfer, institutional strengthening, (ii) field projects to assist members in developing country-specific practical recommendations and national action plans, (iii) organization of international and regional conferences and expert meetings, (iv) training activities and (v) preparation and publication of training materials and tools for information exchange and capacity building. All of these types of assistance have resulted in concrete results such as rapid and cost efficient laboratory methods, accreditation of laboratories, rapid alert systems, implementation of specific GAP-HACCP plans, improved institutional set up and infrastructure, development of national action plans, contamination monitoring programs, regional and national workshops, and publications.

Some examples of FAO projects

FAO sponsors projects in individual countries, in particular regions, and globally, for countries with common interests/concerns. Program content varies, but ranges from policy advice on food control systems and the development of GAP and HACCP quality assurance management systems along the chain, to upgrading, equipping and making accreditation of laboratories possible and providing critical training for those working throughout the food chain in both the public and the private sector. In addition, direct support is provided through emergency projects with consultants and policy advice on urgent situations such as the aflatoxicosis outbreak in Kenya. Horizontal projects also study the social and economic impacts of the application of sanitary and phytosanitary SPS measures on rejection of food products due to mycotoxin levels. Project sustainability is ensured by: (i) identification of country needs and development of feedback mechanisms, (ii) monitoring of progress towards project objectives and anticipated outputs, (iii) commitment to project follow-up at the national level, and (iv) inclusion of project activities complementary to other development activities in the country.

EMRISK: Identification of emerging risks

FAO is a partner in the European Commission-sponsored EMRISK project (http://www.efsa.europa.eu/EFSA/ScientificPanels/efsa_locale-1178620753812_EMRISK.htm) to identify food-related emerging risks, which utilizes FAO case studies on mycotoxins as its basis. The goal is to identify emerging risks, *e.g.*, mycotoxins, associated with feed and food production in a global setting. It considers the increasing complexity of feed and food production systems, trade globalization, novel foods and new food processing technologies. Emerging risks are identified from a combination of knowledge of the food supply chain and the production environment. Five past food incidents were used to identify signals that indicate the possibility of an emerging hazard. For mycotoxins, the emphasis is on the occurrence of conditions for fungal growth and mycotoxin production. The main indicators that contribute to the mycotoxin contamination identified were: (i) food available to the population, (ii) the lack of application of good practices, (iii) erratic or atypical weather patterns conducive to higher mycotoxins production, and (iv) health status of the population. Thus, both technical aspects and information on human behavior and environment are very important and must be considered simultaneously and weighed appropriately. The system can identify an emerging hazard but it does not quantify the associated risk.

Forecasting deoxynivalenol (DONcast)

In most developing countries an early warning system is needed for mycotoxin contamination alerts. In 2001-2002 a *Fusarium* outbreak in the wheat crop in Uruguay resulted in serious economic losses. An electronic deoxynivalenol rapid-alert forecasting system based on a Canadian model that links weather patterns, rainfall, field trials on deoxynivalenol content and crop information (Hooker *et al.*, 2002) was tested in Uruguay as part of an FAO project. DONcast forecasts have two uses: to assist with fungicide application decisions and as a preharvest warning for the grain trade and millers. To be successful, sampling and deoxynivalenol analysis at harvest and the timely provision of near real-time weather data by region is required. DONcast had > 90% accuracy in determining which fields not to spray with fungicide because the risk of deoxynivalenol is low. This project, if extended to regional forecasting, could lead to regional solutions to *Fusarium* epidemics and could be applicable to other areas facing similar situations.

Acute aflatoxicosis in Eastern Kenya

Since May 2004, acute aflatoxin poisoning in Kenya has claimed the lives of over 100 people, with the every-day levels of aflatoxins far exceeding maximum safe consumption levels. FAO coordinated a training of "trainers" for extension staff to raise awareness of mycotoxin contamination and to develop postharvest practices for aflatoxin control. To ensure sustainability, trainees sensitized farmer groups and helped develop action plans for aflatoxin control at the farm level. The action plans were intended to serve as pilot projects that could be replicated in other communities. A second phase of the project stressed raising awareness and pilot drying methods and storage at the community level. A primary factor in the Kenyan outbreak was the availability of food at the time of the outbreak, with many people forced to either eat the moldy grain or to starve, lack of application of Good Agricultural Practices, the weather patterns, and the fragile health status of the population. Future similar events may be avoided if an integrated approach to mycotoxin management is applied, entailing:

- Locally adapted preharvest and postharvest prevention of fungal contamination.
- GAP, GMP and GSP control measures with communication in local languages.
- Intervention in key areas such as seed material, hygiene, cultivation practices, fungicide treatment, limiting plant stress (*e.g.* insect damage and drought), correct harvest date and practices, grain sorting, drying to completeness on an appropriate surface, and proper storage.

The key indicators to monitor include temperature, humidity/moisture/water activity, rainfall, water stress and the presence of insects. The problems that occurred in Kenya could be minimized or avoided if a GAP/HACCP system and its related critical control points is applied, and a forecasting system for acute cases of mycotoxin contamination is developed.

Global enhancement of coffee quality through prevention of mold formation

Persistent low coffee prices have led to the suspension of established agricultural practices and processing techniques. The challenge now is to develop accessible and sustainable recommendations for small-scale growers that have a positive social impact but do not increase production costs. The FAO global project for the Enhancement of Coffee Quality through the Prevention of Mold Formation (FAO/CFC/ISIC, 1999) began in December 2000 in Brazil, Colombia, Ivory Coast, India, Indonesia, Kenya, and Uganda, with ongoing collaboration from the Centre de Coopération Internationale en Recherche Agronomique pour le Développement (CIRAD) and the Institute for Scientific Information on Coffee (ISIC).

The overall scope of the project was to develop solutions and prevention strategies for coffee farmers, buyers and exporters, and to emphasize capacity building while applying internationally accepted food hygiene principles and a HACCP based food safety approach and identifying the critical control points that minimize possible contamination. Underlying the capacity building efforts were recommended measures/good practices, scientific research and field studies and appropriate surveillance and monitoring programs.

The general objective was to increase national awareness of the importance of fungal contamination and ochratoxin A in coffee production on the health of the population and the need to implement effective prevention and control systems along the production and trading chain. More specific objectives included: (i) improvement of analytical methodology and sampling for detection of ochratoxin A and the establishment of good laboratory practices, (ii) preparation of a national action plan for the control of toxigenic fungi and the reduction of ochratoxin A with practices/measures/actions integrated into a unified prevention and control system, (iii) preparation of publications outlining the national action plan, and (iv) improving the quality and value of all coffee varieties. Country needs assessment and country validation-feedback were performed to ensure sustainability.

An electronic training resource (web and CD-ROM based) on good hygiene practices along the coffee chain was developed to guide authorities in the development of national training programs.

Impact of food rejections due to mycotoxins

Case studies of the social and economic impact of the application of sanitary and phytosanitary measures of the WTO SPS Agreement on the rejection of food products due to mycotoxin levels also are being conducted in several developing countries.

Conclusions

Mycotoxins affect a wide range of agricultural products, including most grain crops, which are the agricultural backbone of most developing economies. There are numerous sources of contamination of these commodities, as a result of environmental conditions in the field, as well as improper harvesting, storage and processing. Poor infrastructure and managerial and economic constraints can worsen already difficult situations and increase mycotoxin levels even further.

FAO uses various approaches to help countries address the mycotoxin challenge. Key problems can be grouped into three major areas: inadequate food control systems; lack of prevention and control policies and strategies; and insufficient awareness. The main obstacles to diminishing the mycotoxin problem are: insufficient government and private resources, outdated Food Control Systems, few, if any, action plans for prevention and control, and lack of awareness and communication. Each country's mycotoxin situation is unique, and action plans depend on existing infrastructure, expertise, production protocols and regulations along the food chain. Main general recommendations would be to promote harmonized regulatory framework, modernize food control systems, implement coordinated and integrated action plans along the entire food chain, such as the GAP/HACCP system, conduct proper risk assessment, provide sector specific guidance, educate/train stakeholders and strengthen communication all throughout. Specific technical and managerial approaches will have to be decided and applied on a country by country basis. FAO is instrumental in providing up to date information and technologies, and disseminating effective strategies to developing countries for the prevention and control of mycotoxin contamination. The improvement of the mycotoxin problem in developing countries awaits careful assessment of each country's mycotoxin situation, with subsequent formulation of action plans for prioritization and implementation of defined technical and organizational strategies. It cannot be addressed in one single way; the strategies will depend on the specific developmental situation of the countries involved and need to be integrated into the entire food chain and with a holistic vision (Piñeiro, 2003, 2004*a*).

Further information about the work of FAO in Capacity Building for Food Safety is available at: www.fao.org/ag/AGN. Specific queries can be addressed to food.quality@fao.org.

References

Codex Alimentarius Commission. (2003*a*) *Procedural manual, 13th ed. Joint FAO/WHO Food Standards Program*; http://www.codexalimentarius.net/procedural_manual.stm.

Codex Alimentarius Commission. (2003*b*) Draft working principles for risk analysis for application in the framework of the *Codex Alimentarius*. Appendix IV. ALINORM 03/33a; http://www.codexalimentarius.net/reports.asp.

Codex Alimentarius Commission. (2003*c*) Report of the 34th Session of the Codex Committee of Food Additives and Contaminants (Rotterdam, the Netherlands, March 2002); ftp://ftp.fao.org/codex/alinorm03/Al03_12e.pdf.

Codex Alimentarius Commission. (2003*d*) Report of the 24th Session of the Codex Committee on Methods of Analysis and Sampling (Budapest, Hungary, November 2002); ftp://ftp.fao.org/codex/ alinorm03/A103_23e.pdf.

Codex Alimentarius Commission. (2005) Report of the 28th Session of the Codex Alimentarius Commission; ftp://ftp.fao.org/codex/cac/cac28/if28_05e.pdf.

European Commission. (2002) General Food Law Regulation (EC) No 178/2002 of the European Parliament and of the Council of 28 January 2002 laying down the general principles and requirements of food law. *Official Journal of the European Communities* L31: 1.2.2002, p. 5.

FAO. (2001) *FAO/IAEA Manual on the Application of the HACCP System in Mycotoxin Prevention and Control – FAO Food and Nutrition Paper 73.* FAO, Rome, Italy.

FAO. (2004) *Worldwide Regulations for Mycotoxins in Food and Feed in 2003 – FAO Food and Nutrition Paper 81*. FAO, Rome, Italy.
FAO/CFC/ISIC. (1999) Project on improving quality control through prevention of mold formation. GCP/INT/7634/CFC, FAO, Rome, Italy.
FAO/WHO. (2003) *Assuring Food Safety and Quality: Revised Guidelines for Strengthening Food Control Systems – FAO Food and Nutrition Paper 76*. FAO, Rome, Italy.
FAO/WHO. (2006) *Food Safety Risk Analysis: An Overview and Framework Manual.*
FAO/WHO/UNEP. (1999) Third Joint International Conference on Mycotoxins (Tunis, Tunisia, March 1999); http://www.fao.org/es/ESN/mycoto/mycoto.htm.
Hooker, D.C., Schaafsma, A.W. and Tamburic-Ilincic, L. (2002) Using weather variables pre- and post-heading to predict deoxynivalenol in winter wheat. *Plant Disease* 86, 611-619.
IARC. (1993) Some naturally occurring substances: Food items and constituents, heterocyclic aromatic amines and mycotoxins. *IARC-WHO Monographs on the Evaluation of Carcinogenic Risks to Humans* 56, 445-466.
JECFA. (2001) 56th meeting: Mycotoxins – Summary and conclusions (Geneva, Switzerland, February, 2001); http://www.fao.org/es/ESN/Jecfa/jecfa56.pdf.
Miller, J.D. (1998) Global significance of mycotoxins. In: Miraglia, M., van Egmond, H., Brera, C. and Gilbert, J. (eds.), *Mycotoxins and Phycotoxins – Developments in Chemistry, Toxicology and Food Safety*. Alaken Inc., Fort Collins, Colorado, USA, pp. 3-15.
Piñeiro, M. (2000) Global approaches to mycotoxin risk assessments and standards: Mercosur and Latin American standards for mycotoxins. *Proceedings of the IFT Annual Meeting.* pp. 50-51.
Piñeiro, M. (2003) Mycotoxins: Natural toxins of agricultural and health significance. In: Elmadfa, I., Anklam, E. and Konig, J. (eds.) *Proceedings of the International Congress of Nutrition 2001, Forum of Nutrition Vol. 56*, S. Karger, Switzerland, pp. 409-412.
Piñeiro, M. (2004*a*) Mycotoxins: Current issues in South America. In: Barug, D., van Egmond, H., Lopez-Garcia, R., van Osenbruggen, T. and Visconti, A. (eds.) *Meeting the Mycotoxin Menace*. Wageningen Academic Publishers, Wageningen, The Netherlands, pp. 49-68.
Piñeiro, M. (2004*b*) Mycotoxin issues in developing countries. *Proceedings of the 4th Asian Conference on Food and Nutrition Safety*. Bali, Indonesia.
Pitt, J.J. (1995) Estimation of the direct cost of aflatoxins. *Australian Mycotoxin Newsletter* 6 (1), 12.
Rosner, H. (1998) Mycotoxins: Limits in the European Union and effects on trade. In: Miraglia, M., van Egmond, H., Brera, C. and Gilbert, J. (eds.), *Mycotoxins and Phycotoxins – Developments in Chemistry, Toxicology and Food Safety*. Alaken Inc., Fort Collins, Colorado, USA, pp. 203-212.

34

Mycotoxin Research in USAID's CRSP Programs

John M. Yohe* and Jonathan H. Williams

Abstract

Nine Collaborative Research Support Programs (CRSPs) were established by the U.S. Agency for International Development (USAID) in response to Public Law 94-161, Title XII Legislation: Famine Prevention and Freedom from Hunger Act, which was originally approved in December 1975 and amended and signed by the President in October, 2000 (Public Law 106-373). Two of the CRSPs have significant research activities on the impact of mycotoxins on human health and agricultural productivity. Impaired immunity and malnutrition increases both human susceptibility to HIV infection and the vulnerability to its various impacts. Mycotoxins can significantly reduce human health, especially amongst the poor in developing countries and can become significant non-tariff trade barriers. The greatest cost imposed by mycotoxins is on human health. Aflatoxins are synergistic with hepatitis B and are associated with at least 44% of the disease burden in developing countries. The effects of mycotoxins as market barriers are important to production, socioeconomics and utilization. The Peanut CRSP is involved with mycotoxins because of the vulnerability of this crop to aflatoxin contamination. The Sorghum/Millet CRSP is involved in studies of mycotoxins because mycotoxigenic fungi appear to be less important factors on sorghum and millet than they are on maize in Africa. Although human health concerns are the primary focus, the role of mycotoxins in agricultural trade, especially the export of peanuts, also is important.

Introduction

A wide range of mycotoxins are known food contaminants that can affect exposed persons and animals in various ways. Of particular interest to us are the antinutritional mycotoxin contaminants of peanuts, maize, sorghum and pearl millet that can cause various diseases in humans, livestock and domesticated animals throughout the world. According to Williams *et al.* (2004), as many as 4.5 billion persons living in developing countries are chronically exposed to largely uncontrolled amounts of aflatoxins.

Peanuts play an important role in the diets of rural populations, particularly children, because of their high content of protein (21-30%), fats (41-52%) and carbohydrates (11-27%). They also are rich in calcium, potassium, phosphorus, magnesium and vitamin E. Protein meal, a by-product of oil extraction, is an important ingredient in livestock feed. Peanut haulms are nutritious and widely used for feeding livestock.

Aflatoxin contamination of peanuts is a major hazard to human and animal health and is one of the most important constraints to peanut trade. Aflatoxins are considered to be an important cause of hepatocellular carcinoma, one of the most common cancers in developing tropical countries. Studies in the Ivory Coast, Mozambique, Sudan, Swaziland, China and Korea all have reported relatively high frequencies of liver cancer. A common factor is the presence of toxins – including aflatoxin B_1 produced by fungi such as *A. flavus* and *A. parasiticus* – in peanuts and peanut-based foods.

Sorghum and millets are traditional African cereal grains that have gradually been displaced by maize over the past 150-200 years. Recently the rate of displacement has increased in Africa, often with the encouragement of international agricultural research institutes and aid programs. A major concern for the consumption of maize, especially as a high proportion of the diet (Marasas *et al.*, Chapter 4), is that it may be easily contaminated with aflatoxin and fumonisin mycotoxins produced by *Aspergillus* and *Fusarium* spp., respectively. These toxins are associated with cancer, neural tube defects, immune system depression and acute toxicity in humans. Symptoms depend on the level of the toxin in the diet and the length of the exposure. International Sorghum, Millet and Other Grains CRSP (INTSORMIL)-sponsored research (Bandyopadhyay *et al.*, 2007) has demonstrated that maize is significantly more heavily colonized by aflatoxin-producing *Aspergillus* than is either sorghum or millet, and that overall aflatoxin levels are higher in maize than they are in sorghum or millet from plants grown on-farm by subsistence farmers in Nigeria.

Importance of mycotoxins to economies and health

The wide range of mycotoxins recognized as potential food contaminants are discussed elsewhere in this volume as are details of the regulations and requirements for trade in mycotoxin-prone commodities (FAO, 2004). We focus on the consequences of these food contaminants both in developing and developed countries. The obvious consequences, and those that have attracted the most attention and action, are related to trade and the potential of these compounds to be non-tariff trade barriers in international trade. Recent papers, *e.g.*, Otsuki *et al.* (2001*a,b*) and Wilson and Otsuki (2003), pointing to the trade-off between health gains in Europe and trade consequences for Africa of the new levels of allowable aflatoxin are demonstrative of this focus on trade, and the resulting controversy is indicative of this attention (Coulibaly *et al.*, Chapter 7). For most African countries, reduction in agricultural trade is the least important of the plethora of problems associated with these toxins because only a few countries export the majority of these commodities that they produce. Instead, most of these agricultural products are consumed domestically, which means that the major economic cost of mycotoxin contamination is that required to assure safe food, *i.e.*, the developed country scenario, or to deal with the alternative health costs of failing to do so in developing countries.

The importance of mycotoxins is much greater than is generally recognized. Failure to satisfy mycotoxin standards can result in significant monetary loss as the producers and processors of foods are accountable for the quality of their produce. Thus, they must develop ways to meet the standards, with the additional costs of achieving compliance passed on to the consumer as the cost of assuring a healthy food supply. The additional cost of managing mycotoxins across the full value chain for major commodities may be as much as 10-15% of the final price. For example, aflatoxin contamination in peanuts increases if drought occurs during the last six weeks of the crop's growth. Many farmers have irrigation to avoid this risk, even though simulations of response to irrigation in Florida (Williams and Boote, 1995) show that significant production losses attributable to drought at this stage of production seldom occur. These farmers are responding to the market's requirement for food low in aflatoxins and expect a quality premium on their produce in return. Similarly, there are one-time fixed costs of infrastructure for, and recurring costs associated with, taking samples and analyzing them for toxins. Investments in drying and electronic sorting after harvest to segregate potentially contaminated grains all increase the economic burden imposed by mycotoxins. Despite these investments there are times when a significant fraction of agricultural production is lost in developed countries because it is contaminated. In the United States these losses can be as high as $1.5 billion across all the range of mycotoxins (Robens and Cardwell, 2003). The use of contaminated produce in animal feeds also is recognized as a major source of economic loss (Fink-Gremmels, Chapter 14). Although acute poisoning is relatively rare, the consequences of chronic mycotoxin exposure can be very significant – lowering feed conversion rates, increasing reproductive failures and veterinary medicine costs and escalating the use of antibiotics in feed to protect animals with weakened immune systems.

When there are no investments to assure that foods are not contaminated by mycotoxins, a different importance of these toxins is evident but seldom recognized. In this regard the range of toxicities that may occur must be considered. Exposure to many of these mycotoxins has carcinogenic and teratogenic consequences, so the consequences of exposure need not be immediately connected with exposure and often are not considered because they are deferred or attributed to more obvious causal agents. For example, many mycotoxins have significant immune suppressing capabilities, but mycotoxins are only rarely connected with their impacts on disease epidemiology. As the toxins exert their effects through immunity and nutrition the problems they cause often are missed since current medical training and philosophy focus on responding to symptoms and curing diseases rather than preventing occurrence. Consequences similar to those observed for animal production on farms are observed in human populations in developing countries.

Two independent studies have estimated that the burden of disease connected with aflatoxin alone is > 40% of the disease burden in developing countries (Miller, 1996; Williams *et al.*, 2004), even though there is no way to determine exactly what fraction of this disease burden is attributable to mycotoxins. Based on epidemics in farm animals, the impact of mycotoxins on nutrition and immunity probably is significant. Other data indicate that the potential cost of even marginal changes in the health situation is enormous. As little as a 1% change in the burden of disease in response to managing only aflatoxin would likely increase economic growth rates significantly. The basis for this projection is derived from the data for malaria (which contributes 3% of the burden of disease on a world scale (http://www.globalhealthreporting.org/malaria.asp?id=62) where a 10% change in the incidence of malaria in endemic countries would increase their economic growth on a per/capita annual basis by 1.3% (Gallup and Sachs, 2001). Another example is that there are significant statistical relationships between aflatoxin exposure and HIV incidence that are supported by epidemiological mechanisms (Williams *et al.,* 2005); and there is little doubt of the economic costs of HIV/AIDS.

An analysis of the WHO health risks (Rodgers *et al.*, 2002) combined with the known immunotoxicities and nutritional interferences of mycotoxins suggests that the seven most important identified risk factors (malnutrition of children, sexually transmitted diseases, unsafe water, smoke from open fires, Zn deficiency, Fe deficiency, and vitamin A deficiency) are all subject to modulation by mycotoxin exposure. This fact alone makes health a critical objective for agricultural production and research. Across the world there is a growing realization, as is evidenced by the HarvestPlus program (http://www.harvestplus.org/about.html) that agricultural technologies that can build on the established connections of good nutrition with health-based immune competence and cognitive development are very cost effective ways to increased public health. Mycotoxins are part of that opportunity and might be the single most important part.

How do mycotoxins fit within USAID's priorities?

USAID efforts to address mycotoxins are found in the agricultural and trade sectors – being the support for the CRSPs addressing staple commodities and the CGIAR center programs on maize and peanut. The primary justification for these investments is the importance of the mycotoxins to trade and access to the markets of Europe, Japan and North America. In contrast there is no articulated effort to address these toxins from the health, child survival, HIV and nutrition perspectives. USAID's Environmental Health program is committed to preventing suffering, the saving of lives and the creation of a brighter future for families in the developing world. This commitment includes addressing such issues as maternal and child health, reduction of infectious diseases, *e.g.*, HIV/AIDS, malaria and tuberculosis, environmental issues associated with smoke and water and micronutrient availability. These issues are clearly important risks to health and productivity in developing countries; and the close relationship between the WHO assessment of risk and the USAID priorities is clear. What is not clear, however, is the extent of these problems that is attributable to the anti-nutritional effects of mycotoxins and how much is due to inadequate supply or other environmental factors. From an epidemiological perspective the crucial question is what proportion of these health risks can be attributed to the *prima facie* causal agents and what proportion can be attributed to the indirect effects of impaired immunity and lowered nutritional status from contaminated foods.

A major problem is that the model used is one for commercial large-scale food systems that can assure safe food and meet the standards for *Codex Alimentarius*. Unfortunately, solutions that are effective in developed countries often are impractical in developing countries, where relatively little of the food consumed passes through formal quality control systems, and where food insecurity continues to ensure that even food known to be contaminated is consumed because there are few, if any alternatives. Until the proportion of the burden of disease presently attributed to a health risk can be partitioned into that due to the mycotoxins, and into that due to the direct risk, this perception will continue. However, given the fraction of the burden of disease that is modulated by nutrition and immunity (43%) it is possible that the cost of addressing the direct risk is many times higher because neither developing countries nor donors have considered the indirect role of these toxins when making health estimates or setting health policy(ies). Many of the present investments in programs designed to deliver vitamins and drugs probably could be reduced if the mycotoxin levels in the foods were adequately managed.

Mycotoxin research sponsored by the Sorghum/Millet CRSP

Aflatoxin and fumonisin hazards are associated with cancer, liver disease, immune suppression, retarded growth and development, nutritional interference, acute toxicity, neural tube defects, and death depending on the amount of toxin exposure and its duration. The symptoms depend upon the level of the toxin in the grain and the length of the exposure. The general perception is that sorghum is a safer grain to consume than is maize because the fungi that produce these toxins are rarer on sorghum and more common on maize (Leslie and Marasas, 2002; Leslie *et al.*, 2005). INTSORMIL has worked with the International Institute of Tropical Agriculture (IITA) and the PROMEC unit of the Medical Research Council (MRC) in South Africa to test the hypothesis that the relative mycotoxigenicity of maize and sorghum grain differ.

INTSORMIL CRSP/IITA/Kansas State University (KSU) collaborative research has shown that locally grown maize is significantly more heavily colonized by aflatoxin-producing *Aspergillus* spp. than is either locally grown sorghum or millet, and that the average level of aflatoxin contamination is higher in maize than it is in the other two crops. The average aflatoxin contamination level in maize (36 ng/g) was four to eight fold higher than the contamination level in either sorghum (8.8 ng/g) or pearl millet (4.6 ng/g). The median amount of aflatoxin per sample, however, was similar for all three grains (4.2 ng/g, 5.0 ng/g and 4.4 ng/g respectively) suggesting that the major problem is with the more heavily contaminated samples. Of the 23 maize samples in the test, four exceeded the 20 ng/g internationally accepted guideline (FAO, 2004) recommended by *Codex Alimentarius*, as did two of the 40 sorghum samples but none of the pearl millet samples. In addition to having a higher proportion of samples that exceeded the guidelines, the maize samples that exceeded the guidelines also contained higher levels of aflatoxin than did the non-conforming sorghum samples.

The most effective way to reduce aflatoxin contamination is to prevent it from ever forming on foodstuffs. Maize planted in marginal regions usually encounters both heat and drought stress, two factors that predispose the grain to higher levels of aflatoxin. Cultivating indigenous sorghum or pearl millet crops, which are well adapted to both drought and heat stresses, in marginal environments would disproportionately reduce aflatoxin exposure, since maize with extreme levels of contamination should be encountered less frequently and since sorghum and pearl millet both have a much smaller risk of being contaminated with aflatoxin. The *Fusarium* spp. that colonize maize, sorghum and pearl millet also are known to differ now that the former "*Fusarium moniliforme*" is being broken down into biologically meaningful species (Seifert *et al.*, 2003). Species common on maize, *e.g.*, *F. verticillioides* and *F. proliferatum*, typically produce fumonisins (Leslie *et al.*, 1992*a*,*b*), while species from sorghum and pearl millet, *e.g.*, *F. thapsinum*, *F. andiyazi* and *F. pseudonygamai*, are more likely to produce moniliformin (Leslie *et al.*, 1996; Fotso *et al.*, 2002), a toxin that has not been well-studied and has not been associated with any disease outbreaks in humans or domesticated animals (Desjardins, 2006).

Mycotoxin research sponsored by the Peanut CRSP

Ever since the Peanut CRSP began, aflatoxin has been a consistently high priority for the program. Early research focused on managing aflatoxin to decrease the exposure of peanut consumers in developing countries. The first initiatives (prior to 2000) focused on decreasing the levels of contamination through research on improved management techniques and decontamination. A sustained basic research effort was implemented to prevent/reduce the production of the toxin in peanuts, either by providing resistance to the causal fungi, or by developing ways to block toxin production by the fungus. Genes with this potential were identified and transferred into peanut, but since the techniques used included transformation of host plants with exogenous DNA the resulting lines have not yet been tested in the field. Associated with this research were education/extension activities to transfer to local industries the management techniques being developed or being applied in developed countries.

Peanut CRSP research and outreach has been effective in allowing formal commercial industries to manage aflatoxin and provide safe and exportable products (Galvez *et al.*, 2002*a*,*b*). However, there has not been matching success in the small scale and informal sectors. For example, research in Ghana demonstrated that if available natural products, *e.g.*, leaves of trees and spices, were included with stored peanuts then aflatoxin contamination could be decreased. These results have not been adopted despite extension efforts. Economic research determined that the failure of farmers and marketers to adopt aflatoxin management practices was associated with the lack of incentives to invest in these efforts. Peanuts commonly sell for the same price regardless of quality in many developing countries. We think that quality control is impractical in the informal, small-scale peanut sector of many developing countries. Thus, other options for protecting people by using a food additive approach have become more important.

Part of the earlier research sponsored by the Peanut CRSP focused on the decontamination of peanut oil produced in villages with primitive presses. Research at Texas A&M University found that a hydrated bentonite clay was particularly effective in binding aflatoxin (Huebner and Phillips, 2003). Although the porosity of this product makes it impractical as a technique to decontaminate oil, the properties of this clay led to a new approach to solving the problem of aflatoxin contamination. In animal studies with the clay as a feed additive, the clay provided excellent protection of the animals even at the highest levels of contamination known to occur naturally (7,000 ng/g).

Since 2001 the Peanut CRSP has continued with its efforts to manage aflatoxin in peanuts and has tested the additive for human use in an enterosorption strategy, *e.g.*, Afriyie-Gyawu *et al.* (Chapter 25). This effort has become a high priority for the Peanut CRSP, particularly given the evidence now available for immune suppression and nutritional interference (Williams *et al.*, 2004; Gong *et al.*, Chapter 6; Jolly *et al.*, Chapter 5).

The future

The future activities of the CRSPs in relation to mycotoxins will depend on sustained support by USAID. Assuming that these CRSPs continue, the mycotoxins will remain a high priority, particularly for the Peanut CRSP, where the long term investment in preventing contamination by genetic means could be taken to the field during the next ten years. Other management options also seem likely to be adopted as the broader implications of immune suppression and nutritional interference become better known and consumers begin to create the economic incentives required for the supply chain to justify production of foods with lesser amounts of aflatoxin (Williams, Chapter 30).

Although a lower risk of mycotoxins in sorghum and millet is an important reason to increase the use of these commodities, the present trend to replace them with maize will continue until the hazards of mycotoxins are better understood by the consumers. Changing this perception requires additional research to document the importance of mycotoxins to the epidemiology of priority health risks and to support the current preliminary conclusions regarding the increased safety of sorghum and millet. We believe that USAID has the opportunity to provide world leadership by making the health of local populations at least as important as international trade in agricultural commodities.

The research to date on the use of enterosorption to reduce human aflatoxicosis indicates that such an effort could be easily justified. Virtually everyone living in West Africa carries biomarkers of chronic exposure to aflatoxin. Such chronic exposure interferes with both general nutrition (Gong *et al.*, Chapter 6), micronutrition (Turner *et al.*, 2003) and immunity (Jolly *et al.*, Chapter 5). We also know from the studies done both on animals and humans that there are (as yet) no known risks to the use of this clay as a dietary supplement. One of the major advantages of this approach is that it is effective for the toxin in multiple foods, *i.e.*, it is toxin-specific rather than commodity specific. At the very least this technology enables epidemiological studies that can establish the extent to which aflatoxins contribute to presently identified risks. Such knowledge is needed to identify future options, to develop research agendas, and to define intervention strategies. The use of enterosorption could be justified on a broad basis until management advances have been made and adopted by small-scale farmers and informal food system participants.

When considering aflatoxins produced by *Aspergillus* and fumonisins produced by *Fusarium* in cereals such as sorghum, millets and maize and in peanuts, an ounce of prevention is worth a pound of cure. Thus, the primary management strategy should be to prevent mycotoxins from ever forming on foodstuffs. Because of the susceptibility of maize to mycotoxin load, sorghum and millet should be grown instead of maize, especially in marginal environments where the drought and heat stress increase mycotoxin levels in maize. Genetic resistance is another approach. Genes for resistance to *Aspergillus* and *Fusarium* have been identified in cereals and peanuts. Deployment of resistant varieties holds promise as a management strategy and needs increased research effort. Also, research has identified natural products that decrease mycotoxin levels when mixed with peanuts in storage. Hydrated bentonite clay can bind aflatoxins in peanut oil. However, these strategies are not being adopted by farmers and thus there remains a need to develop practical methods that are farmer acceptable and effective, economical and easy to use.

Acknowledgments

This chapter was made possible by the support of the following: USAID grants: LAG-G-0096-90013-00 to the University of Georgia and LAG-G-00-96-90009-00 to the University of Nebraska and subgrant recipients at Texas A&M University, Kansas State University, IITA and the PROMEC Medical Research Council, Republic of South Africa.

References

Bandyopadhyay, R., Kumar M. and Leslie, J.F. (2007) Relative severity of aflatoxin contamination of cereal crops in West Africa. *Food Additives and Contaminants* 24, 1109-1114.

Desjardins, A.E. (2006) Fusarium *Mycotoxins: Chemistry, Genetics and Biology*. APS Press, St. Paul, Minnesota.

FAO. (2004) *Worldwide regulations for mycotoxins in food and feed in 2003. Food and Nutrition Paper* 81. Rome, Italy.

Fotso, J., Leslie J.F. and Smith, J.S. (2002) Production of beauvericin, moniliformin, fusaproliferin, and fumonisins B_1, B_2 and B_3 by *ex*-type strains of fifteen *Fusarium* species. *Applied and Environmental Microbiology* 68, 5195-5197.

Gallup, L.L. and Sachs, J.D. (2001) The economic burden of malaria. *American Journal of Tropical Medicine and Hygiene* 64, 85-96

Galvez, F.C.F, Francisco, M.L.D.L, Lustre, A.O, and Resurreccion, A.V.A. (2002*a*). *Control of Aflatoxin in Raw Peanuts Through Proper Visual Sorting*, USAID-Peanut CRSP Monograph Series No. 3, 49.

Galvez, F.C.F., Francisco, M.L.D.L., Lustre, A.O., and Resurreccion, A.V.A., (2002b). *Control of Aflatoxin in Raw Peanuts Through Proper Manual Sorting*, USAID-Peanut CRSP Monograph Series No. 3, 1-41

Huebner, H.J. and Phillips, T.D. (2003) Clay-based affinity probes for the selective cleanup and analysis of aflatoxin B_1 utilizing nanostructured montmorillonite on quartz. *Journal of the Association of Official Analytical Chemists International* 86, 534-539.

Leslie, J.F. and Marasas, W.F.O. (2002) Will the real *Fusarium moniliforme* please stand up! In: Leslie, J.F. (ed.) *Sorghum and Millets Diseases*. Iowa State Press, Ames, Iowa, pp. 201-209.

Leslie, J.F., Doe, F.J., Plattner, R.D., Shackelford, D.D. and J. Jonz, J. (1992*a*) Fumonisin B_1 production and vegetative compatibility of strains from *Gibberella fujikuroi* mating population "A" (*Fusarium moniliforme*). *Mycopathologia* 117, 37-45.

Leslie, J.F., Plattner, R.D., Desjardins, A.E. and Klittich, C.J.R. (1992*b*) Fumonisin B_1 production by strains from different mating populations of *Gibberella fujikuroi* (*Fusarium* section *Liseola*). *Phytopathology* 82, 341-345.

Leslie, J.F., Marasas, W.F.O., Shephard, G.S., Sydenham, E.W., Stockenström, S. and Thiel, P.G. (1996) Duckling toxicity and the production of fumonisin and moniliformin by isolates in the A and F mating populations of *Gibberella fujikuroi* (*Fusarium moniliforme*). *Applied and Environmental Microbiology* 62, 1182-1187.

Leslie, J.F., Zeller, K.A. Lamprecht, S.C. Rheeder J.P. and Marasas, W.F.O. (2005) Toxicity, pathogenicity and genetic differentiation of five species of *Fusarium* from sorghum and millet. *Phytopathology* 95, 275-283.

Miller, J.D. (1996). Mycotoxins. In: Cardwell, K.F. (ed.) *Proceedings of the Workshop on Mycotoxins in Food in Africa (6-10 November 1995, Cotonou, Benin)*. International Institute of Tropical Agriculture, Ibadan, Nigeria, pp. 18-22.

Otsuki, T., Wilson, J.S. and Sewadeh, M. (2001*a*) What price precaution? European harmonization of aflatoxin regulations and African groundnut exports. *Agricultural Economics* 28, 263-284.

Otsuki, T., Wilson, J.S. and Sewadeh, M. (2001*b*) Saving two in a billion: Quantifying the trade effect of European food safety standards on African exports. *Food Policy* 26, 495-514.

Robens, J. and Cardwell, K. (2003) The costs of mycotoxin management to the USA: Management of aflatoxins in the United States. *Toxin Reviews* 22, 139-152.

Rodgers, A., Vaughan, P., Prentice, T., Edejer, T.T., Evans, D. and Lowe, J. (2002) The World Health Report 2002: Reducing risks, promoting healthy life. In: Campanini, B. and Haden, A. (eds.). *Defining and Assessing Risks to Health*. World Health Organization, Geneva, Switzerland, pp. 9-25.

Seifert, K.A., Aoki, T., Baayen, R.P., Brayford, D. Burgess, L.W., Chulze, S., Gams, W., Geiser, D., de Gruyter, J., Leslie, J.F., Logrieco, A., Marasas, W.F.O., Nirenberg, H.I., O'Donnell, K., Rheeder, J.P., Samuels, G.J., Summerell, B.A., Thrane U. and Waalwijk, C. (2003) The name *Fusarium moniliforme* should no longer be used. *Mycological Research* 107, 643-644.

Turner, P.C., Moore, S.E., Hall, A.J., Prentice, A.M. and Wild, C.P. (2003) Modification of immune function through exposure to dietary aflatoxin in Gambian children. *Environmental Health Perspectives* 111, 217-220.

Williams, J.H. and Boote, K.J. (1995) Physiology and modeling: Predicting the unpredictable legume. In: Pattee, H.E. and Stalker, H.T. (eds.). *Advances in Peanut Science,* American Peanut Research and Education Society, Stillwater, Oklahoma, pp. 301-353.

Williams, J.H., Phillips, T.D., Jolly, P.E., Stiles, J.K., Jolly, C.M. and Aggarwal, D. (2004) Human aflatoxicosis in developing countries: A review of toxicology, exposure, potential health consequences and interventions. *American Journal of Clinical Nutrition* 80, 1106-1122.

Williams, J.H., Aggarwal, D., Jolly, P.E., Phillips, T.D. and Wang, J.-S. (2005) Connecting the dots: Logical and statistical connections between aflatoxin exposure and HIV/AIDS. http://168.29.148.65/ pdfs/connectionAFwithHIV.pdf

Wilson, J.S. and Otsuki, T. (2003) Food safety and trade: Winners and losers in a non-harmonized world. *Journal of Economic Integration* 18, 266-287.

35

CGIAR Research-for-Development Program on Mycotoxins

Rodomiro Ortiz*, Tomohiro Ban, Ranajit Bandyopadhyay, Marianne Banziger, David Bergvinson, Kerstin Hell, Braima James, Dan Jeffers, P. Lava Kumar, Abebe Menkir, Jiro Murakami, Shyam N. Nigam, Hari D. Upadhyaya and Farid Waliyar

Abstract

The major mycotoxins studied at the Consultative Group on International Agricultural Research (CGIAR) institutes are aflatoxins in maize, peanut, sorghum and cassava, *Fusarium* toxins in maize, wheat and sorghum, and ochratoxin in cocoa and cashew. Genetic enhancement (both through plant breeding and biotechnology), biological control, habitat management, risk assessment, institutional capacity building and public awareness are among the tools in the "CGIAR research-for-development kit" to fight mycotoxins worldwide. A holistic approach should be pursued to deal with mycotoxins that includes the following elements: i) an integrated crop management package that combines mycotoxin-resistant germplasm, biological control, habitat control and soil-amendments; ii) low-cost mycotoxin detection technology for rapid appraisal that also should facilitate trade; iii) a participatory process for mycotoxin assessment in commercially important crops; and iv) a high-level panel composed of scientists, NGOs, farmers, traders, consumers, health officers and policy makers to monitor mycotoxin intervention strategies and to organize awareness campaigns.

Introduction

Food quality and safety are important traits used by people to select their diet. Internationally these standards are agreed through the *Codex Alimentarius* with national laws also having an important role (FAO, 2003). Contamination of staple foods is widespread in some locations, particularly in the developing world, and can occur at all levels of crop production: preharvest, harvest, and storage.

The safety and nutritional quality of food often is compromised by mycotoxins, which are metabolites produced by a few fungi that colonize both staple agricultural products and export crops from countries in the developed and developing world. The best known mycotoxin is aflatoxin, which is produced by a few species of the fungus *Aspergillus*, and commonly occurs in maize, peanut, sorghum, and some root and tuber crops. *Fusarium* spp. also produce mycotoxins such as fumonisins, zearalenone, and deoxynivalenol. Mycotoxins are regulated because they are hazardous to health. In addition to their carcinogenic properties, many mycotoxins are anti-nutritional factors that result in poor growth and immune suppression in young animals and children.

In the developed world, regulatory standards control exposure of humans and animals to dietary mycotoxins. These food safety regulations reduce the risks of morbidity and mortality associated with the consumption of contaminated food. In the developing world, particularly in Sub-Saharan Africa, monitoring and enforcement of standards are rare. Mycotoxins also may form non-tariff trade barriers. The European Union has recently reduced regulatory limits for aflatoxin to 4 ng/g compared to the *Codex Alimentarius* Commission recommended standard of 20 ng/g in peanuts. This dichotomy in legislation could cost some developing nations several hundred million dollars in export losses (Wu, 2004). The costs of food safety regulation includes the cost of production, compliance, and administration, and the deadweight loss associated with these costs. In countries with widespread aflatoxin occurrence, the best quality foods are exported and the poorer quality foods are consumed locally and harm the local population. Thus, mycotoxins degrade food quality, can be barriers to international trade, pose serious risks to health, and are directly and indirectly responsible for human deaths in Africa and Asia. Exposure to aflatoxin increases the incidence of acute toxicosis, liver cancer, and morbidity in children suffering from kwashiorkor.

The centers of the Consultative Group on International Agricultural Research (CGIAR) recognize mycotoxins as one of the most important constraints to the goal of improving human health and well-being through agriculture. The CGIAR centers pursue various strategies for the management of mycotoxins from "field to fork". For example, aflatoxin management practices in farmers' fields and stores have been developed and are being implemented through national partners represented in several chapters elsewhere in this volume. Work continues on the dissemination of management practices, biological control through competitive exclusion strategies, and breeding for resistance. The level of fumonisin contamination in field and stored maize has been surveyed, and management processes that can affect these levels have been identified. Fungal-insect relationships, both for *Aspergillus* spp. and *Fusarium* spp. in the field and in storage, have been investigated. Several aspects of mycotoxin research-for-development need further attention. These include food basket survey, strategies to reduce the impact of mycotoxins on trade, bio-ecological aspects of mycotoxin production, biological control, resistance breeding, and the impact of mycotoxin management options and nutritional improvement on children's growth and health in high-risk zones.

Regional research efforts on mycotoxins in the developing world for information exchange, transfer, and eventual implementation of tested mycotoxin management strategies by various regions need coordination for synergies to develop. A research-for-development program network may be required to deal with mycotoxins, food safety, and trade (Fig. 1). An important step already made by the CGIAR is the development of inexpensive ELISA kits that use monoclonal and polyclonal antibodies to detect aflatoxins, fumonisins and ochratoxins in various crops, food (milk, confectionary, processed meals) and feed samples (Reddy *et al.*, 2002). These assays enable rapid screening of samples for mycotoxins and speed the screening of breeding lines leading to quicker development of resistant cultivars. The cost-effective diagnostic kits provided a new impetus to the research to mitigate aflatoxin contamination, with high-throughput diagnostic labs established by the International

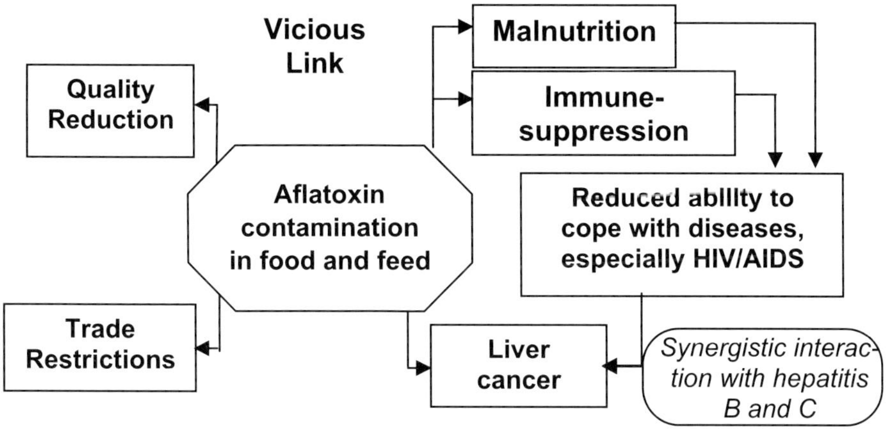

Figure 1. Effects of aflatoxin on trade and human health

Crop Research Institute for the Semi-Arid Tropics (ICRISAT) in Malawi, Mozambique and India. Up to 300 samples per day can be analyzed for mycotoxins by ELISA in these labs.

Controlling mycotoxins in maize

One of the most important socioeconomic changes in the savannas of Africa has been the increasing production of maize in areas that previously were planted to millets and sorghum. The greater use of higher-yielding crop cultivars, increases in the availability and use of pesticides, and deregulation of the market for cereals also have changed these areas dramatically. Maize has essentially become a cash crop. Much of the increase in maize production has occurred in areas at significant risk of attack by pests of stored maize, *e.g.*, fungi that produce mycotoxins.

Aflatoxin contamination is widespread in Africa: in Benin and Togo, aflatoxin levels in maize averaged five times the safe limit of 20 ng/g in up to 50% of the household grain stores surveyed (Egal *et al.*, 2005; Gong *et al.*, 2002). As a result, people, especially children (Gong *et al.*, 2002), are being exposed to high levels of mycotoxins, often in mixtures, and the consequences have been largely ignored. For example, 99% of fully weaned children had ~2-fold higher aflatoxin-albumin adduct levels in their blood than do those receiving a mixture of breast milk and solid foods (Gong *et al.*, 2003, 2004). Surveys also indicate that *Fusarium* infection is prevalent in field and stored maize at many African locations. *Fusarium* spp. are found in all agroecological zones of Benin, but their prevalence is higher in the South than the North. The incidence of *Fusarium* infection is higher in the field than in storage (K. Hell, unpublished). *Fusarium* infection usually is reduced during storage. The most common species found were *F. verticillioides* and *F. proliferatum*. Fumonisins were found in the maize samples with levels often exceeding the limit of 4 μg/g recommended by the U.S. Food and Drug Administration, especially in villages in southern Benin: Yé (12 μg/g), Lainta (7 μg/g), Adjohoun (6.7 μg/g) and Kpomé (4.7 μg/g).

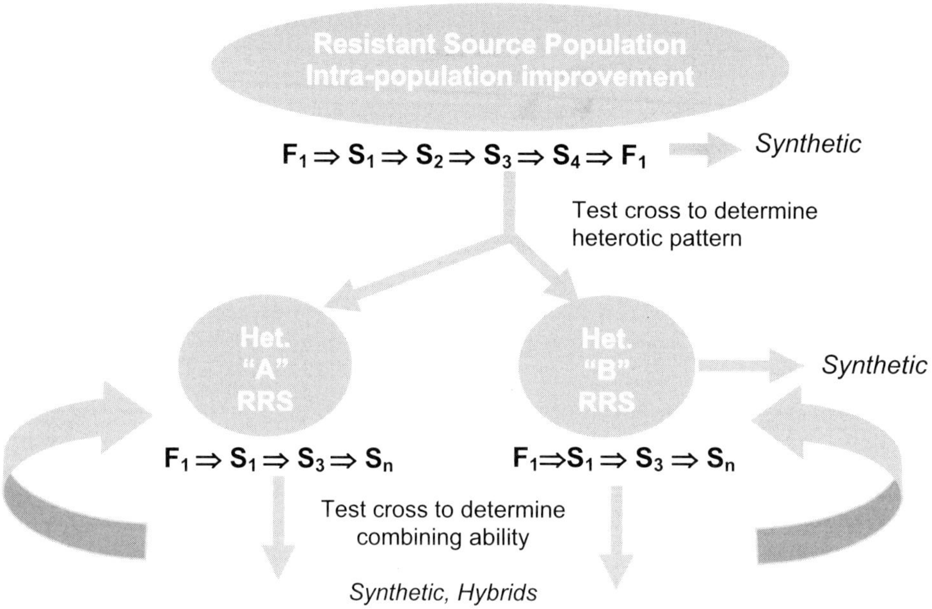

Figure 2. A comprehensive breeding strategy for developing biotic stress resistant maize germplasm. RRS = reciprocal recurrent selection.

Similarly, nearly 35% of maize kernel samples collected from several villages during the rainy and post-rainy seasons of 2004-2005 in Andhra Pradesh (India) contained 1 to 20 ng/g aflatoxin (F. Waliyar, unpublished). Six percent of the rainy season samples and 7.6% of the post-rainy season samples contained > 20 ng/g aflatoxins soon after harvest, whereas 20% of the rainy season crop stored for up to 4 months had > 20 ng/g aflatoxins. This survey suggests that even though toxin levels can be low at the time of harvest, they probably increase during grain storage.

Many insect species, drought and other environment factors, *e.g.*, nutrient stress, or pathogens are positively correlated with aflatoxin content in stored maize. These interactions between biotic and abiotic stresses not only reduce yield but also enable saprophytic fungi, such as *Aspergillus flavus*, to colonize the grain of stressed maize plants. In this regard, researchers at CIMMYT have been working to combine biotic and abiotic stress resistance, and to identify stress-tolerant lines or hybrids that have a reduced incidence of *A. flavus*. Line recycling, *i.e.*, targeting crosses based on known traits in the parental lines, has used sources such as 'La Posta Sequía' for drought tolerance, lines extracted from the 'Multiple Insect Resistance Tropical' (MIRT) population for stem borer and armyworm resistance, stunt resistant populations (mainly 'P73' but also 'P76' and 'P79') and lines resistant to foliar diseases and ear rots (D. Bergvinson and D. Jeffers, unpublished). This broader approach to reducing the mycotoxin load has been successful and has led to efforts to breed source populations and synthetics against the maize weevil and larger grain borer, insects that serve as vectors for fungi and that breach the plant's external integrity allowing fungi to enter and colonize the plant. A comprehensive breeding strategy (Fig. 2) for developing biotic stress resistant maize germplasm has resulted in genetically enhanced lines with abiotic or biotic resis-

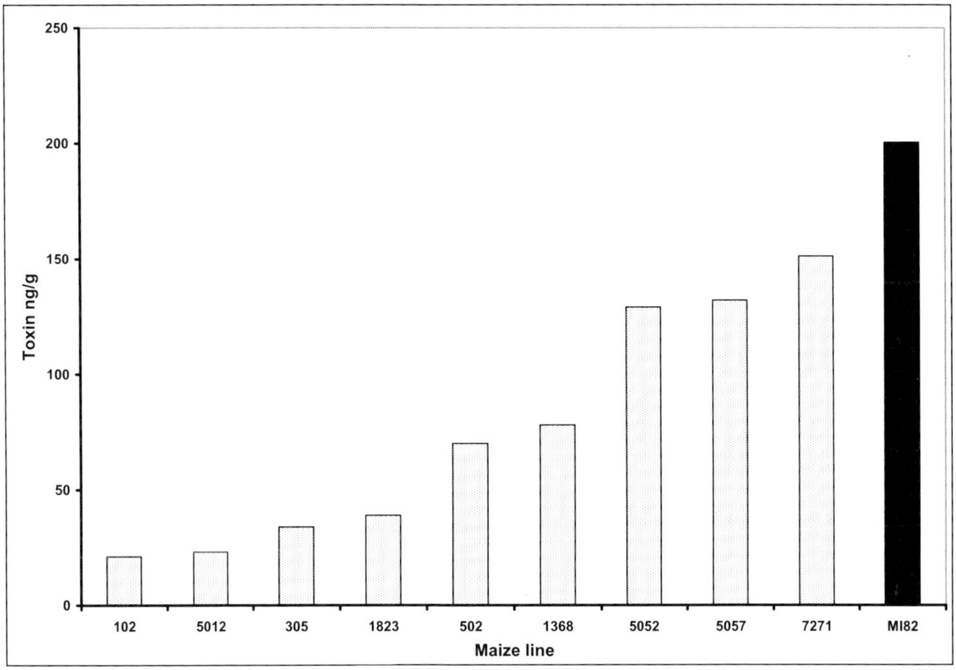

Figure 3. Aflatoxin accumulation in selected IITA maize inbred lines (░) tested with a kernel screening assay. A susceptible hybrid (■) from the United States had 5300 ng/g [after Brown *et al.* (2001)].

tance to storage pests and aflatoxigenic fungi. Valuable sources for resistance to aflatoxin accumulation have been identified in several elite CIMMYT maize lines (Jeffers *et al.*, 2005). These and other maize germplasm sources developed by IITA (Fig. 3) are being incorporated into both tropical and southern United States maize breeding materials (Brown *et al.*, 2001).

Aflatoxin screening in maize kernels needs to be simple and cost-effective. Fluorescent screening uses a black light assay to observe fluorescence from kojic acid, a secondary metabolite observed in colonized grain. Under black light, uninfected kernels are opaque and those infected are bright. This mass preliminary screening of breeding materials, allows selection of parental sources for further population improvement of maize.

Competitive exclusion (one strain competing to exclude another) is a relatively new, but very promising, biological control strategy for aflatoxin management in Africa. This control option leads to the best-adapted fungal strain being dominant in a given environment, and is a promising strategy for replacing toxigenic strains of *Aspergillus* with atoxigenic forms of same fungus. Several strains of *A. flavus* have been isolated and are being tested in Nigeria and Benin, with the goal of using atoxigenic strains to reduce aflatoxin contamination. Systematic knowledge of pre- and postharvest practices can lead to complementary management of aflatoxins through cultural and storage practices (Turner *et al.*, 2005). For example, lodged maize plants, drought predisposition, high grain moisture and grain damage at harvest all increase the risk of aflatoxin production in storage. Thus, farmers need to follow good management practices at harvest time, and in drying, including using an appropriate storage structure and controlling insects. Management options include preventing rain exposure of harvested cobs, storing maize in non-plastic bags, and sorting out kernels with insect damage and/or discoloration. Likewise, drying on black plastic sheets or cemented dry areas can reduce moisture content to safe levels after ~5 days, while drying maize cobs on the ground requires a minimum of 10 days.

Traditional maize processing also may help reduce mycotoxin levels; *e.g.*, aflatoxin levels were reduced by 99% in Benin during the preparation of fermented mawè, and by 79% and 43% when preparing fermented ogi and owo (or "paté de mais"), respectively. Roasting and treatment with alkali also reduces the level of aflatoxin present, whereas boiling and soaking of maize grain in lime-water can eliminate or greatly reduce the levels of aflatoxin in the final product. Last but not least, selective removal or isolation of contaminated portions of the food commodity remains the most widely used physical method for aflatoxin decontamination. Awareness campaigns are needed to sensitize the population to the risks posed by aflatoxins and to popularize management options to minimize its effects. For example, due to civil society public action, more than 10 million people in Benin, Togo and Ghana became aware of the dangers posed by aflatoxin-contaminated feeds and foods.

Aflatoxin management in peanut

Aflatoxins B_1 and G_1 are the most commonly produced forms in peanut. These toxins are involved in several human diseases, particularly liver cancer and growth defects in children. Aflatoxin interactions with Hepatitis B and C viruses result in relatively high levels of primary hepatocellular carcinoma. Aflatoxins also are toxic to livestock, including ruminants, poultry, birds and fish, when contaminated meal is used in their feed. Due to its human and livestock health implications, aflatoxin contamination has become a major issue in the international trade of peanuts and can directly impact the lives of poor farmers by reducing their income.

Infection of peanut by *Aspergillus* spp. can occur both pre- and postharvest. Preharvest infection by *A. flavus* and consequent aflatoxin contamination is important in crops grown under rain-fed conditions in the semi-arid tropics. End-of-season drought and damage to peanut pods by soil pests increases the preharvest aflatoxin levels. Mechanical damage during harvest and postharvest practices, *e.g.*, heaping, increase toxin levels in warm, humid areas. Poor harvest and storage practices may lead to rapid development of the fungi and consequently to higher production of the toxin. Aflatoxin contamination occurs frequently in peanut seeds, with very high toxin levels found in immature and small seeds. Small pods remaining in haulms, damaged and immature seeds often are used as cattle feed. Milk from cattle fed such contaminated fodder contains high levels of aflatoxin M_1.

Field and greenhouse screening methods have been used to increase the efficacy of evaluation of aflatoxin resistance in peanuts. Sick plots with highly aggressive, toxigenic strains of *A. flavus*, and laboratory inoculation methods for selecting individual resistant seeds now enable the screening of large amounts of germplasm. As a result, sources of resistance to seed infection and aflatoxin production have now been identified and used to breed high-yielding lines with resistance to seed infection and aflatoxin contamination that have been registered and shared with national agricultural research systems (NARS) for further use in their programs, *e.g.*, ICGV 88145, 89104, 91278, 91283 and 91284 in Asia and ICGV 87084, 87094 and 87110 in West Africa.

The estimated heritability for seed colonization ranged from 0.55 to 0.79, for seed infection from 0.27 to 0.87, and for aflatoxin production from 0.2 to 0.47. Thus, the levels of resistance in available sources and in the peanut breeding lines are not very high and do not suffice to effectively protect the crop from aflatoxin contamination under all conditions. Further, the diversity of these lines is very narrow. Hence, ICRISAT researchers have developed protocols for the transformation of peanut to produce transgenic plants with anti-fungal genes, *e.g.*, chitinases, that may increase the resistance to *A. flavus* (K.K. Sharma, personal communication).

Other options for aflatoxin control in peanut include the use of isolates of *Trichoderma* and *Pseudomonas*, which provide biological control of *Aspergillus* in both field and greenhouse trials, and cultural practices that reduce aflatoxin contamination, *e.g.*, the application of farmyard manure, lime, gypsum or cereal crop residues to the soil. Treatments including lime and farmyard manure can reduce aflatoxin contamination up to 90% in a highly susceptible cultivar such as "Fleur 11". Harvesting pods at the proper maturity, exclusion of damaged and immature pods, improved harvesting practices, the use of mechanical threshers, and proper seed storage bins are other cultural practices that help reduce aflatoxins in peanuts.

Fusarium head blight host-plant resistance in wheat

Fusarium graminearum is the main pathogen causing scab, or head blight, of wheat. Other species involved, depending on the climate and crops grown in rotation with wheat, are *F. culmorum*, *Microdochium nivale*, *F. avenaceum*, *F. poae*, and *F. sporotrichioides*. These pathogens affect grain yield and quality due to their ability to produce mycotoxins. *Fusarium graminearum* and *F. culmorum* produce the trichothecene mycotoxins, deoxynivalenol and 15-acetyl-deoxynivalenol, and zearalenone in North America, or nivalenol, zearalenone, 3-acetyl-deoxynivalenol, 15-acetyl-deoxynivalenol and fusarenon-X in Asia. Deoxynivalenol is associated with plant disease aggressiveness and may have been responsible for large-scale human poisonings in the last century in China and India. This toxin also causes vomiting and feed refusal in domestic animals and immunosuppression in mice. The closely related nivalenol toxin is toxic to bone-marrow in experimental animals. Zearalenone, a chemically unrelated compound has estrogenic effects in domestic pigs and experimental animals.

Fusarium head blight negatively affects wheat grain quality due both to lower weight of the affected grain and to the reduction in quality that accompanies mycotoxin contamination, and may result in significant economic losses directly to the farmers. CIMMYT, through funding from special grants from the Government of Japan and other donors, provides a global platform for international collaboration on scab research by facilitating the sharing of knowledge and genetically enhanced wheat germplasm and other breeding materials and tools. This global platform capitalizes on the knowledge accumulated on both host plant resistance and genetic improvement of the wheat crop against scab. For example, DNA markers are being mapped and used to incorporate three different types of resistance to the pathogen. These are: (I) resistance to initial infection or penetration, (II) resistance to fungal spread within plant tissues, and (III), degradation of mycotoxins. Chromosome 2D carries quantitative trait loci (QTLs) for type I and type II resistance, which are in the same region as QTLs for heading date and spike length. Although there are a few markers in this chromosomal region, new DNA markers associated with toxin tolerance are being mapped by *in silico* expressed sequence tag mapping that takes advantage of the synteny of the short arm of wheat chromosome 2D with that of rice chromosomes 4 and 7 (T. Ban, unpublished).

Scab screening with a spike inoculation test remains complex, unstable and low throughput. Easy, stable assessment methods for wheat breeders are being developed at CIMMYT that use the primary leaf. When a drop of a conidial suspension is placed on the wounded portion (~1 mm in diameter) of a leaf, the pathogen can infect and produce an oval lesion. This assay can distinguish resistant and susceptible cultivars (J. Murakami, unpublished). This new screening method, when coupled with the advances in genetic enhancement, should lead to novel resistance sources that carry genetically characterized R-loci. It also should be possible to assess transgressive R-segregating genotypes that combine distinct resistance genes, with the aim of pyramiding disease resistance genes in locally adapted wheat germplasm.

Towards a CGIAR-facilitated food safety program

The main role of science in agriculture has been to propel the evolutionary process that allows increased production with less land and less effort (Douthwaite and Ortiz, 2001). Who benefits from these advances depends on who controls the technology, who innovates, how selection decisions are made, and how innovations are enacted. We advocate a "research-for-development end-user-driven" approach that replaces the disconnected concept of research *and* development, in which researchers deal with technology generation and developers test this technology with potential end-users (Ortiz and Hartmann, 2003). Research-for-development needs society-conscious, committed scientists who are willing to transform into developers, by bringing a technology focus to their work. The research products resulting from this work are demand- not supply-driven, by end-users and not by "ivory tower" scientists. Hence, this approach closes the gap between research and development, and ensures that from the start of the research process, *i.e.*, its planning, that development goals are driving the agenda. Two metaphors: "from thinking to acting" and from "research to decision" define this new research-for-development approach, in which advanced research institutes, development organizations, the private sector, development investors and national governments are all partners and share the responsibility for accelerated agricultural diversification and commercialization for the small-scale agricultural sector. Research-for-development, keeping in mind the end-users, operates within a continuum that uses a "means" (research) for an "end" (development), thereby leading to impact on both people's livelihoods and science. With this approach, a working culture evolves in which management rewards internally the top performers following this framework, and externally encourages staff to broaden alliances or partnerships for development in their community of practice. Networking is a *must*, because organizations, which do not always have the same goals, see the advantage of teaming-up to successfully meet the objectives in a target area.

Due to the complex nature of agricultural problems in the developing world, solutions cannot be based on a "one size fits all approach." Research is required to develop decision-making processes that take natural resource fragility, community vulnerability, risk profiles, asset resilience, market options, service provision capacity and competitive advantage into account when designing solutions for specific client needs. Researchers must offer a broad array of products, because low input environments require yield-stabilizing technologies, and high-yield potential technologies should be developed for high-input environments.

Such moving targets must be addressed by a heterogeneous, but dynamic moving strategy, which may change at any given point of time. Researchers following this trajectory must be able to use all of the available research tools for development.

Increasing productivity per unit area results in more food to consume or sell and may diversify the crops being planted. Similarly, higher and more stable yield potential and profitability permits poor farmers to invest in inputs for producing more food and income. High yields also may lead to reduced food prices for the urban and rural poor and to monetarization of rural areas, whose inhabitants may prefer "money in the pocket" (income generation) rather than "a meal on the table" (food security). High yielding crops also may provide employment opportunities for poor people throughout the trade chain (from harvest to processing). Thus, the outputs from research-for-development efforts should be linked to a well-resourced capacity-building program so that farmers will be equipped with plant or animal genetic resources to cope with changing environments and the entrepreneurial skills to assess and take advantage of agricultural market opportunities.

Researchers, farmers, and policymakers should remember that economic phenotype performances (P) are influenced by many factors and their interactions, *i.e.*,

P = Genotype × Environment × Crop Management × Policy (affecting both people and markets) × Institutional Arrangements × Social Demographics.

Decentralized (through networking) end-user participatory research with local partners may provide a means for working in marginal, low input, stressful environments. Such decentralization rearranges priorities as local research partners target crop and resource management and as other responsibilities, *e.g.*, technology testing and the development of new materials through research or selection, shift from a central research station to become local targets. In this way, individual research programs, irrespective of their size, deliberately maintain diversity across locations. Such an approach should be driven by the needs of the rural poor to ensure that the work has a positive impact on their livelihood.

The major mycotoxins studied at the CGIAR institutes are aflatoxins in maize, peanut, sorghum and cassava, *Fusarium* toxins in maize, wheat and sorghum, and ochratoxin in cocoa and cashew. Genetic enhancement (both through plant breeding and biotechnology), biological control, habitat management, risk assessment, institutional capacity development and public awareness are among the tools in the "CGIAR research-for-development kit" to fight mycotoxins worldwide. These "tools" resulted from strategic, applied and adaptive research by scientists at the international agricultural research centers in partnership with their counterparts in the national agricultural research systems and advanced research institutes. Decades of research-for-development by the CGIAR centers suggest that a holistic approach should be pursued to deal with mycotoxins and should include the following elements:

- An integrated crop management package that combines mycotoxin resistant germplasm, biological control, habitat control and soil-amendments.
- Low-cost mycotoxin detection technology for rapid appraisal that also facilitates trade.
- Participatory process for mycotoxin assessment in commercially important crops.
- A high-level panel composed of scientists, NGO staff, farmers, traders, consumers, health officers and policy makers to monitor mycotoxin intervention strategies and to organize awareness campaigns.

In this way, the CGIAR addresses the Millennium Development Goals; *i.e.*, mycotoxin-free food is key for better health, and is especially important for the health of post-weaning children, particularly in sub-Saharan Africa, and to cope with malnutrition and HIV/AIDS.

References

Brown, R.L., Chen, Z.-Y., Menkir, A., Cleveland, T.E., Cardwell, K.F., Kling, J. and White, D.G. (2001) Resistance to aflatoxin accumulation in kernels of maize inbreds selected for ear rot resistance in West and Central Africa. *Journal of Food Protection* 64, 396-400.

Douthwaite, B. and Ortiz, R. (2001) Technology exchange. *Electronic Journal of Biotechnology* 4(2) http://ejb.ucv.cl/content/issues/02/index.html.

Egal, S., Hounsa, A., Gong, Y.Y., Turner, P.C., Wild, C.P., Hall, A.J., Hell, K. and Cardwell, K.F. (2005) Dietary exposure to aflatoxin from maize and groundnut in young children from Benin and Togo, West Africa. *International Journal of Food Microbiology* 104, 215-224.

FAO. (2004) *Worldwide regulations on mycotoxins for food and feed in 2003. FAO Food and Nutrition Paper No. 81*. FAO, Rome, Italy.

Gong, Y.Y., Cardwell, K.F., Hounsa, A., Egal, S., Turner, P.C., Hall, A.J. and Wild, C.P. (2002) Dietary aflatoxin exposure and impaired growth in young children from Benin and Togo: Cross sectional study. *British Medical Journal* 325, 20-21.

Gong, Y.Y., Egal, S., Hounsa, A., Turner, P.C., Hall, A.J., Cardwell, K.F. and Wild, C.P. (2003) Determinants of aflatoxin exposure in young children from Benin and Togo, West Africa: The critical role of weaning. *International Journal of Epidemiology* 32, 556-562.

Gong, Y.Y., Hounsa, A., Egal, S., Turner, P.C., Sutcliffe, A.E., Hall, A.J., Cardwell K.F. and Wild, C.P. (2004) Post-weaning exposure to aflatoxin results in impaired child growth: A longitudinal study in Benin, West Africa. *Environmental Health Perspectives* 112, 1334-1338.

Jeffers, D., Krakowsky, M.D., Williams, W.P. and Bertrán, J. (2005) Response to aflatoxin of CIMMYT germplasm in Southern USA. *Proceedings of the Multicrop Aflatoxin and Fumonisin Elimination and Fungal Genomics Workshop (23-26 October 2005, Raleigh, NC)*, p. 82.

Ortiz, R. and Hartmann, P. (2003) Beyond crop technology: The challenge for African rural development. In: *Building Livelihoods Through Integrated Agricultural Research for Development – Securing the Future for Africa's Children. Vol. 2. Reference Material of the Sub-Saharan Africa Challenge Program*. Forum for Agricultural Research in Africa (FARA), Accra, Ghana, pp. 39-72; http://www.rimisp.org/isc/documentos/beyondcroptechnology.pdf .

Reddy, D.V.R., Thirumala-Devi, K., Reddy, S.V., Waliyar, F. Mayo, M.A., Rama Devi, K., Ortiz, R. and Lenné, J.M.. (2002) Estimation of aflatoxin levels in selected foods and feeds in India. In Hanak, E. Boutrif, E., Fabre, P. and Piñeiro, M. (eds.) *Food Safety Management in Developing Countries – Proceedings of the International Workshop (CIRAD, Montpellier, France, 11-13 December 2000)*. CIRAD, Montpellier, France, pp. 1-4.

Turner, P.C., Sylla, A., Gong, Y.Y., Diallo, M.S., Sutcliffe, A.E., Hall, A.J. and Wild, C.P. (2005) Reduction in exposure to carcinogenic aflatoxins by simple post-harvest intervention measures in West Africa. *Lancet* 365, 1950-1956.

Wu, F. (2004) Mycotoxin risk assessment for the purpose of setting international regulatory standards. *Environmental Science and Technology* 38, 4049-4055.

Index

A

B

C

D

E

F

G

H

I

M

N

O

P

Q

R

S

T

U

V

X

Y

Z